T0212457

Intelligent Internet of Things

Farshad Firouzi • Krishnendu Chakrabarty
Sani Nassif

Editors

Intelligent Internet of Things

From Device to Fog and Cloud

 Springer

Editors
Farshad Firouzi
Department of ECE
Duke University
Durham, NC, USA

Krishnendu Chakrabarty
Department of ECE
Duke University
Durham, NC, USA

Sani Nassif
Radyalis LLC
Austin, TX, USA

ISBN 978-3-030-30369-3   ISBN 978-3-030-30367-9 (eBook)
https://doi.org/10.1007/978-3-030-30367-9

This Springer imprint is published by the registered company Springer Nature Switzerland AG.
The registered company address is: Gewerbestrasse 11, 6330 Cham, Switzerland

Farshad Firouzi:
    "To my beloved mother, father, brother,
and sisters;
    my beloved family;
    and Sied Mehdi Fakhraie"

Krishnendu Chakrabarty:
    "To all my wonderful students over the
years, who have been my best teachers and
my constant inspiration"

Sani Nassif:
    "For Khawla"

Preface

Greetings from a bygone era! We are writing this preface in May 2019, in the year that fully autonomous and self-driving vehicles are not yet available. That being said, we expect this book to be suitable for review and for classroom discussion topics, as well as for practicing engineers and business managers who need to obtain a deep technical understanding, even in 2025, when over 70 billion connected devices are forecast. In 2019, the Internet of Things (IoT) tsunami has already affected all aspects of our lives, from smart homes, smart cars, and smart city to smart health and smart environment. People benefit from automatically controlled room temperature/light to face recognition-based payment systems to various digital voice assistants. But we expect that even bigger waves are coming as the result of the convergence of IoT, artificial intelligence, machine learning, deep learning, distributed ledger technology, big data, and 5G. What we see in Sci-Fi movies will materialize to tangible and affordable technologies being tacitly integrated with our daily routine. This transformation can be truly seen, heard, and touched by everyone.

In this technology revolution, we can never ignore the vast infrastructure change beneath. The IoT wave is built upon rapidly growing and prevailing cloud technologies, evolution of edge technologies, miniaturization of devices, and powerful acceleration of computing capability in edge devices. Furthermore, machine learning and data science, as a core of artificial intelligence, empower the device, edge, and cloud layers to become smarter. In light of the above, this book presents a timely and comprehensive look at cutting-edge IoT technologies. It guides the reader through the fundamentals, principles, architectures, applications, challenges, and promises of the Internet of Things. This book is organized into two parts. The first part addresses the details of IoT and its underlying technologies from embedded systems to cloud computing and big data analytics. The second part discusses the interaction of IoT and healthcare which heralds a paradigm shift in this important area by providing many advantages, including availability and accessibility, ability to personalize and tailor content, and cost-effective health delivery.

Part I: Chapter 1 introduces the Internet of Things with several definitions and discusses the benefits and challenges of establishing the IoT. It reviews IoT reference models, state-of-the-art IoT platforms, and the applications of IoT in vertical markets. In addition, this chapter helps you make sense of the business-related aspects of IoT technology. Chapter 2 addresses "IoT/devices" and its fundamental building blocks including microcontrollers, interfaces, sensors, and actuators. In Chap. 3, the details of networking technologies and protocols are covered. Chapter 4 looks into IoT cloud covering the fundamentals of cloud computing, data ingestion, data processing, data storage, and data visualization. Chapter 5 addresses topics relating to machine learning techniques such as regression, classification, clustering, deep learning, and convolutional neural network. Chapter 6 is a definitive guide to the Hadoop ecosystem. Chapter 7 is an introductory text on cyber-physical systems in the IoT era. Distributed ledger technologies (DLT) and blockchain for IoT are presented in Chap. 8. Emerging hardware technology for processing IoT data at the edge or in the cloud including alternative computing, approximate computing, and in-memory computing is discussed in Chap. 9. Finally, end-to-end IoT security is covered in Chap. 10.

Part II: The second part of this book consists of three chapters. Chapter 11 addresses all important aspects of IoT technologies for smart healthcare: wearable sensors, body area sensors, advanced pervasive healthcare systems, and big data analytics. Chapter 12 covers biomedical engineering fundamentals to enable the building of wearable healthcare devices or to understand biomedical data easily. Finally, Chap. 13 provides a snapshot of the state-of-the-art research in data-driven e-Health studies that leverage artificial intelligence technologies for making sense of personal health data, as well as for delivering situational, actionable insights in care flows.

Durham, USA
Durham, USA
Austin, USA

Farshad Firouzi
Krishnendu Chakrabarty
Sani Nassif

Acknowledgments

The authors realize that this book could not be completed without the support of many people. We sincerely appreciate and gratefully recognize their support here. We would also like to express our appreciation and acknowledge Amanda Knight and Salah Fakhouri for their proofreading and writing assistance.

Contents

Part II IoT Technologies for Smart Healthcare

Part I
IoT Building Blocks

Chapter 1
IoT Fundamentals: Definitions, Architectures, Challenges, and Promises

Farshad Firouzi, Bahar Farahani, Markus Weinberger, Gabriel DePace, and Fereidoon Shams Aliee

All compromise is based on give and take, but there can be no give and take on fundamentals. Any compromise on mere fundamentals is a surrender. For it is all give and no take.

Mahatma Gandhi

Contents

F. Firouzi (✉)
Department of ECE, Duke University, Durham, NC, USA

B. Farahani · F. S. Aliee
Shahid Beheshti University, Tehran, Iran

M. Weinberger
Aalen University, Aalen, Germany

G. DePace
University of Rhode Island, Kingston, RI, USA

© Springer Nature Switzerland AG 2020
F. Firouzi et al. (eds.), *Intelligent Internet of Things*,
https://doi.org/10.1007/978-3-030-30367-9_1

3

1.1 What Is IoT

By now, everyone has heard of the Internet of Things (IoT). Internet of Things has been defined as the next logical stage of the Internet and its extension into the physical world. It is the broad connection of devices that can interact with each other and share data to a larger network, where the shared data can be leveraged to extract value. All devices must have unique identifiers and use embedded technologies to sense and gather data about themselves and their environment and transfer that data to other devices or other hosts. Then these data must be correlated and analyzed to inform more intelligent decisions. The technical challenges are appealing in themselves, but from an industrial and business perspective, IoT presents a grand opportunity to leverage previously unknown information and insight to transform and create industrial processes and business models. This reality is a much greater opportunity than a simple connection. Several companies have defined the Internet of Things in their own terms, and it is instructive to examine these terms to see the similarities and differences.

- IBM defines the Internet of Things as "the concept of connecting any device (physical object) to the Internet and to other connected devices" [1]. IBM also writes that IoT refers to "the growing range of Internet-connected devices that capture or generate an enormous amount of information every day" [1].
- SAP defines the Internet of Things as "the vast network of devices connected to the Internet, including smartphones, and tablets and almost anything with a sensor on it – cars, machines in production plants, jet engines, oil drills, wearable devices, and more. These things collect and exchange data" [2].
- Gartner says "IoT is the network of physical objects that contain embedded technology to communicate and sense or interact with their internal states or the external environment" [3].

- The Bosch corporation defines the Internet of Things as file sharing, e-commerce, social media, and the glue that connects things and devices. The devices can range from sensors and security cameras to vehicles and production machines. The connection of devices results in data that opens up new insights, business models, and revenue streams. The insights can lead to new services complementing conventional product business [4].
- Oxford Dictionary summarizes IoT as "a proposed development of the Internet in which everyday objects have network connectivity, allowing them to send and receive data."
- Finally, IDC defines the Internet of Things as "a network of networks of uniquely identifiable endpoints that communicate without human interaction using IP connectivity (local or globally)" [5].

As you noticed, due to rapid emergence and convergence of technologies, the definition of IoT is evolving, and thus there are several definitions of IoT from different points of view. However, all of them have the following fundamental characteristics:

- *Things or Devices* – Things in IoT (also known as intelligent objects, smart objects, IoT devices, or IoT endpoints) are connected objects that can sense, actuate, and interact with other objects, systems, or people. In order to be a device on the Internet of Things, the device must have a processing unit, power source, sensor/actuator, network connection, and a tag/address so that it can be uniquely identified.
- *Connectivity* – Connectivity empowers the Internet of Things by enabling IoT things to be connected to the Internet or other networks. This implies that there must be a connectivity module in each IoT device as well as an appropriate communication protocol that the network and the device can both understand.
- *Data* – There is no IoT without ("big") data collected from IoT things and indeed "data is the new oil." Data is the first step toward action and intelligence. Sent information from IoT devices most often include environmental data, diagnostic, location data, or report on their status. The data also flows back to the device, for example, a command to tell it to sleep, or decrease power consumption.
- *Intelligence* – Intelligence is the key to unlock IoT potentials because of its ability to extract insights from IoT data. For example, the combination of artificial intelligence (AI), machine learning, data analytics, and IoT data can avoid unplanned downtime (i.e., predictive maintenance), increase operational efficiency, enable new and improved products and services, and enhance risk management [6].
- *Action* – Actions are the consequence of intelligence. It refers to the automated actions to be taken by the device or on the device, but also includes action from the stakeholders in the IoT ecosystem.
- *Ecosystem* – IoT has to be seen and analyzed through an ecosystem perspective. IoT things themselves, the protocols they use, the platforms on which they run, the communities interested in the data, as well as the goals and aims of interested parties all form the ecosystem.

- *Heterogeneity* – The Internet of Things is expected to be made up of heterogeneous devices, working on different platforms on different networks. Therefore, all the components should be interoperable, i.e., they must be able to connect, exchange, and present data in a coordinated manner based on a common reference model.
- *Dynamic Changes* – The state of devices, the contexts in which they operate, the number of connected devices, and the data they transmit and receive are all expected to change dynamically.
- *Enormous Scale* – The number of connected devices will be at least an order of magnitude more than current connections. This means there will be a commensurate increase in the amount of data generated by the devices, which in turn must be transferred and analyzed to be leveraged.
- *Security and Privacy* – Security and privacy are an intrinsic part of IoT. These issues are critical as personal data will be available online (e.g., in a healthcare system, IoT devices could be charting and sharing heart rate, blood glucose levels, sleep patterns, and personal well-being). This demands data sovereignty, secure networks, secure endpoints, and a scalable data security plan to keep all of this information safe.

The Internet of Things exists in an ecosystem, all the components and the environment that supports IoT and its aims. In an IoT ecosystem, there are four major components: *things*, *data*, *people*, and *process*. Let us examine each in turn (see Fig. 1.1) [7]. Of course, all four components, things, data, people, and process, must work in concert in order to achieve the promises of a more connected world.

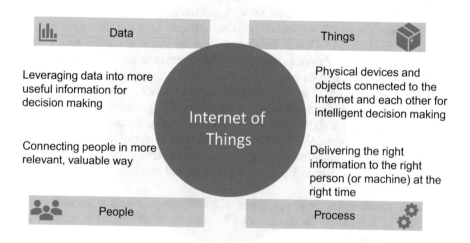

Fig. 1.1 IoT: The networked connection of people, things, data, and process

- *Things* – Things refers to the physical devices that operate as part of the Internet of Things. Each device must have the ability to connect to other devices or the network in general. This could be with a specialized communication protocol such as Zigbee or Bluetooth or the more general Internet Protocols (IP). The device needs the energy and processing power to handle that communication. Also, to be an IoT device, there must be some data to communicate. Most frequently, this is sensor data that is collected by the device itself. Some examples include image data from a security camera, temperature data from a thermometer, humidity or pressure data from a sensor on an industrial manufacturing machine, and so on. The thing or device may also be commanded to perform some action, perhaps sending specific data or moving an actuator or some other control motor. The device must be able to acknowledge these commands, perform these actions, and confirm with the remote controller that the desired action is performed. Routers, switches, and gateways are considered as part of the network but may also be classified as things. Devices must be equipped to survive the environmental conditions in which they are installed and have the necessary power, sensors, and communications to fulfill these roles.
- *Data* – The data component has been partially defined already, as those sensor data being sent from things as well as any commands being issued to the things. With the huge number of things producing data so often, it is easy to understand that the size of the data itself will be enormous. The raw data must be cleaned, that is, checked for errors and formatted, and then either stored for analysis or analyzed immediately. This task can be done at the edge of the network, close to the devices, or the data can be communicated to a more central collection point (e.g., cloud) where it is analyzed. The cost, relevance of the time of data to required actions, and communication barriers are some factors that determine the configuration of data processing. That being said, big data, collected from several IoT things, is most often stored and processed in the cloud.
- *People* – People are affected by the Internet of Things in at least two ways: as the agent of change who must work to make IoT function and as the beneficiary of its outcomes. Typically, people work in their own domain as a specialist at their job. With IoT, however, there is a much broader sense of interconnection between functions, and so people are increasingly finding themselves interacting with people in other business sectors. Sometimes this is a counterpart more or less with a similar function, or at a similar level, what we call horizontally located in the business, but other times it is a more vertical relationship, someone that operates at a lower or higher level. People must be interfacing in order to make sense of the data being collected and to determine the proper interpretation of the outcomes of the analysis of that data. Ultimately, it is people who create and maintain the Internet of Things, and their actions which can derive the most advantage from what IoT has to offer. The other side is the impact that the consumer sees from the IoT, meaning more informed decisions and targeted services from companies. People must also be aware of their personal data, who is collecting it and what is happening to it. Who owns this data? This is a question that has a complex and evolving answer.

- *Process* – The final component of the IoT ecosystem is process and that is where the benefits of intelligent automation, informed decision-making and control, and efficient procedures are realized. All of the methods, techniques, and processes currently used in vertical industries (e.g., manufacturing, logistics) can be made more efficient with the right information at the right time. Analyzing the data gathered from sensors and delivering this information to the appropriate stakeholders is the main idea of the process of IoT.

1.1.1   Internet of Things Terms and Acronyms

In this section, we review some fundamentals terms and explain how they relate to the Internet of Things.

- *Machine to machine communication (M2M)*: M2M is network communication between devices using any channel. Originally, it was used in an industrial context, but has come to mean that communication used to transmit data to personal appliances. Internet of Things is also communications between devices, but is used to also refer to vertical software stacks that automate and manage communications between multiple devices, and therefore refers to communication on a larger scale. Table 1.1 highlights the key differences between IoT and M2M.
- *Cyber-physical systems (CPS)*: The National Institute of Standards and Technology (NIST) has the following definition for CPS: "*Cyber-Physical Systems comprise interacting digital, analog, physical and human components engineered for function through integrated physics and logic. These systems will provide the foundation of our critical infrastructure, form the basis of emerging and future smart services, and improve our quality of life in many areas.*" Many manufacturing processes rely on cyber-physical systems as part of manufacturing. A cyber-physical system can also be found beyond manufacturing, for example, in the Smart Grid or in Smart Cities [8].

Table 1.1 The key differences between IoT and M2M

IoT	M2M
Devices communicate using IP networks, varying communications protocols possible	Point to point communications – embedded in hardware at the customer site
Data delivery is relayed through a middle layer in the cloud	Many devices use these protocols, over cellular networks or wired networks
Active Internet connection required	Not necessarily an Internet connection
Integration options are more varied, but management is necessary	Limited integration options; devices must have communications standards

- *Internet of Everything*: Cisco invented this term to mean the "people, process, data and things to make networked connections more relevant and valuable, turning information into actions' that improve everything." This terminology was abandoned sometime in 2017.
- *Social Internet of Things (SIoT)*: SIoT refers to an IoT in which things are able to create a network of social relationships with one another independent of human intervention. Objects are able to begin constructing social relationships based on the object's profile, interests (i.e., applications deployed, services used), and activities (i.e., movements). These social relationships can also be organized around events causing their creation. For example, a co-work relationship can be created between objects that work together to generate a common IoT application, such as objects that cooperate with each other to provide telemedicine or emergency response. A parental relationship may exist between objects that have the same manufacturer, are of the same model, or were constructed within the same period because they are part of the same batch. Social relationships are created between objects that are in contact occasionally or continuously because the object owners are in contact, and a co-ownership relationship may be created between heterogeneous objects that are owned by the same user. Adoption of the SIoT model offers many advantages [9]:

  – The social network created by the SIoT objects can be shaped as needed to ensure network navigability, the effective discovery of objects or services, and scalability similar to human social networks.
  – Trustworthiness can be created to balance the level of interaction among objects that are friends.
  – Models created to study social networks can be utilized to address IoT issues related to large networks of interconnected objects.

1.1.2 Impact of IoT

The estimated future impact of the Internet of Things is staggering. At the time of writing, it has been estimated that there are about 14 billion devices connected to the Internet. According to a Gartner forecast, it will go up to 25 billion by 2021. Cisco predicts that by 2020, that number will increase to 50 billion things. The government of the United Kingdom speculates twice that number, upward of 100 billion things. With this increased amount of connectivity, the way we interact with everyday objects will fundamentally shift. More of our choices can be driven by data instead of guesswork or habit. In our businesses, data-driven decision-making will prove more efficient and profitable. In our industries, processes and systems will be better managed and monitored, making us safer. Our quality of life will increase as these optimizations save us time, money, and energy. New services can be innovated from the data-rich environment, further improving our well-being.

1.1.3 Benefits of IoT

An organization that embraces the Internet of Things can expect greater safety, comfort, and efficiency. Hazardous environments and workspaces can be more carefully measured and the dangers more readily managed. The increased information about working conditions allows for decisions to improve comfort and consequently productivity, for example, a more localized thermostat can show the differences in the temperatures in specific offices. Adjustments for only those occupied spaces in temperature or lighting can lead to controlled energy costs and greater efficiency. Monotonous tasks can be automated, reducing downtime and yielding faster, more accurate, and greater results. Leveraging benefits like these can make the workplace more rewarding, and that improves employee satisfaction and retention and ultimately improved profits and reduces the necessary investment incurred by employee turnaround.

Organizations can also benefit from more information with which to make business decisions. Using large trends in empirical data means that fewer assumptions need to be made. It becomes possible to be more responsive to emerging trends. From a manufacturing standpoint, there is increased visibility into system behaviors. This can lead to shortened testing cycles and a more optimized production process. Revenue can also be increased or new streams can be realized by improving current procedures or making new ones from the increase in available information. IoT is a unique strategic advantage that early adopters will have over competitors who choose not to pursue digitalization. A few more benefits of IoT can be listed as below:

- *Efficiency*: More information about work/operation processes and rich data sets obtained from connected sensors leads to process streamlining. IoT enables great data sharing, and then manipulating the data as needed helps systems to work more efficiently and make smarter, more informed decisions in real time.
- *Transparency*: IoT digitizes every process and enables physical objects to remain connected, providing greater transparency. For example, IoT sensors can identify the status of the products in a production line or the location of assets in a field and track inventory and parcels.
- *Automation and control*: IoT enables the connection and digital control of physical objects, requiring extensive automation and control within the network. Without requiring human involvement, machines communicate with one another, resulting in more time-efficient output. Automation also ensures uniform completion of tasks and the quality of services provided. Human intervention may only be required in the case of an emergency.
- *Accuracy*: Monotonous tasks are automated, reducing downtime and errors.
- *Monitoring*: IoT provides the advantage of monitoring capabilities. Tracking supply quantities for business or monitoring the air quality of a home is easily accomplished and provides extensive information otherwise not easily obtained.

For example, knowing that the printer is almost out of paper or that you are running low on coffee can enable a user to consolidate shopping and avoid extra trips to purchase supplies. In addition, monitoring product expiration dates provide increased safety.

- *Information*: Access to additional information enables improved decision-making in a diverse array of areas, from everyday decisions like choosing what to purchase at the market to determine if a business has enough inventory. In each situation more knowledge gives the user greater power.
- *Time*: The integration of IoT has the potential to save large amounts of time, which is valuable to everyone.
- *Safety and comfort*: It can be difficult to imagine managing and monitoring hazardous environments requiring the consideration of multiple factors including human safety and optimizing the environment for productivity and comfort. Mundane tasks can be automated resulting in energy savings. For example, smart assembly lines can operate without human intervention and report errors immediately, resulting in greater productivity and less downtime. Automating monotonous tasks would also enable employees to engage in more rewarding work, resulting in increased employee satisfaction/retention and wider profit margins.
- *Security*: Security sensors (e.g., camera) as well as location-based sensors (such as GPS) have a significant ability to enhance security.
- *Cost/money*: The greatest advantage of IoT is the amount of money saved. Fewer errors, higher employee retention, improved processes, and energy-efficient behavior all reduce costs. IoT will be more widely utilized as long as the cost of monitoring equipment is less than the potential cost savings. IoT integration is proving highly useful in daily life as appliances communicate with one another, conserving energy and reducing costs.
- *Industry-specific view:* IoT can revolutionize several industries, for instance:

  - *Targeted marketing*: Greater information leads to individualized experiences, improving the interactions of customers with the company and bringing the company message to those more likely to become customers.
  - *Supply chain enhancements*: Asset tracking and management, security, optimized logistics, and transport all reduce costs of lost inventory, waiting times, and inventory mismatches.
  - *Health*: Individuals can get more information about their own bodies (heart rate, hours of sleep, etc.) to help in maintenance or identifying health problems.
  - *Smart building*: Workplace temperature, lighting, and air quality feedback can ensure a pleasant working environment, increasing satisfaction and productivity. In terms of security, connected cameras can detect the presence of unauthorized individuals.

1.1.4 IoT Challenges

Certainly, any changes bring not only benefits but also challenges that must be overcome. For any organization it is essential to determine which departments are responsible for which changes that must be made. Who will purchase and configure the needed IoT hardware (devices, gateways, etc.)? Who will install and run the needed software and troubleshoot the hardware and software? Who is responsible for networking? Which department will perform the analytics and deliver the reports and findings? There are also issues of what to do with legacy devices and other specialized solutions that will need to be found and addressed. Any potential solution must also be able to scale, to handle current needs but also those of the immediate future as the organization continues to adapt and grow. Through it all, ownership is necessary to maintain an adequate level of production quality, especially as several teams are usually called upon to work together. These are major issues that demand sound leadership in order to meet the challenges of implementing IoT for any organization.

Other challenges are more technical in nature. First of all, scalability and heterogeneity are intrinsic parts of IoT, which should be addressed via appropriate technologies. In this context, the necessary technology standards must be developed or updated including the network protocols and data aggregation standards. As mentioned before, a new connection paradigm will be needed and possibly described by these new protocols. At every stage from gathering data, to transmitting it, to storing and analyzing it, interoperability must be considered. Cloud Services are nonstandardized and non-unified, meaning that changing providers could incur the undue expense. There is currently no consensus on machine to machine protocols, and existing equipment uses a variety of operating systems and firmware technology. The surest way to mitigate these differences is to move computing tasks to the edge of the cloud and take advantage of fog computing models and IoT hubs. This will leave the cloud servers and services to handle the analytical and processing tasks for which they are best suited. Business must be prepared to handle these challenges with adequate planning and a solid business model. The revenue and profits will provide the motivation to invest in IoT and expand into vertical markets, horizontal markets, and consumer markets. If done properly, a market bubble will be avoided as well as regulatory and legal battles.

The final, and perhaps the most important, hurdle will be solving the issues associated with security. There have already been successful hacks, or unauthorized access to several devices on the Internet of Things. Since IoT will become a larger part of our daily lives, it should be obvious that the security of our sensitive information is becoming vital. Losing control of the radio in a car, or the transmissions of a baby monitor, or the home security cameras in a dwelling make for a frightening and compromising future. Controlling access to these and many other devices is a growing concern and is already being addressed.

In summary, the main challenges of IoT can be listed as below:

- *Scale*: Connecting to billions of active connected IoT devices is a big challenge, and the current communication models and technologies should be adjusted to address scalability challenges. In this context, emerging IoT technologies such as decentralized IoT network (e.g., edge/fog computing), peer-to-peer communications, and blockchain can be helpful.
- *Heterogeneity*: IoT in its nature consists of a plethora of devices with different interfaces and communication protocols, and thus there is a necessity to form a common way to abstract the underlying heterogeneity.
- *Privacy*: All the collected data must be kept secure and anonymous when necessary.
- *Data ownership*: Who is the owner of machine-generated data (MGD)? The entity that owns the IoT device or the manufacturer of the device (e.g., in connected cars)?
- *Cybersecurity*: Defeating attackers who seek to control, steal, or mislead is vital.
- *Legal liability*: Who is responsible when something goes wrong with an algorithm or an automated decision?
- *Sensors*: Technically, sensors must be inexpensive, accurate, and energy efficient.
- *Networks*: Transferring data and commands must be secure, reliable (correct and timely), and robust, despite operating in a noisy, busy, dangerous, or harsh environment.
- *Big data*: Connected devices continuously and simultaneously generate large volume and different varieties/forms of data, and thus IoT should be able to address time, resources, and processing capabilities.
- *Analysis*: The data must be properly interpreted and analyzed with fidelity to its meaning, especially if automated actions are taken based on data outcomes.
- *Interoperability*: There is a fierce competition to lead this burgeoning field, and all players must work together to be functional and to protect investments and must do so with fairness and integrity.

1.1.5 IoT and Big Data

Data coming from the Internet of Things is unlike data from the past in at least two important dimensions. First, the large amounts of data being generated demand a new data management approach. Traditional methods need to be adapted or entirely new approaches need to be discovered to handle diverse data constantly streaming from many sources. The second dimension is the nonuniformity of the data. Often the raw data is unstructured, or may come in several different formats, or may even change depending on the context. The new data management techniques must cope with these challenges. Up until now the discussion has been about big data without formally defining it. Big data is a large set of structured, unstructured, and

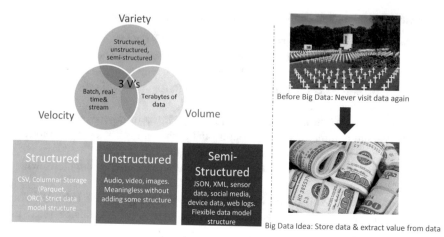

Fig. 1.2 The definition of big data

semi-structured data and the results of analyzing that data to gain insights. Doug Laney defined big data as the three V's (see Fig. 1.2):

- *Volume* – Storing large amounts of data.
- *Velocity* – The rate at which data is generated is high, so it must be stored or processed quickly.
- *Variety* – There are many possible formats of the data, from structured numeric to text, e-mails, video, audio, and so on.

Turning big data into tangible business insights is one of the major benefits of IoT. Most well-known approaches for dealing with IoT data include:

- *Analyzing data*: Before data becomes useful in making decisions, it must be analyzed. Traditional manual analytics, though powerful and informative, simply will not be practical in the face of the staggering amount of data that IoT will generate. Therefore, some automated analytics must be employed. These analytics need to provide descriptive reports of the environment, visualizations, dashboards, trigger alerts from data sources, and automated actions to be taken based on the data. They will also be used to detect patterns in the data, predict outcomes, and detect anomalies. There are open-source frameworks currently available for performing automated analytics. The two main approaches are to process the data in batches or to analyze the data as it is generated in real time. Which technique to use depends on the context of the problem as well as the resources available. The analytics can be run in a distributed fashion, also called in the cloud or at the edge, in servers nearer the sensors. First, the data is preprocessed, that is, duplicates are filtered out, and the data is possibly reordered, aggregated, and most likely normalized. These and other similar preprocessing tasks can be performed on the IoT device itself or on a gateway device before it is sent upstream. The most common automated analytics performed now are machine learning algorithms.

- *Machine learning (ML)*: Traditional mathematical statistical models analyze data by fitting the data to a model. Then the model is used to make predictions. This is a difficult process to follow especially when the data is dynamic or has many variables or the important points of the data are unknown. Machine learning is an algorithmic approach where the important parameters are extracted from the data in a process called learning. The data itself provides the structure of the mathematical model. Machine learning techniques can be applied to historical data or data taken in real time. The main way to think of it is that machine learning finds patterns or relationships or key variables in data. The model that is learned can be updated over time as more data is collected. One of the important applications of machine learning in the context of IoT is about finding patterns in the data, so that anomalies can be found quickly. Traditionally, anomalies were detected when certain values crossed thresholds. Machine learning allows for more complex patterns in the data to be identified as anomalous, therefore increasing speed and accuracy in detecting problems. The machine-learning-driven intelligence in IoT can be used for predictive analytics (what will happen), prescriptive analytics (what should we do next), and adaptive or continuous analytics (how can we adapt to the latest changes).
- *Edge analytics*: When analytics is applied at the edge of the network close to the IoT devices that generate the input data, it is referred to as edge analytics. Since network traffic is reduced, this is an attractive approach to reduce bandwidth and the latency from data gathering to a useful result. One drawback is that more processing power is needed in the devices and close to them, and cost or the particulars of the environment or device may make this prohibitive. On the other hand, sending large amounts of data across a network into the cloud may also be too expensive. Often a hybrid of edge and upstream analytics in the cloud is used to mitigate these costs.
- *Real-time analytics*: Any time that data is collected and immediately analyzed is known as real-time analytics. This is the best choice when a delay in the results of the analysis would reduce the value of the data. Time series data, rolling metrics, running averages, and any other occasion where the window of time analysis needs to be controlled are also good candidates for real-time analytics. Some real-time analytics frameworks available include Apache Storm, Apache Spark, and Flink frameworks.
- *Distributed analytics*: When the data sets are particularly large, too large to be handled by a single node (server), then distributed analytics can be used. As the name implies, the analysis tasks can be broken up and spread out to several compute nodes, possibly across multiple databases. If the data allows, it could be bucketed by time period and thereby effectively split up in order to make it more manageable. This is also an example of batch processing. Hadoop provides an ecosystem of frameworks for performing analytics. Apache Hadoop is used for batch processing and uses the MapReduce engine to process distributed data. Hadoop is a good open-source framework and one of the first to become available. It is used successfully for historical data analytics.

Data storage, besides the data processing, is another challenging issue in the era of big data. As more and more devices are connected to the Internet of Things, the amount of data they generate will drastically increase. They will be sending messages with their status, sensor outputs, metadata, and other messages. Despite the large amounts of data, it still must be stored. Two common methods are listed here, NoSQL databases and time series databases. Traditional techniques (SQL databases) are usually not feasible because of the amount of data being stored, with its varied and often unstructured nature. NoSQL databases offer high throughput and low latency of storage and retrieval. Since there is no schema, dynamic new data types are allowed. Couch Base, Apache Cassandra, Apache Couched, MongoDB, and Apache HBase (Hadoop) are examples of frameworks that use NoSQL. There is also a NoSQL in the cloud solution offered by IBM's Cloudant (a distributed database) and AWS' DynamoDB. A time series database can also be a NoSQL database or even a relational database. The indexing and queries are all based on timestamps in the data. Some frameworks using time series databases are InfluxDB, Prometheus, and Graphite.

1.1.6   IoT and Cloud Computing

The two worlds of IoT and Cloud experienced swift and independent progress. However, the complementary features of IoT and big data generated many new opportunities and advantages. Cloud computing is the solution to the increased demand for storage and processing. The cloud is defined as a group of servers and computers connected over the Internet in a large, distributed infrastructure. The concept is to deliver on-demand services over the Internet. The model is typically based on pay for the usage consumed (metered service), with the ability to scale up and down as needed (elastic resources). Amazon, Microsoft, and Google are dominating this *Infrastructure as a Service* (IaaS). They also provide *Platform as a Service* (PaaS) and *Software as a Service* (SaaS). The advantage to consumers is a lowered computation cost versus purchasing the hardware and then paying to operate and support it in-house. In summary, the main drivers for integration of IoT and Cloud are listed below [10]:

- *Device lifecycle management*: As the Internet of Things grows in size, the number of devices that need to be registered, managed, and updated while maintaining security requirements also grows and must be accommodated. It is possible for tools to configure and update firmware and software over the air (FOTA). The cloud platforms enable device lifecycle management, so devices can be connected, registered, on-boarded, updated remotely, and even remotely diagnosed should something need to be fixed. This reduces the operation and support cost of the devices. That means the enterprise Internet of Things is remotely managed, with minimal time and a reduced cost of ownership. In other words, a 360-degree view of the IoT devices is possible via the cloud.

- *Communication*: Cloud platform can be leveraged with the help of IoT to deliver scalable domain-independent services by providing appropriate service-oriented domain mediators.
- *Resource pooling*: Physical resources of IoT can be integrated into the cloud resource pool enabling us to allocate and share them on demand like regular Infrastructure as a Service (IaaS).
- *Storage*: IoT drives a real tsunami of big characterized by volume, variety, and velocity. In this context, IoT benefits from large-scale and long-lived storage of the cloud.
- *Computation*: Data processing is typically a very resource-hungry task. Therefore, IoT can benefit from virtually unlimited processing resource of cloud to aggregate data and execute batch and/or real-time analytics on the collected data.
- *Device shadowing or digital twin:* Another benefit available through cloud computing is device shadowing. The concept here is to have a backup of running applications and devices also running in the cloud. Any time there is a fault or failure in the original device or application; the twin can be examined to extract the result or to help in diagnosing the problem. System availability can be increased by using the digital twin as software redundancy. If the original system needs to be taken offline for maintenance, the twin can continue the operation uninterrupted; it can also provide system behavior statistics and behavior profiles for the original system at decreased risk.

1.1.7 IoT and Digitalization

Gartner defines *"digitalization"* as leveraging digital technologies to change business models and provide new revenue and value-producing opportunities. The process of updating a business to digital technologies is an evolutionary one and indeed has been happening for decades. The process is enabled by increased interoperability, information transparency across departments and industries, automated assistance and support, and a trend toward decentralized decision-making. In this context, IoT is considered as the major pillar for digitalization. The other important pillars are blockchain, big data, and machine learning.

1.1.8 IoT and Industry 4.0

The phrase "Industry 4.0" is rooted in a high-tech, German government research and development project in the manufacturing industry. It was initially coined at the Hannover Fair in 2011. Although there is some difference of opinion around the definition of historical industrial revolutions, Industry 4.0 is considered as the fourth industrial revolution. The initial industrial revolution occurred in the late 1800s and is responsible for mechanizing the power of steam and water. The second industrial revolution began in the early 1900s and was characterized by the use of electricity to drive mass production through assembly lines and a reorganization of labor. The 1970s brought the third industrial revolution which utilized computers to automate production and processes. It is predicted that the coming fourth industrial revolution will fully utilize digital manufacturing in smart factories. Industry 4.0 is propelled by the merging of technologies such as:

- Industrial Internet of Things (IIoT) and extensive sensor use
- Analytics and big data
- Machine learning and artificial intelligence (AI)
- The convergence of IT/OT
- Augmented reality (AR)
- Advanced robotics
- Additive manufacturing

Benefits of Industry 4.0 Industry 4.0 will generate benefits in many areas. Product development will move more quickly due to analytics, and original equipment manufacturers (OEMs) will utilize analytics to understand better how consumers actually use products compared to a product's anticipated use. Sensor data will be used to optimize production through constant status updates that are compared to a digital twin (i.e., a perfectly efficient simulation which creates a virtual and digital replica of the target physical product/entity or process) to predict the physical counterpart's performance characteristics and guide corrective action and predictive maintenance needs. Additive manufacturing will become highly profitable based on highly flexible, small production capabilities. Augmented reality will drive learning and efficiency, and machines will assist humans with dangerous or complicated tasks as they gain autonomy. Many of these technological advancements are already occurring on a smaller scale. However, the guiding vision of Industry 4.0 is to revolutionize manufacturing and its connected industries. The main goal of the Industry 4.0 vision is to help manufacturing and its connected industries to evolve away from a logistics or end product focus. This revolution seeks to help these fields move toward an efficient customer-responsive business model that generates innovative revenue sources. Industry 4.0 also has the potential to revolutionize cities and utilities on a larger scale.

Industrial IoT The Industrial Internet of Things (IIoT) uses actuators and sensors to improve industrial and manufacturing processes. The IIoT is vital in many industries such as oil and gas, logistics, manufacturing, energy/utilities, transportation, resource mining, and aviation as well as other industrial fields or use cases common to these industries. However, there are some companies and professional researchers who consider Industry 4.0 and IIoT to be equivalent.

1.2   Architectures and Reference Models of IoT: A Layard View

1.2.1   IoTWF Reference Model of IoT

There are several standardizations in the IoT ecosystem. The IoT World Forum (IoTWF) is an exclusive annual industry event hosted by Cisco. As an outcome of their collaboration, they published a Standardized Architecture in 2014. The committee was comprised of Cisco, IBM, Rockwell Automation, and others. The proposed IoT Architecture is a seven-layer reference model, with control originating from the center (e.g., cloud) to the endpoint devices. Generally, data is gathered at the endpoint devices and is sent toward the center. The central processing can, in fact, be decentralized and implemented as a cloud service. The purpose of such a model is to give a common understanding of how the problem of creating IoT can be divided. With the different goals of each layer identified, and the interfaces specified, different companies can contribute pieces that will interoperate. Security can also be enforced at each layer of the model. These seven layers include (see Fig. 1.3) [11]:

1. *Physical devices and controllers (things)* – These are the physical devices, sensors, actuators, and controllers that form the Internet of Things. Their primary function is to collect data to transmit upstream, but they should also be capable of receiving commands, e.g., power down, etc.
2. *Connectivity (networking)* – This is the layer that serves as the medium to bring the sensor data from the devices to the upper layers where that data is cleaned and analyzed. The chief responsibility here is for reliable, secure, and timely delivery of data. This includes any switching or routing that is necessary as well as translation between protocols if necessary.

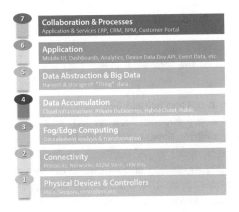 

Fig. 1.3 IoT reference model by IoTWF

3. *Edge computing (data element analysis and transformation)* – This is also known as the fog layer because it is the layer where data cleaning, aggregation, and processing begin. It is the responsibility of this stage to prepare the data for analysis and storage. One of the methods to mitigate the enormous data flows is to start the analysis as early as possible. Data is evaluated, possibly reformatted or reordered, filtered, and checked for warning thresholds. Edge/fog computing facilitates data processing, data storage, and networking services between endpoint IoT devices and the center (e.g., cloud or data centers). The key idea of edge/fog is to process the data and make actions closer to where the data is created. This can ultimately result in reducing the traffic between IoT devices and real-time actions (i.e., respond fasters to events).

4. *Data accumulation (storage)* – This is the layer which prepares data to be stored in a database, whatever that format is. The key here is that after this layer, the data is expected to be able to be retrieved based on queries.

5. *Data abstraction (aggregation and access)* – Another layer that deals with the data. At this layer, the data is consistent, complete, and validated. In practice, data is often stored across multiple databases; the task of this layer is to ensure that the data is able to be queried and a unified, reliable result is returned.

6. *Application (reporting, analytics, control)* – This is where individual software applications can query the data to perform specific functions, such as reporting, monitoring, control of devices, visualizations, and analytics.

7. *Collaboration and processes (involves people and business processes)* – This is the layer that makes use of the outputs from the software applications of the previous layer. Data and conclusions from that data are shared with other entities or applications. The collaboration of several data sources illuminates new business practices, makes existing processes more efficient, and opens the doors to innovation. This is where the benefits of the Internet of Things are largely realized.

1.2.2 Simplified Reference Model of IoT

A simplified IoT architecture is comprised of the following layers (see Fig. 1.4) [12, 13]:

- *IoT Things Layer* – Consists of all IoT sensors and actuators.
- *IoT Network Layer* – Includes network components such as IoT gateways, switches, and routers responsible for transmitting data in a timely and dependable fashion. This layer also includes fog/edge nodes to perform data analysis and transformation and information processing as quickly and closely to the things as possible. This is very helpful in real-time applications such as IoT healthcare to be able to provide low-latency and faster responses to emergencies.

Fig. 1.4 Simplified IoT architecture

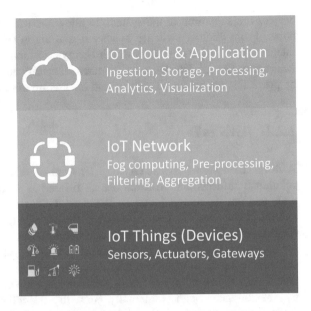

- *IoT Cloud and Application Layer* – Manages and processes IoT devices, as well as data created by the other two layers. It is also responsible for data ingestion, data interpretation through software applications, as well as integration with other platforms to improve business value.

1.3 IoT Frameworks and Platforms

1.3.1 FIWARE

FIWARE is funded by the European Union (EU) to be an open-source middleware platform. This means that it specifies interfaces for application programmer interfaces, allowing anyone to be able to connect devices to a catalog hosted in the cloud. The idea is to simplify the task of integrating devices into IoT and enable an economy based on data. Since it is a standard, it relies on participation and adoption, but the promise of interoperability is real and appealing. To that end there is an active and well-funded community surrounding the platform and encouraging participation [14].

1.3.2 SmartThings

SmartThings is a cloud-based platform offered by Samsung that focusses on building and running an IoT-driven smart home. The application management system is used to process subscriptions from device type handlers. Over 300 different devices are supported in the system, to allow the user control over objects in the home. From dimmer switches to sensors and alarm systems, SmartThings offers integration of varied devices with third-party assistants such as those produced by Amazon or Google. The system can increase security and convenience in any home by providing a common connection and integration of devices found there.

1.3.3 AWS IoT

With Amazon Web Services (AWS) IoT, Amazon offers a managed, cloud-based solution. Platforms and software are all offered as a service, with the ability to scale and use its analytics tool on IoT data. Things can be registered as devices, and the architecture features a Message Broker, Thing Registry, Thing Shadows (Digital Twins), and Rules Engine in addition to Security and Identity components. AWS is aimed at home users as well as industrial users with its mix of device software, control, and data services. Amazon machine learning provides analytics and visualization tools as a service. Users can make use of the same technology used by Amazon data scientists internally, but with a friendlier wizard-style interface to begin. Getting started building and performing IoT tasks or data science functions are painless. The services scale as your needs or businesses grow, and stopping is just easy since no capital investment is required [15].

1.3.4 Microsoft Azure IoT

The Azure Internet of Things (IoT) is a collection of services that is capable of connecting, controlling, and tracking billions of IoT devices. Available services in Microsoft IoT include [16]:

- Azure Internet of Things (IoT) Hub
- Azure IoT Edge
- Azure Stream Analytics
- Azure Machine Learning
- Azure Logic Apps

1.3.4.1 Azure Internet of Things (IoT) Hub

The Azure IoT hub is a cloud-hosted service that functions as a centralized, bidirectional message hub for an IoT application and its connected devices. It can be used to build dependable and secure communications among millions of IoT devices and back-end solutions hosted by the cloud. Almost any device can be virtually connected to the IoT hub, which supports communication coming from the device to the cloud and vice versa. IoT hub is able to support different message patterns used to manage devices including file uploads from devices, device-to-cloud telemetry, and request-reply methods. IoT hub monitoring is useful for supporting solution health because it monitors events including device connections, failures, and connections. The IoT hub provides a secure channel for devices to communicate and send data [16]:

- Individual device authentication allows each device to connect to the hub securely and to be controlled securely.
- The IoT hub provides total control over device access and can manage each per-device connections.
- When a device initially boots up, the *IoT Hub Device Provisioning Service* automatically provisions devices to the correct IoT hub.
- Various device capabilities are supported by multiple authentication types:

  - SAS Token-Based Authentication
  - Individual X.509 Certificate Authentication
  - The X.509 CA Authentication IOT hub connects devices using the following protocols: AMQP, AMQP over WebSocket, HTTPS, MQTT, MQTT over WebSocket

IoT hub also includes built-in message routing which provides the flexibility to create a rules-based, automated message fan-out. Additionally, the IoT hub can be combined with additional Azure services to create comprehensive solutions such as:

- *Azure Logic Applications* – Business process automation
- *Azure Machine Learning* – Adds AI models and machine learning to solutions
- *Azure Stream Analytics* – Provides real-time data analytics on data streaming from devices

There are two available Software Development Kit (SDK) categories used with the IoT hub:

- *IoT Hub Device SDKs* – Allow one to create IoT applications to be executed on IoT devices. These applications can send telemetry to the IoT hub and include the option to receive messages, method, job, or updates from the hub. Compatible languages include Python, Node.js, Java, C#, and C/C++.
- *IoT Hub Services SDKs* – Allow a developer to create back-end applications that manage the hub and schedule jobs, send messages, invoke other functions, or send updates to IoT modules or devices.

1.3.4.2 Azure IoT Edge

Edge enables an organization to focus on business insights rather than focusing on data management by transferring cloud analytics and some business logic from cloud to edge. Azure IoT Edge has three main components (see Fig. 1.5) [16]:

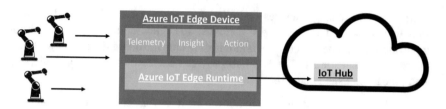

Fig. 1.5 The architecture of Azure IoT Edge

- *IoT Edge Modules* – These are the fundamental execution units that run the business logic of the system at the edge. These modules are implemented as Docker-compatible containers. There is a possibility to create more complex data processing pipeline by connecting several containers to each other. IoT Edge allows you to create custom modules or bundle different Azure services into modules able to extract insights from IoT data offline at the edge.
- *IoT Edge Runtime* – It is located in the edge and provides cloud and custom business logic for IoT Edge. In addition, it performs communication and management operations including:
  - Manages workload installation and updates
  - Manages Azure IoT Edge Security Standards
  - Ensures IoT Edge Modules are running
  - Monitors and reports module health remotely
  - Manages communication and handles communication between downstream endpoint IoT devices and IoT Edge, between modules, and between the cloud and IoT Edge devices
- *IoT Cloud Interface* – It sits in the cloud and allows remote management and monitoring of IoT Edge devices from the cloud.

1.3.4.3 Azure Stream Analytics

As an event-processing engine, Azure Stream Analytics enables you to monitor large-volume streaming data coming from IoT devices as well as data from social media feeds, applications, web sites, etc. You can also use Azure Stream Analytics to visualize relationships and find patterns in streaming data. Once identified, data patterns can be used to drive downstream actions like sending information to reporting tools, storing data, or creating data alerts [16].

Azure Stream Analytics utilizes a source of streaming data that is ingested into the Azure IoT hub, Azure event hub, or from Azure storage. To evaluate the data streams, you must create an analytics job that identifies the input data stream source and uses a transformation query to determine how to search for data relationships or patterns. When analyzing incoming data is done, you are able to identify the desired output and then determine how to respond to the analyzed information. For example, you can take follow-up actions including:

- *Trigger Alerts/Customized Workflows* – Triggers a specific process or function in response to an input pattern.
- *Visualize Data* – Data is sent to a Power BI (a business intelligence framework) dashboard to allow real-time data visualization.
- *Store Data* – Utilizes Azure storage system to store the data; therefore, you can perform batch analytics or train holistic machine learning models based on historical data.

1.3.4.4   Azure Machine Learning

Azure Machine Learning is a cloud-based service supportive of open-source technology and useful for large-scale training, deploying, automating, and managing machine learning models. Azure Machine Learning enables the user to access thousands of open-source Python packages that include machine learning components such as PyTorch, Scikit-learn, and TensorFlow. Microsoft also offers another framework called Azure Machine Learning Studio, a drag-and-drop, a collaborative area that allows you to create, test, and deploy machine learning solutions without writing code. This workspace also provides preconfigured and pre-built algorithms as well as data management modules that make experimenting with machine learning modules quick and uncomplicated. Azure Machine Learning Service is beneficial instead of Azure Machine Learning Studio, when greater control over the details of the machine learning algorithms is needed or you need the flexibility to utilize open-source machine learning libraries [16].

1.3.4.5   Azure Logic Apps

Azure Logic Apps is a cloud service used to arrange or automate tasks, workflow, or business processes when data, applications, systems, or services must be integrated across large enterprises. One of the main benefits of Azure Logic Apps is that it makes the designing and implementation of scalable data integration, applications, and other system solutions including business-to-business (B2B) communication within the cloud or on premises (or both) easier and more straightforward. Below you will find examples of workloads that is possible to automate using Azure Logic Apps [16]:

- *Event Processing* – Events can be processed and routed across cloud services and on-premises systems.
- *Email Notification* – Email notification can be automatically sent via Office 365 when an event occurs in an app, service, or system.
- *File Transfer* – Uploaded files can be transferred from FTP or SFTP servers to Azure storage.
- *Tweet Monitoring* – Tweets can be reviewed by subject or analyzed based on sentiment and alerts or tasks can be created if the additional inspection is required.

1.4   IoT Applications in Vertical Markets

There are many areas where the Internet of Things will have a major impact. What follows is a sampling and a brief discussion of some of these IoT application areas.

1.4.1   Smart Agriculture

Also known as precision farming, since the power of data is brought to bear on agricultural decisions, instead of the traditional wisdom and guesswork.

- *Smart Greenhouses:* An IoT-enabled greenhouse will allow for the finer automation and control of environmental parameters. As expected, all aspects of the greenhouse can be monitored and recorded, including temperature, sunlight, air quality, humidity, and air flow. Adjustments to the environment can be recommended or automatically carried out depending on the recommendations of cloud servers.
- *Livestock Monitoring:* Cattle and other livestock can be monitored with IoT sensors to determine their location and vital signs to determine their health. Those animals with warning signs of sickness can be identified, quarantined to protect the others, and treated to overcome the sickness. This process saves on labor costs, improves the health of the overall herd, and reduces the risk to the animals and farmers.
- *Agricultural Drones:* In addition to placing sensors at key points, farmers can use airborne drones to monitor much larger and widespread areas. These drones can be recruited to plant seeds, spray existing crops, take soil samples, assess the health of crops, or even monitor fields or assets for security purposes simply. Historical records of crops can be more easily kept with drones assisting. They could also be used for integrated GIS mapping and visualization. All of these data points can ease the burdens on farmers, saving time and money and potentially increasing the agricultural output of the farm.

As an example of IoT in agriculture, we can name Cropx company. This company installs sensors to understand better water usage in fields growing crops. The data are used to conserve better and utilize irrigation. The company also advises on the type and use of pesticides and fertilizers to maximize the yields of crops. They do this by collecting data on soil, air quality, crop maturity, and even weather and then using algorithms and machine learning techniques to determine when and where to intervene.

1.4.2   Logistics and Transportation

Logistics is all about moving items from one place to another. Warehouses are used to store the items temporarily until they can be loaded onto vehicles and moved

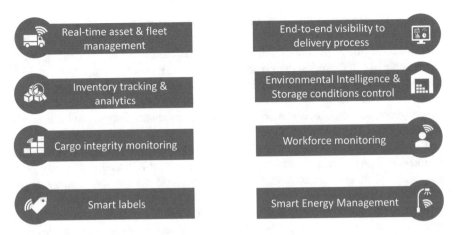

Fig. 1.6 A few use cases of IoT in logistics

further along toward their destination. The vehicles use the transportation network to maneuver between endpoints and warehouses. Generally, suppliers are the ones that dispatch these vehicles to deliver the items to customer locations. To succeed at logistics, it is important to reliably, safely, and predictably deliver items from suppliers to customers. All stakeholders also want to know the status and position of the items in the transportation network and to be able to forecast how long it will take to receive these items through borders, customs, or other checkpoints. This means that a quality logistics operation has mastery over the capacity of each stage in the transport and can optimize vehicle routes between endpoints in an energy-efficient and proactive manner. IoT can assist with all of these goals, from real-time traffic and environmental conditions for vehicles, to real-time monitoring of vehicle and warehouse capacity, to sensors to locate and verify that items are in good condition and route. The value of items is linked to the length of time they are in transit, so minimizing risk in damage or delay is another prime concern, and another one that can be addressed with the data that IoT delivers. The vehicle fleet itself can also be monitored to ensure that timely maintenance is being done, increasing its availability and longevity. Fuel and time can be saved by re-routing around bad weather or accidents, and theft and loss can be prevented and stopped with cargo validation and monitoring. There may be other business innovation opportunities when the IoT is fully leveraged in this field. A few more uses cases of IoT in logistics/transportation are shown in Fig. 1.6.

1.4.3 Smart Grid

The traditional power grid consists of monolithic power generation plants that deliver electricity across transmission lines to power substations where electricity is distributed to customers via distribution lines. The customers had meters to record the use of the electricity, and these meters needed to be visited to be read. The next generation of power grid adds intelligence at the customer end by making those meters able to communicate their readings back to the power company. Further, with advanced metering, they can be updated in real-time to reflect changing tariffs on power based on loading and time factors. Demand can be better understood making power generation timelier and more efficient. Energy spikes, equipment failure, and power failures can be detected more quickly with smart sensors and the response can be more rapid with the automatic dispatch of engineers or even an automated restoration. Power outages and interruptions cost several billion dollars every year, so finding solutions to reduce and eliminate the occurrences improves quality of life and makes financial sense. Generators of electricity can better understand where, how, and how much electricity is used on the grid, enabling them to be more adaptive and responsive. Especially as the move is toward renewable energy sources and decentralization of electricity generation, smart technologies are vital to unlocking wind, solar, and tidal power to its full potential.

On the consumer side, understanding when and how electricity in the home is used can lead to better choices. Home automation can activate appliances during off-peak times, and thermostats can control home heating and cooling depending on the time of day and occupancy to maximize comfort and energy savings. Electric vehicles can serve as power storage for a smart grid, or a micro-grid for the neighborhood, being charged during off-peak times, and returning power to the grid during the times of highest demand. To address the above challenges, research and development to design IoT-driven power grid as a robust, reliable, and secure infrastructure is critical to the future of technological advances, since it powers all the other technologies.

1.4.4 Smart Building

Large buildings are currently outfitted with proprietary solutions to assist in solving the problems faced by facilities managers. They need information about how the building is functioning, the heating, ventilation and air conditioning system, the boilers, the power, the security system, and many other systems and subsystems that make up a modern building. While some management systems do a good job, they are often difficult to integrate with other solutions. Since they are often hardware based, once they become obsolete, it can be costly to update them, making them inflexible. Legacy buildings are a significant contributor to the increase in greenhouse gases in the atmosphere, with some estimates as high as 36 percent of CO2. Forty percent of total energy consumption is from the maintenance of buildings, with as much as 75 percent of current structures being inefficient. In 2016, the Paris Climate Agreement specifically targeted reducing the high energy consumption of buildings as an excellent method for addressing climate change.

IoT will bring greater interoperability to these older, disparate systems and, through the cloud, allow for greater remote management, improving efficiency and response time. The sensors in an IoT-enabled building can collect the traditional information but also many other pieces of information not currently monitored, like air quality and occupancy. Using this data, building services could be improved to make occupants more comfortable and safer and also to use less energy. Workers who have greater peace of mind are able to concentrate better and be more productive. A smart building can offer this increased comfort with targeting thermostats to maintain more consistent temperatures. This results in more employee satisfaction, but also in a reduction in facilities calls to come and adjust the thermostat. Smart lighting can adjust light levels based on time of day and the presence of people needing that light, resulting in added savings. As systems change and adapt, software-based solutions will more easily adapt to them. The output of the sensors and detectors can be collected and visualized for facilities managers, to improve maintenance timing and effectiveness. Several companies are working for IoT-integrated solutions for the smart building, including Intel.

A case study in the possible energy conservation is that of a conference room. These rooms are important locations for productive meetings and a valuable space in which to work, but whether they are actually being used or not, traditionally they receive HVAC services. Intel performed a study using its smart building product and showed in its report that it was able to save 4 percent on HVAC costs in conference rooms in its subject building. With LED lighting and occupancy detection, it was able to reduce lighting wattage used per square foot from 1.09 to 0.39. Through analyzing other data points, there is room to innovate other cost savings in offices and shared spaces in buildings.

Power companies incentivize customers to reduce consumption during peak times by offering lower rates or credits called Automated Demand Response. IoT-driven smart buildings can take advantage of this by knowing the current building power usage and the grid rates and adjusting the load accordingly. There may be some load devices whose use can be delayed until an off-peak time, for example. If more power is demanded, then solar panels on the building, batteries or fuel cells in the building, or even a local power generator (diesel perhaps) could be utilized to make up the shortfall. This process can be automated via IoT solutions to optimize the cost of power to the building.

1.4.5 Smart Factory

Like IoT and other emerging technologies, there is not a unique and universally accepted definition for smart factory. However, smart manufacturing or smart factories can be explained by their main characteristics and core contributing technologies such as IoT, machine learning, 3D printing, cyber-physical system, robotics, big data, and blockchain. In the context of IoT, smart factories are manufacturing plants that incorporate IoT technologies into their processes to improve and optimize each and every aspect of the factory.

1.4.5.1 Current Manufacturing Model

The current manufacturing automation is based on a hierarchical architecture consisting of the following layers:

- *Level 1: Sensor and Actuator Layer* – This is the base level where the devices, sensors, and actuators exist on the plant/shop floor to perform different manufacturing process. This layer is a part of Operational Technology (OT).
- *Level 2: Field Automation Layer* – This layer (mostly based on PLC: Programmable Logic Controller) monitors and controls the devices that are attached to. This layer is also a part of Operational Technology.
- *Level 3: Supervisory and Integration Layer* – This layer mainly addresses supervisory control of the whole production process in the shop floor, shop floor monitoring, data acquisition, and data storage. It also functions as a multi-protocol intermediate gateway between the underlying industrial systems and the upper enterprise systems. This layer is usually implemented by Manufacturing Executive Systems (MESs), and Supervisory Control and Data Acquisition systems (SCADA). Level 3 is also a part of Operational Technology.
- *Level 4: Enterprise Layer* – Finally, decisions at this level concern the business planning, customer orders, material acquisition, and administration. Note that this layer is classified as an Information Technology (IT) layer.

Three important points should be noted here. The first point is that the above layers sometimes melt into each other in a way that some functions can be implemented at multiple levels. Second, many existing factories have not integrated the integration of Information Technology (Layer 4) with Operational Technology (Layer 1–3) yet. The third point is that the above model is rigidly structured to some extent, meaning that in the first two layers, there is almost a strict master-slave communication paradigm, with the master taking charge. According to Industry 4.0, the above model can be evolved toward a more decentralized model, allowing for more autonomy. In Industry 4.0, the decentralization allows for more flexibility, self-governance, self-organization, self-maintenance and self-repair. These are all goals of smart manufacturing, and is not surprising as Industry 4.0 is a much more recent standard.

1.4.5.2 Potential Use Cases

Here are the most popular IoT applications that are reshaping manufacturing and factories:

Operating Efficiency IoT-based smart factories are more responsive to changes in the environment and armed with more detailed and timely data are poised to proactively address potential problems or events as soon as they occur or even before they occur. In addition, traditional manufacturing plants depend on the skill and training of the operators and technicians to produce their output. For decades, manufacturing has worked to increase the amount of automation in the process. Some benefits were realized, but often new technicians were needed to ensure the automation was working optimally. The Internet of Things should improve this situation as automation can be better monitored and controlled. A networked control system can sense, visualize and control every aspect of the manufacturing process even remotely. The smart factory can deliver a cost-effective, efficient, sustainable, and safe manufacturing system.

Real-Time Quality Control Manufacturing business success is dependent upon a rigorous inspection process applied across each production phase. IoT enables manufacturers to program equipment and utilize big data analytic frameworks within factories to effectively monitor the manufacturing line, equipment, raw materials quality, and the quality of completed products at each point in the manufacturing process. Integrating IoT in this manner provides the following benefits to the quality control process:

- Enabling real-time action in alignment with the manufacturing process
- Optimizing in-process manufacturing using production engineering insights
- Continuous adaptation and learning based on production output
- Continuous optimization to address process drift or production variance

Predictive Maintenance The ability to predict difficulties or perform predictive maintenance is an advantage with increased uptime and safety. Predictive maintenance is repairing or replacing equipment or components before predicted failures. Traditionally, historical mean time between failure data was used to schedule this maintenance, but with more accurate and timely data from IoT devices, a more specific time can be found, meaning good parts are not replaced, or unexpected weaknesses can be located and addressed before catastrophic failure. Of course, the data must be analyzed to extract these benefits, using machine learning and other data analytics techniques as mentioned earlier.

Safety Employee safety is another area that can be improved with IoT devices. Workers can be observed to find lapses in focus or other mistakes, and preventative action can be taken. With increased knowledge of activities on the floor, should there be a problem, help can be dispatched more quickly and accurately. When all activities are analyzed, it is possible to discover new processes or methods to use during the manufacturing itself. There is the potential to improve efficiency with these process ideas or with real-time solutions as situations develop in the plant.

Supply Chain Management IoT can help with supply chain management, as sensors track and help manage the location and condition of inventory, management can better plan, and schedules can be adjusted to optimize output. In addition to sensors, IoT devices can be used directly for automation. Integrating robotics can improve worker safety and factory throughput and reduce costs by increasing efficiency.

Machine as a Service (MaaS) This approach will allow updated machines to be deployed from the cloud, with remote configuration, connectivity, and monitoring.

Services such as these will allow for 100% uptime and zero-touch deployment, two desirable goals for manufacturers.

IT/OT Convergence Since the 1970s, there has been an increase in automation in the manufacturing sector. This trend continues as operational technology (OT) and information technology (IT) converge with programmable logic controllers, computers, networking, and connected devices and sensors. IoT brings manufacturing technology and enterprise networks together, eliminating technological silos. These improvements will lower costs with scaled, automated, and platform-based machine connectivity that will increase monitoring and optimization. It can be claimed that the main driving factor in IIoT and IoT in the manufacturing industry is the convergence of IT/OT. There are two terms that must be understood before discussing the convergence of IT/OT:

- *Information Technology (IT)* – Using computers, hardware/software, and other telecommunications devices to complete business operations. IT is mainly linked with the back-end functions required to handle operations including billing, resource planning, asset monitoring, accounts receivable/payable, and maintaining client information.
- *Operational Technology (OT)* – The foundation of modern smart factories. Manages infrastructures powering manufacturing plants and ensures factory lines keep running. The value of OT is amplified as additional machines or components are connected.

IT/OT convergence means operational technologies such as meters, sensors, programmable logic controllers, and SCADA are integrated to work together in near real time or real time with IT systems. The fields of OT and IT have existed side by side since the beginning of modern manufacturing. However, they have been siloed, with minimal interaction, resulting in a lack of understanding about how individual departments fit into the manufacturing process. Before IT/OT convergence, data sharing among departments was guided by the calendar, but the birth of the IIoT has vastly reduced the gap between IT and OT. Therefore, in a post-integration era, both IT and OT can share data in real time. There are several main benefits to IT/OT convergence, including agility, performance, productivity, cost, and agility. Combining IT and OT generates a complete picture of operational improvement opportunities and challenges facing manufacturers. This increased transparency helps IT and OT teams to better define their roles in light of a clearer team goal or purpose.

- *Cost*: The benefit that most often overlies both IT and OT departments is cost. In the area of IT, the cost is tied to predicting or illustrating profitability while the cost is generally linked to reducing production expenses in the area of OT. In both departments, reducing costs is good for the organization's profit margin.
- *Performance and productivity*: The benefits of improved performance and productivity are connected. Businesses can enable IT and OT to collaborate through a common platform to create accurate key performance indicators (KPIs) that

equip both departments to work toward common goals together while improving company-wide visibility.
- *Agility*: When an organization does a better job of controlling costs and analyzing KPIs, it is better able to act with agility to reduce production time and make space for innovation, which was a highly difficult task in a siloed IT/OT environment.

1.4.5.3 Major Challenges

There is the perception of several barriers that must be overcome in order to evolve manufacturing plants to smart factories. A recent survey of manufacturing executives by Cisco ranked these problems for IoT in manufacturing, starting from the most serious [17]:

- Lack of supply chain visibility
- Lack of visibility of plant floor KPIs
- Inability to access data within production
- Lack of common metrics across plants
- Plant floor IT apps in silos
- Employee skills gap
- The complexity of manufacturing operations
- Inflexible automation
- Lack of understanding the plant floor
- Lack of reliable plant floor network
- The process not automated (manual)
- Lack of clear manufacturing strategy
- Unable to justify return on investment (ROI)
- Insufficient investment to modernize
- Security threat or fear

1.4.6 Smart City

People have lived in cities for centuries, but only relatively recently, the mass migration from rural areas to cities has intensified worldwide. In 1950, less than one-third of the world's population lived in cities; that fraction is expected to increase to two-thirds by 2050. In raw numbers, that was fewer than 1 billion people, to upward of 4 billion people. As these populations increase, it puts tremendous pressure on the local environment. The amount of energy consumed, the amounts of food and products that must be brought into the city, and waste that must be removed strain the transportation system and the city itself. The world's cities use 60–80% of the energy used in the world. They also contribute the most to greenhouse gas emissions. Cities consume 60% of potable water in the world, wasting an estimated 20% in leakage. It is important to optimize the use of these critical resources and maximize their conservation [18].

The main reason cities have not been well designed is because they grew organically in response to increases in population. When the population increases rapidly, then urban planning cannot keep pace. The other problem is that city services are independent of each other, not communicating to solve problems together. The way cities are organized prevents collaboration, with each service or department getting their own funding and incentivized to solve their narrow problems. This leads to redundancy, waste, and shortcomings in meeting the needs of the city population. What is needed is a more scalable, collaborative, efficient system of city management and improvement. The Internet of Things can provide a city with more detailed and timely information and facilitate better solutions to the inefficiencies of modern cities.

1.4.6.1 Smart City Layers

As proposed by Cisco, an IoT solution for a smart city can be described with four general layers [17]:

- *Street Layer*: At the base is the street layer. This is where devices and sensors are placed in various parts of the city to collect data and take automated or commanded actions resulting from the analyzed data. The sensors used will depend on the location and function expected of them.

  - Video cameras are currently in widespread use in cities for various reasons. Some are aimed at highway sections, interchanges, or some city street intersections, and these are used to determine and report traffic conditions primarily. Other cameras are mounted at street level and are intended to monitor pedestrian behavior or are used for security purposes. The improvements in video recognition technology mean that these can be automated to perform facial recognition and vehicle recognition and make automated reports for security, traffic, and accidents.
  - Device counters or vehicle detectors are used to count the number of vehicles passing a certain area, or that are parked on streets or in structures. This is another technology that has been in use for many years to great benefit. Its use can be expanded to make parking counts more available to private drivers and their applications to better coordinate parking. They can also be adapted to count other things such as birds behaving as pests in public areas.
  - Magnetic sensors are able to detect the presence of vehicles in specific locations. This is another sensor that can be applied to the parking problem. It can also be used to make traffic lights more responsive.
  - An air quality sensor can be used to measure the amounts of particulate matter present in the atmosphere. This data can be used to give warnings to citizens when air quality is bad or to detect the culprits of high levels of pollution in order to improve air quality.

- – There are other sensors and controllers available and the choice of which one to be used depends on the problem to be solved and the resources available. There are several factors to consider when selecting a sensor. What are its lifetime maintenance costs? Can it be mounted on existing infrastructure? What is the cost of operation? Can it store its own data, or must it be transmitted to the cloud immediately? If such a connection is needed, is it available? How can this sensor interoperate with other such sensors? Can it scale? Once these questions are answered, then a tradeoff analysis can be conducted and the appropriate sensors can be selected. This is another reason why it is more efficient for the different departments of a city to work in concert, as they can leverage sensors and infrastructure to solve multiple problems.

- *City Layer*: The next layer is the city layer. This is above the street layer and provides the connectivity for the myriad devices in use at the lower layer. This means the network routers and switches are at this layer along with the communications protocols that allow the connected devices to exchange data. This is also the edge layer and the start of data processing. Some sensor data will be time sensitive, while others must be cleaned or reordered before transmission to the higher levels. A resilient and reliable network is therefore a necessity at this layer. Often, the networking equipment will be placed outdoors or in a harsh environment and therefore must be designed to work under inhospitable conditions. A malfunction at this layer may cause automated false alarms due to missing or mishandled data.

- *Data Center Layer*: When the data has been collected at the edge and transmitted, possibly over different transport protocols, it is delivered to the next layer up, the data center layer. This is where the final analysis is performed, and the results of these analytics are stored for further use. Therefore, analytics, storage, and some method of making results available are the primary functions at this layer. As previously discussed, the cloud plays a major role at this stage, providing the required storage and processing power.

- *Services Layer*: The services layer is the final layer in the Internet of Things smart city model. At this point, the results of the sensor data are provided to applications that make use of it – for example, a visualization tool to show the real-time status of traffic in the city. City managers, law enforcement, and private citizens should all have access to the data. City managers will want to ensure that the city is running smoothly and could use the data to find opportunities to conserve energy, for instance, or to check on the status of waste removal in a given neighborhood. Law enforcement could be verifying that tolls were paid or the payment made for the use of a public parking space, among other things. A private citizen could be looking for an open parking space or for the speediest path to the other side of the city. Once the sensors are in place and the data made available, then the city is ready to reap the benefits of the Internet of Things.

1.4.6.2 Applications of IoT in Smart City

Here is a sample of some areas where the Internet of Things is making a positive impact on smart cities.

- *Smart Lights* – Public outdoor lighting is beneficial to society as it makes public spaces safer to live and work. Unfortunately, it is also expensive to operate and often wasteful. Many systems merely use a timer to activate and deactivate the lights at certain hours of the day. The Internet of Things can make the system more efficient. Using sensors to monitor usage and activity at the lights, they can be directed to activate the lights only when needed. They can adapt the lighting settings to environmental conditions, such as fog or rain, when visibility has decreased. Lights can also be used to assist emergency responders or law enforcement by providing more lights in high-crime areas or when an accident has occurred. Real-time data about the lights themselves can provide operational status, making maintenance and replacement tasks proactive. This can increase the longevity and operational time of the lights while reducing maintenance costs.
- *Traffic* – Traffic lights can also benefit from the sensing and command possibilities of being connected by the Internet of Things. Real-time traffic data can be used to smooth traffic loads throughout cities. The goals are to reduce idling time, to improve flow and runtimes through the city, and to reduce pollution and fuel use. The data can be collected from cameras and correlated with data from vehicle counters. The ideal system would integrate traffic data with private navigation applications, so that a centralized view of the city can help direct drivers to balance routes. With smart traffic light technology and sensors on roadways, vehicle accidents can be detected more quickly, and assistance can be dispatched to the scene more efficiently. The path of the responders (e.g., police, ambulance) can be expedited, and even emergency rooms can be alerted so they can be prepared to receive victims. These efficiencies will help to make our roads safer.
- *Smart Parking* – Parking in a congested city can be frustrating and wasteful. With access to data about traffic patterns and open parking spots, applications can help citizens to locate and travel to available parking more efficiently. Kansas City in the United States and Paris in Europe have already implemented IoT smart parking solutions.
- *Smart Water* – Every city needs to manage its water supply. Water treatment plants must treat potable water for citizens, and a distribution system must deliver this water to residents. Currently, up to 20% of water is lost from the network because of leaks. It is difficult to predict water demand, and without accurate predictions, treatment plants can run inefficiently. The Internet of Things sensors can be used to improve water metering, leakage detection, planning for increased distribution, and understanding water use. Having better, more accurate data helps when creating water usage models, which improves predictions. More accurate water meters make water bills more accurate and build trust between

city water authorities and customers. Water is a precious commodity for us and it is vital that we manage it in an informed, thoughtful and efficient manner.

- *Smart Waste* – All cities produce waste and managing that waste is a difficult challenge. The most used current solution to the waste problem is to use manual collection based on a schedule set by a waste management company. The schedule may or may not be effective as it depends on the details of the waste management contract. The scope of the waste problem includes the collection, transport, processing, and disposal of the various kinds of waste generated by a city's population. Some waste can be recovered by recycling techniques, but this must be identified and separated and then transported to a recycling facility for processing. The entire process must be managed and monitored all at the cost of time, money, and labor. Improvements in the process can benefit all the stakeholders, the city council, manufacturing plants and other companies, health and safety authorities, and the people themselves. Using the Internet of Things to improve the process involves adding sensors in the waste receptacles and in the waste removal vehicles. These sensors can detect the amounts of garbage and the types of garbage present. In this way, a logistics platform can match the collection agents to the receptacles that are at or near capacity. The routes that collection trucks use can be optimized for efficiency.

1.4.6.3   Examples of Smart City

There are some cities that are already embracing transformative IoT technologies to improve the well-being of their citizens. For example, in Stockholm, a smart management system in conjunction with smart applications has addressed traffic and environmental issues in the city. The city implemented a policy of a shared waste management vehicle fleet that resulted in better waste collection routes and improved waste collection.

In Helsinki, the collective inputs of the citizens were leveraged by making over 1 thousand databases publicly available. The data concerned transport, economics, employment, and overall well-being of the people in the city. This was done via an open urban data platform, called the Helsinki Region Infoshare Project. The project won the European Prize for Innovation in Public Administration for empowering the citizens of the city. One of the chief results was to foster more public involvement in policy- and decision-making in the city.

1.5   IoT Business Implications and Opportunities

Internet of Things is seen as a strategic topic in many industries. For example, in 2018 the number of job postings related to IoT in Germany doubled compared to 2017 [19]. Bain & Company projected the global IoT market to reach $318 billion by 2021. Expectations are high regarding future IoT-based turnover [20]

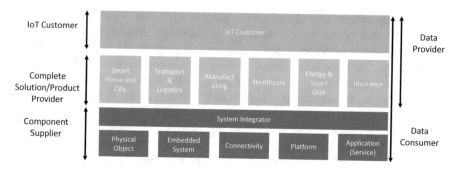

Fig. 1.7 Three basic IoT business opportunities

and emerging new business models [21]. Despite those huge expectations, many companies, especially small- and medium-sized enterprises, struggle to identify promising business models and solid use cases [22]. To be able to design new business models and draft business cases, the key business opportunities that IoT provides for a specific company should be analyzed and evaluated. In this context, we also need to understand the IoT business ecosystem, the stakeholders, as well as their motives. As shown in Fig. 1.7, three business opportunities and stakeholders can and should be distinguished when discussing the strategic impact of IoT from a company's point of view:

- *Complete Product and Solution Provider (Vendor):* These stakeholders aim at creating additional revenue streams from smart products/services which comprise the hybrid value proposition of IoT solutions [23].
- *IoT Customer:* The customer of an IoT solution on the other side is ultimately looking for optimization and cost reductions within its own operations. The IoT solution makes the operations of IoT customers smarter and optimized. The fundamentally different perspectives of IoT provider and IoT customer and their relation can be illustrated better by an example. John Deere is an international corporation that manufactures agricultural, construction, and forestry machinery. The so-called field connect system from John Deere allows for monitoring the moisture levels on various depths of a farmer's field [24]. The IoT provider (John Deere) intends to create additional revenues from an innovative offering based on IoT, which he did not sell before. The farmer (IoT customer) on the other hand invests money in an IoT solution hoping to reduce the cost for monitoring moisture manually on site and waste of water.
- *Component Supplier:* Many companies could leverage the third strategy as well. A component supplier facilitates the design and deployment of the Internet of Things. In this case, no complete IoT solutions are involved, but just IoT components. These might be technical components on a single layer of the IoT technology stack (e.g., IoT device, gateway, connectivity, and cloud platform) or components from two or more layers.

1.5.1 Component Supplier: Component Business

While a complete IoT solution comprises the whole IoT stack, IoT components relate to one or at maximum three layers of the IoT stack. A company could focus on selling such IoT components to complete IoT solution providers who intend to build and release complete IoT solutions. Let us examine each layer of the IoT value stack and their corresponding stakeholders using a connected electric bike (e-bike) example [25]:

- *Physical Object* – The physical object (i.e., e-bike) provides the initial direct benefit for the user. As with traditional bicycles, the e-bike provides an eco-friendly, healthy mode of transportation while also enabling motorized cycling.
- *Embedded System* – In this layer, the physical thing is equipped with a processing unit (e.g., a microcontroller), a connectivity module (e.g., 3G, 4G, NB-IoT), sensors, and actuating components to become smart. These pieces operate locally by gathering data and providing localized benefits. In our example, sensors are responsible for monitoring battery status or sensing when motorization is required. An example of an embedded system provider is Bosch which designs and provides several IoT sensors for years.
- *Connectivity* – In this layer the smart object and its functions/status (e.g., battery status) can be accessed online globally with the help of network providers/operators. Moreover, new services could be added to the system e.g., online location monitoring or theft prevention. In many IoT solutions such as e-bike, we can use SIM cards and mobile networks to be connected to the Internet. Indeed, already in the first quarter of 2016 in the United States, 69% of newly activated SIM cards were related to non-phone devices, like cars, dog collars, etc. Thus, all major mobile network operators aim at selling SIM cards as IoT components.
- *Platform (Cloud)* – Platforms are one of the central foundations of IoT as they unite connectivity, service providers, applications, and embedded systems to create specialized IoT solutions for diverse industries. Platform providers offer data ingestion, data storage, data analytics, data visualization, device/user management, and integration with other third parties through SDK or APIs. In our e-bike IoT solution, this layer enables one to track the movement patterns of e-bike users, study the difficulty levels of specific cycling routes to understand motorized support demand better, or discover the location of stolen e-bikes in real time. An example of the platform layer could be Amazon. With Amazon Web Services (AWS), the company has been successful in the cloud computing business. Indeed, Amazon is now offering an IoT component (i.e., AWS) that can be used to build IoT solutions.
- *Application (Service)* – This final layer combines the options and features provided by the prior layers to structure digital services. Users can receive digital services in appropriate formats that are independent of location via mobile applications or web tool. In our example, this feature enables customers to find e-bikes in the case of theft or provides pertinent location information to law enforcement.

- *System Integration* – The stakeholders of this layer play a large role in the IoT ecosystem because not all IoT components are plug-and-play right out of the box. Therefore, system integrators are needed to enable individual IoT components to collaborate in the best possible way. System integrators should identify a specific niche and then make partnerships with other stakeholders.

1.5.2 Complete Solution and Product Provider: Additional Revenue

IoT solutions are addressing business problems across several vertical markets from health to smart building, transportation/logistics, energy, and manufacturing. In this context, many companies, incumbents and startups, seek to create revenues from smart connected products. This could include enhancing the companies' already existing products with embedded systems (e.g., sensors, connectivity, etc.) in order to enable new features or digital services. But it could also mean developing entirely new connected offerings.

One of the main challenges for such an endeavor lies in handling the rather complex IoT value stack. A company from the digital or Internet world or a startup would need to develop and produce the connected thing, which would mean to enter the hardware world, with comparatively high upfront investments for development and production setup. A manufacturing company needs to complement its hardware expertise with the required skills on the connectivity, analytics, and service layer, which includes user front ends like apps as well. A second aspect which might be new for many manufacturing companies is the fact that servers and their software, as well as apps and other user front ends, need to be operated and maintained throughout the whole usage phase. This poses two challenges. On the one hand, the organization has to bear operating cost over the whole lifetime of the offering. And, on the other hand, the organization needs to have the capability to perform the abovementioned operations. Especially manufacturing companies might need to install new units taking care of those tasks [26].

The upside of smart connected products is new revenue sources which wait to be captured by an appropriate business model [23]. It provides a hybrid value proposition consisting of physical and digital parts. Both parts can be monetized either in a product (one-time payment and transfer of ownership) or service manner (continuous payments and usage rights). This opens up a space of four potential revenue sources. Especially in B2B scenarios, vendors manage to monetize two or more of those revenue sources. But even if only the hardware is being monetized, prices for connected products are, in many cases, much higher compared to similar non-connected products. For example, connected Philips Hue light bulbs sell for much higher prices compared to not connected light bulbs [27].

It should be noted that the creation of IoT products currently is not directed by specific guidelines or a systematic method. The application of a traditional

Fig. 1.8 Phases of IoT
product development

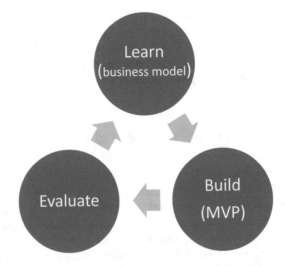

legacy product development paradigm to IoT products is generally ineffective and disadvantageous. Agile methodology and in particular Lean model is typically considered a good fit for many organizations working in the IoT domain. According to this model, when developing an IoT product, there are three main stages that are important to constructing a competitive and sustainable IoT solution (see Fig. 1.8):

- *Learn* – Develop an innovation plan and construct or revise the business model.
- *Build* – Implement and build a minimum viable product (MVP).
- *Evaluate* – Measure and evaluate the product and provide feedback to the first stage.

1.5.3 IoT Customer: Optimization and Cost Reduction

While IoT consumers might buy an IoT solution in order to increase their comfort, for the peace of mind or just for fun, in business to business cases, the customer always calculates a return on investment. For IoT solutions, this usually translates into expected cost reductions that amortize the investment. Among the most popular approaches to realize IoT-based cost reductions is condition monitoring [28]. Critical parameters in the production process, like soil moisture in the John Deere example, are being monitored and optimized in order to reduce waste, or equipment is being monitored with the aim to reduce downtime. Other approaches include optimizing the supply chain. IoT, in terms of RFID technology being applied in a warehouse for example, could lead to a much more detailed picture of the actual inventories of raw material. This in return could allow for reducing the warehouse stocks, which leads to cost reductions as well.

In general, the Internet of Things helps to gather data regarding the status of the physical world. This could be the condition of machines or other equipment, inventories in a warehouse/whereabouts of goods. Those data can be analyzed and leveraged by advanced machine learning algorithms and big data analytics algorithms to optimize a company's operations.

1.5.4 Important Aspects of Implementation

The three IoT-based business opportunities pose different challenges for companies. But in general, they require a significant change in the company's business model. While changing a business model is a serious management challenge already, business model innovation poses various additional challenges. Bilgeri et al. identified 16 barriers to IoT business model innovation. They are distributed along with the following innovation phases: idea generation, concept development and evaluation, technical implementation, and commercialization. Many of these issues are related to organizational questions. As already discussed, IoT solutions require continuous efforts, e.g., in back-end operations, maintenance, and development of new features throughout the whole lifecycle. Most incumbents from the manufacturing industry do not have units for these tasks in their organization yet. To name another example, IoT solutions provide the opportunity to sell services in addition to or rather than products. However, selling services requires different skills as well as controlling and financial mechanisms compared to selling products.

1.5.5 Data Monetization

Transforming IoT data into a marketable product is a fast-growing trend many companies are considering as a secondary revenue source; however, the idea of selling data is not a new one. Gartner has labeled the creation and utilization of data or information with the term, "*infonomics*" [29]. With millions of smart devices connecting to the IoT and collecting data, a new market based on data providers and data customers has been born (see Fig. 1.7). Profiting from IoT data can be approached in two ways [30]:

- *Direct Data Monetization* – Regardless of why you may be willing to offer your raw data, there are probably consumers interested in using and paying for your data. While there are many ways to sell data, a primary means is through a data marketplace. When selling data, direct monetization is generally separated into two categories [30]:

    - *Selling Raw Data* – Direct access to data (i.e., APIs or data sets) is provided in trade for cryptocurrency or money. There are two general marketplaces from which you can choose and the appropriate choice depends on your strategic requirements [30].

- *Centralized Marketplace* – This is a platform owned by one party that serves as a centralized location to exchange multiple kinds of data among diverse participants. In this marketplace, both metadata and raw data are stored.
- *Decentralized Marketplace* – This is a platform where participants are able to exchange data directly in peer-to-peer transactions. In this context, the marketplace only stores the metadata to enable data consumers to find the provider/owner of the data.

- *Selling Data Insights or Analysis* – Performing data analytics on raw data improves the quality of the information being sold. Not all companies have the capability to analyze data, creating an opportunity for monetization that is beneficial for both sides of the transaction. Analysis services can be offered in marketplaces or through other channels.

- *Indirect Data Monetization* – Data can be used to improve business intelligence and function, generate new products or services, and create new business models. Generally, there are two approaches to making good use of your own data [30]:

- *Data-Driven Optimization* – Utilizing data in this way decreases cost and increases the effectiveness and efficiency of business processes. This optimization is applicable across many fields. For example, manufacturing test benches could be optimized by shortening the testing time or field data could be utilized to improve the design of a product.
- *Data-Driven Business Models* – Monetizing by employing this strategy means that process or product data is used to generate new business opportunities or attract new customer groups through the development of new services or products or by improving existing products or services. Building a data-driven business model enables you to uncover innovative, new businesses rather than adjacent businesses. These models are also important for diversifying revenue streams. For example, Bosch makes use of manufacturing data to create customized subscription-based services that monitor the conditions of hydraulic systems.

The market of IoT data will keep growing as companies learn how powerful it can be to provide data to others and how much others are willing to pay to obtain data. The primary challenges around monetizing IoT data include [29]:

- *Ensuring Data Quality* – In order for customers to trust the data provided, it must be of high-quality and complete. The data should also be accurate and timely and have been obtained ethically.
- *Determining Information Type* – Providers of data will need to adapt and flex to customer needs as companies may require IoT data in diverse forms or

may consider data that was not originally fit to their particular business model. Customers may seek additional information for data points that were not initially recorded.

- *Traditional Product Management and Marketing* – IoT data is not like a traditional physical product. Therefore, companies may need to forego the usual activities that help sell physical products such as research, design, development, promotions, packaging, or marketing support.
- *Protecting Against Unlicensed Use* – It is very important to ensure data sovereignty for the creator of the data. It is easy to copy data, and thus it can become difficult to make sure customers are not utilizing data in unintended manners. Therefore, we need to consider contracts that ensure a licensed user understands the appropriate and ethical handling of information products, how to audit usage, etc.

1.5.6 Business Model

It is important to understand the basic business model before attempting to create an IoT Solution. The term *"business model"* was born toward the end of the 1990s when it became a buzzword in popular media. Since that time, it has received significant attention from scholars and business practitioners and currently exists as a clear point of interest in many areas of IoT. Typically, the business model is defined as an analytical model used to determine how a business functions. The available literature regarding the business model has not yet reached an agreement regarding which elements are vital to the creation of a business model. However, two widely known tools currently exist to illustrate business models: *St. Galler Magic Triangle* and *Osterwalder Business Model Canvas*.

The St. Galler Magic Triangle is comprised of four dimensions and is illustrated using a triangle shape (see Fig. 1.9) [31]:

- *Who*

  - Who are the target customers?
  - How can customers be classified into groups?
  - What are the basic demographics and shared characteristics of customers?

- *What*

  - What is the opportunity being offered to the customer?
  - What value is being added for the customer? (value proposition)
  - What combination of services or products make up the opportunity?

- *How*

  - How is the value proposition created, applied, and distributed?

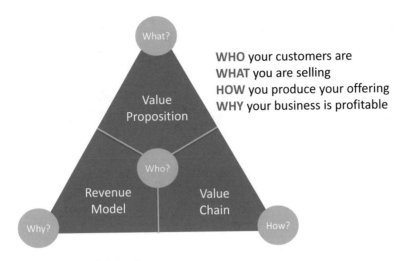

Fig. 1.9 St. Galler Magic Triangle

- How will the activities and processes need to provide the product look?
- What kind of resources will be needed?
- Which IoT business ecosystem stakeholders will be needed and how should they be organized?

- *Revenue*

  - Does it look as though the opportunity will be financially sustainable?
  - What does the cost structure look like?
  - What revenue mechanisms will be applicable?
  - How can the value proposition be monetized?

Thoughtfully answering the questions in each of the four areas noted above creates a solid business model and a foundation for further innovation in IoT ecosystem.

Osterwalder Business Model Canvas, created by Osterwalder in 2010, serves as an alternative method to the St. Galler Magic Triangle for illustrating a business model (see Fig. 1.10) [32]. It provides a well-known guide for explaining a business model in only one page. This model includes the following components [32]:

1. *Key Partners*: Who are the key partners and suppliers?
2. *Key Activities*: What key activities (e.g., marketing, designing, producing) our value propositions, distribution channels, customer relationships, and revenue streams need? What tasks does the company need to perform to fulfill its business purpose [32]. Some typical key activities in IoT business model include *Research & Development, Production, Marketing*, and *Sales & Customer Services*.

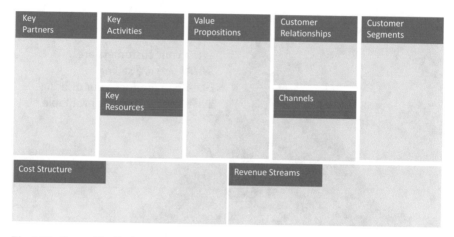

Fig. 1.10 Osterwalder Business Model Canvas

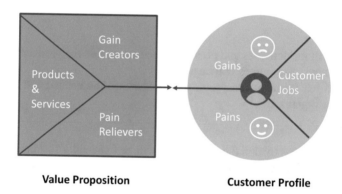

Fig. 1.11 Value Proposition Canvas

3. *Key Resources*: What key resources (e.g., *physical resources, intellectual resources, human resources, financial resources*) do our value propositions, distribution channels, customer relationships, and revenue streams require [32]?

4. *Key Propositions*: What value do we deliver to our customers? What bundles of products/services do we offer? Which problems of the customer are solved by our products/services? To find the key proposition, one can use the Value Proposition Canvas [32]. As shown in Fig. 1.11, the Value Proposition Canvas consists of two building blocks to be able to model and visualize the relationship between product/service and customer/market [32]:

- *Customer Profile*: This shows the task/job a customer needs to get done, potential pains that the customer might face during and after the job, and benefits that a customer expects from the product/service.

- *Value Proposition*: This shows the list of products/services, explains how they can kill the pains of the customer, and demonstrates how the offered products/services can create customer gains.

5. *Customer Relationship*: What type of relationship does each of our customer segments expect us to establish and maintain a long-term relationship with them? The most common types of customer relationships include transactional, personal assistance, self-service, automated services, communities, and co-creation. Note that these types of relationships can coexist in a company's relationship [32].
6. *Channels*: Through which channels (e.g., website, email) do our customer segments want to be reached [32]?
7. *Customer Segments*: From whom are we creating value? Who are our most important customers [32]?
8. *Cost Structure*: What are the most important costs inherent in our business model? Which key resources/activities are the most expensive [32]?
9. *Revenue Streams*: For what value are our customers really willing to pay? For what do they currently pay? How are they currently paying? How would they prefer to pay? How much does each revenue stream contribute to overall revenues [32]?

1.5.7 Minimum Viable Product (MVP)

The concept of a *minimum viable product* (MVP) was first introduced in Eric Ries' popular book, *The Lean Start-Up*, in 2001. The goal of an MVP is to evaluate if the product fits in the market with the smallest possible amount of risk. In this approach, a new product is created with features adequate to satisfy the earliest users. The final features are not developed until feedback from initial users can be evaluated. In short, the main idea is to construct a very simple, testable version of the product. The results of testing can be included in the next stage of development during the scaling phase or for revising the business model. A "build, evaluate, and learn" approach enables the solution provider to build the more important and viable basics into the product as quickly as possible. At the start of the process, there is usually a large-scale, almost unreachable vision of what the finished product will be, and shaping the vision at such a high level can consume large amounts of time and considerable resources. It is important to avoid the pitfall of trying to create a *perfect IoT product*. Instead, one should focus on creating a *possibly viable IoT product* with the potential to focus team creativity and original ideas throughout the process, while addressing the question of whether the product should even be created at all or not. In order to choose the most significant value proposition for creating the MVP, the company must concentrate on the specific intersection of the customer's wants and the value of the product as illustrated in Fig. 1.12. As shown in this figure, to be able to define the list of important features which should be

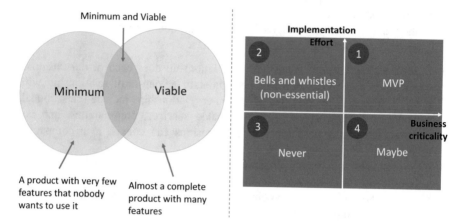

Fig. 1.12 Minimum viable product (MVP)

included in the MVP, we can classify the product features based on two dimensions, namely, implementation effort and business criticality.

- *Quadrant I* – Features in this area are vital to the MVP because they are critical to business success and are usually more straightforward to implement.
- *Quadrant II* – These features are nice, but they are not vital and are still easy to implement. Leaving these out of the MVP saves both time and resources.
- *Quadrant III* – Features in this area are trivial and can be hard to implement. They should be avoided in an MVP and in the following product iterations.
- *Quadrant IV* – These features are business critical, but are also arduous to implement. The elements in this quadrant need a maximum amount of deliberation and thought.

1.6 Summary

This chapter introduced the Internet of Things (IoT) with several definitions and discussed the benefits and challenges of establishing the IoT. There are advantages to be gained in the personal lives of individuals as well as the operations of businesses and manufacturers. This chapter discussed all the promises and challenges of IoT. The complete IoT stack from the sensors and devices, to the fog, and to the cloud has also been explained. There are several commercial frameworks, cloud technologies, and IoT-enabled devices and ecosystem providers, which we presented their offerings briefly. Next, some examples of the applications of IoT technology have been listed with their expected impacts to varied sectors of our society, from agriculture to the cities in which we live. Finally, the details of the business implications, business models, and opportunities of IoT have been addressed.

References

1. IBM. Available from: https://www.ibm.com
2. SAP. Available from: https://www.sap.com/
3. Gartner
4. Bosch. Available from: https://www.bosch.com/
5. IDC. Available from: https://www.idc.com
6. Intelligent IoT. Available from: https://www2.deloitte.com/insights/us/en/focus/signals-for-strategists/intelligent-iot-internet-of-things-artificial-intelligence.html
7. A. Rayes, S. Salam, *Internet of Things—From Hype to Reality* (The road to Digitization. River Publisher Series in Communications, Springer, Denmark, 2017), p. 49. https://www.amazon.de/Internet-Things-Hype-Reality-Digitization/dp/3319448587
8. NIST: National Institute of Standards and Technology. Available from: https://www.nist.gov/
9. P. Raj, A.C. Raman, *The Internet of Things: Enabling Technologies, Platforms, and Use Cases* (Auerbach Publications, 2017)
10. A. Botta et al., Integration of cloud computing and internet of things: a survey. Futur. Gener. Comput. Syst. **56**, 684–700 (2016)
11. Internet of Things World Forum. Available from: https://www.iotwf.com/
12. F. Firouzi et al., Keynote paper: from EDA to IoT eHealth: promises, challenges, and solutions. IEEE Trans. Comput. Aided Des. Integr. Circuits Syst. **37**(12), 2965–2978 (2018)
13. B. Farahani et al., Towards fog-driven IoT eHealth: promises and challenges of IoT in medicine and healthcare. Futur. Gener. Comput. Syst. **78**, 659–676 (2018)
14. FIWARE. Available from: https://www.fiware.org/
15. AWS IoT. Available from: https://aws.amazon.com/iot/
16. Microsoft Azure IoT. Available from: https://azure.microsoft.com/en-us/services/iot-hub/
17. D. Hanes et al., *IoT Fundamentals: Networking Technologies, Protocols, and Use Cases for the Internet of Things* (Cisco Press, 2017)
18. I.A.T. Hashem et al., The role of big data in smart city. Int. J. Inf. Manag. **36**(5), 748–758 (2016)
19. Diese Digitalexperten sind bei deutschen Firmen besonders begehrt. Available from: https://www.handelsblatt.com/unternehmen/beruf-und-buero/digitaler-jobindex/digitaler-job-monitor-diese-digitalexperten-sind-bei-deutschen-firmen-besonders-begehrt/22957856.html?ticket=ST-1817019-o0qDpTQqZsW0sFKadd20-ap
20. Roundup of Internet of Things Forecasts. Available from: https://www.forbes.com/sites/louiscolumbus/2017/12/10/2017-roundup-of-internet-of-things-forecasts/#6a7ab8041480
21. E. Fleisch, M. Weinberger, F. Wortmann, *Business Models and the Internet of Things, Bosch IoT Lab Whitepaper* (Bosch Internet of Things and Services Lab, 2014)
22. Businesses are Expected to Continue IoT Adoption Despite Security Risks, Survey Says. Available from: https://biztechmagazine.com/article/2017/02/businesses-are-expected-continue-iot-adoption-despite-security-risks-survey-says
23. F. Wortmann et al., Ertragsmodelle im Internet der Dinge, in *Betriebswirtschaftliche Aspekte von Industrie 4.0*, (Springer, 2017), pp. 1–28
24. John Deere, John Deere Prescision Ag Technology, Brochure. Available from: https://www.deere.com/assets/publications/index.html?id=004d03e7#36
25. D. Bilgeri, et al., *The IoT business model builder.* A White Paper of the Bosch IoT Lab in collaboration with Bosch Software Innovations GmbH, 2015
26. D. Bilgeri, F. Wortmann, E. Fleisch, How digital transformation affects large manufacturing companies' organization. 2017
27. E. Fleisch et al., *Revenue Models and the Internet of Things? A Consumer IoT-based Investigation* (ETH Zurich, 2016)
28. How the Internet of Things is driving cost-saving efficiencies for manufacturers, The shi blog. Available from: https://blog.shi.com/hardware/internet-things-driving-cost-saving-efficiencies-manufacturers/

29. D.B. Laney, *Infonomics: How to Monetize, Manage, and Measure Information as an Asset for Competitive Advantage* (Routledge, 2017)
30. A guide to data monetization. Available from: https://blog.bosch-si.com/business-models/a-guide-to-data-monetization/
31. O. Gassmann, K. Frankenberger, M. Csik, The St. Gallen business model navigator. 2013
32. Osterwalder Business Model Canvas. Available from: http://alexosterwalder.com/

Chapter 2
The Smart "Things" in IoT

Farshad Firouzi, Bahar Farahani, and Mahdi Nazm Bojnordi

> *Be as smart as you can, but remember that it is always better to be wise than to be smart.*
>
> Alan Alda

Contents

F. Firouzi (✉)
Department of ECE, Duke University, Durham, NC, USA

B. Farahani
Shahid Beheshti University, Tehran, Iran

M. N. Bojnordi
University of Utah, Salt Lake City, UT, USA

© Springer Nature Switzerland AG 2020
F. Firouzi et al. (eds.), *Intelligent Internet of Things*,
https://doi.org/10.1007/978-3-030-30367-9_2

2.1 Definition and Architecture of Smart Things

The phrase "smart things" is sometimes used interchangeably with similar terms including smart object, IoT device, IoT endpoint, endpoint device, smart sensor, intelligent device, intelligent node/thing, or ubiquitous thing. Regardless of the term used, a smart thing is an object with embedded electronics which can exchange data over a network without any human interaction [1]. A typical smart thing is made up of several components, namely, a processing unit, a power source, a communication device, and transducers (see Fig. 2.1) [2].

- *Processing Unit* – Used to obtain data and process and analyze sensor information, synchronizes actuator control signals, and controls smart object functions like power systems and communication. The kind of processing unit may vary based on the application needs; however, the *microcontroller* is the most common type of the processing unit used in IoT things because of being small, easy to program, flexible, power efficient, low cost, and ubiquitous.
- *Transducers* – Transducers are devices that convert/transfer one energy into another. Common energy domains include electrical, fluid, thermal, mechanical, and chemical. Transducers are categorized into two groups, namely, *sensors* and *actuators*. Actuators and sensors enable smart objects to interact with the physical world.

  - *Sensors*: Sensors recognize the existence of energy as well as changes in or transfers of energy (i.e., motion, heat, light, or chemical reaction) and then produce an output that can be understood or read. In general, the output is an electrical signal (analog or digital) that is readable.
  - *Actuators*: Actuators are tasked with utilizing energy to produce motion. Essentially, actuators are devices that change energy into motion or mechanical energy; therefore, an actuator is a kind of transducer. Two basic kinds of motion are rotary motion and linear motion. Linear actuators transform energy into a straight-line motion useful in positioning applications that require a

Fig. 2.1 Block diagram of typical smart things (smart object) in IoT

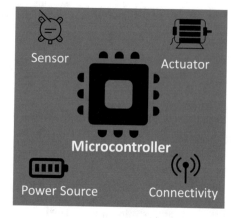

Fig. 2.2 An illustrative example of the hardware/software layers in an IoT thing

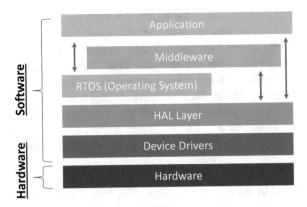

push/pull motion. Rotary actuators transform energy into a rotating motion useful in controlling valves such as a butterfly or ball valve. Actuators can take many forms and vary in size or power configuration, depending on their intended use. The most well-known actuators include:

- *Thermal Actuators* – Often a dual-metallic strip that transforms thermal energy into motion
- *Mechanical Actuators* – Convert mechanical energy input (often rotary) into linear motion (i.e., screw jack)
- *Electrical Actuators* – The most common actuators in the IoT domain that convert electrical energy into motion or mechanical energy (i.e., electrical motors)

- *Communication Device* – A wired or wireless device that connects a smart object to the physical world or other smart objects using a network; most often, smart objects using IoT networks are connected wirelessly due to limited infrastructure, deployment ease, and cost efficiency. The details of communication technologies for IoT application will be discussed in the next chapter.
- *Power Source* – All smart objects contain elements that require power. Generally, the communication device is the component that consumes the most power. Smart objects usually contain limited power, have long deployment periods, and are not easy to access. Because of these factors, especially when a smart object relies on battery power, it is important to design smart objects with efficient power use, connectivity, sleep modes, and low-power components.

As shown in Fig. 2.2, from the hardware/software point of view, smart IoT things consist of six architectural layers.

- *Hardware:* Hardware comprises the physical implementation of various components of the IoT thing, such as the processing units, transducers, power supply system, and the communication interface for connecting the device to the outside world. Various technologies are necessary to make the hardware implementation of an IoT possible. Low-power and often high-performance CMOS devices are commonly used for building the processing units. Various memory types may be necessary for a typical IoT think that rely on both volatile and nonvolatile memories. Power supply and distribution networks are specifically designed and optimized for IoT things to better serve user applications. The transducers and communication devices are modular components that may be included in an IoT thing. The higher-level software may access these devices through device drivers.

- *Device Drivers*: Software that interfaces with and controls hardware allows operating system and other programs to access hardware functions without requiring an understanding of the underlying hardware details
- *Hardware Abstraction Layer (HAL)*: A software layer on top of the device drivers that defines the protocols, tools, and routines needed to interact with hardware. HAL is concerned with creating high-level functions required to enable hardware to function without extensive knowledge of how the hardware works. This is important for developers working with many microcontroller hardware pieces that require port applications to connect platforms. A HAL also enables engineers with less knowledge of lower-level hardware to create useful application code without knowledge of all the small details. The principal difference between a driver and HAL is that HAL is constructed on top of device drivers and is capable of hiding hardware differences from higher software layers. For example, a USB mouse driver and a PS2 mouse driver are very different, but with the help of HAL, you can treat them interchangeably.
- *Real-Time Operating System (RTOS)*: Generally, operating systems multitask to enable many programs to execute simultaneously. The scheduler determines which program should run and in what order. The scheduler then quickly switches between programs so that it looks as all programs are executing simultaneously. For example, a desktop OS such as Windows includes a scheduler that attempts to ensure user responsiveness. In contrast, the scheduler in an RTOS functions toward creating a more predictable pattern of execution where each task must complete within a given time budget. Particularly, this feature is important for IoT systems as they include real-time applications with strict response time requirements.
- *Middleware:* IoT devices are often designed and optimized for a class of target applications using a heterogeneous system on a chip (SoC) with diverse IP cores for processing, sensing, and communication. Middleware eases the development process of such systems by supporting interoperability within the diverse applications and services that compose an IoT device. Along these lines, numerous operating systems have been developed to support the development of IoT middleware. Middleware combined with the HAL and device driver layers provide the necessary functionalities to enable service deployment. Example constituent services of a diverse application domain for IoT systems are wireless sensor networks (WSNs), radio frequency identification (RFID), machine-to-machine communication, and supervisory control and data acquisition (SCADA). Middleware is responsible for managing the interaction between the application layer and a variety of devices in the system. Various functional components are often considered to manage this interaction. Interface protocols oversee

providing technical interoperability. Device abstraction is used to make the interaction among the application components possible. Central control and context detection are used to support context-aware computation for the entities that interact with the system. Application abstraction is to interface the high-level applications and end users with the underlying devices of the system.

- *Application Programming Interface (API):* Defines the tools, protocols, and routines that are required to create an application. APIs are designed to be generic and independent from specific implementations so that an API can be utilized across multiple applications with only slight changes to implementation only (no change to behavior or general interface). While HALs and APIs are similar, they serve two different purposes in software development. HAL is located between low-level drivers and creates a common interface space for software stacks (i.e., RTOS) or middleware components (i.e., Ethernet, USB, file systems). A HAL can also serve as a driver interface or as a wrapper or common interface for higher-level code and current drivers. An API serves as a tool to assist advanced developers in quickly creating application code by providing interface code needed to manage real-time system behavior and allowing access to common components like file access or serial communication.

- *Application*: The application layer is responsible for producing services and defining a set of protocols for communication among different IoT applications, such as logistics, retail, and healthcare. In addition to various applications, numerous application protocols may be found in the literature, for example, extensible message and presence protocol (XMPP), message queue telemetry transport (MQTT), constrained application protocol (CoAP), and representational state transfer (REST). MQTT is a machine-to-machine architecture that enables lightweight connectivity that supports publish and subscribe over TCP. CoAP employs a request and response protocol to enable communication in recourse-constrained environments. XMPP is mainly designed for instant messaging, multiparty chats, and audio calls. Important parameters such as the bandwidth requirement, data latency, reliability, and memory footprint need to be considered to choose an application protocol for an IoT system.

2.2 Sensors

Sensors are used to detect incidents or alterations in the physical world (i.e., sound, temperature, pressure, motion, flow, magnetic qualities, or chemical/biochemical factors), obtain required data, and then send the data to a monitoring system [2, 3]. Outlined below are the different kinds of sensors useful in various IoT applications.

- *Temperature Sensors* – The most common sensor in the IoT domain because most biological/chemical, physical, and electronic/mechanical systems can be altered by temperature. The four main categories of temperature sensors include:

- *Thermocouple Sensors* – They are made of two different metal conductors (e.g., chrome and alumel) that have dissimilar thermal conduction characteristics. The conductors are welded at one end so to generate temperature-dependent voltage through a thermoelectric effect.
- *Resistance Temperature Detector (RTD)* – The level of resistance changes based on the temperature in pure material, such as platinum, nickel, and copper. RTD sensors are based on this phenomenon to measure temperature. Due to their stability, precision, and repeatability, RTD sensors are often used to monitor laboratory or industrial process temperatures.
- *Thermistors* – They employ semiconductor elements to measure temperature. Similar to the RTD sensors, the level of resistance in thermistors changes based on temperature. These sensors exhibit a more nonlinear resistance behavior as the temperature varies.

• *Flow Sensors* – These sensors can measure flow rate and the total volume of gases or liquids in pipes or systems. Such capability is vital for practical everyday applications such as brewing machines as well as scientific endeavors like monitoring the flow of highly pure acids. An example of how important flow monitoring sensors are can be found in the 2014 Flint, Michigan, United States, water crisis. The city of Flint switched its water sourcing from the Great Lakes through Detroit Water to obtaining water through the Flint River. However, officials failed to identify high levels of lead, a very serious danger to public health, in the water source. The very acidic water of the Flint River leeched lead out of the pipes and passed it into the water, resulting in very high levels of heavy metals in the water. Thousands of children who consumed or were exposed to the contaminated water experienced severe health problems. In 2016, criminal charges were filed by Michigan's attorney general against three individuals regarding the crisis.
• *Level Sensors* – These sensors measure fluid levels constantly or at certain points. Level sensors can be used in many ways and are capable of measuring contained fluids or elements in a natural setting such as oil levels at an oil rig site. For example, ultrasonic sensors do not require direct contact and are often utilized in measuring viscous liquids or bulky solid materials. These sensors are also useful in controlling pumps and measuring open channel flow in water treatment plants. Capacitance level sensors measure the existence of liquids and solids using radio frequency signals in the circuit.
• *Imaging Sensors* – This sensor captures and transforms the changing attenuation of waves into signals. Imaging sensors are often found in medical imaging machinery, digital cameras, and night vision paraphernalia.
• *Noise Sensors* – High noise levels from machinery, aircraft, trains, construction, or loud music can cause harm to both humans (i.e., cardiovascular or hearing damage) and animals. Regulatory entities have begun employing noise sensors in order to better measure noise pollution and disturbances that may cause harm. For instance, ambient noise sensors constantly measure environmental noise levels and transmit an electronic signal to an ambient noise system when the noise level

changes. The system can take automatic action such as regulating the music level or notifying authorities.

- *Pressure Sensor* – A typical pressure sensor comprises a pressure-sensitive element to determine the actual pressure applied to the sensor and some components to convert the sensed pressure into an output signal. Examples of pressure-sensitive elements that have been examined in the literature are the metal strain gauges that are glued together, capacitance variable cavity and diaphragm, and silicon diaphragm with integrated strain.
- *Barometer* – Quantifies atmospheric pressure.
- *Altimeter* – Measures altitude of an object above a certain designated level.
- *Acoustic Sound Sensors* – Measure and convert the level of sound into a digital or analog signal, for example, hydrophone, microphone, and geophone.
- *Radiation Sensors* – Measure radiation in the environment through ionization or scintillating detection.
- *Biosensors* – Detect biological components including enzymes, antibodies, organisms, cells, tissues, and nucleic acid.
- *Ohmmeter* – Quantifies resistance.
- *Voltmeter* – Calculates the voltage.
- *Galvanometer* – Determines current.
- *Watt-Hour Meter* – Measures electrical energy provided to and consumed by a business or residence.
- *Oxygen Sensor* – Calculates the percentage of oxygen present in a liquid or gas.
- *Carbon Dioxide Detector* – Recognizes the presence of CO2.
- *Light Sensor (Photodetector)* – Perceives light as well as electromagnetic energy.
- *Photocells (Photoresistor)* – Resistor affected by ambient light intensity changes.
- *Infrared Sensor* – Recognizes infrared radiation.
- *Seismometer* – Measures seismic waves.
- *Acoustic Wave Sensor* – Calculates wave velocity to recognize chemicals present.
- *Air Pollution Sensors* – These sensors generally monitor five main air pollutants including nitrous oxide, ozone, sulfur dioxide, particulates, and carbon monoxide.
- *Infrared Sensors* – Monitor an object's movement by generating and receiving infrared heat waves.
- *Moisture and Humidity Sensors* (Hygrometer Sensors) – Measure and record air humidity using electrical capacitance.
- *Speed Sensors* – Often used to detect vehicle speeds. Examples include laser surface velocimeter, speedometers, Doppler radar, and wheel speed sensors.
- *Accelerometer* – Small device that detects static (i.e., gravity) and dynamic acceleration (i.e., starting/stopping). These sensors are most often used in tilt sensing (or stop sensing) because they react to gravity and tell you how something is oriented in relation to the Earth's surface. For example, Apple's iPhone uses an accelerometer to detect if the phone is being held in landscape or portrait mode. Accelerometers are also capable of sensing motion. For example, Nintendo's Wiimote can sense the imitated forehand or backhand tennis racket motion or the motion of rolling a bowling ball. This device is also able to sense

Table 2.1 Types of proximity sensors

Sensor technology	Range of use	Common use
Inductive sensor	4 mm–40 mm	Close-range detection of ferrous metals such as copper, aluminum, iron, etc.
Capacitive sensor	3 mm–60 mm	Close-range detection of nonferrous materials such as liquid, plastic, or wood
Photoelectric sensor	1 mm–60 mm	Long-range military target recognition
Ultrasonic sensor	3 mm–30 mm	Long-range military target detection with difficulties such as many colors or inconsistent service

the state of free fall. This feature is included in many computer hard drives. If the device senses that the hard drive is falling, the drive automatically turns off to protect against losing data. When choosing an accelerometer, the key characteristic for consideration is number of axes measured. It is actually very simple: There are three possible axes (x, y, z). Of the three, how many can the accelerometer measure? A device that can measure all three axes is usually the best option. These are widely available and are usually no more expensive than accelerometers that measure only one or two axes.

- *Gyroscope* – Measures angular velocity (i.e., how fast an object spins around an axis). When measuring the orientation of a moving object, an accelerometer may not be able to provide orientation-specific information. However, gyroscopes are not affected by gravity and work as a complementary partner with the accelerometer. Angular velocity is measured in units of rotation per minute (RPM) or in degrees per second. The axes of rotation (x, y, and z) are also referred to as roll, pitch, and yaw. Gyroscopes have proven useful in navigating space, controlling missiles, providing underwater navigation, and guiding aircraft. Currently, gyroscopes and accelerometers are being used together in vehicle navigation and motion-capture applications. In general, gyroscopes are not as advanced as accelerometers, and less expensive three-axis gyroscopes have only recently become available on the market, with most commonly available gyroscopes still measuring only one or two axes. When purchasing a gyroscope, it is important to understand which axes are measurable. For example, one two-axis gyroscope could measure pitch and roll, while another two-axis gyroscope measures pitch and yaw.

- *Proximity and Displacement Sensors* – Use electromagnetic fields, sounds, or light to detect the absence or presence of objects. The different types of proximity sensors are each useful in specific environments or scenarios (see Table 2.1) [3]:

  - *Inductive Sensors* – Detect ferrous material at close range
  - *Capacitive Sensors* – Detect nonferrous material at close range
  - *Photoelectric Sensors* – Detect long-range targets
  - *Ultrasonic Sensors* – Detect long-range targets with difficult surfaces

2.3   Actuators

While sensors provide data, actuators are devices responsible for acting or control-ling systems. Actuators take data or sources of energy (i.e., hydraulic fluid pressure) and convert it into motion. The most well-known actuators in IoT applications include *relays* and *electric motors* (e.g., DC motors, servo motors, and stepper motors).

2.3.1   Switches and Relays

Switches control the flow of electrons to open/close the connection in an electric circuit. While a switch is usually a mechanical device, relays are a kind of switch that is controlled electronically. Depending on the operational mode, relays can be categorized as "normally open" or "normally closed." A normally open switch does not conduct electricity until energized and normally closed switches will conduct electricity until energized. Based on the structure of their inputs and outputs, relays can be categorized as follows (see Fig. 2.3) [4, 5]:

- *Single Pole Single Throw Relay (SPST)* – SPST stands for single pole single throw meaning that it has one input which can be connected to a single output. The relay is actuated by a single coil.
- *Single Pole Double Throw (SPDT)* – This relay includes one input and two outputs useful in applications requiring a switch between two circuits. A single coil can actuate the delay.
- *Double Pole Single Throw Relay (DPST)* – This relay includes one coil, two inputs, and two outputs, and it can be considered as two SPST relays. However, it needs just one coil of them. Each input has one corresponding output.
- *Double Pole Double Throw (DPDT)* – This relay includes one coil, two inputs, and two outputs, and it is equivalent to two SPDT switches or relays which is activated by only one coil.

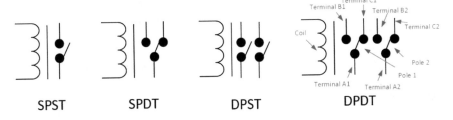

SPST   SPDT   DPST   DPDT

Fig. 2.3 Types of relay

Finally, relays can also be categorized in the following manner [4, 5]:

- *Electromechanical Relays (Armature Relays)* – Electrical switches are made of coils, contacts, and a moveable armature. These relays are mostly used to control high-power electrical devices. An energized coil creates an electromagnetic field which in turn it can open/close the circuit using physical movement of armature contacts (see Fig. 2.4).
- *Reed Relays* – Switches comprised of overlapping ferromagnetic blades and coils wrapped around reed switches. This type of relay utilizes mechanically actuated physical contacts to open or close a circuit. However, Reed relay contacts are smaller and have less mass than electromechanical relay contacts (see Fig. 2.5).
- *Solid-State Relays* – Light, commonly from an encapsulated LED, actuates a photosensitive MOSFET enabling current to flow through (see Fig. 2.6).

Fig. 2.4 Electromechanical relay (armature relay)

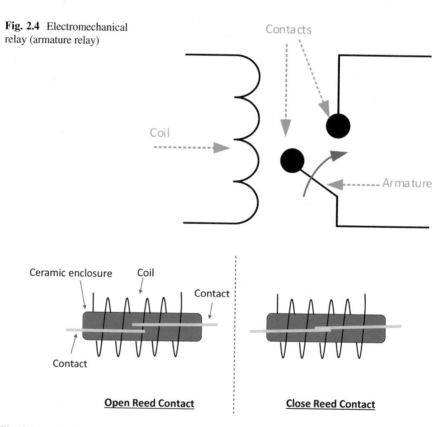

Fig. 2.5 Reed relay

Fig. 2.6 Solid-state relay

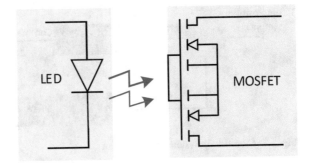

2.3.2 *Electrical Motors*

The common electrical motors in IoT applications include DC motors, servo motors, and stepper motors.

Direct Current (DC) Motors Many movement-based applications utilize DC motors because they are cost-efficient, easily drivable, electric motors. Examples of a DC motor include radio-controlled car wheels or computer cooling fans. Each DC contains two terminals (a ground wire and a power wire), across which voltage is applied. The motor's rotation direction can be adjusted by changing the voltage polarity across the terminals. The motor's speed is proportional to the level of voltage used, and the motor's torque is proportional to the level of current. A DC motor's speed is controlled with pulse width modulation (PWM), a means of quickly pulsing power on and off. The motor speed is determined by the percentage of time spent cycling on and off. For example, the motor will rotate at half the speed of 100% (completely on) if the power is cycled at 50%. However, every pulse is so fast that it looks like the motor is constantly spinning.

Servo Motors These motors are utilized for specific tasks requiring a precisely defined position such as controlling a robotic arm, moving a camera, or adjusting a boat's rudder. Unlike DC motors, servo motors do not rotate freely because the angle of rotation is typically limited to approximately 180 degrees. These motors work based on closed-loop mechanisms able to utilize position feedback to maintain motion and control the position. Servo motors are generally comprised of four elements including a DC motor, control circuit, gearing set, and position sensor encoder. The encoder is often a potentiometer able to generate speed and position feedback. These motors generally contain control, power, and ground wires. Power is continuously supplied and the servo control circuit is responsible for managing the power draw needed to drive the motor. As input, servo motors need a control signal representing the final position. Next, power is applied to the DC motor until the shaft rotates to the appropriate position determined by the position sensor (see Fig. 2.7).

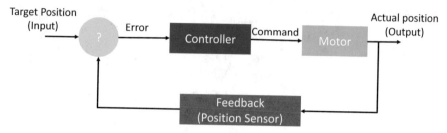

Fig. 2.7 Closed-loop system of servo motor

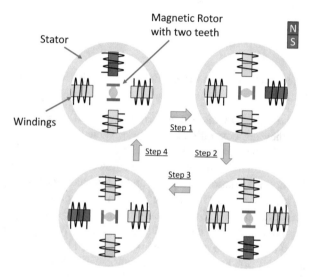

Fig. 2.8 A simple example of stepper motor

Stepper Motor This is an electric DC motor that divides full rotations into equal steps. Stepper motors are often found in 3D printers or similar devices that require very specific positioning. These motors include multiple windings and the voltage is applied in accurate sequences to rotate the motor shaft. Based on the applied voltage, the motor rotates step-by-step incrementally. More precisely, stepper motors define position by using multiple-toothed electromagnets arrayed around a central gear. An electromagnet is powered and attracts the gear's teeth, making the motor shaft rotate. When the teeth are in alignment with the initial electromagnet, it is slightly offset from the second electromagnet. When the second electromagnet is powered, the first one turns off and the gear turns to align with the second electromagnet. This process is repeated to make a complete rotation. Note that each turn is known as a "step" and a complete rotation consists of an integer number of steps (see Fig. 2.8). This process enables the motor to be turned to a precise angle. It is worth noting that stepper motors need a microcontroller or external control circuit to independently power each electromagnet and turn the motor shaft. The primary advantage of a stepper motor as opposed to a servo motor is their ability to control position. While servo motors need a feedback mechanism and supportive circuitry to adjust the position, a stepper motor can control position through the fractional, incremental rotation.

2.4 Processing Unit: Microcontroller

While the terms "microprocessor" and "microcontroller" have often been inter-changed with one another, both were created for use in real-time IoT applications. Although they share many similarities, they also have distinct differences. A microprocessor is an integrated circuit (IC) that includes only a central processing unit (CPU) inside (i.e., Intel's Pentium 1 . . . 4, Core 2 Duo, i3, i5, etc.). Micropro-cessors do not include ROM, RAM, or other peripherals on the chip and target general-purpose applications in which tasks are unspecified (i.e., photo editing, document creation, developing games, software, websites, etc.). In these cases, the input/output relationship hasn't been defined; therefore, large amounts of resources such as ROM, RAM, I/O ports, etc. are required. However, microcontroller units (MCU) include a CPU as well as memory and programmable input and output peripherals. Modern manufacturers (i.e., TI, Microchip, Freescale, ATMEL, Philips, Motorola, etc.) can create microcontrollers with a broad range of features across multiple versions. Microcontrollers are utilized to automatically control devices like remote controls, implantable medical devices, home appliances, power tools, auto-mobile engine control systems, children's toys, and other products with embedded systems. Because applications are highly specific, small resources such as ROM, RAM, input/output ports, etc. are needed and can be included on the same IC, which reduces both cost and size.

2.4.1 Classifications of Microcontrollers

Microcontrollers can be categorized based on bus-width, memory structure, instruc-tion set architecture, and memory map [6].

2.4.1.1 Classification by Bus-Width (Number of Bits)

Microcontrollers can have 8-bit, 16-bit, or 32-bit data unit and register.

- An 8-bit microcontroller includes an 8-bit internal data unit, and the ALU handles 8-bit arithmetic and logic operations. Intel 8080 is the first widely adopted 8-bit microprocessor that has 8-bit data words and 16-bit addresses.

- A 16-bit microcontroller functions with greater performance and precision than its 8-bit counterpart (i.e., extended 8051XA and PIC2x families).
- A 32-bit microcontroller utilizes a 32-bit instruction set to complete both logic and mathematical operations. These MCUs can generally be used to control devices such as office and home appliances, medical devices, systems designed for engine control, and other embedded systems (i.e., PIC3x and Intel/Atmel 251 family). Please note that 32-bit architecture does not mean that the system has an exact 32-bit data path. For example, IBM System/360 is considered a 32-bit architecture; however, it has only an 8-bit native path width, and all 32-bit arithmetic operations are implemented by executing several 8-bit operations at a time.

2.4.1.2 Classification by Instruction Set (RISC vs CISC)

Instruction set architecture (ISA), also known as computer architecture, is an abstract model of the underlying design that serves as an interface for hardware and software. ISA defines several key parameters of the processor including instruction set, addressing modes, I/O, memory (registers, etc.), as well as data types. Note that different processors with different designs (microarchitectures) can share the same ISA. For instance, Intel and AMD both implement x86 ISA while having different microarchitectures.

An ISA can be implemented with different microarchitectures that can vary in cost, physical size, and performance. However, software created to run on one ISA implementation can generally run on other implementations of the same ISA. This flexibility supports binary compatibility among various generations of computers and enables the creation of processor families.

ISAs can be categorized in a variety of ways, but a common means of classification is based on instruction set complexity. Complex instruction set computers (CISC) include several specialized instructions that are infrequently needed in practical programs. Therefore, the instruction set of a CISC is very sophisticated, and typically, instructions have different complexity levels resulting in different execution times. On the other hand, a reduced instruction set computer (RISC) only implements those instructions that are frequently used in applications. In RISC, operations that are rarely needed are designated as subroutines, and the additional processing time needed to execute them is offset by less frequent use. In addition, in RISC, generally instructions can be executed within one clock cycle, but it needs to execute more instructions compared to CISC. Thanks to the simplicity of RISC, it usually results in low-power consumption, which is critical in battery-powered systems.

2.4.1.3 Classification by Memory Structure and Bus Architecture

A modern microcontroller is a structure that includes one or more processing cores, a simple or complex memory hierarchy, and interface controllers to connect with the peripheral devices. The microcontroller architecture is designed after a model proposed by John von Neumann (1903–1957), which consists of a central processor to perform arithmetic and logic operations, a control unit, memory and mass storage, and input and output. This model is sometimes referred to as the Princeton Architecture. The von Neumann model employs the same physical memory for both instruction and data. As the instruction fetch and data access cannot happen simultaneously, two clock cycles may be necessary for executing an instruction. Figure 2.9 demonstrates a general block diagram of the von Neumann model machine. Data and instruction are stored in a unified memory and accessed through the same memory bus. The proportions of data and instruction in the memory may vary from one application to another. The von Neumann model was later enhanced for modern processors by separating instruction and data using dedicated memory units. This new model, which is referred to as Harvard architecture, enables transmission of data and instruction to take place on separate buses simultaneously. As a generally accepted practice, the Harvard architecture utilizes two distinct memories or separate instruction and data cache units connected to a unified memory in high-performance systems. Figure 2.9 illustrates the block diagram of an example of Harvard architecture. The main advantages of adding a second bus to the system include:

- *Pipelining* – Thanks to the dedicated bus for instruction, a second instruction can be obtained from memory as the first instruction is being executed, leading to a greater performance. We will discuss the details of the pipelining concept later in this section.
- *Wider Instructions* – In the Harvard architecture, the size of instruction memory words may be larger from the data word; therefore, more instructions may be fetched to the processor that may lead to performance improvement for compute intensive applications.

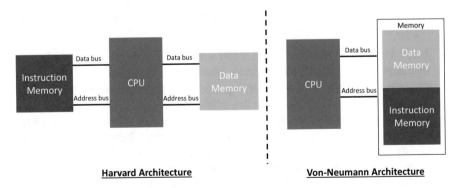

Fig. 2.9   von Neumann and Harvard processor architectures

Table 2.2 Comparison between port-mapped I/O and memory-mapped I/O

Port-mapped I/O	Memory-mapped I/O
Two different address for I/O and memory	Memory and I/O use the same address bus
To access I/O, we need a special class of instruction	Same instructions can access both I/O and memory
Intel x86 is implemented based on port-mapped I/O	Most widely I/O technique in processors

2.4.1.4 Classification by IO

A microcontroller is a group of resources from the perspective of the computer programmer. Every resource is identified using at least one "address" in an "address space." A memory map is a pictorial representation of the way resources related to addresses. Usually, memory maps are based on the structure of hardware as generated by the microcontroller and external devices. Changes cannot be made to the map while a program is being executed. However, in some situations, writing in "special configuration registers" will allow the user to disable or move resources to a new part of the address space. Microprocessor generally connects external devices using two methods (see Table 2.2) [7]:

- *Memory-Mapped Input/Output (MMIO)* – In this method, I/O devices and ROM are mapped into the system memory map together. Accessing hardware requires only reading or writing to their corresponding address using normal memory access instructions. One of the primary benefits of this method is that each instruction with the ability to access memory can be utilized to control input and output devices. One of the main drawbacks is that the whole address bus must be completely decoded for each device. For instance, a system with a 32-bit address bus would need logic gates to process the state of all 32 address lines and to be able to decode the unique address of any device. This may result in several overheads such as higher cost and reduced operating frequency (i.e., higher delay).

- *Port-Mapped I/O (PMIO or Isolated IO)* – This method requires the I/O devices be mapped into distinct address spaces. Most often this is achieved by utilizing a separate set of signal lines to delineate memory access versus IO access. Address lines are generally shared between two address spaces (i.e., memory and IO), but fewer of them are utilized for IO access. For example, a PC using 16 bits of IO address space may have 32 bits of memory address space. The primary advantage of this method is that less logic is required to decode an IO address. When it comes to software development, this method is less advantageous compared to MMIO because a larger number of instructions are needed to complete the same task. For example, testing one bit on a memory a single instruction is required. On the other hand, for IO, we should first read its data to a register before testing the corresponding bit.

2.4.2 Three Main Types of Microcontrollers

The three well-known microcontrollers currently on the market include PIC MCUs, AVR MCUs, and ARM MCUs [6].

2.4.2.1 Peripheral Interface Controller (PIC) Microcontrollers

These microcontrollers are a group of specialized microcontroller chips created in the early 1990s by Microchip Technology in Chandler, Arizona. PIC microcontrollers are created using a Harvard architecture and are utilized in a variety of device families. Basic and midrange families utilize 8-bit data memory, while higher-end device families utilize 16-bit data memory. In 2007, Microchip began producing 32-bit microcontrollers using the MIPS32 Core. The MIPS32 core is a reduced instruction set architecture (RISC) developed by MIPS Technologies, Inc. PICs instruction sets can vary from 35 instructions on the low end to 80+ instructions for higher-end PICs.

2.4.2.2 AVR Microcontrollers

The AVR microcontroller was initially created in 1996 at Atmel Corporation by Vegard Wollan and Alf-Egil Bogen. The AVR is an acronym standing for *A*lf-Egil Bogen and *V*egard Wollan's *R*ISC processor. These microcontrollers utilize a modified Harvard RISC architecture with divided instruction and data memory buses. The AVR performance is generally higher in comparison to a PIC. AVR microcontrollers can be divided into four categories:

- *TinyAVR* – Smaller in size with less memory; utilized for simple applications:

  - .5–32 KB program memory
  - 6–32 pins
  - Limited peripherals

- *MegaAVR* – Highly popular as these provide a larger memory (up to 256 KB) and more in-built peripherals; utilized for modestly to more complex applications:

  - 4–256 KB program memory
  - 28–100 pins
  - Larger instruction set (e.g., multiply instructions)
  - Extensive peripheral set

- *XmegaAVR* – 8-/16-bit AVR XMEGA is utilized in highly complicated or commercial applications that require a larger program memory and higher speed:

  - Compatible with tinyAVR and megaAVR devices
  - 16–384 KB program memory

- 44–64 – 100 pin package known as A4, A3, A1
- 32 pin package (XMEGA-E, XMEGA8E5)
- More extensive performance features including DMA, cryptography support, and event system
- Large peripheral set including ADCs

- *32-Bit AVR* – Atmel developed 32-bit microcontrollers in 2006. This microcontroller included DSP instructions, a 32-bit data path, and additional audio/video processing features. While the instruction set resembled that of other RISC cores, unfortunately, it is not compatible with the initial AVR. AVR32 support has not been provided by Linux since the 4.12 kernel. Recently, Atmel also started to produce microcontrollers based on M variants of the ARM architecture. M variants of ARM will be discussed in the next section.

2.4.2.3 ARM Microcontrollers

ARM (Advanced RISC Machines) is a group of reduced instructions set computing (RISC) architectures designed by Arm Holdings. ARM Holdings is a fabless business in a way that licenses are purchased by other companies that utilize the architecture when designing their own products. The list of notable companies that utilize ARM core processors includes (but is not limited to) Atmel, Cypress Semiconductors, Apple, Analog Devices, Broadcom, Nvidia, Freescale Semiconductors, NXP, Samsung Electronics, Qualcomm, and Texas Instruments.

In comparison to AVR and PIC systems, ARM microcontrollers utilize Synchronous Dynamic RAM (SDRAM) instead of simple SRAM. SDRAM can be synchronized with the processor clock cycle to improve efficiency and allow a greater number of operations to be executed during each clock cycle. ARM processors also include a proprietary technology called the Advanced Microcontroller Bus (AMB). This technology enables hardware developers to more easily design and incorporate additional components into the ARM core.

ARM microcontrollers are a popular option for consumer electronics (i.e., smartphones) because of their storage capability, speed, and cost efficiency when compared to other higher-end microcontrollers. However, longer set up times mean that these microcontrollers are a less likely option when trying to develop a prototype quickly.

2.5 ARM Microcontrollers

2.5.1 Background

The ARM architecture indicates the specifications such as register sets, operation modes, and instruction sets that should be supported by any implementations of the ARM core. Several ARM architectures, ranging in size from 32-bits to 64-bits, have

been crafted over the last several years including ARMv1, ARMv2, ARMv3 ...
ARMv8, etc. On the other hand, the way a given architecture is implemented and
the corresponding blueprint of the transistors is known as a core. It is noteworthy
that a core does not contain input or output ports as well as memory. Memory is
usually provided by the company providing the processor [8, 9].

A central processing unit (CPU), which is also known as a processor, con-
tains one or many cores. Microprocessors or microcontrollers are more than the
CPU/cores. They incorporate additional hardware components and circuitry (see
Fig. 2.12). For example, interconnects connect the cores to a shared cache as
well as the outside world. ARM has created several 32-bit processors over the
last 20 years which can be grouped into families. The ARM families include
ARM1, ARM2, ARM3, ARM6, ARM7, ARM8, ARM9, ARM10, and ARM11.
Figure 2.10 illustrates several of the most well-known members of these ARM
processor families. ARM was determined to increase their share of the market and,
around 2003, created another series of high-performance processors known as the
Cortex family. This family is made up of three subfamilies of processors: Cortex-A,
Cortex-R, and Cortex-M [9].

The ARM architecture has matured over time and the seventh version (ARMv7)
is now the most widely used version. ARMv7 identifies three possible architecture
profiles (see Fig. 2.11) [9]:

- *Microcontroller Profile (M-Profile)* – Low-cost processors which are designed
  based on ARMv6-M architecture (Cortex-M0 and Cortex-M0+) as well as the
  ARMv7-M architecture (Cortex-M3 and Cortex-M4). This architecture relies on
  a minimal instruction set (a subset of the larger A-profile set) and generally has no

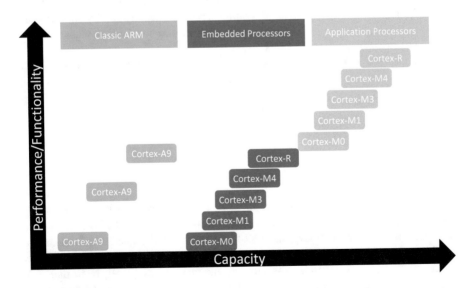

Fig. 2.10 ARM families

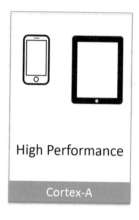

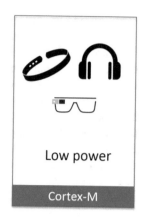

| High Performance | Fast Response | Low power |
| Cortex-A | Cortex-R | Cortex-M |

Fig. 2.11 Applications of ARM Cortex profiles

cache, memory management, or floating-point unit. This profile is utilized with RAM, FLASH, and peripherals in Cortex-M series. Because Cortex-M chips are focused on being cost-efficient rather than high performing, they are a top choice for IoT applications.

- *Application Profile (A-Profile)* – Highest-performance processers built based on ARMv7-A and ARMv8-A architecture, capable of operating at more than 5GHz. It exploits a more expansive instruction set, cache(s), floating-point unity, and memory management. In contrast with the M-Profile, this profile focused on high performance and is not as cost-efficient. These profiles are considered microprocessors and are meant to be utilized in conjunction with off-chip FLASH and RAM. These microprocessors are usually run by an operating system, mostly Linux. Cortex-A is the best option for GPS devices, tablets, gaming systems, or smartphones.
- *Real-Time Profile (R-Profile)* – A trade-off between the A-Profile and the M-Profile, this profile is utilized by the Cortex-R series. R-Profile is generally used in network devices, automotive applications, hard-disk controllers, and high-speed microcontroller applications.

2.5.2 Architecture

In this section, we will briefly review the Cortex-M3 processor, utilized successfully in many smart IoT objects (see Fig. 2.12). This processor includes a 32-bit data path, 32-bit registers, 32-bit memory, and a 32-bit processor all designed within a Harvard architecture. This means that data and instruction access can occur simultaneously because it utilizes a separate data bus and instruction bus [8, 9]. Note that chip manufacturers (e.g., Texas Instrument, Samsung, etc.) to be able to design and produce a microcontroller need to license the ARM cores (processors) and then add

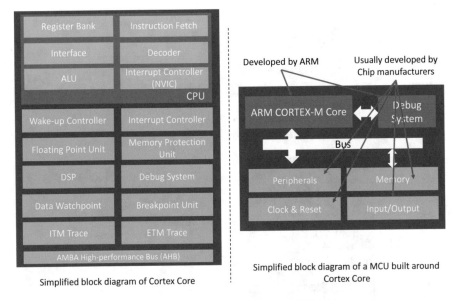

Simplified block diagram of Cortex Core

Simplified block diagram of a MCU built around Cortex Core

Fig. 2.12 A simplified model of an ARM Cortex core and an ARM microcontroller

extra components (e.g., memory, peripherals, etc.) to the chip. Therefore, Cortex-M3-based microcontrollers from different manufacturers (or even from the same manufacturer) can have different specifications and features such as memory sizes, peripherals, etc.

Access Modes in Cortex-M3 It utilizes two access levels/modes (usually user/normal level and privileged level). In the privileged level, applications can access all memory (unless restricted by settings) and can utilize all instructions. The reliability of the system can be increased by the separation of user and privileged levels because configuration registers cannot be accessed or altered by suspicious codes.

Pipeline in Cortex-M3 The ARM architecture is built on pipelining concept, an architecture that allows several instructions to overlap during execution. Each pipeline consists of several different stages that are responsible for executing a piece of instruction in parallel. Each stage is connected to the next stage, forming a pipe structure where instructions enter one end, work through each stage, and then exit the other end. Pipelining increases the instruction throughput, which is computed by the frequency with which instructions leave the pipeline. Simple ARM cores execute instructions in three different stages (see Fig. 2.13):

- *Fetch* – Fetches the instruction from memory and brings to pipeline
- *Decode* – Decodes the instruction
- *Execute* – Executes the instruction (uses arithmetic and logic unit (ALU)) and writes the result in appropriate registers

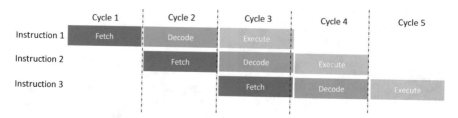

Fig. 2.13 An example of a pipeline of Cortex-M

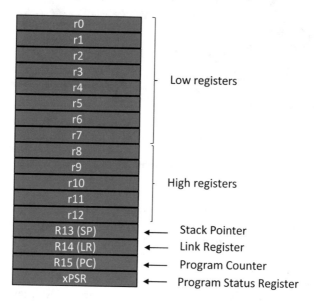

Fig. 2.14 Registers of Cortex-M

Register Registers are dedicated, specialized memory circuit in processors that can be read or written to faster than normal memory. The contents (operands) of registers are directly available as input for the processor. This means that instructions with operands in the registers can be executed more quickly. A register is used to hold information like states of CPU execution or computation results. Therefore, those microcontrollers that are designed for higher speed usually have a larger number of internal registers. Generally, an ARM processor includes the following 32-bit registers (see Fig. 2.14):

- *General-Purpose Registers* – Thirteen 32-bit registers (R0-R12), for general data operations:

  - *Registers R0-R7* – Accessible by 16-bit and 32-bit instructions
  - *Registers R8-R12* – Accessible by all 32-bit instructions that define a general-purpose register but inaccessible to 16-bit instructions

- *Stack Pointer (SP)* – Register R13 stores the address of the last item in a stack.
- *Link Register (LR)* – Register R14 is responsible for holding and returning information needed for function calls, exceptions, and subroutines.
- *Program Counter (PC)* – Register R15 stores the most recent program address.
- *One Program Status Register (PSR)* – This register combines the following components:

  - *Application Program Status Register (APSR)* – Stores condition code flags
  - *Interrupt Program Status Register (IPSR)* – Includes the Interrupt Service Routine (ISR) number of current exception activation
  - *Execution Program Status Register (EPSR)*

2.5.3   GPIOs and Interfaces

2.5.3.1   General-Purpose Input/Output (GPIO)

GPIO stands for general-purpose input/output. GPIOs are standard interface available on every modern microcontroller. These interfaces are used to connect external device, sensors, and actuators to microcontrollers. When functioning as an input port, the GPIO can communicate with the CPU regarding sensor readouts or "on/off" signals received from switches. When functioning as an output port, GPIO can trigger external operations in accordance with CPU instructions and calculations. For example, a GPIO can be used to send control signals of a DC motor or to control (turn on/off) an LED.

GPIOs are general-purpose elements because individual pins can function independently as an input or output based on their application-level configuration. Traditional MCUs included ports that were used solely for input or output. More modern GPIOs are more adaptable. For example, if a GPIO contains eight pins, each can be set to meet customized needs: 7 input/1 output, or 4 input/4 output, etc.

In general, GPIOs are clustered into several ports to be able to manage them simply. In simple words, a port is a group of IO pins that are addressed/configured as one logical entity/channel and all pins in one port work in a similar way. For instance, STM32F411RET6 is an ARM microcontroller with 64 pins. Fifty-two of those pins are available for GPIO. In this specific microcontroller, GPIOs are clustered into five ports in a way that arranges ports A, B, and C to be 16-bits wide, port D to be 1-bit wide, and port H to be 3-bits wide.

It is also important to note that an ARM Cortex-M includes three digital input modes for GPIO:

- *Input with Internal Pull-Up*
- *Input with Internal Pull-Down*
- *Input Floating*

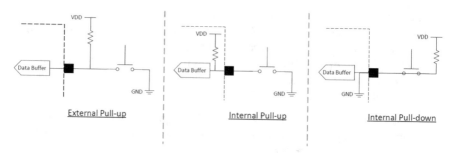

Fig. 2.15 How a simple switch circuit works

Figure 2.15 illustrates the function of a simple switch circuit. When an internal pull-up input mode is utilized, the button circuit is active low, meaning that pushing the button results in a "0" in logic in the input data register. If the input mode is pull-down, a pressed button results in a "1" in logic in the input data register. To utilize input floating, an external pull-up/pull-down resistor is needed. For instance, when an external pull-down resistor is used, the circuit is active high, meaning that pressing the button results in "1" in the corresponding input data register [10].

Other than GPIOs, MCUs also include circuits that handle various peripheral functions, making deployment in diverse settings easier. An MCU generally includes at least one analog/digital converter (ADC) needed to change incoming analog signals into digital values as well as at least one digital/analog converter (DAC) needed to change digital values into outgoing analog signals.

2.5.3.2 Analog Inputs

Because MCUs cannot read non-digital data values, the analog input pin requires an extra element, known as an analog-to-digital converter (ADC), to read analog voltage and then transform it into a digital format. The ADC reads the input voltage and then transforms it into a binary value.

ADCs can be implemented in several ways and their details are beyond the scope of this chapter. However, it is very important to define "ADC resolution." The ADC resolution specifies the number of distinct values generated across the complete range of analog values. In other words, resolution regulates the accuracy of ADC and the magnitude of the quantization error. The ADC resolution is specified by the number of bits it uses to digitize an analog signal. For instance, an 8-bit ADC can encode an analog input to 1 of 256 different levels ($2^8 = 256$). On the other hand, in a 16-bit ADC, the range of the analog signal is represented by 2^{16} (65536) discrete values.

2.5.3.3 Analog Outputs

Analog output is needed for several applications such as controlling motor speed. MCUs are digital by nature and can generate either high or low voltage; however, they are not able to produce a varying voltage (analog value). Therefore, it is necessary to emulate varying voltage (analog value) by generating a series of voltage at regular intervals but with different pulse widths, known as pulse width modulation (PWM). PWM is a method for obtaining analog values utilizing digital means. The voltage produced by PWM is often referred to as pseudo-analog voltage.

To better explain the key idea behind PWM, consider the following example. The graph below illustrates pulsing a pin both high and low for the same amount of time. The amount of time that the voltage is high is known as pulse width. The duty cycle is the ratio between pulse width (high voltage) and the total time (i.e., the amount of time needed for a pulse to move from low to high and back to high again). In this example, the duty cycle is 50%, and therefore the effective voltage (average voltage) is 50% of the total voltage.

Figure 2.16 demonstrates that it is possible to generate a duty cycle at less than 50% by using pulses that are shorter in duration than the length of pauses (i.e., shorter high voltage duration compared to low voltage duration). In this case, the average voltage is less than half of the total voltage.

It is also important to note that PWM output can be transformed to create a genuine analog voltage. In order to achieve this, all that is required is a low-pass filter (created using a ceramic capacitor and resistor) which should be connected to the PWM output.

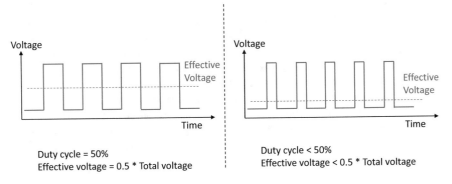

Fig. 2.16 How pulse width modulation (PWM) works

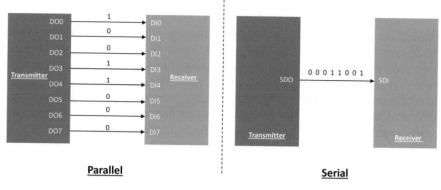

Parallel   **Serial**

Fig. 2.17 Parallel and serial communication

2.5.3.4 Parallel Interfaces vs Serial Interfaces

Circuits rely on communication protocols to be able to exchange data among themselves. While there are hundreds of communication protocols to support the exchange of data, they can all be classified into two categories – *serial* or *parallel* (see Fig. 2.17).

Parallel interfaces move many bits simultaneously and generally need data buses with a size of 8, 16, or more bits. In contrast, serial interfaces stream data one bit at a time, thereby it can function with only 1 wire, and generally do not use more than 4 wires. Serial interfaces can also be categorized into two groups, namely, synchronous and asynchronous [10].

- *Synchronous*: Synchronous serial interfaces match data line(s) to a clock signal so that all devices share a mutual clock on the synchronous serial bus. This results in an uncomplicated, quick, serial transfer; however, it also means at least one additional wire is needed to connect communicating devices. SPI and I^{2}C are two common synchronous serial communication interfaces.
- *Asynchronous*: An asynchronous interface exchange data without the aid of an external clock signal, which reduces the number of wires and I/O pins needed. However, guarantee the reliability of data transmitting/receiving needs extra effort. A universal asynchronous receiver/transmitter (UART) is an example of a synchronous interface.

2.5.3.5 Universal Asynchronous Receiver/Transmitter (UART)

UART systems support reliable, reasonably speedy, full-duplex (two-way) communication using three signals: Rx (received serial data), Tx (transmitted serial data), and a ground. Note that the Tx of one device should be linked to the Rx of the other device and vice versa. Note that UART does not need a clock signal because it is asynchronous (see Fig. 2.18) [11].

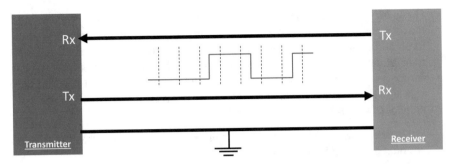

Fig. 2.18 Configuration of universal asynchronous receiver/transmitter with one transmitter and one receiver

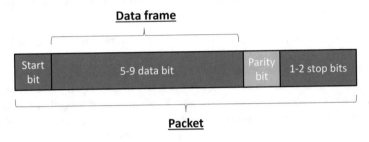

Fig. 2.19 The structure of UART packets

Although there is not any explicit external clock signal between two devices in UART, both the transmitter and receiver contain their own internal clock. However, this clock is not transmitted between devices. Instead, the internal clock of the transmitter regulates how fast logic levels are created on the Tx side, and the internal clock of the receiver regulates the sampling frequency on the Rx side. If the receiver and transmitter are designed to receive or send data on different transmission frequencies, UART communication will not work. In addition, their internal clock signals have to be accurate in relationship to the target data-transmission frequency and adequately stable in the presence of temperature variations over time.

UART data is arranged into packets for transmission. As shown in Fig. 2.19, every packet includes 1 starting bit, 5–9 bits for data, an optional parity bit, and 1–2 stop bits [11]:

- *Start Bit* – Initial bit of each packet. This bit shows that the state of the data line is changing from idle state to active state. Since typically the idle state is represented by active high, the start bit is active low. Although a start bit is mandatory in UART to be able to start the data-transmission process, it does not actually transfer useful data; therefore, it is an overhead bit.
- *Stop Bit* – Final bit of each packet. This bit is logic high and similar to start bit can be considered as an overhead bit.

- *Parity Bit* – A bit that detects errors in each set of transmitted bits. In "odd parity," the total complete number of 1s in the transmit data when added together is equal to an odd number. On the other hand, in "even parity," the number of 1s in the transfer data including the parity bit itself should be an even number.
- *Bit Rate* – In literature, bit rate is also referred to as total physical layer bit rate, data signal rate, raw bit rate, or total data transfer rate. Bit rate indicates the total number of transmitted *bits per second* (bps).
  - *Gross Bit Rate* – This parameter captures both protocol overhead (e.g., start and stop bit) and useful data.
  - *Net Bit Rate* – Unlike the gross bit rate, this parameter indicates a digital communication channel's capacity and excludes overhead protocol. Net bit rate is also known as information rate, useful bit rate, net data transfer rate, payload rate, effective data rate, and wire speed.
- *Baud Rate* – This parameter also referred to as "symbol rate." Baud rate shows the number of symbol or signal changes occurring per second. A symbol can be a change in phase, voltage, or frequency. Each symbol can represent one or more bits. Based on this, the baud rate is always lower than or equal to the bit rate. The baud rate is never higher than a bit rate. For example, if a symbol rate is 4800 baud with each symbol representing two bits, then the total bit rate is 9600 bps. In the case of UART, each symbol represents only one bit; therefore, the baud rate and gross bit rate are equal.

UART protocols follow the following steps to exchange data between two devices:

1. UART module in the transmitter device receives parallel data from the data bus.
2. UART modules create the UART packet including start, parity, stop bits, as well as the payload (actual intended message).
3. The whole data packet is sent via the bus in series to receiving UART module in the receiver device.
4. Receiving UART module samples bus at a preset baud rate.
5. Receiving UART module removes start, parity, and stop bits and extracts the payload.
6. Receiving UART module formats the received serial data (payload) and publishes it to the data bus of the receiver device.

2.5.3.6   Serial Peripheral Interface (SPI)

Serial Peripheral Interface (SPI) is often used to transmit data between small peripherals (e.g., sensors and SD cards) and their corresponding microcontrollers. While asynchronous serial communication can also be used to address this situation, it requires complicated hardware for data transmission, and also it includes significant overhead because of extra start and stop bits that are packaged with each payload (actual intended message). On the other hand, SPI is a synchronous data

Fig. 2.20 SPI configuration
with one master and one slave

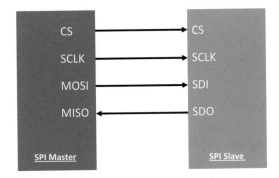

bus requiring both data line and a clock line to maintain synchronization between both devices. The clock is a precise periodic signal that indicates when the receiver samples the data line to obtain the corresponding data bit. Note that the sampling is done side at the rising or falling edge of the clock.

Figure 2.20 illustrates connections between master and slave via the SPI. In an SPI, only one of the sides provides the clock signal. The clock is known as SCK (serial clock) or CLK. The side responsible for managing the clock is referred to as the "master," while the corresponding side that receives the clock is known as the "slave." In any SPI there is only one master; however there can be more than one slave. An SPI usually consists of four wires:

- Clock (CLK, SCLK)
- Chip Select (CS)
- Master Out, Slave In (MOSI)
- Master In, Slave Out (MISO)

The master device can select and enable any of the slaves via the corresponding chip select signal. The chip select signal is typically active low meaning that when the signal is high, the slave disconnects from the SPI bus. The MOSI data line sends data from the master to the slave and the MISO data line sends data from the slave to the master. To start SPI communication, the master activates the clock and enables the slave via chip select signal. Because the chip select is active low, the master puts logic 0 on chip select line to be able to select the appropriate slave. As a full-duplex interface, in SPI, data can be transmitted simultaneously via the MISO and MOSI data lines. In other words, data is simultaneously sent serially to the MOSI data bus and received on the MISO bus.

As stated earlier, multiple slaves can be used in conjunction with one master using one of the following two modes (see Fig. 2.21) [12]:

- *Regular Mode* – Master needs a dedicated chip select line for each slave device. When one slave is selected by its chip select line, then the clock and MISO/MOSI data lines are accessible to the enabled slave. If more than one chip select is selected, the MISO line will malfunction because the master cannot determine which slave is sending the data.

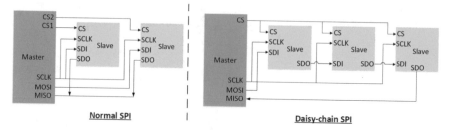

Fig. 2.21 SPI configuration with one master and several slaves

- *Daisy-Chain Mode* – In regular mode, the number of chip select lines increases in correlation to the number of slaves, rapidly increasing the IO pins required from the master. Considering the limited number of SPI pins in each master, this approach will limit the number of slaves that can be connected to one single master. Daisy-chain mode can address this issue because slaves are designed in a way that the chip select signal and clock signal for all salves are the same while the data are flowing from one slave to the next slave. In this approach, the master sends the data to the first slave, and the first slave propagates it to the second slave, the second slave transmits it to the third one, and so on. As data is sent from one slave to another, the number of clocks needed to send data to a particular slave is a linear function of the target slave's position in the chain.

2.5.3.7 I2C (Inter-integrated Circuit)

As previously mentioned, the hardware overhear of UART is high. In addition, UART is innately designed to facilitate communication between only two devices and the data transfer rate can be problematic. SPI solves some of these issues; however, a drawback to SPI is the number of required pins. This is rooted in the fact that connecting a master to a slave with an SPI demands 4 lines, and each additional slave also needs another chip select. This issue makes SPI less advantageous in scenarios where many devices act as slaves to the same master. In addition, SPI is not a multi-master interface meaning that there must be one and only one master with one or several slaves.

I2C, which was initially developed by Philips in 1982, is the Best of Both Worlds as it makes use of the best of what UART and SPI have to offer. I2C is also designed based on master and slave concept. In I2C, there is a possibility to have multiple masters and multiple slaves. In other words, each device in I2C can be a transmitter, a receiver or both. Each I2C device is specified by a unique 7-bit or 10-bit address. Every I2C data bus includes two bidirectional lines, namely, an SCL (serial clock) line and an SDA (serial data) line. Each master generates its own clock and data is changed only when the clock is logic low. Since there may be several masters within the I2C, we need an arbitration process to determine which master can use the bus to communicate with its slaves. To implement the arbitration process, each device must constantly monitor SDA and SCL for start and stop conditions to figure out when the bus is idle/available or busy. By this approach, a master can realize when another master is active and using the bus, so it can immediately stop its transfer.

SDA and SCL lines are logic high in normal state. Then the master node can start the communication. To do so, the master produces a start condition and then specifies the slave device address. If bit 0 of the address byte is designated 0, then the master will write to the slave. If not, the master realizes that it should read from the slave. When all data have been read, or written, the master sends the stop signal, indicating that the bus is now free and available for use by other devices.

I2C sends data in the form of messages and these messages are separated into two frame types (see Fig. 2.22) [10]:

- *Address Frame* – Master determines which slave the message is sent to.
- *Data Frame* – One or more (8-bit data messages) are transferred from the master to the slave and vice versa.

Each message in I2C consists of one start and one stop condition:

- *Start* – In this condition, SCL is high and SDA has transmission from high to low. After a successful start, the bus is busy and other masters cannot use the bus.
- *Stop* – In this condition, SCL is high and SDA has transmission from low to high. After a successful stop, the bus is free and other masters can start using the bus.

After the start condition, the first part of any new communication series is always the address frame. The master sends the 7-bit address of the slave (starting from most important bit (MSB)) following by a R/W bit. The R/W indicates if the operation is a write (0) or a read (1). Write means that the master writes to slave and read means that the master reads from slave. The 9th bit of any data or address frame is always known as the NACK/ACK bit. When the initial 8 bits of the frame have been sent by the master, the slave device gains SDA control. If the slave does not pull the SDA low prior to the 9th clock pulse, it means that the slave did not receive the message or was not able to parse it; therefore, the message exchange stops and the master must determine the next step (see Fig. 2.23) [10].

When the address frame has been successfully received by the slave, the actual data transmission can begin. The master continues to send clock pulses at regular intervals. Depending on if the R/W designates a read or write operation, the slave or master publishes data on the SDA. It should be noted that there is no limitation and

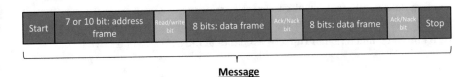

Message

Fig. 2.22 The structure of I2C packets

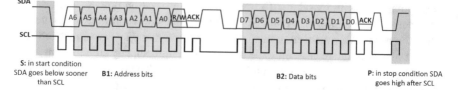

Fig. 2.23 The signaling mechanism in I2C between one master and one slave

upper bound on the number of data frames that can be transmitted between master and slave. Finally, when all data frames have been dispatched, the master creates a stop condition.

I2C has its own advantages and drawbacks. The main benefits of I2C are:

- *Minimal Wires* – I2C needs only two wires.
- *Greater Flexibility* – Unlike SPI, several masters can be connected to the same I2C bus concurrently.
- *ACK/NACK Bit* – This bit can be used to confirm the successful transfer of each frame.
- *Simplicity* – Hardware is not as complicated as with UART.
- *Universality* – It is widely used by vendors to connect peripherals to microcontrollers.

The main drawbacks of I2C can be summarized as below:

- *Slower Data Transfer* – Data transmits at a slower rate than SPI.
- *Smaller Data Frames* – Data frame size is limited to no more than 8 bits.
- *Complicated Hardware* – I2C requires more complex hardware than SPI.

2.5.3.8 Universal Synchronous Asynchronous Receiver Transmitter (USART)

USARTs can function asynchronously in the same manner as a UART (asynchronously) as well as acting synchronously. The main difference between a USART and a UART is the manner in which serial data is clocked. In UARTs each counter generates its own data clock internally. Since there is not any incoming clock signal to the receiver, it must know what the communication baud rate is in advance of receiving the data stream. On the other hand, a USART can run synchronously. A sending peripheral creates a clock that the receiver can obtain from the data stream without advance knowledge of the baud rate. An external clock enables the USART data rate to be much higher (up to 4 Mbps) than the data rate of a UART.

Fig. 2.24 Configuration of
RS232

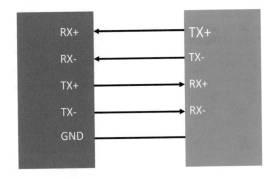

2.5.3.9   RS232 and RS422

In microcontrollers, UARTs generally operate at 5 V TTL or 3 V levels. Transistor-transistor logic (TTL) refers to a logic family created using bipolar junction transistors. A TTL signal is "low" if it is between 0 V and 0.8 V, and the signal is "high" if it is between 2 V and 5 V. RS232 fundamentally operates like a UART, but with the addition of a new line driver to improve signal strength. Thanks to the new line driver, RS232 is capable of functioning between $+/-$ 15 V. In other words, RS232 data is bipolar. Typically, "logic 1" is determined by $+3$ V to $+12$ V, whereas "logic 0" is indicated by -3 V to -12 V. This feature enables the signal to transfer data in a cable with a length of up to 10 meters. Note that like UART, RS232 also has two data lines, namely, TXD and RXD (see Fig. 2.24).

The RS422 design also utilizes the same general UART but includes a different line driver IC. The line driver generates a different signal from the original signal. Therefore, there are four data lines in RS422: TXD+, TXD-, RXD+, and RXD-. Utilizing a differential signal is beneficial because it gives the system greater immunity to noise and allows for longer cables.

2.5.4   Clock Tree

A clock distribution network inside MCUs is known as a clock tree and includes several circuitries and interconnects from the clock source to the destination. There are two system clock (SYSCLK) sources available in the Cortex-M series, namely, the external clock and the internal clock (see Fig. 2.25). The external clock is usually created using one of the following ways [9]:

- *High-Speed External (HSE)* – Generated by an external resonator device, crystal, or external clock signal.
- *Low-Speed External (LSE)* – Generated by an external crystal which then feeds into the internal real-time clock (RTC). Note that RTC functions as a 24-hour clock and/or a Julian calendar that monitors the current time.

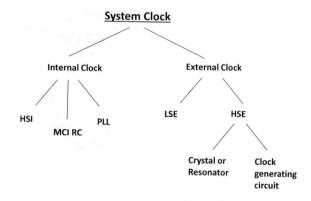

Fig. 2.25 The general structure of the clock tree of Cortex-M series

An internal clock can be created using the following:

- *High-Speed Internal (HSI)* – A precise, RC-based, high-speed internal clock with a manufacturer set tolerance level of at least 1%.
- *Multi-Speed Internal (MSI) RC* – An RC-based clock source generating clock in various frequencies that can be trimmed using software. Typically, MSI is capable of providing about 12 different clock frequencies in most Cortex-M MCUs.
- *Phase-Locked Loop (PLL)* – PLL receives data from the HSE or MSI clocks and uses it to create different system clocks.

2.5.5 *Interrupts*

A simple analogy is best to help explain the basic concept of how things work with interrupts and without interrupts. When boiling eggs for 10 minutes, one might simply glance at the clock every few minutes to determine when 10 minutes have passed. This is also true with embedded systems. When waiting for a specific status change to happen before taking action, then one could check the state periodically. Similarly, if the program is pending for a GPIO input level to move from 1 to 0 before advancing to the next step, then you could repeatedly test the GPIO value. This approach is called "polling."

Polling is an uncomplicated manner of confirming status; however, it does have a drawback. If the status check interval is too long, you risk missing the status change entirely because the state could revert back again before the next checkpoint. A shorter interval can provide quicker, dependable detection but also requires longer processing time and higher power consumption because many status checks will return a negative result.

An alternative to polling is the utilization of interrupts. To revisit our boiling egg analogy, when cooking the eggs, you could set a timer for 10 minutes and then carry on with another activity until the timer goes off, telling us to come back to the

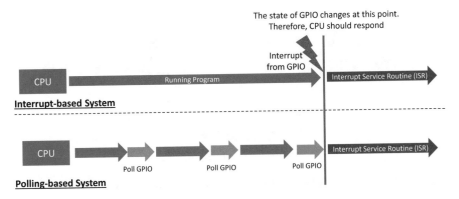

Fig. 2.26 How polling and interrupt work

stove. The timer acts like an interrupt and removing the eggs from the stove would be the corresponding processing. When using the interrupt method, a state change creates an interrupt signal telling the processor to pause a current operation and save the present state before executing the processing connected to the interrupt. Then, the previous state is restored and operations continue from the point of interruption. Figure 2.26 compares the polling-based system with the interrupt-based system [13].

Interrupts can initiate from internal MCU devices or external devices. An interrupt that originates from an external sensor or switch is often referred to as an "attached interrupt," because it is created by an external device connected to an IRQ (interrupt request) pin in the MCU. When the appropriate status change is achieved, an external device sends an IRQ to the pin. Next, the pin creates a notification that is sent to the MCU's interrupt controller. On the other hand, interrupts from peripherals on the on-chip peripherals such as GPIO ports, internal times, etc. are known as "peripheral interrupts." These interrupts send a notification directly to an interrupt controller without aid from the IRQ pin.

Interrupt controllers are responsible for moving IRQs to the CPU in a harmonious manner. In the case of multiple interrupts, the controller must transmit them to the CPU in the correct order, according to their respective priorities. The controller must also perceive which interrupts are disabled (or masked) to disregard those interrupts entirely. When the processor receives an IRQ from the controller, it suspends its current program operations and saves all important information of the processor, so that it can continue work after processing the interrupt. Next, the processor loads and executes the associated processing program that matches the IRQ received. This associated program is called ISR. An ISR (Interrupt Service Routine, sometimes referred to as Interrupt Service Procedure) is a subroutine that the processor executes to respond to a unique event/interrupt/exception. Once the IRQ processing is finished, the CPU restores the saved work information and begins working from the previous stopping point. The saving and the resuming process are managed automatically by the processor (see Fig. 2.27) [13].

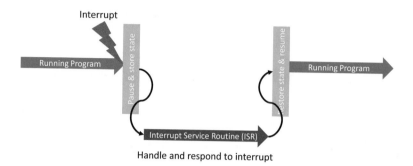

Fig. 2.27 How an interrupted is addressed in a processor by Interrupt Service Routine (ISR)

Vector tables and interrupt vectors are important to understanding both software and hardware interrupts. Interrupt vectors are actual addresses that notify the interrupt handler where the ISR can be found. For example, an ARM Cortex-M3 includes 255 interrupt vectors. Generally, a processor utilizes a vector table to hold ISR addresses for each interrupt, and when an interrupt occurs, the processor obtains the appropriate address from the vector table.

There are several processor architectures that facilitate nesting interrupts, meaning that when a low priority ISR is executed, a higher priority service is able to preempt and suspend the ISR. The low priority ISR then resumes after the higher priority ISR has resolved. The regulations governing a nested interrupt system include [14]:

- Prioritization of all interrupts.
- An interrupt may occur at anytime, anywhere.
- If a higher priority ISR interrupts a lower priority ISR, the higher priority is completed first.
- If a lower priority ISR interrupts a higher priority ISR, the higher priority finishes execution prior to the execution of the lower priority.
- ISRs of equal priority level are executed in time order.

The NVIC (Nested Vector Interrupt Controller) found within the Cortex-M family is an example of an interrupt controller that is highly flexible when it comes to managing interrupt priorities. Nested Vector Interrupt Controllers allow several interrupts to be defined and individual interrupts are given a priority, with "0" recognized as the highest priority level. It is very important to remember that the Cortex-M family utilizes a reverse numbering system for interrupts with higher numbers indicating a lower priority value. It is also important to remember that the priority level is programmable. The Cortex-M series offers several kinds of interrupts (see Fig. 2.28) [8]:

- *Interrupt Requests (IRQs)* – Interrupts that are asynchronous and not connected with the code currently in execution by the processor (i.e., ADC converter completing conversion).

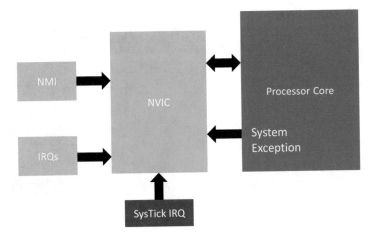

Fig. 2.28 Nested Vector Interrupt Controller

- *Non-maskable Interrupt (NMI)* – Like IRQs, except that they cannot be disabled (not maskable).
- *SysTick Timer Input* – SysTick is a timer with 24 bits counting from a preloaded value to zero. When the counter transitions to zero, it can generate an interrupt.
- *Exceptions, Faults, and Software Interrupts* – These are synchronous and are the result of executing a unique instruction (e.g., encountering an undefined instruction or overflow due to the execution of a specific instruction).

2.5.6   Addressing Modes

Addressing modes are an integral part of instruction set architecture and define how instructions can identify the operand of each instruction. The addressing mode rules how an instruction address field is interpreted or modified before the instruction is executed.

Immediate Addressing Also known as "literal addressing" uses the data in the address field of instruction as an operand. In other words, the data is in the address field of instruction.

Direct Addressing Mode Memory location's address is directly given and specified in the instruction. This kind of instruction can only be utilized for special function registers (SFR) or internal RAM (see Fig. 2.29).

Register Indirect Addressing Mode The memory location address is specified indirectly in a register (see Fig. 2.30).

Register Relative Indirect Addressing Mode Effective memory location is calculated by adding the content of a register and an immediate value (see Fig. 2.31).

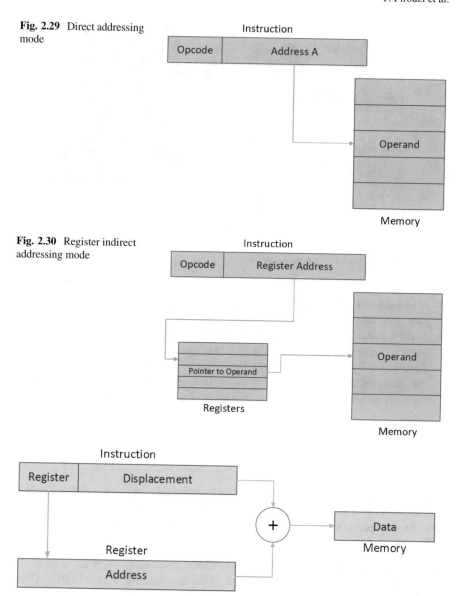

Fig. 2.29 Direct addressing mode

Fig. 2.30 Register indirect addressing mode

Fig. 2.31 Register relative indirect addressing mode

Base-Indexed Indirect Addressing Mode Effective address in memory is calculated by combining information from two registers (sometimes with an immediate value) as illustrated in Fig. 2.32.

Base-Scaled Register Addressing Mode Memory address is calculated by adding the content of a register with the content of a second register shifted left.

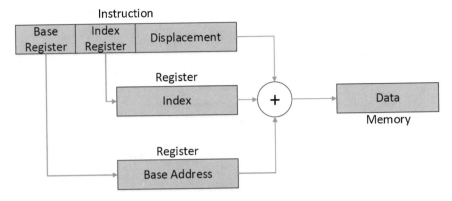

Fig. 2.32 Base-indexed indirect addressing mode

2.5.7 Timers

In any microcontroller, the timer is one of the most important features. Timers have many use cases. For example, they can be utilized to generate accurate time intervals enabling users to control code and complete tasks that are dependent on time. Timers are also used to create delays, change servo shaft angles, set baud rates, sample analog signals, or turn a device on or off after a given period of time. In general, you can find a timer subsystem managing a broad range of functions in all microcontrollers, e.g.:

- Producing accurate time intervals
- Measuring the length of external events
- Counting events

Microcontrollers contain exclusive timers or general-purpose timers that handle the following functions:

- A real-time clock
- Creation of pulse width modulated (PWM) signals
- As watchdog to automatically reset the system if the main program neglects to periodically service

The most important time in Cortex-M is SysTick timer. SysTick is a 24-bit system timer that is part of standard hardware for the ARM Cortex-M series. Figure 2.33 illustrates the block diagram of a typical system timer. As depicted in this figure, system timers include three registers:

- SYST_CVR: This register stores the current SysTick counter value.
- SYST_RVR: This register indicates the starting value of the counter which should be loaded to SysTick.
- SYST_CSR: This register manages and controls the SysTick features.

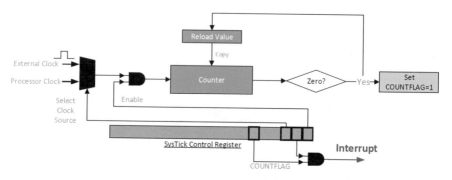

Fig. 2.33 How the system timer works and generates an interrupt

The SysTick timer counts down from a preloaded starting value to zero. SYST-CVR register holds current SysTick counter value. When the counter reaches zero, a new starting value will be loaded to the timer. Note that the starting value is stored in a register called SYST_RVR. SysTick counts down from a continuously lower starting point as it starts at an incrementally lower point for each subsequent clock cycle. The counter can be disabled by writing a zero value to the SYST-RVR register. When the counter reaches zero, the COUNTFLAG status bit is designated at 1. COUNTFLAG is one bit of SYST_CSR which controls the status of the timer. Note that this special bit can trigger an interrupt in the system. Reading of the SYST-CSR returns the value of COUNTFLAG to zero again (see Fig. 2.33) [15].

2.5.8 Low-Power Modes

In many IoT applications, it is very important that the deployed microcontroller supports sleep mode capabilities to be able to suspend many or all operations to reduce energy use, enabling it to run for many years with limited energy. ARM microcontrollers generally use low-leakage technology and advanced design to reach extremely low current consumption, making them an ideal option for battery-operated or energy-harvesting IoT applications. Making the most of low-power capabilities in IoT devices requires understanding the low-power modes currently available. Typically, the following techniques are used to activate a low-power mode in the Cortex-M series [8, 16]:

- *Wait for Interrupt (WFI) Instruction* – The low-power mode is utilized by this instruction. In this low-power mode, the device will sleep until an interrupt or exception occurs which wakes the device. Keep in mind that one must first enable interrupts using a Nested Vectored Interrupted Controller (NVIC).
- *Wait for Event (WFE) Instruction* – Fundamentally, this instruction works similar to WFT; however, it is more flexible than the WFT. Because of executing this instruction, the device enters low-power mode. The main difference between WFE and WFI is that WFE instruction allows the device to be woken up by

interrupts disabled in the NVIC as well, if those interrupts are enabled in the related peripheral control register.

- *SLEEPONEXIT* – The other way to begin using low-power mode is by enabling the *SLEEPONEXIT* bit found in the System Control Register (SCR). When this bit is designated at 1, it allows the processor to immediately go to low-power mode once the execution of interrupt/exception is complete, before the program resumes execution.

The low-power modes enabled on a device are determined by the specific implementation within the processor family series. The most common low-power modes in the Cortex-M series include Sleep Mode, Low-Power Run Mode, Stop Mode, Low-Power Sleep Mode, and Standby Mode.

The main differences between the modes noted above typically can be represented by three important parameters, namely, *wake-up time*, *power consumption*, and *performance*. The table below compares low-power modes discussed above in terms of these three parameters. To make this comparison clear, each parameter uses a ranking scale of #1 to #5 with "#1" being the best possible ranking and "#5" being the worst. For example, "performance = 1" indicates the highest performance, and "performance = 5" indicates the worst performance. In general, as power consumption reduces, performance decreases, and wake-up time increases (see Table 2.3) [16].

- *Low-Power Run Mode* – System clock frequency is decreased; however, the core does not stop.
- *Sleep Mode* – Only the core stops and all peripherals keep running.
- *Low-Power Sleep Mode* – A combination of low-power run mode and sleep mode; core stops and system clock frequency decreases.
- *Stop Mode* – In this mode, core and external high-speed clocks stop, while internal clocks and low-speed external clocks work in a limited capacity. This strategy allows a reduction in the power consumption on the order of nano-amps, while SRAM, registers, as well as a few peripherals remain functional. For example, UART and I2C can receive data, when it is needed.
- *Standby Mode* – In this low-power mode, the entire chip stands by. There are limited options for exiting this mode because only a wake-up pin, a reset signal, or a real-time clock wake-up event will wake the device. Note that unlike the other lower power modes, standby mode does not maintain the content of SRAM and registers. In other words, in the case of wake-up, the whole system will be reinitialized.

Table 2.3 Low-power modes of Cortex-M based on various parameters

	Performance	Power consumption	Wake-up
Low-power run mode	1	5	2
Sleep mode	2	4	1
Low-power sleep mode	3	3	4
Stop mode	4	2	3
Standby mode	5	1	5

2.5.9 Programming and Debugging Techniques

There are a few main techniques available to program a microcontroller, namely, *JTAG/SWD* and *Bootloader*. Outlined below are the benefits and drawbacks of each method [17].

2.5.9.1 JTAG/SWD

Joint Test Action Group (JTAG) is an industry standard initially designed for electrical testing of integrated circuits after manufacturing by means of boundary scanning mechanisms. Since then, JTAG has evolved into a common interface utilized to control the entire microcontroller and is currently used to program, test, and debug almost all embedded devices. A complete JTAG interface needs the five pins indicated below:

- *Test Mode Select (TMS)* – This signal is used to determine the next state of the system.
- *Test Clock (TCK)* – Clock of the system during test/debug mode.
- *Test Data Input (TDI)* – The input data that is shifted into the device.
- *Test Data Output (TDO)* – The output data that is shifted out of the device.
- *Test Reset (TRST)* – Rest pin to reset the JTAG controller system.

Figure 2.34 illustrates a typical JTAG connector needed to program a microcontroller.

Serial Wire Debug (SWD) is an ARM-specific protocol used in ARM chips. This method includes two pins (SWDIO/SWCLK). SWD is pin-compatible with JTAG

Fig. 2.34 20Pin JTAG adapter board kit

and requires fewer wires. However, note that the adapter and ARM microcontroller support both SWD and JTAG.

Utilizing JTAG and/or SWD requires the use of an adapter (i.e., external board connecting the computer to the microcontroller). This adapter can also be thought of as a "USB to JTAG adapter." JTAG/SWD is the most definitive method used in programming microcontrollers because it enables powerful features including built-in debugging functions. Although JTAG/SWD can be more complicated and expensive due to the need for an adapter, this method is dependable and works consistently because it is implemented at the hardware level.

2.5.9.2   Bootloader

In the context of embedded systems and microcontrollers, a bootloader is a small piece of code that is located at the start of the microcontroller's memory to:

- Program the microcontroller with a compiled program (i.e., user code) that is received from a communication port (i.e., USB, serial port)
- Execute the received program

The bootloader enables programming of the microcontroller without an adapter, whereas the bootloader is just meant to load a user code, and it is not as powerful as a JTAG/SWD adapter. The bootloader does not allow step-by-step debugging (e.g., single step through code, examine memory). In addition, the bootloader must be written into the microcontroller once via an adapter. Note that usually in many microcontrollers which can be purchased, the bootloader has already been installed.

2.5.10   Real-Time Operating System (RTOS)

It is very important to answer the following question: "Is an RTOS really necessary in my IoT applications"? The answer is Yes and No. In most of IoT applications, RTOS is not mandatory, and most smart objects (IoT things) are implemented on "bare metal" without any RTOS. Generally, one should consider an RTOS if the product needs any of the following components:

- TCP/IP or other complex networking stack
- Complex GUI
- File system
- Multitasking

Some real-life IoT applications run multiple, independent tasks on the same processor. While CPUs can execute only one task at a time, it appears that multiple tasks are being executed simultaneously because the processor quickly exchanges control among multiple tasks. This rapid exchange of control is managed by a scheduler. Scheduling is a foundational element in RTOSs. In general, there are two different techniques for scheduling:

- *Non-preemptive Scheduling* – When a process begins running and the resources (CPU) are allocated to it, it keeps running and holds all the resources until the process is completed or the process switches to waiting state.
- *Preemptive Scheduling* – In this technique, a process can be preempted. In other words, a task given to a CPU can be removed. Preemptive scheduling can be implemented in several ways including:

  - *Round Robin (Cyclic Executive) Scheduling* – This scheduling utilizes time sharing by assigning a time slot or quantum to jobs using a circular pattern without designating any particular job as a priority.
  - *Interrupt-Driven Scheduling* – In this scheduling technique, tasks can be prioritized and the most urgent task gets the resources (CPU) first. If a new task with a greater priority is received, the scheduler ends the current task to start running the new highest priority task.

2.6　Summary

In this chapter, the "Things" in IoT and its main building blocks (i.e., processing unit, sensors, actuators, power source, and connectivity) have been presented. Next, we provided the details of the main technologies of sensors and actuators. Finally, we described the details of a microcontroller (ARM Cortex-M) covering all the details of registers, addressing mode, interrupts, GPIOs, parallel/serial interfaces (e.g., UART, SPI, I2C), timers, clock tree, low-power modes, as well as programming and debugging techniques.

References

1. IBM. Available from: https://www.ibm.com
2. D. Hanes et al., *IoT Fundamentals: Networking Technologies, Protocols, and Use Cases for the Internet of Things* (Cisco Press, 2017)
3. A. Rayes, S. Salam, *Internet of Things—From Hype to Reality* (The road to Digitization. River Publisher Series in Communications, Denmark, 2017), p. 49
4. openlabpro. Available from: https://openlabpro.com/guide/relays-and-actuators/
5. National Instruments. Available from: http://www.ni.com/product-documentation/3960/en/
6. Microcontrollers – Types & Applications. Available from: https://www.elprocus.com/microcontrollers-types-and-applications/
7. Memory-mapped IO vs Port-mapped IO. Available from: https://www.bogotobogo.com/Embedded/memory_mapped_io_vs_port_mapped_isolated_io.php
8. ARM. Available from: https://www.arm.com/
9. D. Ibrahim, *ARM-Based Microcontroller Projects Using Mbed* (Newnes, 2019)
10. Sparkfun. Available from: https://learn.sparkfun.com/tutorials

11. The Universal Asynchronous Receiver/Transmitter (UART). Available from: https://www.allaboutcircuits.com/technical-articles/back-to-basics-the-universal-asynchronous-receiver-transmitter-uart/
12. Introduction to SPI Interface
13. Interrupts. Available from: https://www.renesas.com/br/en/support/technical-resources/engineer-school/mcu-programming-peripherals-04-interrupts.html
14. A Software Approach to Using Nested Interrupts. Available from: https://www.nxp.com
15. Y. Zhu, *Embedded Systems with Arm Cortex-M Microcontrollers in Assembly Language and C* (E-Man Press Llc, 2017)
16. Low-Power Modes on the STM32L0 Series. Available from: https://www.digikey.com/eewiki/display/microcontroller/Low-Power+Modes+on+the+STM32L0+Series
17. JTAG/SWD vs Bootloader. Available from: https://libtungsten.io/tutorials/jtag_vs_bootloader

Chapter 3
Engineering IoT Networks

Enrico Fraccaroli and Davide Quaglia

> *Successful engineering is all about understanding how things break or fail.*
>
> Henry Petroski

Contents

E. Fraccaroli (✉) · D. Quaglia
University of Verona, Verona, Italy
e-mail: enrico.fraccaroli@univr.it; davide.quaglia@univr.it

© Springer Nature Switzerland AG 2020
F. Firouzi et al. (eds.), *Intelligent Internet of Things*,
https://doi.org/10.1007/978-3-030-30367-9_3

3.1   IoT Network Scenarios

Figure 3.1 shows the general network structure of a typical IoT application. There
are two main sets of networks, i.e., the *Internet* and the *Edge Network*. They are
different in the objective and the standards. The everyday's objects that become
connected "things" constitute the Edge Network. They can be found in houses,
buildings, factories, and open environments (e.g., cities or agricultural fields). They
can be vehicles in a smart transport network. In the factory automation scenario, we
also use the term *Industrial IoT (IIoT)*.

The quality of service parameters of the Edge Network is strictly related to the
application requirements and constraints. The "things" usually require low data rate,
but video surveillance cameras are an example of exception. Data collection usually

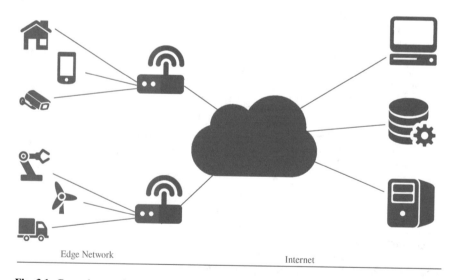

Fig. 3.1 General network structure of an IoT application

tolerates high and variable delay, but for closed-loop control applications and alarm notification, the delay should be kept small and constant. The error rate is usually not negligible in case of wireless channels, and this aspect should be considered in the deployment of the application. Network standards for the Edge Network will be presented in the main body of this chapter.

The Edge Network is connected to the Internet through nodes acting as *gateways* between the two worlds. Internet architecture and protocols are not the topics of this book even if some of them will be found in the next sections since they are also used in the Edge Network. The Internet moves data to/from user's computer (e.g., for data visualization) and data centers for storage as well as processing (e.g., with machine learning algorithms); the use of data centers is related to the concept of *Cloud* [1]. Gateways play an important role in the IoT scenario of Fig. 3.1. Their basic function consists in moving data between Edge Network protocols and Internet protocols. Since they are more standard computational devices and are usually powered by stable sources, they can also perform data processing, thus moving "intelligence" from the cloud closer to the "things." This technique may reduce response delay and is related to the concept of *Fog Computing* or *Edge Computing* [2].

This chapter focuses on the red part of Fig. 3.1. It has some distinguishing features:

- strict dependence between application and communication aspects;
- system-of-systems nature;
- strict relationship with the environment.

IoT applications feature a *strict dependence between application requirements as well as constraints and communication aspects*. While in the traditional Internet, communication requirements are quite uniform for the users, in IoT applications there is a large variability of requirements for data rates, delay, and error rates. For instance, agricultural sensor and gas meters produce meager data rates, while video-surveillance cameras generate a large amount of data. Furthermore, sensors can be very far from the power grid, and, therefore, the corresponding protocols should consume a small amount of energy to allow long autonomy with batteries or the use of environmental energy sources. In factory automation, level and variability of delays are crucial for the correct behavior of closed-loop control algorithms.

Many IoT applications can be regarded as a *system-of-systems* since even if the various nodes can independently operate, they interact together to achieve the *good behavior of the global application* [3]. For instance, in a building automation application, the final objective may be to achieve reasonable control of the temperature, and it does not matter the set of nodes that provides such functionality, as long as the global application behavior satisfies design objectives. Thus, these applications pose new questions to designers, traditionally mainly interested in the specification of each single network node as done for Internet servers and clients. Most relevant issues are:

- finding the optimal number of nodes to achieve the common mission;

- finding the best assignment (according to given metrics) between software tasks and hosting nodes by taking into account tasks' requirements and nodes' capabilities;
- finding the best set (according to given metrics) of network protocols by taking into account communication requirements and the presence of a legacy network infrastructure.

The last distinguishing feature is a *strict relationship with the environment*. In many IoT applications, the position of "things" is a constraint for the network. For instance, networked sensors and actuators should be placed where data should be acquired, or action should be performed, respectively. This fact affects the communication architecture significantly. For instance, the position of smart meters requires the use of radio-frequency bands that can propagate through thick walls. Finally, the number and position of nodes affect the communications among them and application performance.

3.2 The Simplified ISO/OSI Reference Model and IoT

This section is devoted to introducing the main telecommunication concepts. It starts by defining some terms that will be used in the following text. Then, the telecommunication layered architecture is described with the various communication functions. Finally, the main standardization bodies are introduced.

3.2.1 Fundamental Terminology

3.2.1.1 Network Nodes

A *network* consists of *nodes* exchanging data over *links*. Nodes that host applications are called *end nodes*, while nodes that connect links to create the network are called *intermediate systems*. Nodes are connected to link through *interfaces*. In IoT applications, the term *Machine-to-Machine (M2M) communications* is very common to emphasize that communications takes place between unmanned "things" and not people using computers.

3.2.1.2 Links and Topologies

Each link is made of a *physical medium*, e.g., a radio-frequency band, a copper wire, or an optical fiber. The physical medium and, optionally, some protocols over it represent the so-called channel which can be considered an abstract view of the link. A link can move data in a single direction (*unidirectional link*), in both directions

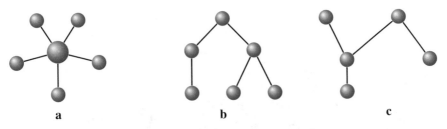

Fig. 3.2 Main network topologies: (**a**) star, (**b**) tree, (**c**) mesh

alternatively (*half-duplex link*) or simultaneously (*full-duplex link*). The data flow exiting the user's interface is named *uplink channel*, while the opposite one is named *downlink channel*.

Nodes and links can be arranged in different ways known as *topologies* as depicted in Fig. 3.2. The simplest topology is the *point-to-point network*. A slightly more complex (and useful) topology is the *star* in which a central node is an intermediate system that connects peripheral nodes. More stars can be joined into a *tree*. Stars and tree are very common topologies in the Edge Network of IoT applications. Stars are usually employed when several sensors collect data and transmit them to the Internet. In this case, the maximum distance between the sensor and the gateway is given by the range of every single link. Trees are used to extend the range of the network by using *multi-hop communications*. In *stars and trees*, there are no multiple paths between nodes, thus simplifying the routing of data but also sacrificing redundancy. In *mesh networks*, the interconnection graph contains cycles which provide more routing opportunities and thus more reliability.

3.2.1.3 Quality of Service

The quality offered by a channel is named *quality of service – QoS –* and consists of three main parameters, i.e., *capacity*, *delay*, and *bit error rate*. Capacity is the maximum number of bits that can be delivered reliably over the channel. The delay is the difference between the time of leave of a bit from the transmitter node and the time of its arrival at the receiver node. The error rate is the percentage of bit values that are erroneously read by the receiver considering all the transmitted bits. It is worth noting that QoS depends on the characteristics of the physical medium as well as of all protocols used over it. A physical medium may have a significant bit error rate (e.g., a radio link), but a retransmission technique provided by a protocol over it can reduce such rate. The term *bandwidth* refers to the frequency interval (reported in Hertz) used to transmit bits at the physical level. This feature is used to allocate various communications of the electromagnetic spectrum. People may use this term with the same meaning as capacity, but this practice is formally wrong since bandwidth is measured in Hertz while capacity is measured in bit/s. Actually, at the physical level, the relationship between bandwidth and capacity is given by the *modulation scheme*, and, given it, the higher is the bandwidth, the higher is the capacity.

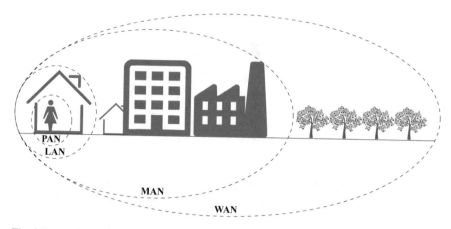

Fig. 3.3 View of an IoT network at different physical sizes

If we observe communications from node's perspective, the *physical bitrate* is the number of bits injected into the physical link in the time unit. Instead, the amount of data transferred in the time unit from the user's perspective is named *throughput* or *data rate* and is usually lower because of the overhead of protocols. The physical bitrate is also known as *wire speed* even if in wireless communications, there is no wire.

3.2.1.4 Network Size

The traditional classification among personal area network (PAN), local area network (LAN), metropolitan area network (MAN), and wide area network (WAN) should be specialized in case of IoT applications (Fig. 3.3). PAN involves nodes less than one meter apart and can be deployed around a body or a machine. LAN involves nodes spanning a private premise over a maximum range of 2–3 km; it is used for ambient intelligence or factory automation. MAN regards nodes spanning over a public city area and can be used to implement smart city applications. WAN regards nodes distance of several kilometers in a public area. Usually PAN and LAN use protocols optimized for short-range communications and exploiting unlicensed radio frequencies, while MAN and WAN use protocols allowing long-range communications, most of them exploiting cellular network standards. In the specific context of IoT, some special names have been derived from the previously described acronyms. Low rate wireless PAN (LR-WPAN) is a term used in the specific context of IEEE 802.15.4 3.3.4. Low-power WAN (LP-WAN) is a term used in the context of LoRaWAN 3.3.9, LTE-M 3.3.17.5, and Sigfox 3.3.10.

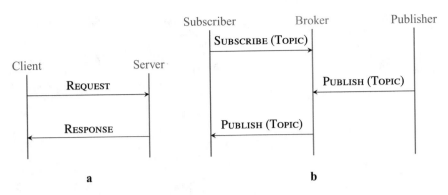

Fig. 3.4 Main communication patterns in IoT applications: (**a**) client/server, (**b**) publish/subscribe

3.2.1.5 Communication Patterns

Network nodes communicate each other by following predefined patterns. The most used pattern is called *client/server*. As depicted in the message sequence chart of Fig. 3.4a, the Client starts the interaction by asking something to the Server that should be in listening state. Then the Server answers and the interaction ends. This pattern requires that the Client knows the address of the information provider (i.e., the Server), and this is not always the case, especially in IoT networks. For instance, one may be interested in the temperature of the room without knowing the address of each temperature sensor inside it. Therefore a more complex pattern can be used, named *publish/subscribe*. As shown in Fig. 3.4b, the device that needs information, whose role is named *Subscriber*, is assumed to know a third-party device, whose role is named *Broker*, that receives data from various sources named *Publishers* and forwards it to various Subscribers. Subscription and publishing are performed for a specific *topic* of interest to avoid broadcasting undesired data.

3.2.2 The ISO/OSI Layers

In general, data transfer between two or more end nodes of the network is a complex task, and the usual engineering approach consists in decomposing it into smaller and simpler problems. In telecommunication, decomposition is depicted as a set of layers, from the user down to the physical medium. As depicted in Fig. 3.5, there are seven layers in the original ISO/OSI Reference Model [4] and five layers in the most used version named *Simplified ISO/OSI Model* since the upper three layers of the original ISO/OSI model are merged into a single layer in the Simplified ISO/OSI Model.

As depicted in Fig. 3.6, at each layer there are two *communication entities* (denoted by rounded rectangles) that interact each other by exchanging data

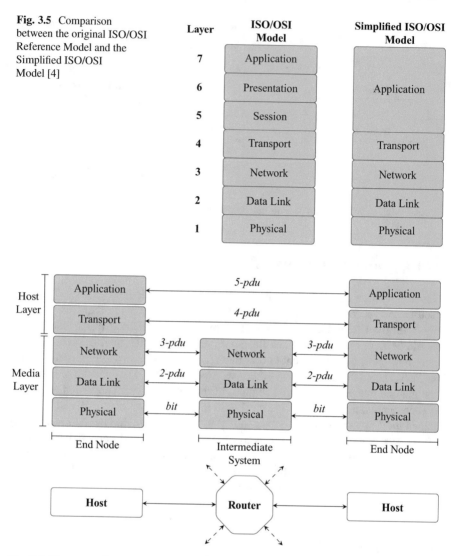

Fig. 3.5 Comparison between the original ISO/OSI Reference Model and the Simplified ISO/OSI Model [4]

Fig. 3.6 Upper part: PDU exchange at the different layers of the Simplified ISO/OSI Model: two end nodes and an intermediate system are represented. Lower part: in the IP context end nodes are called hosts and intermediate systems are called routers

messages named Protocol Data Unit (PDU). The noun PDU is the ISO/OSI term, while the following synonyms are often used in standard-specific contexts: *packet, datagram, frame,* and *segment.*

Figure 3.7 details PDU exchange from node's perspective. User's data comes from the upper layer of the transmitting end node and goes down through the layers. At each layer the corresponding communication entity performs a communication

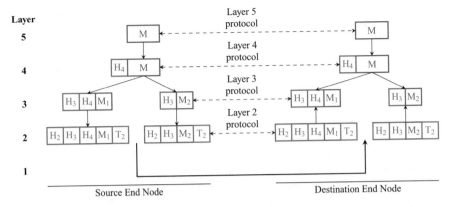

Fig. 3.7 Protocols, PDU and headers in the Simplified ISO/OSI Model [4]

function through the implementation of a *protocol* and adds a *header* in front of data to control such protocol in the interaction with the corresponding counterpart at the receiving end node; header information and data constitute the PDU. Then, such entity calls services provided by the lower-layer entity to transmit the resulting PDU. At Layer 2, also a trailer is added at the end of the PDU to delimit it. At the lowest layer, all these bits are physically transmitted on the medium.

A protocol, and thus the corresponding entities implementing it, can provide different kinds of services:

- **Unacknowledged connectionless service**: it just provides the PDU transmission without any guarantee on reception;
- **Acknowledged connectionless service**: PDU reception is guaranteed by using acknowledge and retransmission which increases the *reliability* of the communication;
- **Connection-oriented service**: PDU flow is preceded by the establishment of the so-called *connection*, i.e., the logic grouping of the PDUs, so that acknowledges and state information can be used to maximize reliability and achieve a given QoS.

PDUs generated by a node can be of three different types:

- **Unicast**. Their destination is a specific interface of a specific node. This is the usual case in which a user wants to send data from node A to node B. In this case the PDU header contains the address of the destination.
- **Broadcast**. Their destination is *all* the interfaces of the network. This mechanism is used when the transmitting node wants to either announce or ask something to all neighbors without the need to repeat the message for each different destination. In this case the PDU header contains a special address representing all interfaces.
- **Multicast**. Their destination is a *subset* of the interfaces of the network. This mechanism is used to reach destinations that satisfy a given property. In this case

the PDU header contains a special address identifying a group of interfaces. Each receiver interested in this kind of communication should listen to PDUs reporting this *group address*.

The Simplified ISO/OSI Model is shown in Fig. 3.6, and its layers are described below.

3.2.2.1 Application Layer

It is responsible for defining how application data are exchanged between the end nodes, the message types that implement the application logic (e.g., the format of the message used to transfer a temperature), and the semantic of application data (e.g., a temperature value with respect to a humidity value). In the Simplified ISO/OSI Model, this layer includes aspects that are split in different layers in the full ISO/OSI Model, i.e., *Presentation Layer* devoted to data representation and encryption and *Session Layer* devoted to session establishment, management, and recovery.

3.2.2.2 Transport Layer

It performs *protocol multiplexing*, i.e., addressing the different processes running at the Application Layer. It is worth referring to well-known network architecture, i.e., TCP/IP, to better understand the role of this layer. In the Transport Layer of the TCP/IP architecture, there are two alternative protocols, i.e., Transmission Control Protocol (TCP) and User Datagram Protocol (UDP). TCP is a connection-oriented protocol; therefore, it provides a reliable and byte-oriented transport service. About 80% of the IP PDUs belonging to a TCP stream do not deliver actual data but are devoted to connection management and acknowledgment. Furthermore, connection establishment and retransmission of lost PDUs take time, and they are not suitable for real-time communications. UDP just provides encapsulation of application data into IP PDUs without any effective error control. It only addresses the basic functionality of protocol multiplexing. In IoT applications, UDP is usually the best choice [5] since:

- devices usually transmit small messages (e.g., a temperature sample) that do not need to be fragmented and re-assembled as well as acknowledged;
- TCP connection overhead is not acceptable for low power wireless transmissions;
- devices may frequently go into sleep mode due to energy constraints; thus, it is infeasible to maintain a long-lived TCP connection;
- some applications (e.g., device actuation) may have a low-latency requirement, which may not tolerate the delay caused by TCP connection establishment.

However, TCP is still used in IoT applications when web services are exploited (see Sect. 3.4.3) since they are conveyed by HTTP/HTTPS application protocols which are delivered over TCP connections.

3.2.2.3 Network Layer

It is responsible for routing PDUs in tree and mesh topology thus increasing network coverage with respect to star topology.

The most common protocol at the Network Layer is the *Internet Protocol (IP)*. There are two options for IP protocol, i.e., IP Version 4, the traditional one, and the new IP Version 6 which is spreading over the Internet. The former is not well-optimized for low data rate transmissions, while the latter is very flexible, and one of its flavors, named 6LoWPAN and described in Sect. 3.3.14, has been specifically designed for wireless IoT transmissions.

IP provides an unacknowledged connectionless service. Each network interface has an IP address (32 bits in IPv4 and 128 bits in IPv6) consisting of a *network prefix* and a remaining part which is interface-specific. Each PDU contains the source and destination address for the delivery. All the interfaces on the same physical medium have the same network prefix thus simplifying the routing process. IP networks are created by connecting end-nodes (named *hosts* in IP terminology) to intermediate systems named *routers* which handle PDU delivery to the destination interface in a *hop-by-hop* way. Figure 3.6 reports the Simplified ISO/OSI Model for two hosts connected through a router. Layer 4 and 5 PDUs go unchanged through the router as if it were not present (for this reason, Layer 4 and 5 are grouped under the name *host layer*). The router handles the three lowest layers (grouped under the name *media layer*). In particular, forwarding decision is performed at Layer 3 according to routing rules.

IoT applications that use IP protocol are widespread since, in this case, the inter-operation with traditional Internet applications is straightforward.

3.2.2.4 Data Link Layer

It deals with the communication on each shared physical medium by identifying the nodes' interfaces and regulating the access to the medium. Since the simultaneous transmission of two or more messages on the same physical medium corrupts all of them, an arbitration policy should be adopted to share a common medium as the wireless channel. There are two main approaches, i.e., Carrier Sense Multiple Access (CSMA) born with Ethernet 40 years ago and Time Division Multiple Access (TDMA) taken from telephone networks. In CSMA, a node willing to send a message checks whether the medium is free and in that case starts transmitting. Multiple transmitters recover from collisions by resending the message after a random delay. In TDMA, a master node keeps synchronized all the other nodes by sending a periodic broadcast message (usually denoted as *beacon*) thus defining a periodic time interval named *superframe* divided into time slots. Each node is only allowed to transmit in a specific time slot to avoid collisions. CSMA is simple thus reducing the software size but may lead to non-deterministic delays and waste of energy to sense the channel. TDMA leads to deterministic periodic transmissions but requires the presence of a master which may become a point of failure. CSMA is preferred where transmission patterns are not regular (e.g., messages triggered by asynchronous events) and variable delays are not an issue. TDMA is preferred in closed-loop control systems where messages are scheduled.

3.2.2.5 Physical Layer

It deals with the transmission of bits on the medium. Many IoT applications exploit radio transmissions by using specific frequency intervals (named *frequency bands*) of the electromagnetic spectrum. Transmissions over different frequency bands do not interfere each other and can go in parallel, but two transmissions cannot be performed on the same frequency band simultaneously. For this reason, in general, the allocation of transmissions in the electromagnetic spectrum is regulated by the government. Frequency bands assigned by the government to specific bodies (usually telecommunication providers) are named *licensed bands*. They are used in *cellular networks*. Other frequency bands can be freely used without asking the government for a permit provided that the transmitted radiated power is kept below a given threshold and the number of sent PDU per time unit is kept under a given threshold; such bands are named *unlicensed bands*. Traditionally, local area networks (and their evolution as personal area network) use unlicensed bands, while commercial telecommunication networks use licensed bands.

There are unlicensed bands in various positions of the spectrum. Some properties of electromagnetic signals depend on frequency value. Bands below 1 GHz are named *sub-gigahertz bands*. Their propagation properties are similar to old FM transmissions, i.e., radio signals go around obstacles and can propagate for tens of kilometers. Vice versa at frequencies above 1 GHz (e.g., around 2.4 and 5 GHz as in Wi-Fi) radio signals behave similarly to light rays and require Line-of-Sight (LoS) propagation between transmitter and receiver. However physical bitrate is proportional to frequency. Communication standards use such different bands according to their application field. For instance, network architectures for agricultural monitoring and smart meter reading use sub-gigahertz bands for their excellent propagation capability considering that the required bitrate is low. Vice versa 2.4/5 GHz bands are used when the required data rate is high, and deployment place does not create LoS issues.

3.2.3 Standardization Bodies

Standardization bodies are organizations that coordinate the development of new standards. Standardization bodies can cover specific areas, like the Institute of Electrical and Electronics Engineers (IEEE) which gather partners interested mainly into electrical and electronics engineering and computer science, or a wider selection of topics, like the International Organization for Standardization (ISO).

Nowadays, there are mainly three types of standards [6]:

- *de facto*: those standards that are followed by informal convention or dominant usage.
- *de jure*: those standards that are part of legally binding contracts, laws or regulations
- *voluntary*: those standards which are published and available for people to consider for use.

3GPP

The 3rd Generation Partnership Project (3GPP) unites seven telecommunications standard development organizations (ARIB, ATIS, CCSA, ETSI, TSDSI, TTA, TTC), known as "Organizational Partners" and provides their members with a stable environment to produce the Reports and Specifications that define 3GPP technologies [7]. The project covers cellular telecommunications network technologies, including radio access, the core transport network, and service capabilities – including work on codecs, security, and quality of service – and thus provides complete system specifications. The specifications also provide hooks for non-radio access to the core network and for interworking with Wi-Fi networks.

ITU

International Telecommunication Union (ITU) is a specialized agency of the United Nations (UN) that is responsible for issues that concern information and communication technologies [8]. ITU coordinates the shared global use of the radio spectrum, promotes international cooperation in assigning satellite orbits, works to improve telecommunication infrastructure in the developing world, and assists in the development and coordination of worldwide technical standards [9].

IEEE

Institute of Electrical and Electronics Engineers (IEEE) is a professional association with a special interest in electrical and electronic engineering, telecommunications, computer engineering, and related disciplines [10]. All working groups address standardization. The most relevant working group for this chapter is 802 and, in particular, 802.11 (Wi-Fi) and 802.15 (personal area networks).

ISO

The International Organization for Standardization (ISO) is an independent, non-governmental organization promoting the development of proprietary, industrial, and commercial standards [11]. It is the world's largest developer of voluntary international standards and facilitates world trade by providing common standards between nations [12]. ISO standards help businesses increase productivity while minimizing errors and waste. By enabling products from different markets to be directly compared, they facilitate companies in entering new markets and assist in the development of global trade regularly. The standards also serve to safeguard consumers and the end users of products and services, ensuring that certified products conform to the minimum standards set internationally.

ETSI

The European Telecommunications Standards Institute (ETSI) is an independent, not-for-profit, standardization organization in the telecommunications industry in Europe, headquartered in Sophia Antipolis, France, with worldwide projection [13]. ETSI deals with telecommunications, broadcasting, and other electronic communications networks and services. ETSI is a European Standards Organization supporting European regulations and legislation through the creation of harmonized European standards. Only standards developed by the three ESOs (Cen, Cenelec, and ETSI) are recognized as European Standards (ENs). ETSI is a member of 3GPP which develops specifications for advanced mobile communications including 5G as well as oneM2M which develops specifications for the most efficient deployment of machine-to-machine communications systems.

IETF

The Internet Engineering Task Force (IETF) is an open organization which develops and promotes voluntary Internet standards, in particular, the standards that comprise the Internet protocol suite (TCP/IP) [14]. The IETF is organized into a large number of working groups and informal discussion groups, each dealing with a specific topic and operates in a bottom-up task creation mode, largely driven by these working groups [15]. IETF standards are named *Request For Comments (RFC)* or *Internet Draft*.

3.2.4 IoT Network Standards and the Simplified ISO/OSI Model

In recent years, many standardization bodies and industrial alliances have proposed new standards or adapted a previous one for IoT scenarios. A standard is a set of recommendations regarding communication functions belonging to one or more layers explained in Sect. 3.2.2. Therefore a possible way to present and compare them is showing their coverage of the Simplified ISO/OSI Model as done in Fig. 3.8.

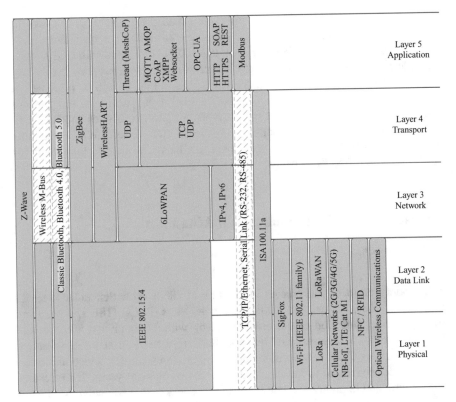

Fig. 3.8 Mapping of IoT network standards on the Simplified ISO/OSI Model

From Fig. 3.8, the potentiality of each standard can clearly be identified. For instance, both IEEE 802.15.4 and Wi-Fi cover Physical and Data Link Layers. Therefore they, like all the standards covering the same layers, can be seen as different alternatives to connect nodes on a wireless medium. ZigBee standard covers all layers over Data Link Layer [16], and therefore it is not an alternative to IEEE 802.15.4.

From the same figure, we can see standards that do not cover the Network Layer. They cannot be used, alone, to create tree and mesh networks since routing is implemented at the Network Layer. A particular case is given by Bluetooth that does not support routing in the core set of protocols but rather in a specific Mesh Profile in Bluetooth 4.0 Low Energy [17]. Theoretically, we can say that routing is implemented at Application Layer in this case.

A set of protocols belonging to different vertically adjacent layers constitutes a *protocol stack*. From the same figure we can see that Bluetooth and Z-Wave standards provide a full protocol stack, i.e., these specifications cover communication functions belonging to all layers. In all other cases, more standards can be combined to create a full protocol stack to be used in actual applications. For instance, IP

Protocol (either version 4 or version 6) can be used over IEEE 802.15.4, Wi-Fi, LoRaWAN, and all cellular standards. Some combinations are mandatory (e.g., ZigBee or 6LoWPAN over IEEE 802.15.4, ISA100.11a or Thread with UDP), while others are optional (e.g., IP over LoRaWAN or over cellular networks).

In the rest of the chapter, the standards reported in Fig. 3.8 will be presented in details, but it is worth referring to Fig. 3.8 to recall their position in the protocol stack. Standards that cover the Physical Layer fall into two classes depending on the type of adopted frequency bands, i.e., licensed or unlicensed. Cellular networks described in Sect. 3.3.17 use licensed frequency bands, while all other standards adopt unlicensed frequency bands.

3.3 IoT Network Technologies and Standards

3.3.1 Modbus

The Modbus protocol was created in 1979 by Modicon (now Schneider Electric) to share data between programmable logic controllers (PLCs) [18]. The open and royalty-free specification of the protocol, along with its simplicity, allowed it to become the first widely accepted de facto standard for industrial communication. Modbus enables client/server communication between a master device and many slave devices connected to the same network, for example, a system that measures temperature and humidity and communicates the results to a computer. Modbus is often used to connect a supervisory computer with a remote terminal unit (RTU) in supervisory control and data acquisition (SCADA) systems. There are many variants of Modbus protocol:

- **Modbus RTU** It uses serial communication standards such as RS-232 and RS-485. Devices can be connected in a linear tree topology also known as *daisy chain* in which each node act as both end node and intermediate system relaying messages for the other nodes. It uses a compact, binary representation of the data for protocol communication. The RTU format follows the commands/data with a cyclic redundancy check checksum as an error check mechanism to ensure the reliability of data. Modbus RTU is the most common implementation available for Modbus. A Modbus RTU message must be transmitted continuously without inter-character hesitations. Modbus messages are separated by idle periods. .
- **Modbus ASCII** It is used in serial communication and makes use of ASCII characters for protocol communication. The ASCII format uses a longitudinal redundancy check checksum. Modbus ASCII messages are framed by leading colon (":") and trailing newline (CR/LF).
- **Modbus Plus (Modbus+, MB+ or MBP)** It is proprietary to Schneider Electric, and unlike the other variants, it supports peer-to-peer communications between multiple masters. It requires a dedicated coprocessor to handle fast HDLC-like token rotation. It uses twisted pair at 1 Mb/s and includes transformer isolation

at each node, which makes it transition/edge-triggered instead of voltage/level-triggered. Special hardware is required to connect Modbus Plus to a computer, typically a card made for the ISA, PCI or PCMCIA bus.

- **Modbus TCP/IP or Modbus TCP** It is a Modbus variant used for communications over TCP/IP networks, connecting over port 502. It does not require a checksum calculation, as lower layers already provide checksum protection.
- **Modbus over TCP/IP or Modbus over TCP or Modbus RTU/IP.** It is a Modbus variant that differs from Modbus TCP in that a checksum is included in the payload as with Modbus RTU.
- **Modbus over UDP** Some have experimented with using Modbus over UDP on IP networks, which removes the overheads required for TCP.

The last three variants were developed to take advantage of the benefits of the TCP/IP architecture over Ethernet networks. The IP layer provides addresses and routing functionality, and Ethernet implements medium access control and physical transmission.

The core part of all Modbus messages contains a device identifier as destination address, a function code (basically to denote read or write operations), and a data field to handle data or command values. Depending on the specific Modbus variant, such fields are preceded and followed by delimiters and an error checking field.

3.3.2   Near-Field Communication (NFC)

Near-field communication (NFC) is a set of communication protocols that enable two electronic devices to establish communication by bringing them within 4 cm (1.6 in) of each other. This is sometimes referred to as NFC/CTLS (Contactless) or CTLS NFC [19, 20]. The standardization body is NFC Forum.

NFC devices are used in contactless payment systems, similar to those used in credit cards and electronic ticket smartcards and allow mobile payment to replace or supplement these systems. NFC-enabled devices can act as electronic identity documents and key cards. NFC is also used for social networking, for sharing contacts and small files. Even if NFC offers a low-speed connection, its setup is simple, and it can be used to bootstrap more capable wireless connections.

NFC is a particular case of *radio-frequency identification (RFID)* technology which employs short-range electromagnetic induction between two loop antennas as a communication channel. NFC operates within the globally available unlicensed radio-frequency ISM band of 13.56 MHz on ISO/IEC 18000-3 air interface at rates ranging from 106 to 424 kb/s. NFC always involves an initiator and a target; the initiator actively generates an RF field that can power a passive target. This enables NFC targets to take very simple form factors such as unpowered tags, stickers, key fobs, or cards. NFC peer-to-peer communication is possible, provided both devices are powered. Secure communications are available by applying encryption algorithms as is done for credit cards.

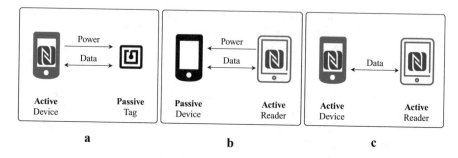

Fig. 3.9 NFC communication models: (**a**) reader/writer, (**b**) card emulation, (**c**) peer-to-peer

As shown in Fig. 3.9, each full NFC device can work in three modes:

- NFC *reader/writer*: it enables NFC devices to read information stored on inexpensive NFC tags embedded in labels or smart posters;
- NFC *card emulation*: it enables NFC devices such as smartphones to act like smart cards, allowing users to perform transactions such as payment or ticketing.
- NFC *peer-to-peer*: it enables two NFC devices to communicate with each other to exchange information in an ad hoc fashion.

NFC tags are passive data stores which can be read and under some circumstances written to, by an NFC device. They typically contain data (as of 2015 between 96 and 8,192 bytes) and are read-only in normal use, but may be rewritable. Applications include secure personal data storage (e.g., debit or credit card information, loyalty program data, personal identification numbers, contacts).

3.3.3 Bluetooth

Bluetooth is a wireless technology standard for exchanging data over short distances using radio waves in the ISM band from 2.400 to 2.485 GHz from both fixed and mobile devices, and building personal area networks. It was initially conceived as a wireless alternative to short-range data cables [21–23], but now the Bluetooth Special Interest Group (SIG), which manages the standard, has more than 30,000 member companies in the areas of telecommunication, computing, networking, and consumer electronics. The IEEE standardized Bluetooth as IEEE 802.15.1, but no longer maintains the standard. The Bluetooth SIG oversees the development of the specification, manages the qualification program, and protects the trademarks. A manufacturer must meet Bluetooth SIG standards to market it as a Bluetooth device. A network of patents applies to the technology, which is licensed to individual qualifying devices.

Bluetooth standard covers Physical Layer and Data Link Layer while the so-called Profiles contain aspects that cover Transport and Application Layers.

Table 3.1 Transmission power and range as a function of the Bluetooth class

| Class | Max permitted power | | Typical range |
	(mW)	(dBm)	(m)
1	100	20	100
1.5 (BT 5 Vol 6 Part A Sect 3)	10	10	20
2	2.5	4	10
3	1	0	1
4	0.5	-3	0.5

Network Layer functionality is not addressed in Bluetooth since topology is kept mainly point-to-point. One exception is given by Mesh Profile in Bluetooth 4.0 which calls low-level services to move data over point-to-point links as in a mesh. Regardless of version, Bluetooth classifies transmitters as a function of emitted power. Transmission power affects range and energy consumption that determines battery lifetime or enables energy-harvesting operations. Table 3.1 reports transmission power and range as a function of the Bluetooth class.

3.3.3.1 Bluetooth Versions

All versions of the Bluetooth standard support downward compatibility. Therefore the latest standard covers all older versions. The main milestones in Bluetooth development are:

- Bluetooth 1.x. This set is also denoted as *Bluetooth Basic Rate (BR)*. It contains the first versions of the standard supported by IEEE 802.15.1 working group. Physical bitrate was up to 721 kb/s. Its target was the basic replacement of cables and audio transmissions.
- Bluetooth 2.x. Physical bitrate was increased to 3 Mb/s, thanks to Enhanced Data Rate (EDR) technique. Security was enforced.
- Bluetooth 3.0. Alternative MAC/PHY was introduced to use 802.11 as a high-speed transport network. Unicast Connectionless Data mode was introduced. Power control was improved.
- Bluetooth 4.0. It introduces the so-called Bluetooth Low Energy or Bluetooth LE or simply BLE. The features introduced in the previous versions of the standards are still maintained and referred to by using the single-word term "Bluetooth" or "Classic Bluetooth" or "Bluetooth Basic Rate/Enhanced Data Rate (BR/EDR)." Table 3.2 reports a comparison between Classic Bluetooth and Bluetooth Low Energy. Compared to Classic Bluetooth, Bluetooth Low Energy is intended to provide considerably reduced power consumption and cost while maintaining a similar communication range. Maximum physical bitrate is 1 Mb/s. New profiles and protocols have been introduced to support IoT applications such as beacon advertising. Mesh topology was introduced.

Table 3.2 Comparison of Bluetooth versions

Aspect	Classic Bluetooth	Bluetooth low energy 4.x
Range	100 m	100 m
Data rate	1–3 Mb/s	1 Mb/s
Throughput	0.7–2.1 Mb/s	0.27 Mb/s
Security	56-/128-bit key	128-bit AES with Counter Mode CBC-MAC
Robustness	Frequency hopping, FEC, fast ASK	Frequency hopping, message integrity check
Latency	100 ms	6 ms
Voice capability	yes	no
Topology	star, tree	star, tree, mesh
Power consumption	1 W	0.01–0.5 W
Main use cases	Headset, automotive, PC, audio	Headset, automotive, PC, audio, healthcare, sport & fitness, retail

Credit to https://cdn.everythingrf.com/live/Bluetooth-A_636234469544772000.jpg

- Bluetooth 5.x. With respect to BLE, it provides options that can double the speed (2 Mb/s burst) at the expense of range, or up to fourfold the range at the expense of data rate, and eightfold the data broadcasting capacity of transmissions, by increasing the packet lengths. The increase in transmissions could be important for IoT devices, where many nodes connect throughout the same environment. Bluetooth 5 adds functionality for connectionless services such as location-relevant navigation of low-energy Bluetooth connections. Bluetooth 5.1 introduces direction-finding to improve localization accuracy.

To specify interoperability between Classic Bluetooth and BLE, Bluetooth 4.0 introduces two different configurations:

- *Single-mode (BLE, Bluetooth Smart) device.* A device that implements BLE and can communicate with single-mode and dual-mode devices, but not with devices supporting Classic Bluetooth only.
- *Dual-mode (BR/EDR/LE, Bluetooth Smart Ready) device.* A device that implements both Classic Bluetooth and BLE and can communicate with any Bluetooth device.

Figure 3.10 in three different columns shows logos, protocol stacks, and interoperation modes of Classic Bluetooth, Bluetooth Smart Ready, and Bluetooth Smart.

3.3.3.2 Bluetooth Protocols and Profiles

Figure 3.10 shows the protocol stack of the various Bluetooth configurations. They share the same separation of:

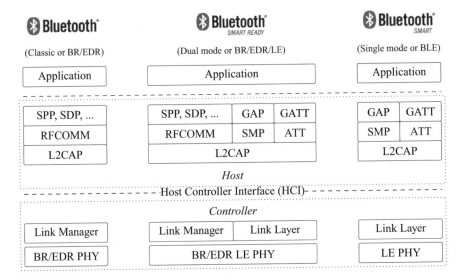

Fig. 3.10 Bluetooth protocol stack and interoperation between Bluetooth versions and device types [24]

- **Application**. It contains the logic, user interface, and data handling of everything related to the actual use-case that the application implements. The architecture of an application is highly dependent on each particular implementation.
- **Host**. It defines and manages services and attributes for each application domain and implements the security functionality.
- **Controller**. It implements the Physical and Data Link Layers.

The Controller stack is generally implemented in a low-cost silicon device containing the Bluetooth radio and a microprocessor. The Host stack is generally implemented as part of an operating system, or as a package on top of an operating system. For integrated devices such as Bluetooth headsets, the host stack and controller stack can be run on the same microprocessor to reduce mass production costs; this is known as a hostless system. The upper layer of the Host part of Fig. 3.10 is occupied by the various Bluetooth profiles that provide definitions and semantics needed to build user applications.

In the following text, a list of basic concepts to understand how Bluetooth works:

- **Physical Layer**. Bluetooth uses a radio technology called frequency-hopping spread spectrum. Bluetooth divides transmitted data into packets and transmits each packet on one of 79 designated Bluetooth channels. Each channel has a bandwidth of 1 MHz. It usually performs 1,600 hops per second. Two modulations are available, i.e., *basic rate (BR)* and *Enhanced Data Rate (EDR)* allowing an instantaneous physical bitrate of either 1 Mb/s or 2–3 Mb/s, respectively. The

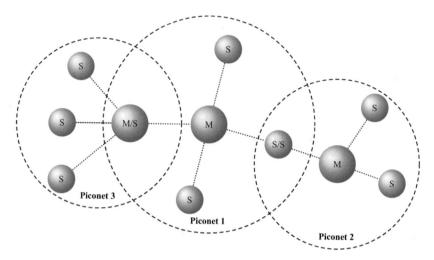

Fig. 3.11 Topologies in Bluetooth networks: three piconets forming a scatternet. Nodes' roles are: Master (M), Slave (S), Master/Slave (M/S), and Slave/Slave (S/S)

modulation rate for Bluetooth Low Energy is fixed at 1 Mb/s to reduce power consumption.

- **Data Link Layer**. Bluetooth is a packet-based protocol with a master/slave architecture. All devices share the master's clock. Packet exchange is based on the basic clock, defined by the master, which ticks at 312.5 μs intervals. Two clock ticks make up a slot of 625 μs, and two slots make up a slot pair of 1250 μs. In the simple case of single-slot packets, the master transmits in even slots and receives in odd slots. The slave, conversely, receives in even slots and transmits in odd slots. Packets may be 1, 3, or 5 slots long, but in all cases, the master's transmission begins in even slots and the slave's in odd slots. BLE has only one packet format and two types of packets, i.e., advertising and data packets. Advertising packets broadcast data for applications that do not need the overhead of a full connection establishment and discover slaves to connect to them. As shown in Fig. 3.11, Bluetooth basic topology is a tree named *piconet* that can be connected to other trees forming a *scatternet*. A piconet is an ad hoc network that links a wireless user group of devices using Bluetooth technology. A piconet starts with one master connecting up to seven slave devices. The devices can switch roles, by agreement, and the slave can become the master (e.g., a headset initiating a connection to a phone necessarily begins as master, i.e., initiator of the connection, but may subsequently operate as the slave). At any given time, data can be transferred between the master and one other device (except for the little-used broadcast mode). The master chooses which slave device to address; typically, it switches rapidly from one device to another in a round-robin fashion. Since it is the master that chooses which slave to address, whereas a slave is (in theory) supposed to listen in each receive slot, being a master is a lighter burden than being a slave. Being a master of seven slaves is possible;

being a slave of more than one master is possible. A scatternet is a number of interconnected piconets. Scatternets can be formed when a member of one piconet (either the master or one of the slaves) elects to participate as a slave in a second, separate piconet. The device participating in both piconets can relay data between members of both ad hoc networks. Using this approach, it is possible to join together numerous piconets into a large scatternet and to expand the physical size of the network beyond Bluetooth's limits in both communication range and number of devices.

- **Host Controller Interface (HCI)**. It provides standardized communication between the host stack (e.g., a PC or mobile phone OS) and the controller (the Bluetooth chip). This interface allows the host stack or controller chip to be swapped with minimal adaptation. There are several HCI Transport Layer standards, each using a different hardware interface to transfer the same command, event, and data packets. The most commonly used are USB (in PCs) and UART (in mobile phones and PDAs).
- **Logical link control and adaptation protocol (L2CAP)**. It is the lowest layer of the Host part and its functions include:

  - Multiplexing data between different higher layer protocols.
  - Segmentation and reassembly of packets.
  - Providing one-way transmission management of multicast data to a group of other Bluetooth devices.
  - Quality of service (QoS) management for higher layer protocols.

  In basic mode, L2CAP provides packets with a payload configurable up to 64 KB, with 672 bytes as the default size and 48 bytes as the minimum mandatory supported size. In retransmission and flow control modes, L2CAP can be configured for reliable or asynchronous data per channel by performing retransmissions and CRC checks.
- **Radio-frequency communication (RFCOMM)**. It is a simple set of transport protocols, made on top of the L2CAP protocol, providing emulated RS-232 serial ports (up to 60 simultaneous connections to a Bluetooth device at a time). The protocol is based on the ETSI standard TS 07.10. RFCOMM is sometimes called serial port emulation. The Bluetooth serial port profile is based on this protocol. RFCOMM provides a simple reliable data stream to the user, similar to TCP. It is used directly by many telephony-related profiles as a carrier for AT commands, as well as being a Transport Layer for OBEX over Bluetooth. Many Bluetooth applications use RFCOMM because of its widespread support and publicly available API on most operating systems. Additionally, applications that used a serial port to communicate can be quickly ported to use RFCOMM.
- **Service discovery protocol (SDP)**. It is used to allow devices to discover what services each other support and what parameters to use to connect to them. For example, when connecting a mobile phone to a Bluetooth headset, SDP will be used to determine which Bluetooth profiles are supported by the headset (headset profile, hands free profile, advanced audio distribution profile, etc.) and the protocol multiplexer settings needed to connect to each of them. Each service

is identified by a Universally Unique Identifier (UUID), with official services (Bluetooth profiles) assigned a short form UUID (16 bits rather than the full 128).

- **Low Energy Attribute Protocol (ATT)**. It is similar in scope to SDP but specially adapted and simplified for Bluetooth Low Energy. It allows a client to read and/or write certain attributes exposed by the server in a non-complex, low-power friendly manner.
- **Low Energy Security Manager Protocol (SMP)**. It is used by Bluetooth Low Energy for pairing and transport specific key distribution.
- **Serial Port Profile (SPP)**. It emulates a serial communication interface (like RS-232 or a UART). It sends bursts of data between two devices as if there were RX and TX lines connected between them.
- **Generic Access Profile (GAP)**. It dictates how Bluetooth LE devices interact with each other. GAP can be considered to define the BLE topmost control layer, given that it specifies how devices perform control procedures such as device discovery, connection, security establishment, and others to ensure interoperability and to allow data exchange to take place between devices from different vendors [24].
- **Generic Attribute Profile (GATT)**. It establishes in detail how to exchange all profile and user data over a BLE connection. In contrast with GAP, which defines the low-level interactions with devices, GATT deals only with actual data transfer procedures and formats [24]. The GATT Profile specifies the structure in which profile data is exchanged as depicted in Fig. 3.12. This structure defines basic elements such as services and characteristics, used in a profile. The top level of the hierarchy is a profile. A profile is composed of one or more services necessary to fulfill a use-case. A service is composed of characteristics or references to other services. Each characteristic contains a value and may contain optional information about the value. The service and characteristic and the components of the characteristic (i.e., value and descriptors) contain the profile data and are all stored in Attributes on the server [25].
- **Mesh Profile**. It is used by Bluetooth Low Energy devices to communicate with other Bluetooth Low Energy devices in the network. Each device can pass the information forward to other Bluetooth Low Energy devices creating a "mesh" effect.

There are many other profiles in today's Bluetooth. Some of them, related to healthcare, sporting, and fitness, boosted the spreading of IoT applications based on wearable devices, e.g., to monitor blood pressure or physical activity.

Proximity sensing applications have been enabled by the long battery life of low-energy Bluetooth devices. Manufacturers of iBeacon devices implement the appropriate specifications for their device to make use of proximity sensing capabilities supported by Apple's iOS devices [26]. Relevant application profiles include:

- FMP, the "find me" profile. It allows one device to issue an alert on a second misplaced device.

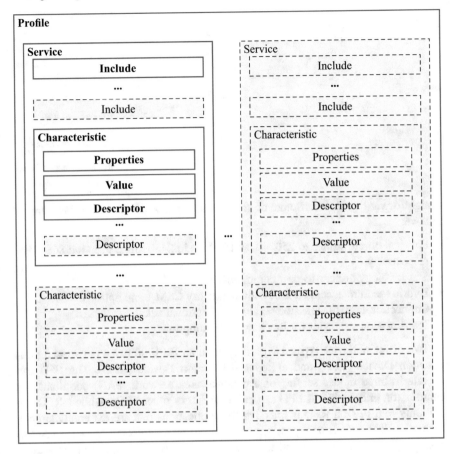

Fig. 3.12 GATT data hierarchy [25]

- PXP, the proximity profile. It allows a proximity monitor to detect whether a proximity reporter is within a close range. Physical proximity can be estimated using the radio receiver's RSSI value, although this does not have absolute calibration of distances. Typically, an alarm may be sounded when the distance between the devices exceeds a set threshold.

3.3.4   IEEE 802.15.4

IEEE 802.15.4 specifies the Physical and Data Link Layer for the so-called Low-Rate Wireless Personal Area Networks (LR-WPANs) [27]. The main objectives of the standard are ease of installation, reliable data transfer, short-range operation, extremely low cost, and reasonable battery life while maintaining a flexible and

Fig. 3.13 Star topology in IEEE 802.15.4 networks

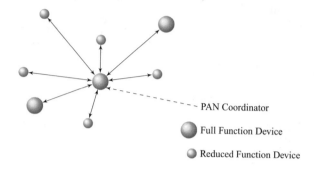

PAN Coordinator

Full Function Device

Reduced Function Device

straightforward protocol. Some of the characteristics of an LR-WPAN are as follows:

- Physical bitrate values of 250, 40, and 20 kb/s in 2450 MHz, 915, and 868 MHz, respectively.
- Star topology or peer-to-peer communications.
- Hybrid medium access control with mandatory CSMA and optional TDMA.
- Connection-less transmission service with optional acknowledgment.
- Transmission strategies for low power consumption.

Two different device types can participate in an IEEE 802.15.4 network: a Full-Function Device (FFD) and a Reduced-Function Device (RFD). The FFD can operate in three modes serving as a personal area network (PAN) coordinator, a coordinator, or a device. A FFD can talk to RFDs or other FFDs, while a RFD can talk only to a FFD. a RFD is intended for applications that are extremely simple, such as a light switch or a passive infrared sensor; they do not have the need to send large amounts of data and may only associate with a single FFD at a time. Consequently, the RFD can be implemented using minimal resources and memory capacity.

In the star topology (Fig. 3.13), the communication is established between peripheral RFD nodes and a single FFD node acting as central controller, called the PAN coordinator. The PAN coordinator typically has some associated application and is either the initiation point or the termination point for network communications. The PAN coordinator is the primary controller of the PAN; it generates a unique PAN ID value for the network and keeps track of all peripheral end nodes that should explicitly associate to it to operate. All devices operating on a 802.15.4 network shall have unique 64-bit addresses. This address may be used for direct communication within the PAN before network association. During association, the PAN coordinator assigns short 16-bit addresses to end nodes to reduce the number of bits transmitted in the PDU. The PAN coordinator might often be mains powered, while a battery or a natural source will most likely power the end s through the so-called *energy harvesting* approach.

For this reason, end nodes are supposed to sleep for a significant part of their life. Since sleeping nodes cannot hear messages, IEEE 802.15.4 adopts asymmetric

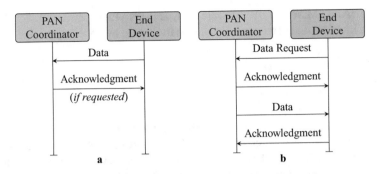

Fig. 3.14 Asymmetric data transmission in IEEE 802.15.4 networks: (**a**) from end nodes to the coordinator, (**b**) from the coordinator to end nodes

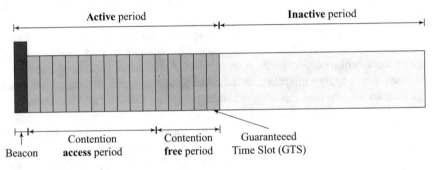

Fig. 3.15 Superframe structure for hybrid medium access control in IEEE 802.15.4 networks

data transmission, as shown in Fig. 3.14. End nodes willing to send data to the PAN coordinator can always do it since the coordinator never sleeps. Vice versa, data messages for a given end node should be stored in the coordinator until the end node asks whether there are messages for it.

Concerning medium access control, IEEE 802.15.4 defines a flexible solution that merges CSMA and TDMA approaches and ensures further power saving. If properly configured, the PAN coordinator sends periodic messages named *beacons*. The time between two beacons is named *superframe*, and its structure is depicted in Fig. 3.15. It can be divided into three parts. The first part is named *Contention Access Period* and implements CSMA policy. The second part is named *Contention Free Period* and is divided into a number of time slots in which only a node can transmit. In the third part no node is allowed to transmit, and therefore they all can sleep to save power. The presence and length of these parts is decided by the designer for the best trade-off between CSMA and TDMA operations (see Sect. 3.2.2) and sleeping time. Such structure is described in the beacon message so that all the end nodes are synchronized and informed.

IEEE 802.15.4 is a continuously evolving standard [28] and provides Physical and Data Link Layers for several network architectures such as ZigBee, ISA100.11a,

WirelessHART, MiWi, 6LoWPAN, Thread, and SNAP. In particular, 6LoWPAN defines a low data rate version of IPv6 and is itself used by upper layers like Thread.

3.3.5 ZigBee

ZigBee is an IEEE 802.15.4-based specification for a suite of high-level communication protocols used to create personal area networks with small, low-power digital radios, such as for home automation, medical device data collection, and other low-power low-data-rate needs, designed for small-scale projects which need wireless connection [29]. Hence, ZigBee is a low-power, low data rate, and close proximity (i.e., personal area) wireless ad hoc network [30, 31]. It is worth noting that documents describing ZigBee standard do not contain the description of IEEE 802.15.4 whose definition is outside the scope of ZigBee.

The technology defined by the ZigBee specification is intended to be simpler and less expensive than other wireless personal area networks, such as Bluetooth or more general wireless networking such as Wi-Fi. Applications include wireless light switches, home energy monitors, traffic management systems, and other consumer and industrial equipment that requires short-range low-rate wireless data transfer.

The ZigBee constraint on power consumption limits transmission distances to 10–100 m line-of-sight, depending on power output and environmental characteristics. ZigBee devices can transmit data over long distances by passing data through a mesh network of intermediate devices. ZigBee is typically used in low data rate applications that require long battery life and secure networking (128-bit symmetric encryption keys secure ZigBee networks). ZigBee features a physical rate of 250 kb/s, best suited for intermittent data transmissions from a sensor or input device.

ZigBee was conceived in 1998, standardized in 2003, and revised in 2006 as well as 2008 (ZigBee PRO). The name refers to the waggle dance of honey bees after their return to the beehive.

3.3.6 ZigBee IP

ZigBee IP is the first open standard for an IPv6-based full wireless mesh networking solution and provides seamless Internet connections to control low-power, low-cost devices [32]. It connects dozens of different devices into a single control network. ZigBee IP was designed to support ZigBee 2030.5 (formerly known as ZigBee Smart Energy 2). It has been updated to include 920IP, which provides specific support for ECHONET Lite and the requirements of Japanese Home Energy Management systems. 920IP was developed in response to Japan's Ministry of Internal Affairs and Communications (MIC) designation of 920 MHz for use in HEMS and Ministry of Economy, Trade and Industry (METI) endorsement of

ECHONET Lite as a smart home standard. 920IP is the only standard referenced by the Telecommunications Technology Committee (TTC) which supports multi-hop mesh networking.

The ZigBee IP specification enriches the IEEE 802.15.4 standard by adding network and security layers and an application framework. ZigBee IP offers a scalable architecture with end-to-end IPv6 networking, laying the foundation for an Internet of Things without the need for intermediate gateways. It offers a cost-effective and energy-efficient wireless mesh network based on standard Internet protocols, such as 6LoWPAN, IPv6, PANA, RPL, TCP, TLS, and UDP. It also features proven, end-to-end security using TLS1.2 protocol, Link Layer frame security based on AES-128-CCM algorithm and support for critical public infrastructures using standard X.509 v3 certificates, and ECC-256 cipher suite. ZigBee IP enables low-power devices to participate natively with other IPv6-enabled Ethernet, Wi-Fi, and HomePlug devices.

From this foundation, product manufacturers can use the ZigBee Smart Energy version 2 standard to create multi-vendor interoperable solutions. As with any ZigBee Alliance specification, custom applications, known as manufacturer-specific profiles, can be developed without multi-vendor interoperability.

Characteristics of ZigBee IP include:

- Global operation in the 2.4 GHz frequency band according to IEEE 802.15.4 Regional operation in the 915 MHz (Americas), 868 MHz (Europe) and 920 MHz (Japan)
- Incorporates power saving mechanisms for all device classes
- Supports development of discovery mechanisms with a full application confirmation
- Supports development of pairing mechanisms with a full application confirmation
- Multiple star topology and inter-personal area network (PAN) communication
- Unicast and multi-cast transmission options
- Security key update mechanism
- Utilizes the industry standard AES-128-CCM security scheme
- Supports Alliance standards or manufacturer-specific innovations

3.3.7 WirelessHART

WirelessHART is a wireless extension of the Highway Addressable Remote Transducer Protocol (HART) [33] used in factory automation and process control. In April 2010, WirelessHart was approved by the IEC as the wireless international standard IEC 62591-1. Figure 3.16 shows a complete example of WirelessHART architecture which includes:

- **field devices** performing field sensing or actuating functions;

- **router devices**, i.e., all devices that have the ability to route packets in the wireless mesh network;
- **adapters** that bind wired HART devices into the wireless mesh network;
- **handheld devices** carried by mobile users such as plant engineers and service technicians;
- a single **gateway** (may be redundant) that functions as a bridge to the host applications;
- **access points** that connect wireless mesh network to the gateway;
- a single **Network Manager** (may be redundant) that may reside in the gateway device or be separate from the gateway;
- **a Security Manager** that may reside in the gateway device or separate from the gateway.

In WirelessHART, communications are precisely scheduled based on TDMA and employ a channel hopping scheme for added system data bandwidth and robustness. The vast majority of communications are directed along graph routes in the wireless mesh network. Graphs are a routing structure that creates a connection between network devices over one or more hops and one or more paths. Scheduling is performed by a centralized Network Manager which uses overall network routing information in combination with communication requirements that devices and applications have provided. The schedule is translated into transmit and receive slots and transferred from the Network Manager to individual devices; devices are only provided with information about the slots for which they have to transmit or receive requirements. The network manager continuously adapts the network graphs and network schedules to changes in the network topology and communication demand [33].

Scaling WirelessHART to service large numbers of wireless devices and high network data rates can be accomplished in a number of ways. One of the ways to do this is to use multiple access points as shown in Fig. 3.16. This architecture allows for a WirelessHART centralized network management of the wireless communications and has the following advantages:

- it coordinates the wireless resources to prevent islands that overlap in the RF space from interfering;
- it reuses wireless resources in non-overlapping islands to scale the network to large number of devices and higher system throughput;
- it provides multiple backbone access points for higher throughput to the backbone network (each access point has the potential throughput of 100 packets per second);
- it provides access points to connect to backbones that go to different plant organizations and separate plants;
- WirelessHART islands may represent different parts of a plant like separate operations or separate geographic regions.

Figure 3.17 shows the protocol stack of WirelessHART. Physical and Data Link Layers are based on the IEEE 802.15.4-2006 standard using 2.4 GHz band

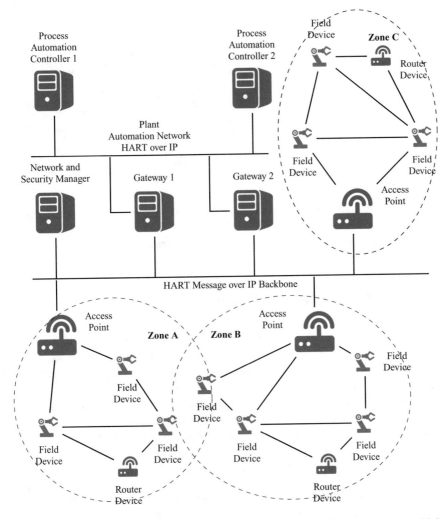

Fig. 3.16 Example of WirelessHART architecture using a single backbone to connect multiple wireless zones

and beacon-based mode. Upper layers have been designed specifically for WirelessHART. The Network Layer is responsible for several functions, the most important of which are routing and security within the mesh network. Whereas the Data Link Layer moves packets between devices, the Network Layer moves packets end-to-end within the wireless network. The Network Layer also includes other features such as route tables and time tables. Route tables are used to route communications along graphs. Time tables are used to allocate communication bandwidth to specific services such as publishing data and transferring blocks of data. Network Layer security provides end-to-end data integrity and privacy across

Fig. 3.17 WirelessHART
protocol stack

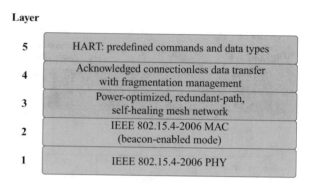

Layer	
5	HART: predefined commands and data types
4	Acknowledged connectionless data transfer with fragmentation management
3	Power-optimized, redundant-path, self-healing mesh network
2	IEEE 802.15.4-2006 MAC (beacon-enabled mode)
1	IEEE 802.15.4-2006 PHY

the wireless network. Transport Layer provides an acknowledged, connectionless delivery service to the Application Layer. Application Layer is HART which defines data types and commands in factory automation domain.

3.3.8 Wi-Fi (IEEE 802.11 Family)

IEEE 802.11 is part of the IEEE 802 set of LAN protocols and specifies the set of Physical and Data Link Layer protocols for implementing wireless local area network (WLAN) Wi-Fi computer communication in various frequencies, including but not limited to 2.4, 5, and 60 GHz frequency bands [34, 35].

It is the world's most widely used wireless computer networking standard, used in most home and office networks to allow laptops, printers, and smartphones to talk to each other and access the Internet without connecting wires. It was created and maintained by the Institute of Electrical and Electronics Engineers (IEEE) LAN/MAN Standards Committee (IEEE 802). The base version of the standard was released in 1997 and has had subsequent amendments. The standard and amendments provide the basis for wireless network products using the "Wi-Fi" brand. While each amendment is officially revoked when it is incorporated in the latest version of the standard, the corporate world tends to market to the revisions because they concisely denote capabilities of their products. As a result, in the marketplace, each revision tends to become its own standard.

3.3.9 LoRaWAN

LoRaWAN specification allows creating low-power wide area networks (LP-WANs) with large geographical coverage [36]. It consists of a Physical Layer named LoRa and a Data Link Layer protocol named LoRaWAN MAC. LoRa is a patented wireless communication technology developed by Cycleo (Grenoble, France) and

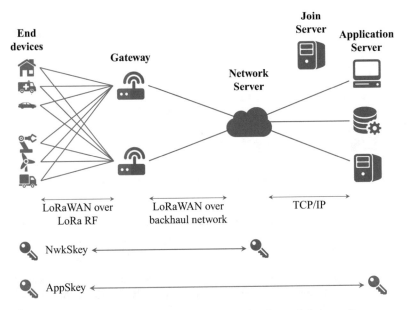

Fig. 3.18 LoRaWAN topology with the various network roles and the security coverage of LoRaWAN keys

acquired by Semtech in 2012. It provides very long-range transmission, thanks to sub-gigahertz frequency bands and chirp spread-spectrum modulation. LoRaWAN is the Data Link Layer protocol over LoRa. Version 1.0 of the LoRaWAN specification was released in June 2015. LoRaWAN is also responsible for managing the communication frequencies, data rate, and power for all devices [37, 38].

The upper part of Fig. 3.18 shows the LoRaWAN topology with the various network roles. LoRaWAN architecture is deployed in a star topology in which *gateways* relay messages between *end devices* and the central *network server*. The gateways are connected to the network server via a backhaul network (e.g., standard IP). Each of them acts as a transparent bridge, simply converting RF packets into IP packets and vice versa. LoRaWAN links consist of the *uplink channel* from the end device to the network server and the *downlink channel* from the network server to the end device. Each end device can transmit to one or many gateways which should be kept synchronized. A stable source usually powers each gateway. Differently from Wi-Fi and IEEE 802.15.4, a gateway is not the coordinator of the network and does not manage join operations of end devices. The *network server* is the coordinator, whereas, association management is performed by the *join server*. Application logic is split between end devices and the *application server*. Network Server, Join Server, and Application Server may reside in the same physical node. Figure 3.19 shows the various LoRaWAN protocol stacks as a function of the network role. The "transparency" of the gateway ensures efficient operations since:

- the gateway is kept simple;

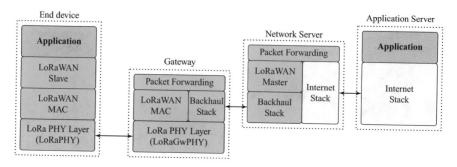

Fig. 3.19 LoRaWAN protocol stack in the various network roles (extension of a figure in [39])

- messages can go through different gateways to allow mobility (no handover is needed) and reliability (messages can go through multiple paths);
- new gateways can be added when the number of end devices increases.

All communication packets between end devices and gateways also include a variable data rate (DR) setting. The selection of the DR allows a dynamic trade-off between communication range and message duration. Also, due to the spread-spectrum technology, communications with different DRs do not interfere with each other and create a set of virtual "code" channels increasing the capacity of the gateway. LoRaWAN network servers manage the DR setting and RF output power for each end device individually by using an adaptive data rate (ADR) scheme, which maximizes both battery life of end devices and the overall network capacity. LoRaWAN physical bitrate ranges from 250 b/s to 50 kb/s. The use of multichannel multi-modem transceiver in the gateway is recommended to increase the efficiency of the gateway by working on different frequency bands.

Data rate is changed by acting on the *spreading factor (SF)* of chirp modulation. LoRa operates with spread factors from 7 to 12. SF7 is the shortest time on air, SF12 will be the longest. Each step-up in spreading factor doubles the time on air to transmit the same amount of data. With the same bandwidth, longer time on air results in fewer data transmitted per unit of time. LoRaWAN uses a different configuration of frequencies, spreading factors, and data rates depending on where the devices are located in the world. Table 3.3 reports such data for some common bands, i.e., 868 and 433 MHz in Europe, 780 MHz in Canada, and 923 MHz in Asia. It is worth noting that DR7 uses FSK modulation instead of chirp modulation.

LoRaWAN uses unlicensed sub-GHz bands that are regulated in all countries they can be used. The rules are based on two restrictions:

- Transmission power: it is the maximum power an emitter can use on the channel when it is communicating. 25 mW is the typical power the device uses for communicating.
- The duty cycle – it is defined as the maximum ratio of time on the air per hour. For instance, 1% means a device can transmit 36 s per hour, not more. Duty Cycle is usually applicable for each sub-band.

Table 3.3 LoRa data rates

Data rate type	Configuration	Physical bitrate (b/s)	Max payload size (byte)
DR0	SF12/125 kHz	250	59
DR1	SF11/125 kHz	440	59
DR2	SF10/125 kHz	980	59
DR3	SF9/125 kHz	1,760	123
DR4	SF8/125 kHz	3,125	230
DR5	SF7/125 kHz	5,470	230
DR6	SF7/250 kHz	11,000	230
DR7	FSK: 50 kb/s	50,000	230

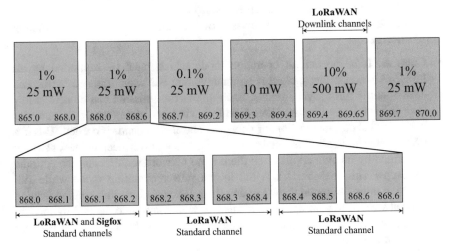

Fig. 3.20 Organization of 868 MHz band in Europe

In Europe, LoRaWAN shares the 868 MHz unlicensed band [40] with Sigfox (discussed in Sect. 3.3.10). This band is regulated by different norms like ERC-REC-70-3E [41] and have national norms in relation. Basically, the 868 MHz band ranges from 865 to 870 MHz and is split in six different sub-bands where different rules apply. The six channels are defined as in Fig. 3.20. The first channel (865.0–868.0) is a 25 mW/1% channel, it is a large area to add LoRa channels, but it is also a zone used by RFIDs. The more interesting channel is the second one, from 868.0 to 868.6; on these 600 kHz, we have the 2000 Sigfox channels and the three standard LoRaWan channels (868.1, 868.3, and 868.5 MHz). As a consequence the duty cycle has to be divided by the number of channels in the same band. As the standard configuration has three channels in the same sub-band, the duty cycle of each of the channel is 0.33%. But, if you allocate some channels on other bands (like 869.7–870), you can set a 1% duty cycle more on this one. So your device can be able to communicate 3 × 0.33% on a band plus 1% on the other band. Basically you can communicate 2% (up-to 3%) with this mechanism. The 868.7 to 869.2 sub-band is a 25 mW area, but the duty cycle is 0.1%; this zone can be interesting to communicate when an object is emitting once a day: the risk of collision is really lower and the number of time you will have to re-emit is, as a consequence, lower, so in this sub-band, you can expect to preserve your energy. The 869.4 to 869.65 zone is particularly interesting because you can communicate at 500 mW with a 10% duty cycle. An end device would not be able to exploit such resource when running on battery but a gateway can use it: the higher power allows to communicate far away and be heard over the local noise; the larger duty cycle allows the gateway to communicate with many devices or send a larger amount of data. The last zone 869.7 to 870 is the last 25 mW/1% zone where you can deploy extra LoRaWAN channels. It is worth noting that this regulation on 868 MHz band applies in Europe, but similar regulations exist for all other countries.

LoRaWAN has three different classes of operations to address the different application needs:

- **Class A**. It is the default operation mode which must be supported by all LoRaWAN end devices; class A communication is always initiated by the end device and is fully asynchronous. Each uplink transmission can be sent at any time and is followed by two short downlink windows, giving the opportunity for bi-directional communication, or network control commands if needed. This is a CSMA-type protocol. The end device can enter low-power sleep mode for as long as defined by its own application: there is no network requirement for periodic wake-ups. This makes class A the lowest power operating mode while still allowing uplink communication at any time. Because downlink communication must always follow an uplink transmission with a schedule defined by the end device application, downlink communication must be buffered at the network server until the next uplink event.
- **Class B**. Devices are synchronized to the network using periodic beacons and open downlink "ping slots" at scheduled times. This provides the network the ability to send downlink communications with a deterministic latency, but at the expense of some additional power consumption in the end device. The latency is programmable up to 128 s to suit different applications, and the additional power consumption is low enough to still be valid for battery powered applications.
- **Class C**. It further reduces latency on the downlink by keeping the receiver of the end device open at all times that the device is not transmitting (half duplex). Based on this, the network server can initiate a downlink transmission at any time on the assumption that the end device receiver is open, so no latency. The compromise is the power drain of the receiver (up to 50 mW) and so class C is suitable for applications where continuous power is available. For battery-powered devices, temporary mode switching between classes A and C is possible and is useful for intermittent tasks such as firmware over-the-air updates.

LoRaWAN defines two layers of cryptography, i.e., a unique AES 128-bit Network Session Key (NwkSKey) shared between the end device and network server and a unique AES 128-bit Application Session Key (AppSKey) shared end-to-end at the application level (lower part of Fig. 3.18). As depicted in Fig. 3.21, NwkSKey

Fig. 3.21 Use of LoRaWAN keys in the protection of the message

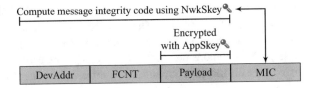

is used to provide authentication and integrity of packets, while AppSKey provides end-to-end encryption of the application payload. The keys can be Activated By Personalization (ABP) on the production line or during commissioning, or can be Over-The-Air Activated (OTAA) in the field. OTAA allows devices to be re-keyed if necessary. These two security levels, together with "transparent gateways," allow to implement multi-tenant or public shared networks without the network operator having visibility of the users payload data.

3.3.10 Sigfox

Sigfox is a French network operator founded in 2009 that created and currently manages a technology and a worldwide infrastructure for wireless LP-WAN based on unlicensed frequencies. It aims to connect low-power objects, e.g., electricity meters and smartwatches, which need to be continuously on and emitting small amounts of data [42–44]. As of October 2018, the Sigfox IoT network has covered a total of 4.2 million square kilometers in a total of 50 countries and is expanding. Sigfox has partnered with a number of embedded systems providers such as Texas Instruments, Silicon Labs, and ON Semiconductor. Sigfox technology is mainly focused on the access network from "things" to the base station, and therefore it covers the Physical and Data Link Layers.

At the Physical Layer, Sigfox employs the differential binary phase-shift keying and the Gaussian frequency shift keying modulations over the unlicensed bands of 868 MHz in Europe and 902 MHz in the United States. Signals in these frequencies pass easily around solid objects and can cover large areas even reaching underground objects while requiring little energy. The physical bitrate ranges from 100 to 600 b/s, depending on the region. Regional regulations also determine the permitted transmission power and the maximum number of messages per day. For instance, in ETSI regions, the maximum transmission power is 14 dBm per device for a maximum transmission time of 2 s and a total of 140 messages per day. In FCC regions, the maximum transmission power is 22 dBm, for a maximum transmission time of 0.346 s, with 140 maximum messages per day. To overcome these limitations, devices transmit at an initial random frequency and then send two duplicates of the same data at two other randomly frequencies at different time-slots.

At the Data Link Layer, the network is based on star topology with a large set of Sigfox-managed base stations which monitor the spectrum looking for messages to be relayed to the cloud through a third-party telecommunication network. A device is not attached to a specific base station unlike Bluetooth, IEEE 802.15.4, Wi-Fi, or cellular protocols. The broadcast message is received by any base station in the range (three in average, according to Sigfox). The spatial distribution of base stations, where the same signal can be received under different conditions, together with message repetition on different frequencies, contributes to increase reliability considering that unlicensed bands are very crowded. The uplink messages are 26 bytes long with a 12 bytes data payload. Therefore they take a maximum of 2 s to be transmitted over the air. The payload allowance in downlink messages is 8 bytes. Sigfox technology was mainly conceived for upload operations and the fact that listening intervals are very limited increases security. Security is mainly enforced in the core Sigfox network, among base stations and towards the cloud. Low bit rate and simple radio modulation enable a 163.3 dB link budget for long-range communications. Sigfox nodes are designed to transmit less than 1 min per day. This feature together with the absence of a tight connection with the base station leads to low transmission and computation overhead to maximize the autonomy of devices.

3.3.11 Z-Wave

Z-Wave is a wireless communications standard used primarily for home automation [45]. It features a mesh network using low-energy radio waves to communicate from appliance to appliance, allowing for wireless control of domestic appliances and other devices, such as lighting control, security systems, thermostats, windows, locks, and door openers. Like other protocols and systems aimed at the home and office automation market, a Z-Wave system can be controlled via the Internet from a smartphone, tablet, or computer and locally through a smart speaker, wireless key fob, or wall-mounted panel with a Z-Wave gateway or central control device serving as both the hub controller and portal to the outside. Z-Wave provides Application Layer interoperability between home control systems of different manufacturers that are a part of its alliance. Z-Wave Plus is the latest certification standard that enforces security and automatic configuration. Z-Wave classic is also fully interoperable but may require more configuration effort.

Z-Wave is a proprietary standard which operates in sub-GHz radio-frequency bands, at 868.42 MHz in Europe and 908.42 MHz in the United States. Z-Wave employs the Frequency Shift Keying (FSK) and the Gaussian Frequency Shift Keying (GFSK) modulations. While Z-Wave has a range of 100 m in open air, building materials reduce that range, and therefore it is recommended to have a Z-Wave device roughly every 10 m, or closer for maximum efficiency. Physical bitrate ranges from 40 to 100 kb/s. Z-Wave is based on a mesh network topology. This means each (non-battery) device installed in the network can become a signal repeater. Each Z-Wave network can support up to 232 devices.

Table 3.4 Standard operation modes of Wireless M-Bus

Mode	Frequency (MHz)	Description
S (Stationary)	868	Meters that send data few times a day
T (Frequent transmit)	868	Meters that send data several times a day
C (Compact)	868	Higher data rate version of Mode T
N (Narrowband)	169	Long range, narrow band system
R (Frequent Receive)	868	Collector that reads multiple meters on different frequency channels
F (Frequent Tx and Rx)	433	Frequent bi-directional communication

3.3.12 Wireless M-Bus

Wireless M-Bus or Wireless Meter-Bus is the European standard (EN 13757-4) that specifies the communication between utility meters and data loggers, concentrators, or smart meter gateways [46]. It was developed as the European standard for the networking and remote reading of utility meters.

The Physical Layer derives from the Konnex industrial standard (KNX-RF) [47] originally using frequency shift keying (FSK) modulation on 868 MHz unlicensed frequency band with 25 mW max power. Then, several operation modes have been added to the standard as reported in Table 3.4. Modes S, T, C, and N are most commonly used with Mode N gaining popularity in the 169 MHz un-licensed frequency band. Modes F, P, Q, and R are less common. These modes have unidirectional and bidirectional sub-modes. Devices are arranged in a simple star topology. No retransmission and channel arbitration procedures are specified in the standard. KNX-RF applies a very simple listen-before-talk mechanism for channel arbitration by measuring the received channel power. The Application Layer is user-defined and may follow well-known standards for smart metering, e.g., Open Metering System (OMS) [48], Dutch Smart Meter Requirements (DSMR) [49], and DLMS/COSEM [50].

3.3.13 Optical Wireless Communications

Optical Wireless Communications (OWCs) are a form of optical communication in which unguided visible, infrared, or ultraviolet light is used to carry a signal [51]. OWC systems operating in the visible band (390–750 nm) are commonly referred to as Visible Light Communication (VLC) [52]. VLC systems take advantage of light emitting diodes (LEDs) which can be pulsed at very high speeds without noticeable effect on the lighting output and human eye. VLC can be possibly used in a wide range of applications including wireless local area networks, wireless personal area networks, and vehicular networks, among others. On the other hand, terrestrial point-to-point OWC systems, also known as the free space optical systems, operate at the near-infrared frequencies (750–1600 nm). These systems typically use laser transmitters and offer a cost-effective protocol-transparent link with high data rates, i.e., 10 Gbit/s per wavelength and provide a potential solution for the backhaul bottleneck. There has been also a growing interest in ultraviolet communication as a result of recent progress in solid-state optical sources/detectors operating within the solar-blind ultraviolet spectrum (200–280 nm). In this so-called deep ultraviolet band, solar radiation is negligible at the ground level, and this makes possible the design of photon-counting detectors with wide field-of-view receivers that increase the received energy with little additional background noise. Such designs are particularly useful for outdoor non-line-of-sight configurations to support low-power short-range communications such as in wireless sensor and ad hoc networks.

IEEE started the standardization of short-range OWC in IEEE 802.15.7, while IEEE 802.11bb working group also started to address optical communications for wireless LAN. The IEEE 802.15.7 standard defines the Physical Layer (PHY) and Media Access Control (MAC) Layer. The standard can deliver enough data rates to transmit audio, video, and multimedia services. It takes into account optical transmission mobility, its compatibility with artificial lighting present in infrastructures, and the interference which may be generated by ambient lighting. The MAC layer permits using the link with the other layers as with the TCP/IP Protocol [53]. The standard defines three PHY layers with different rates:

- The PHY 1 was established for outdoor application and works from 11.67 to 267.6 kbit/s.
- The PHY 2 layer permits reaching data rates from 1.25 to 96 Mbit/s.
- The PHY 3 is used for many emissions sources with a particular modulation method called color shift keying (CSK). PHY III can deliver rates from 12 to 96 Mbit/s.

The modulation formats recognized for PHY 1 and PHY 2 are on-off keying (OOK) and variable pulse position modulation (VPPM). The Manchester coding used for the PHY 1 and PHY 2 layers includes the clock inside the transmitted data by representing a logic 0 with an OOK symbol "01" and a logic 1 with an OOK symbol "10", all with a DC component. The DC component avoids light extinction in case of an extended run of logic 0's.

3.3.14   6LoWPAN

6LoWPAN is an acronym of IPv6 over low-power wireless personal area networks. It is an IETF proposal to bring the benefits of IP and, in particular, IPv6 networking into low-power, low data rate, low-cost wireless personal area networks [54–56]. The IETF 6LoWPAN group defined encapsulation and header compression mechanisms that allow IPv6 packets to be sent and received over IEEE 802.15.4-based networks. The problem statement was given in RFC 4919 [57], while the base specification was developed in RFC 4944 [58] (updated by RFC 6282 with header compression, and by RFC 6775 with neighbor discovery optimizations). Even if

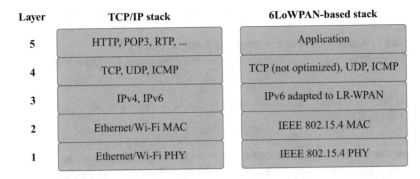

Layer	TCP/IP stack	6LoWPAN-based stack
5	HTTP, POP3, RTP, ...	Application
4	TCP, UDP, ICMP	TCP (not optimized), UDP, ICMP
3	IPv4, IPv6	IPv6 adapted to LR-WPAN
2	Ethernet/Wi-Fi MAC	IEEE 802.15.4 MAC
1	Ethernet/Wi-Fi PHY	IEEE 802.15.4 PHY

Fig. 3.22 Example of 6LoWPAN-based protocol stack compared with the traditional TCP/IP stack

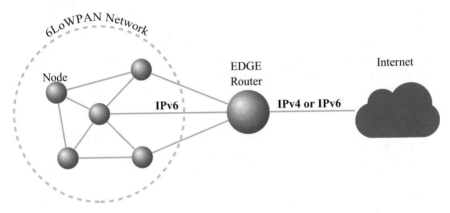

Fig. 3.23 Example merging IPv4, IPv6 and 6LoWPAN networks through the presence of the Edge Router

6LoWPAN was originally though for IEEE 802.15.4, a solution for Bluetooth Low Energy (BLE) was also defined in RFC 7668 [59].

Figure 3.22 shows an example of 6LoWPAN-based protocol stack compared with the traditional TCP/IP Protocol stack. Most protocol stacks based on 6LoWPAN only supports UDP at Transport Layer since 6LoWPAN provides header compression specifically for UDP even if it does not prohibit the use of TCP if needed. It is worth noting that documents describing 6LoWPAN standard do not contain the description of the other protocols involved in the depicted stack. The definition of such protocols is outside the scope of 6LoWPAN.

The 6LoWPAN concept originated from the idea that the same protocol (i.e., IP Protocol) should be applied to all actors of an IoT application, from the traditional hosts and cloud servers even to the smallest low-power devices with limited processing capabilities, thus avoiding complex protocol translations. This objective can be achieved by using *Edge Routers* [60] mixing IPv4, IPv6, and 6LoWPAN as shown in Fig. 3.23. The edge router has three objectives:

1. the data exchange between 6LoWPAN devices and the Internet (or other IPv6 network);
2. local data exchange between devices inside the 6LoWPAN;
3. the generation and maintenance of the radio subnet (the 6LoWPAN network).

3.3.15 Thread

Thread is an IPv6-based networking protocol targeted for use in low-power, embedded consumer, and commercial IoT devices. Thread's original contribution in the IoT protocol context is a secure and reliable mesh network with no single point of failure. [61]. Figure 3.24 shows a Thread-compliant protocol stack. It shows that Thread standardization group merged several well-known protocols belonging to different layers to create a complete communication solution. Up to the Transport Layer, Thread adopts the well-known 6LoWPAN platform based on IPv6, IEEE 802.15.4, UDP, and ICMPv6. As explained in 6LoWPAN RFC, TCP can be used even if its support is not optimized. Between UDP and application protocols, Datagram Transport Layer Security (DTLS) protocol is adopted to introduce the same security properties of TLS over UDP [62]. At Application Layer, Thread adopts well-known industrial standards to interact with sensors and actuators:

- Dotdot: ZigBee-based ontologies for different application domains such as home, building, industrial, retail, health, and energy [63];
- Weave: a secure, reliable communications backbone for Google Nest products [64];
- KNX: an open standard for commercial and domestic building automation;
- CoAP: detailed in Sect. 3.4.6.

At Application Layer, there are also two important service protocols to manage and use mesh topology, i.e., Mesh Commissioning Protocol (MeshCoP) which was specifically developed in Thread and Mesh Link Establishment (MLE) which is an IETF Internet Draft [65].

Fig. 3.24 Thread-compliant protocol stack; MeshCoP is highlighted since it was specifically introduced by Thread's

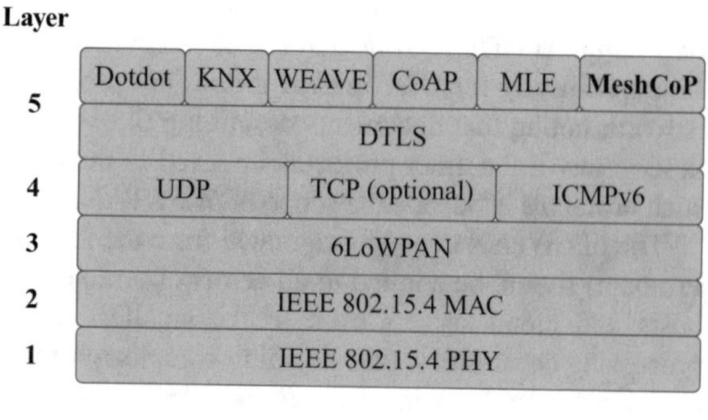

Layer						
5	Dotdot	KNX	WEAVE	CoAP	MLE	**MeshCoP**
	DTLS					
4	UDP		TCP (optional)		ICMPv6	
3	6LoWPAN					
2	IEEE 802.15.4 MAC					
1	IEEE 802.15.4 PHY					

Thread guarantees a high level of security. Only devices that are specifically authenticated can join the network. All communications through the network are secured with a network key. Thread states that there is no single point of failure in its system. However, if the network is only set up with one edge router, then this can become a single point of failure. The edge router or another router can assume the role of leader for specific functions. If the leader fails, another router or edge router will take its place. This is the main way that Thread guarantees no single point of failure.

3.3.16 ISA100.11a

ISA100.11a is a competitor of WirelessHART in wireless networks for process control. It was developed by the International Society of Automation (ISA) and standardized as IEC 62734 [33] in 2014.

The ISA100.11a Physical Layer is taken from the IEEE 802.15.4-2006 2.4 GHz DSSS Physical Layer. The Data Link Layer is unique to ISA100.11a and uses a non-compliant form of the IEEE802.15.4 MAC. It implements graph routing, frequency hopping and time-slotted time domain multiple access features. The forwarding of messages within the wireless network is performed at the Data Link Layer by using a mesh approach. Since the TDMA slot size and the ACK format are not fully specified, it is possible that two ISA100.11a devices may not be able to communicate. ISA100.11a leverages 6LoWPAN protocol and addressing for routing messages outside the wireless domain. The ISA100.11a Transport Layer consists in a connectionless service based on UDP with an enhanced message integrity check and end-to-end security. Currently no process control Application Layer is specified.

ISA100.11a and WirelessHART share some similarities and exhibit some differences [33]. Each of the technologies uses the IEEE 802.15.4 standard at Physical Layer. WirelessHART also uses the IEEE802.15.4-2006 Data Link Layer. ISA100.11a uses a modified, non-compliant version of the same layer. Both have similar mechanisms for forming the wireless mesh network and transporting data to and from the gateway. These networks are very low power. The radio spectrum used in each is in the 2.4 GHz band and does not require licensing. The radio technology utilizes a combination of channel hopping and Direct Sequence Spread Spectrum to achieve coexistence with other users of the same spectrum. Networks can occupy the same physical space and radio spectrum without blocking one another. Both specifications use similar graph routing, source routing, security, and centralized network management functions.

The major differences between WirelessHART and ANSI/ISA100.11a-2011 can be directly traced to the differences in the goals of each standard. Whereas WirelessHART is focused on providing a wireless medium for HART protocol, ISA100.11a is designed to provide flexibility by offering a variety of build options to the manufacturer and run-time options for customizing the operation of the system. This flexibility can be a source of interoperability. Furthermore, ISA100.11a exploits all the potentialities of 6LoWPAN instead of designing a custom Network Layer protocol.

3.3.17   Cellular Network Standards

Wireless devices such as mobile phones or data transmitters receive and send radio signals with any number of cell site base stations fitted with microwave antennas [66]. These sites are usually mounted on a tower, pole, or building, located throughout populated areas, then connected to a cabled communication network and switching system. The phones have a low-power transceiver that transmits voice and data to the nearest cell site, normally not more than 8 to 13 Km away. In areas of low coverage, a cellular repeater may be used, which uses a long-distance high-gain antenna to communicate with a cell tower far outside of the normal range, and a repeater to rebroadcast on a small short-range local antenna that allows any cellphone within a few meters to function correctly.

When the mobile phone or data device is turned on, it registers with the mobile telephone exchange, or switch, with its unique identifier, and can then be alerted by the mobile switch when there is an incoming telephone call. The handset always listens for the strongest signal being received from the surrounding base stations and can switch seamlessly between sites. As the user moves around the network, the "handoffs" are performed to allow the device to switch sites without interrupting the call.

Cell sites have relatively low-power (often only one or two watts) radio transmitters which broadcast their presence and relay communications between the mobile handsets and the switch. The switch, in turn, connects the call to another subscriber of the same wireless service provider or to the public telephone network, which includes the networks of other wireless carriers. Many of these sites are camouflaged to blend with existing environments, particularly in scenic areas.

The dialogue between the handset and the cell site is a stream of digital data that includes digitized audio. The technology that achieves this depends on the system which the mobile phone operator has adopted. The technologies are grouped by generation. The first-generation systems started in 1979 with Japan, are all analog and include AMPS and NMT. Second-generation systems, started in 1991 in Finland, are all digital and include GSM, CDMA, and TDMA.

In an effort to limit the potential harm from having a transmitter close to the user's body, the first fixed/mobile cellular phones that had a separate transmitter, vehicle-mounted antenna, and handset (known as car phones and bag phones) were limited to maximum 3 watts Effective Radiated Power (ERP). Modern handheld cellphones which must have the transmission antenna held inches from the user's skull are limited to a maximum transmission power of 0.6 watts ERP. Regardless of the potential biological effects, the reduced transmission range of modern handheld phones limits their usefulness in rural locations, and handhelds require that cell towers are spaced much closer together to compensate for their lack of transmission power.

3.3.17.1 Second Generation (2G)

2G (or 2-G) is short for second-generation cellular technology [67]. Second-generation (2G) cellular networks were commercially launched on the GSM standard in Finland in 1991. Three primary benefits of 2G networks over their predecessors were that phone conversations were digital. 2G systems were significantly more efficient on the spectrum enabling far higher wireless penetration levels. The most common 2G technology was the TDMA-based GSM, originally from Europe but used in most of the world outside North America. Over 60 GSM operators were also using CDMA2000 in the 450 MHz frequency band.

2G introduced data services for mobile, starting with Short Message Service (SMS), i.e., text messages as well as Multimedia Message Service (MMS), i.e., pictures. All text messages sent over 2G are digitally encrypted, allowing the transfer of data in such a way that only the intended receiver can read it. With General Packet Radio Service (GPRS), 2G offers a theoretical maximum transfer speed of 50 kbit/s (40 kbit/s in practice). With EDGE (Enhanced Data Rates for GSM Evolution), there is a theoretical maximum transfer speed of 1 Mbit/s (500 kbit/s in practice).

3.3.17.2 Third Generation (3G)

3G, short for third generation, is the third generation of wireless mobile telecommunications technology [68]. It is the upgrade for 2G and 2.5G mainly for faster data services. This is based on a set of standards that comply with the International Mobile Telecommunications-2000 (IMT-2000) specifications by the International Telecommunication Union. 3G finds application in wireless voice telephony, mobile Internet access, fixed wireless Internet access, video calls, and mobile TV.

3G telecommunication networks support services that provide an information transfer rate of at least 0.2 Mb/s. Later 3G releases often denoted 3.5G and 3.75G also provide mobile broadband access of several Mb/s to smartphones and mobile modems in laptop computers. The first 3G networks were introduced in 1998.

3.3.17.3 Fourth Generation (4G)

4G is the fourth generation of broadband cellular network technology, succeeding 3G [69]. A 4G system must provide capabilities defined by ITU in IMT Advanced. Potential and current applications include amended mobile web access, IP telephony, gaming services, high-definition mobile TV, video conferencing, and 3D television.

The first 4G networks were introduced in 2008. The first-release Long Term Evolution (LTE) standard was commercially deployed in Oslo, Norway, and Stockholm, Sweden, in 2009 and has since been deployed throughout most parts of the world. It has, however, been debated whether first-release versions should be considered 4G LTE, as discussed in the technical understanding section below.

As opposed to earlier generations, a 4G system does not support traditional circuit-switched telephony service, but all-Internet Protocol (IP)-based communication such as IP telephony. As seen below, the spread-spectrum radio technology used in 3G systems is abandoned in all 4G candidate systems and replaced by OFDMA multi-carrier transmission and other frequency-domain equalization (FDE) schemes, making it possible to transfer very high bit rates despite extensive multi-path radio propagation (echoes). Smart antenna arrays further improve the peak bit rate for Multiple-Input Multiple-Output (MIMO) communications.

3.3.17.4 NB-IoT

Narrowband IoT (NB-IoT) is a low-power wide area network radio technology standard developed by 3GPP to enable a wide range of cellular devices and services [70]. The specification was frozen in 3GPP Release 13 (LTE Advanced Pro), in June 2016. Other 3GPP IoT technologies include eMTC (enhanced Machine-Type Communication) and EC-GSM-IoT [37].

NB-IoT focuses specifically on indoor coverage, low cost, long battery life, and high connection density. NB-IoT uses a subset of the LTE standard but limits the bandwidth to a single narrow-band of 200 kHz. It uses orthogonal frequency-division multiple access (OFDMA) modulation for downlink communication and a couple of options for uplink communication (for more information on the uplink options, refer to the 3GPP specification TR 36.802.). OFDMA is a modulation scheme in which individual users are assigned subsets of subcarrier frequencies. This enables multiple users to transmit low-speed data simultaneously.

As depicted in Fig. 3.25, NB-IoT can operate in three different modes:

- **Standalone**: A GSM carrier is used as an NB-IoT carrier, enabling reuse of 900 MHz or 1800 MHz.
- **In-band**: Part of an LTE carrier frequency band is allocated for use as an NB-IoT frequency. The service provider typically makes this allocation, and IoT devices are configured accordingly. You should be aware that if these devices must be deployed across different countries or regions using a different service provider, problems may occur unless there is some coordination between the service providers, and the NB-IoT frequency band allocations are the same.
- **Guard Band**: An NB-IoT carrier is between the LTE or WCDMA bands. This requires coexistence between LTE and NB-IoT bands.

The link budget of NB-IoT is 164 dB. The GPRS link budget is 144 dB, used by many machine-to-machine services. The additional 20 dB link budget should guarantee better signal penetration in buildings and basements while improving

Fig. 3.25 NB-IoT modes of operation [71]: (**a**) Standalone, (**b**) In Band, (**c**) Guard Band

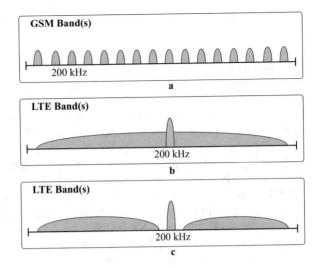

battery life. At Layer 1, the maximum transport block size (TBS) for the downlink is 680 bits, while uplink is 1000 bits. At Layer 2, the maximum payload size is 1,600 bytes. NB-IoT operates in half-duplex frequency-division duplexing mode with a maximum uplink data rate of 60 kb/s and downlink of 30 kb/s.

3.3.17.5 LTE Cat M1

LTE category M1 (LTE Cat M1 or simply LTE-M) is 4G profile specifically designed for IoT and M2M communications. It features a transmit power of 20 dBm and provides an average upload bitrate between 375 kb/s and 1 Mb/s. Main advantages of this technology are: extended battery lifecycle, excellent in-building range, and support of voice functionality through Voice over LTE (VoLTE).

LTE-M supports two duplex modes:

- Full-duplex: data can be downloaded and uploaded simultaneously at a rate of 1 Mb/s;
- Half-duplex: data can travel only in one direction at a time at a rate of 375 kb/s.

The advantage of the second mode is that it requires less power, which makes it suitable for scenarios where there is no need to both send and receive at the same time.

It is worth noting that both NB-IoT and LTE-M were designed to support a large number of sensors for environmental monitoring or fleet tracking. LTE-M requires an LTE infrastructure, while NB-IoT can be hosted in GSM infrastructure. With low data rate applications, NB-IoT consumes less energy than LTE-M but at the cost of a lower bitrate.

3.3.17.6 Fifth Generation (5G)

5G (from "5th generation") is the latest generation of cellular mobile communications. 5G targets high data rate, reduced latency, energy saving, cost reduction, higher system capacity, and massive device connectivity. The first phase of 5G specifications in Release 15 has been completed by April 2019 to accommodate the early commercial deployment. The second phase in Release 16 is due to be completed by April 2020 for submission to the ITU as a candidate of IMT-2020 technology.

The ITU IMT-2020 specification demands speed up to 20 Gb/s, achievable with wide channel bandwidths and massive MIMO. 3GPP is going to submit the 5G NR (NR = New Radio) standard proposal. 5G NR can include lower frequencies, below 6 GHz, and mmWave, above 15 GHz. However, the speeds and latency in early deployments, using 5G NR software on 4G hardware (non-standalone), are only slightly better than new 4G systems, estimated from 15% to 50% better.

The key trends which must be accommodated by hardware designers include:

- Increased data rate for Enhanced Mobile Broadband (eMBB) and other applications, specifically driving the instantaneous available data rate at 10x current rates [72]. Furthermore, deployment of 5G will also be staged depending on frequency band, sub-6GHz will be deployed first, followed by the contiguous bands at mmWave frequencies enabling more key eMBB applications at a later stage. Simulation of standalone eMBB deployments showed improved throughput by $2.5\times$ below 6 GHz and by nearly $20\times$ at millimeter waves.
- Connectivity to many more devices will happen because expectations are that there will be 50 billion connected devices within 2 years. This is partly addressed by existing standards but will also be encompassed by the current specification of Massive Machine-Type Communications (mMTC) in Release 16 of 3GPP.
- New usage models, exerting new requirements onto mobile devices and the cellular infrastructure that they connect to. Good examples include low data rate, low power requirements for connecting battery-powered IoT end-points within mMTC. High reliability, low latency cellular for vehicle-to-vehicle and vehicle-to-infrastructure connectivity (C-V2X) to complement existing V2X solutions like collision detection. Low latency support for new and emerging applications like remote surgery and augmented/virtual-reality. The second two examples will be addressed by the upcoming 3GPP standard for Ultra-Reliable, Low Latency Connectivity (URLLC).

3.4 Application Layer Protocols

This section describes several Application Layer protocols to build IoT applications. The starting point is the well-known HyperText Transfer Protocol (HTTP) which is also the main building block of the World Wide Web. Many IoT-specific Application Layer protocols have been inspired by HTTP because of its spread.

3.4.1 HyperText Transfer Protocol (HTTP)

The HyperText Transfer Protocol (HTTP) is for the World Wide Web as the water for the ocean, and therefore all its definitions are likely to seem restrictive. HTTP is a textual protocol to build distributed, collaborative, hypermedia information systems. Even if its origin comes back to the invention by Tim Berners-Lee at CERN in 1989, HTTP is still updated by IETF. In the last years, privacy, authentication, caching, and connection persistence have been added to fulfill the requirements of the ever-increasing types of applications conveyed over HTTP.

HTTP is an Application Layer protocol. Its definition presumes an underlying and reliable Transport Layer protocol, and Transmission Control Protocol (TCP) is commonly used. However, HTTP can be adapted to use unreliable protocols such as the User Datagram Protocol (UDP), for example, in HTTPU and Simple Service Discovery Protocol (SSDP). HTTP functions as a request/response protocol in the client/server computing model. A web browser, for example, may be the client and an application running on a computer hosting a website may be the server. The client submits the HTTP request message to the server. The server, which provides resources such as HTML files and other content, or performs other functions on behalf of the client, returns a response message to the client. The response contains completion status information about the request and may also contain requested content in its message body. A web browser is an example of client. Other types of client include the indexing software used by search providers (web crawlers), voice browsers, mobile apps, and other software that accesses, consumes, or displays web content.

HTTP resources are identified and located on the network by Uniform Resource Locators (URLs), using the Uniform Resource Identifiers (URIs) schemes named "http" and "https". An example, including all optional components, is reported in Fig. 3.26. URIs are encoded as hyperlinks in HTML documents, so as to form interlinked hypertext documents. In HTTP v1.0 a separate connection to the same server is made for every resource request. HTTP v1.1 can reuse a connection multiple times to download images, scripts, stylesheets, etc. after the page has been delivered. HTTP/1.1 communications therefore experience less latency as the establishment of TCP connections presents considerable overhead.

HTTP is designed to permit intermediate network elements to improve or enable communications between clients and servers. High-traffic websites often benefit from web cache servers that deliver content on behalf of upstream servers to improve response time. Web browsers cache previously accessed web resources and reuse them, when possible, to reduce network traffic. HTTP proxy servers at private

Fig. 3.26 Example of Uniform Resource Identifier for HTTP scheme [73]

network boundaries can facilitate communication for clients without a globally routable address, by relaying messages with external servers.

3.4.2 WebSocket

WebSocket is an Application Layer protocol, providing a full-duplex communication channel over a TCP connection originally created for an HTTP connection. The IETF standardized the WebSocket protocol as RFC 6455 in 2011 [74], and the World Wide Web Consortium (W3C) is standardizing the WebSocket API in Web Interface Definition Language [75]. WebSocket and HTTP are located at the Application Layer in the ISO/OSI model and, as such, depend on TCP at Transport Layer. A WebSocket connection originates from an HTTP connection by using the *HTTP Upgrade header* as shown in the message sequence chart depicted in Fig. 3.27. The WebSocket protocol enables a peer-to-peer interaction between the agent previously acting as HTTP client (e.g., a web browser or other client application) and the one previously acting as HTTP server so that the latter can send content to the former without being first requested by it as in HTTP. WebSocket protocol also allows to keep the connection open and support TLS-based security. Since the WebSocket connection uses the previously created HTTP connection, it is compliant with NAT-based and firewall-protected networks. In summary, WebSocket protocol allows two-way conversation without recurring to complex workarounds and thus facilitating real-time data transfer.

Fig. 3.27 Example of WebSocket life cycle

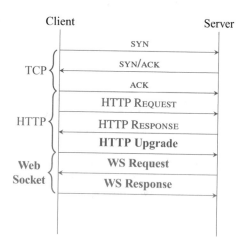

3.4.3 Web Services and Representational State Transfer (REST)

The term *web service* is a service offered by an electronic device to another electronic device, communicating with each other via the World Wide Web technology [76]. In a web service, Web technologies such as HTTP or HTTPS, initially designed for human-to-machine communication, is utilized for machine-to-machine communication, more specifically for transferring machine-readable file formats such as XML and JSON. Web services are usually not very optimized for low-power and low data rate communications since they rely on HTTP/HTTPS and TCP, but they have the advantage of being compliant with firewall-protected networks.

Web resources were first defined on the World Wide Web as documents or files identified by their URLs. However, today they have a much more generic and abstract definition that encompasses every thing or entity that can be identified, named, addressed, or handled, in any way whatsoever, on the web. For instance, a web service can consist of an object-oriented web-based interface to a database server utilized by several sensor devices to store sensed data or to a regulator server managing a set of actuation devices.

A first technology to implement web services was based on the *Simple Object Access Protocol (SOAP)* which uses XML to encode request and response messages [77]. Such encoding mechanism revealed to be heavy even for high data rate networks. Furthermore, according to this paradigm, the server should implement application logic as a set of running objects whose state information should be stored for a long time. This feature leads to high memory overhead on the server. Both drawbacks make SOAP a non-scalable technique for large M2M applications.

A recent web service technique is called *Representational State Transfer (REST)*. It is a software architectural style that defines a set of constraints to implement M2M communications by using basic HTTP operations [78]. Web services that conform to the REST architectural style, termed RESTful web services, allow the requesting systems to access and manipulate textual representations of web resources by using a uniform and predefined set of stateless operations. By using a stateless protocol and standard operations, RESTful systems aim for fast performance, reliability, and the ability to grow, by re-using components that can be managed and updated without affecting the system as a whole, even while it is running.

The term "representational state transfer" was introduced and defined in 2000 by Roy Fielding in his doctoral dissertation. Fielding's dissertation explained the REST principles based on the "HTTP object model." The term is intended to evoke an image of how a well-designed Web application behaves: it is a network of Web resources (a virtual state-machine) where the user progresses through the application by selecting links and operations such as GET or DELETE, resulting in the next resource (representing the next state of the application) being transferred to the end user.

In a RESTful web service, requests made to a resource's URI will elicit a response with a payload formatted in HTML, XML, JSON, or some other format.

The response can confirm that some alteration has been made to the stored resource, and the response can provide hypertext links to other related resources or collections of resources.

It the very common case in which HTTP/HTTPS is used, the REST operations are mapped onto HTTP methods [79] as follows:

- **GET method**. It is used to read a representation of a resource. If the path is correct, GET returns a representation in XML or JSON and an HTTP response code of 200 (OK). In an error case, it most often returns a 404 (NOT FOUND) or 400 (BAD REQUEST). According to the design of the HTTP specification, GET (along with HEAD) requests are used only to read data and not change it. Therefore, when used this way, they are considered *safe*, i.e., they can be called without risk of data modification or corruption. Additionally, GET (and HEAD) is *idempotent*, which means that making multiple successive identical requests ends up having the same result as a single request.
- **POST method**. It is most often utilized to create new resources. In particular, it is used to create subordinate resources, i.e., subordinate to some other (e.g., parent) resource. In other words, when creating a new resource, POST to the parent and the service takes care of associating the new resource with the parent, assigning an ID (new resource URI). On successful creation, return HTTP status 201, returning a Location header with a link to the newly created resource with the 201 HTTP status. POST is neither safe nor idempotent. It is therefore recommended for non-idempotent resource requests. Making two identical POST requests will most likely result in two resources containing the same information.
- **PUT method**. It is most often utilized to update capabilities, i.e., to put to a known resource URI with the request body containing the newly updated representation of the original resource. However, PUT can also be used to create a resource in the case where the resource ID is chosen by the client instead of by the server, i.e., if the PUT is to a URI that contains the value of a nonexistent resource ID. Again, the request body contains a resource representation. On successful update, PUT call returns 200 (or 204 if not returning any content in the body). If PUT is used to create, it returns HTTP status 201 on successful creation. A body in the response is optional and a waste of bytes since the client already knows the resource ID. PUT is not a safe operation, in that it modifies (or creates) state on the server, but it is idempotent. In other words, if you create or update a resource using PUT and then make that same call again, the resource is still there and still has the same state as it did with the first call. It is recommended to keep PUT requests idempotent and to use POST for non-idempotent requests.
- **PATCH method**. It is used to modify capabilities. The PATCH request only needs to contain the changes to the resource, not the complete resource. This resembles PUT, but the body contains a set of instructions describing how a resource currently residing on the server should be modified to produce a new version. This means that the PATCH body should not just be a modified part of the resource but in some patch languages like JSON Patch or XML Patch. PATCH is neither safe nor idempotent. However, a PATCH request can be issued

in such a way as to be idempotent, which also helps prevent adverse outcomes from collisions between two PATCH requests on the same resource in a similar time frame. Collisions from multiple PATCH requests may be more dangerous than PUT collisions because some patch formats need to operate from a known base-point, or else they will corrupt the resource. Clients using this kind of patch application should use a conditional request such that the request will fail if the resource has been updated since the client last accessed the resource. For example, the client can use a strong ETag in an If-Match header on the PATCH request.

- **DELETE method**. It is used to delete a resource identified by a URI. On successful deletion, it returns HTTP status 200 (OK) along with a response body, e.g., the representation of the deleted item or a wrapped response. DELETE operation is idempotent regarding the deleted resource but not regarding the return code. Calling DELETE on a resource a second time will often return 404 (NOT FOUND) since it was already removed and therefore is no longer available. This, by some opinions, makes DELETE operations no longer idempotent; however, the end-state of the resource is the same. Returning a 404 is acceptable and communicates the status of the call accurately.

3.4.4 Message Queuing Telemetry Transport (MQTT)

Message Queuing Telemetry Transport (MQTT) is an ISO standard (ISO/IEC PRF 20922) describing a publish/subscribe messaging protocol [80, 81]. It works on top of the TCP/IP Protocol even if MQTT-SN is a variation of the main protocol aimed at embedded devices on non-TCP/IP networks, such as ZigBee.

3.4.4.1 How MQTT Works

An MQTT session starts with a client creating a TCP/IP connection with the broker by using either a standard port or a custom port defined by the broker's operators. An MQTT connection is established by using the following standard ports: 1883 for non-encrypted communication and 8883 for encrypted communication using SSL/TLS. The client validates the server certificate to authenticate the server during the SSL/TLS handshake. The client may also provide a client certificate to the broker during the handshake, which the broker can use to authenticate the client. Whilst not part of the MQTT specification, it has become customary for brokers to support client authentication with SSL/TLS client-side certificates. This protocol is designed to be low-demanding in terms of resources especially for IoT applications. As a consequence, relying on SSL/TLS might not be an optimal solution. In these cases, authentication is done by sending non-encrypted username and password as part of the CONNECT/CONNACK packet sequence. This is the case with *public*

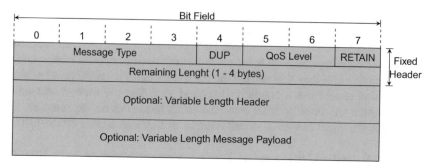

Fig. 3.28 MQTT packet structure

brokers which are configured to accept anonymous clients (i.e., username and password are simply left blank).

MQTT is designed to be a lightweight protocol where all the messages have a small code footprint. The structure of an MQTT message is shown in Fig. 3.28. Each MQTT message consists of a 2 byte fixed header, followed by an optional variable header, a message payload that is limited to 256 MB of information. Fourteen types of MQTT messages can be exchanged [82]:

1. CONNECT: After a Client establishes a Network Connection to a Server, the first Packet sent from the Client to the Server MUST be a CONNECT Packet.
2. CONNACK: Is sent by the Server in response to a CONNECT Packet received from a Client.
3. PUBLISH: Is sent from a Client to a Server or from Server to a Client to transport an Application Message.
4. PUBACK: Is the response to a PUBLISH Packet with QoS level 1.
5. PUBREC: Is the response to a PUBLISH Packet with QoS 2. It is the second packet of the QoS 2 protocol exchange.
6. PUBREL: Is the response to a PUBREC Packet. It is the third packet of the QoS 2 protocol exchange.
7. PUBCOMP: The PUBCOMP Packet is the response to a PUBREL Packet. It is the fourth and final packet of the QoS 2 protocol exchange.
8. SUBSCRIBE: The SUBSCRIBE Packet is sent from the Client to the Server to create one or more Subscriptions. Each Subscription registers a Client's interest in one or more Topics. The Server sends PUBLISH Packets to the Client in order to forward Application Messages that were published to Topics that match these Subscriptions. The SUBSCRIBE Packet also specifies (for each Subscription) the maximum QoS with which the Server can send Application Messages to the Client.
9. SUBACK: A SUBACK Packet is sent by the Server to the Client to confirm receipt and processing of a SUBSCRIBE Packet. A SUBACK Packet contains a list of return codes that specifies the maximum QoS level that was granted in each Subscription that was requested by the SUBSCRIBE.

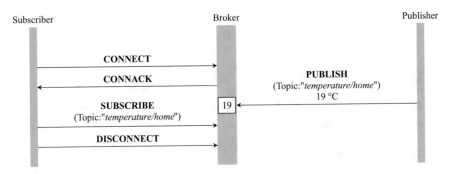

Fig. 3.29 Example of MQTT protocol operation at QoS Level 0

10. UNSUBSCRIBE: An UNSUBSCRIBE Packet is sent by the Client to the Server to unsubscribe from topics.
11. UNSUBACK: The UNSUBACK Packet is sent by the Server to the Client to confirm receipt of an UNSUBSCRIBE Packet.
12. PINGREQ: The PINGREQ Packet is sent from a Client to the Server. It can be used to:

   - Indicate to the Server that the Client is alive in the absence of any other Control Packets being sent from the Client to the Server.
   - Request that the Server responds to confirm that it is alive.
   - Exercise the network to indicate that the Network Connection is active.

13. PINGRESP: The Server sends a PINGRESP Packet to the Client in response to a PINGREQ Packet. It indicates that the Server is alive.
14. DISCONNECT: The DISCONNECT Packet is the final Control Packet sent from the Client to the Server. It indicates that the Client is disconnecting cleanly.

MQTT supports three Quality of Service (QoS) levels, which determine the protocol's behavior and how it manages the communication. The first QoS level is shown in Fig. 3.29 and is referred to as *at most once*, or *fire and forget*. The publisher sends a message to the broker one time and then deletes the sent data. The broker receives the data and sends it to the subscribers one time. The second QoS level is shown in Fig. 3.30 and is referred to as *at least once*. In this case, each publish action from the publisher and the broker, and from the broker to the subscriber, is followed by an acknowledge. If an acknowledge is not received promptly, the packet is sent again. The third QoS level is shown in Fig. 3.31 and is referred to as *exactly once*. In this third scenario, the mechanism requires two pairs of messages to be sent. The first pair is *publish/pubrec*, and the second is *pubrel/pubcomp*. These two pairs of messages ensure that, even with multiple retries, subscribers receive the data only one time.

One of the benefits of the MQTT protocol is that it preserves battery power and efficiently delivers messages. There are other Application Layer protocols with a similar publish/subscribe paradigm, e.g., the Advanced Message Queuing Protocol

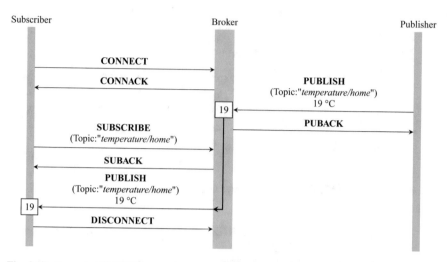

Fig. 3.30 Example of MQTT protocol operation at QoS Level 1

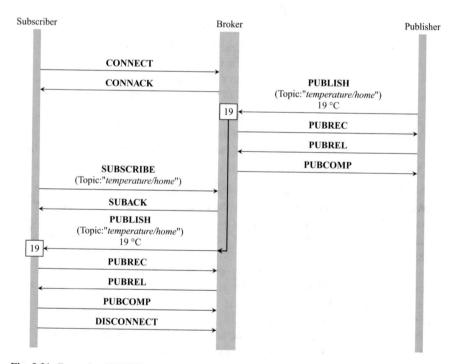

Fig. 3.31 Example of MQTT protocol operation at QoS Level 2

(AMQP) and the Constrained Application Protocol (CoAP); the former has been traditionally used in normal Internet, while the latter has been specifically designed for IoT applications. They are discussed in the next sections.

3.4.5 Advanced Message Queuing Protocol (AMQP)

The Advanced Message Queuing Protocol (AMQP) is an open standard Application Layer protocol for message-oriented middleware [83]. AMQP is a binary, Application Layer protocol, designed to efficiently support a wide variety of messaging applications and communication patterns. The defining features of AMQP are message delivery (both client/server and publish/subscribe), queuing, reliability, and security. It provides flow-controlled, message-oriented communication with message-delivery guarantees such as at most once (where each message is delivered once or never), at least once (where each message is certain to be delivered, but may do so multiple times), and exactly once (where the message will always certainly arrive and do so only once). Authentication and encryption is based on Simple Authentication and Security Layer (SASL) and Transport Layer Security (TLS). It assumes an underlying reliable Transport Layer protocol such as TCP. It provides a symmetric and asynchronous flow of messages whose format can be extended by the application designer.

The main difference between AMQP and MQTT is that the former has been designed as a general-purpose middleware for traditional computers, while the latter has been optimized for constrained devices. In fact, AMQP provides more features and allows both client/server and publish/subscribe communication patterns at the cost of an increased overhead. Both end points of the data transfer can be either client or server so that communication can start from any direction. MQTT is less symmetric and allows only the publish/subscribe communication pattern with a message header size of just 2 bytes. Therefore, MQTT is more suitable for IoT applications.

3.4.6 Constrained Application Protocol (CoAP)

Constrained Application Protocol (CoAP) is a specialized Internet Application Layer protocol for constrained devices with limited computation and communication resources, e.g., nodes with 8-bit microcontrollers with small amounts of ROM and RAM, connected to wireless network with high packet error rates and a typical throughput of 10s of kb/s. As defined in RFC 7252 [84], it enables those constrained nodes to communicate with the broader Internet using a protocol similar to HTTP/HTTPS that can be easily translated for interoperation. CoAP is designed for use between devices on the same constrained network (e.g., low-power, lossy networks), between constrained devices and traditional Internet nodes, and between devices on different constrained networks linked by Internet. CoAP is also being used via other mechanisms, such as SMS on cellular networks [85]. CoAP was

Table 3.5 Comparison between HTTP and CoAP architecture

Feature	HTTP	CoAP
Transport protocol	TCP	UDP
Network protocol	IP	6LoWPAN
Multicast support	No	Yes, through UDP
Client/server mode	Yes	Yes
Publish/subscribe mode	No	Yes
Synchronization requirement	Yes	No
CPU overhead	Large	Small

originally designed as a client/server protocol, but in 2019 IETF published an Internet Draft describing a publish/subscribe mechanism [86].

Table 3.5 compares HTTP and CoAP. Like HTTP, CoAP is an Application Layer protocol. A client requests a resource at a Uniform Resource Identifier (URI) and the server responds. Morever, RESTful principles are followed; verbs GET, POST, PUT, DELETE are used. Protocol is indicated with `coap://`. Where CoAP differs from HTTP is that UDP is used for transport instead of TCP. UDP handshaking is lighter and easier to implement on microcontrollers. CoAP header is only 4 bytes. CoAP can also use UDP's broadcast and multicast features. Since there's no TCP, CoAP takes care of message acknowledgments, retries with congestion control, and duplicates detection. With a very simplified reasoning, we can say that CoAP, MQTT, and AMQP exhibit an increased message overhead with higher consumption of computational and communication resources as well as energy.

Figure 3.32 shows an example of CoAP/HTTP interoperation. A REST client, implemented on a constrained node, calls services provided by a REST server implemented on a powerful Internet node. REST messages flow seamlessly between client and server as they both were on traditional Internet. On the constrained network, HTTP, TLS, and IP are replaced by CoAP, DTLS, and 6LoWPAN, respectively. A proxy node is also shown. It has in charge the mapping between CoAP messages into HTTP messages and vice versa.

CoAP supports four different message types:

- confirmable (CON);
- non-confirmable (NON);
- acknowledgment (ACK);
- reset (RST).

A *confirmable message* is considered as a reliable message. Using this kind of message, the client can be sure that the message will arrive at the server because it is repeatedly sent until the receiver answers with an *acknowledge* message. The *acknowledge* message must contain the same ID of the *confirmable* message. Figure 3.33 shows an example of a confirmable message exchange.

Whenever the server is unable to manage the incoming request, it can answer to the client with a *reset* message (RST) instead of the *acknowledge* message (RST). Figure 3.34 shows an example where the server answer with a *reset* message.

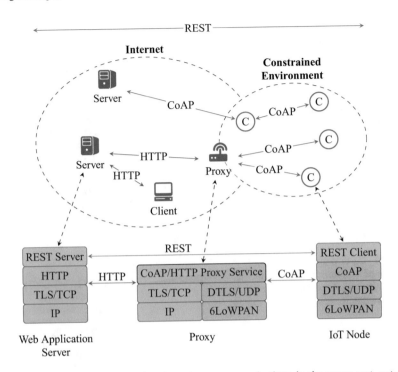

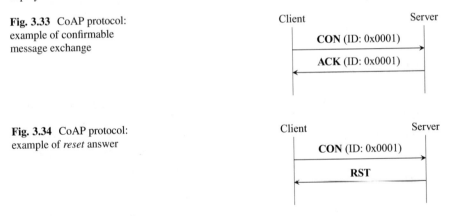

Fig. 3.32 CoAP/HTTP interoperation through a proxy mechanism: in the upper part, network deployment and interconnection, in the lower part, protocol stacks of the involved nodes

Fig. 3.33 CoAP protocol: example of confirmable message exchange

Fig. 3.34 CoAP protocol: example of *reset* answer

The *non-confirmable* (NON) messages instead do not require an *acknowledge* by the server. They are unreliable messages, which usually do not contain critical information that must be absolutely delivered to the server. Examples of *non-confirmable* messages are those that contain values read from sensors. Even if these messages are considered unreliable, they still need to have a unique ID associated. Figure 3.35 shows an example of a *non-confirmable* message.

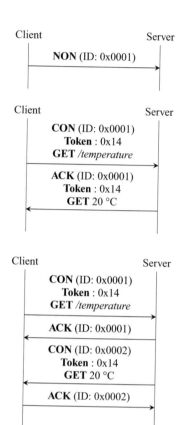

Fig. 3.35 CoAP protocol: example of *non-confirmable* message

Fig. 3.36 CoAP protocol: example of *request/response* communication, where the server can answer immediately

Fig. 3.37 CoAP protocol: example of *request/response* communication, where the server is initially busy

3.4.6.1 CoAP Request/Response Model

The CoAP *request/response model* consists of *confirmable* and *non-confirmable* messages. As shown in Fig. 3.36, if the request is carried using a *confirmable* message and the server can answer immediately, the server sends back to the client an *acknowledge* message containing the response or an error code. The *confirmable* message also contains a Token, which is different from the ID, and it is used to match the request and the response.

As shown in Fig. 3.37, if the server cannot answer to the request immediately, then it sends an *acknowledge* message with an empty response. As soon as the response is available, then the server sends a new *confirmable* message to the client containing the response. Then, the client sends back an *acknowledge* message.

Figure 3.38 shows the structure of a CoAP message. It is worth noting that the CoAP protocol is meant for constrained-size messages, and to avoid fragmentation, a message occupies exactly the data section of a UDP packet. The CoAP message consists of:

- Ver (2 bits): Indicates the CoAP version number.

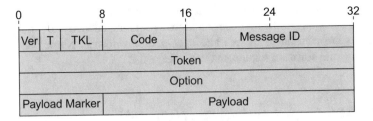

Fig. 3.38 CoAP message structure

- T (2 bits): Indicates if this message is of type confirmable (0), non-confirmable (1), acknowledgment (2), or reset (3).
- TKL (4 bits): It is the token length and indicates the length of the variable-length token field, which may be 0–8 bytes in length.
- Code (8 bits): It is the CoAP request/response code. The three most significant bits form a number known as the "class," which is analogous to the class of HTTP status codes. The five least significant bits form a code that communicates further detail about the request or response.
- Message ID (16 bits): It is used to detect message duplication and to match messages of type acknowledgment/reset to messages of type confirmable/non-confirmable.
- Option (32 bits): First 4 bits determine which option, the next 4 bits determines the option length, and the remaining bits is the option itself.
- Payload Marker (8bit): The marker has value 0xFF and indicates the end of options and the start of the payload.
- Payload (32bit): Contains the transmitted data.

3.4.7 *Extensible Messaging and Presence Protocol (XMPP)*

Extensible Messaging and Presence Protocol (XMPP) is a communication protocol for message-oriented middleware based on XML [87] promoted by the XMPP Standards Foundation (formerly the Jabber Software Foundation) and standardized by IETF [88]. It enables the near-real-time exchange of structured yet extensible data between any two or more network entities. Originally named Jabber, the protocol was developed by the homonym open-source community in 1999 for near real-time instant messaging (IM), presence information, and contact list maintenance. Designed to be extensible, the protocol has been used also for publish-subscribe systems, signaling for VoIP, video, file transfer, gaming, IoT applications (e.g., smart grids), and social networking services [88, 89].

Unlike most instant messaging protocols, XMPP is defined as an open standard and uses an open systems approach of development and application, by which anyone may implement an XMPP service and interoperate with other organizations' implementations. Because XMPP is an open protocol, implementations can be developed using any software license and many servers, clients, and library implementations are distributed as free and open-source software. Numerous freeware and commercial software implementations also exist.

3.4.8 OPC Unified Architecture (OPC-UA)

OPC Unified Architecture (OPC-UA) is a machine-to-machine Application Layer communication protocol for industrial automation developed by the OPC Foundation [90]. The idea behind OPC-UA is providing a platform-independent communication standard with a service-oriented architecture, where devices communicate using specific messages of request and response. OPC-UA addresses a wide range of target applications, i.e., Supervisory Control and Data Acquisition (SCADA), Human-Machine Interface (HMI) (or, more generally, Human-System Interface (HSI)), distributed control systems, and management of programmable logic controllers. Sources or destinations of OPC-UA data belong to the so-called shop floor of the smart factory, e.g., they may be milling machines, 3D printers, robots, conveyor belts, etc.

Figure 3.39 shows the OPC-UA reference software architecture [91]. At the lowest level, the fundamental components of OPC-UA are transport mechanisms and data modeling. The transport mechanisms do not depend on a specific protocol mapping and allow adding new protocols in the future. The first version of OPC-UA defined an optimized binary TCP protocol for high-performance intranet communication as well as a mapping to accepted Internet standards like web services, XML, and HTTP for firewall-friendly Internet communication. Both transports are using the same message-based security model known from web services.

The OPC-UA standard can be used to build two main communication scenarios, which are exemplified in Fig. 3.40, i.e., client/server, and publish/subscribe. The

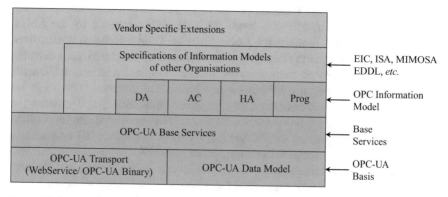

Fig. 3.39 OPC-UA reference software architecture

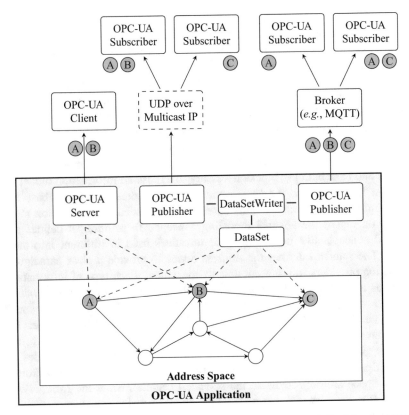

Fig. 3.40 Delivery of data through the OPC-UA architecture. Data, defined in the OPC-UA Address Space, can be delivered in three different modes, i.e., client/server, broker-less publish/subscribe, and broker-based publish/subscribe

latter one can be further subdivided into two more scenarios, i.e., broker-less and broker-based. In the *client/server mode*, an OPC-UA client requests information and receives a response from an OPC-UA server. Each device can host multiple clients and servers. Each client can concurrently interact with more than one server, and vice-versa a server can concurrently interact with more than one client. This communication mode is simple and does not require gateway, but client and server should know each other, and sharing information may lead to message repetition. The *publish/subscribe mode* (also known as *PubSub*) allows to handle efficiently one-to-many and many-to-many communications by using a message-oriented middleware. Information providers and consumers are not required to know each other. In the *broker-based* scenario, the publisher sends a message to a broker, which then distributes the messages to the various subscribers. OPC-UA does not define the message-oriented middleware; instead, it leverages existing technologies, e.g., MQTT. The broker is an active actor which sends messages to devices that subscribed on a specific *topic*. In the *broker-less* scenario, OPC-UA relies on the *multicast* function of the network infrastructure to deliver messages to one or more receivers. In this scenario topics are mapped on multicast addresses, and interested nodes should interact with the low-level network infrastructure to join a multicast group. Broker-less PubSub mode seems the optimal solution since message repetitions are avoided without the need of an additional device acting as a broker. However, broker-less PubSub mode requires connection-less transport protocols, such as UDP, since TCP cannot be conveyed over multicast transmissions. Therefore many useful features of TCP (e.g., loss recovery) should be moved inside UPC-UA implementation in the involved devices.

Data are at the heart of an OPC-UA application. The data modeling defines the rules and base building blocks necessary to expose an information model with OPC UA. It defines also the entry points into the address space and base types used to build a type hierarchy. This base can be extended by information models building on top of the abstract modeling concepts. In addition, it defines some enhanced concepts like describing state machines used in different information models. The standard defines the *Address Space* to provide a clear paradigm for information providers to represent data to consumers. Each type of information is defined as a *node* in the Address Space. Each node has its own set of *attributes*. There are different types of nodes, called *NodeClass*, based on their specific purpose, i.e., Variable, Object, Method, View, DataType, VariableType, ObjectType, and ReferenceType. Depending on the NodeClass, a node can have different attributes, plus a set of attributes that are common to all the types of nodes. The Address Space is a network of nodes interconnected by *references*. Data from the Address Space are collected into the so-called DataSet, while the *DataSetWriter* prepares *DataSetMessages* starting from the *DataSets* and makes them available for publishing.

The OPC-UA Services are the interface between servers as supplier of an information model and clients as consumers of that information model. The Services are defined in an abstract manner. They are using the transport mechanisms to exchange the data between client and server.

This basic concept of OPC UA enables an OPC UA client to access the smallest pieces of data without the need to understand the whole model exposed by complex systems. OPC UA clients also understanding specific models can use more enhanced features defined for special domains and use cases.

To cover all successful features known from Classic OPC, information models for the domain of process information are defined by OPC-UA on top of the base specifications. There are four main *Information Models* defined in OPC-UA:

- **OPC Data Access (DA)** It gets data out of the control systems into other systems on the shop floor. Each information about a specific tag or data point contains some information about it. First you have the data itself, and that is called Value and of course the Name of it. To that comes a number of other pieces of information that describes the information; the first is the Timestamp that gives you the exact time when the value was read. This timestamp can be taken either directly from the underlying system or assigned to it when the data is read in the

OPC server. The last piece is called Quality which gives a basic understanding if the data is valid or not.

- **OPC Alarm & Events (AE)** It is fundamentally different from the DA model simply due to the fact that events do not have a current value. This means that this protocol always is a subscription-based service where the clients gets all the events that come in. In terms of data that comes with the event, there is no tags and therefore not any name and quality, but there is of course a Timestamp. But like in the case with DA, there is no store in the server, and once the event is transferred, the server forgets it was ever there.
- **OPC Historical Analysis (HDA)** The difference between DA, AE, and HDA is that HDA contains historical data and you can call for a large amount of past data. The model therefore supports long record sets of data for one or more data points. It was designed to provide a unified way to get out and distribute historical data stored in SCADA.
- **Programs (Prog).** It specifies a mechanism to start, manipulate, and monitor the execution of programs.

Other organizations can build their models on top of the UA base or on top of the OPC information model, exposing their specific information via OPC-UA. Examples for standards already working on mappings to OPC-UA are Field Device Integration (FDI) combining Electronic Device Description Language (EDDL) and Field Device Tool (FDT) both used to describe, to configure, and to monitor devices, and PLCopen, a standard for PLC programming languages. Additional vendor-specific information models will be defined using directly the OPC-UA base, the OPC models, or other OPC-UA-based information models.

3.5 IoT Network Design Methodology

In Sect. 3.1 we described the general structure of an IoT application and its specific characteristics:

- strict dependence between application and communication aspects;
- system-of-systems nature;
- strict relationship with the environment.

Because of these characteristics, the design of IoT network structure can be a complex task, and the huge amount of opportunities provided by standard technologies (as the ones described in previous sections) can be an issue. Furthermore IoT applications can be heterogeneous, i.e., they may involve more than one technology in the same scenario. For instance, a LR-WPAN technology can be used inside some buildings that are then connected together by using a LP-WAN technology. Research work addressed this problem. In [92] a *communication-aware design flow* is proposed together with a *communication-aware formal specification of the whole distributed IoT application* to formulate and solve the design problem by using an optimization approach. Such formal specification considers the following aspects:

- **Energy consumption and type of energy source**. Energy consumption is proportional to the throughput of the node. Therefore, it can be reduced by lowering the physical bitrate or introducing sleeping periods. High energy consumption requires to connect the device to a stable energy source such as the power grid or a large battery to be recharged periodically by a person. Devices that exhibit such requirements cannot be unattended. This is the case of gateways, coordinators, network servers, routers. Medium energy consumption allows the use of batteries that have autonomy of at least 2–3 years. In this case, the replacement period is acceptable and can be combined with other activities already required in that application context. This is the case of beacons that highlight the presence of an object (e.g., in a museum). Low energy consumption allows the use of *energy-harvesting* (or *energy scavenging*) techniques to obtain a small amount of energy from the surrounding environment [93]. Well-known examples are NFC passive tags that use radio-frequency energy and sensors supplied by photovoltaic cells.
- **Frequency band**. There are two main sets of frequency bands, i.e., sub-GHz and millimeter-wave. Sub-GHz bands allow low-power and long-range operation as well as obstacle avoidance at the cost of data rates usually far below 1 Mb/s. Millimeter-wave bands (e.g., 2.4 and 5 GHz bands) allow higher data rates but requires "line-of-sight" transmission. It is worth noting that frequency bands that are blocked by walls (as in the case of 5 GHz band) allow intrinsic security with respect to eavesdropping. Furthermore, as reported below, the choice of an unlicensed or licensed frequency standard has an impact on telecommunication cost.
- **Quality of service**. This is the general term that encompasses some different performance aspects of the communication. One aspect is the *data rate*, which denotes the amount of information transferred in the time unit. Multimedia data (e.g., the output of a camera) usually require a higher data rate than physical sensing (e.g., light intensity). If data from several sensors are merged, the resulting stream may require a high data rate. It is worth noting that data processing at the edge of the network can help reduce the required data rate; for instance, in a video surveillance application, object recognition performed by the camera allows to replace video transmission with a simple alert notification. Another aspect is *delay* which denotes the time to move a message from one node to another of the network. Delay constraints are important in control applications. The absolute value of delay affects the promptness in device actuation. In closed-loop control applications, the delay variation is very important since it affects the stability of control. Finally, the third aspect is *error rate* which denotes the fraction of bits whose value is erroneously received at the destination. Bit errors, if detected, decrease data rate since the message is discarded; otherwise, they can compromise data processing at the destination.
- **Transmission range**. The right distance between two communicating devices is an application requirement. For some applications, e.g., precision agriculture

and smart metering, long distance values are desirable to avoid the need for intermediate systems to re-launch the signal. Vice versa, in proximity services and electronic payments, keeping the maximum distance very short is a requirement. It is worth noting that large distance values increase power consumption, electromagnetic noise, and risk of eavesdropping.

- **Mobility.** There are two kinds of mobility. In case of *low mobility*, users are interested in avoiding the presence of cables to reduce cost or to be able to change the position of nodes if needed. In the case of *high mobility*, devices are transmitting while moving. This feature requires to adapt the transmission to the changing environmental condition, e.g., distance from the counterpart, reflections over obstacles, and presence of other electromagnetic sources in the same frequency range.
- **Scalability.** Except for mesh networks, usual IoT infrastructures are based on star or tree topology with intermediate systems connecting a set of end nodes. In these cases the maximum number of end nodes supported by an intermediate system is very important. Small values indicate low scalability of the technology which leads to high cost to deploy intermediate systems.
- **Security.** Data confidentiality and integrity will become ever more important in IoT applications according to their diffusion. Security at Data Link and Network Layer should be directly enforced by the protocol standard. Security at Application Layer can be easily provided by recurring to the traditional TCP-based mechanisms (e.g., TLS/SSL). If TCP is not supported by the protocol stack, then custom mechanisms should be implemented by the application designer.
- **Cost.** The cost of an IoT network infrastructure depends on different aspects. The most evident aspect is the *hardware cost* of the various devices involved in the application. Most of these devices are end nodes (e.g., sensors), but the number of intermediate systems (e.g., gateways) can be significant especially if the technology is not scalable. We can include in this category also the cost for the use of cloud services. The second aspect is the *telecommunication cost* due to subscription fees for the use of licensed frequency standards. The third aspect is the *energy cost*, which encompasses energy bill, purchase of new batteries, and disposal of the exhausted ones, as well as personnel cost for battery replacement.

Table 3.6 compares some of the standard technologies described before according to such aspects. It is interesting the comparison between standards based on unlicensed frequencies (i.e., IEEE 802.15.4, Bluetooth, ZigBee, WirelessHART, Z-Wave, IEEE 802.11, LoRaWAN and Sigfox) and cellular standards (i.e., NB-IoT and LTE-M). Regarding the first set, the use of the radio channel is free of charge but the noise level due to its shared nature leads to a decrease of the QoS and scalability. To cope with crowded channels and provide a wide coverage, more gateways are needed thus increasing deployment cost. Vice versa, in cellular networks, end nodes can be directly connected to the telecom infrastructure which natively guarantees a higher QoS and coverage (e.g., for NB-IoT) at the cost of subscription fees. The

Table 3.6 Comparison of IoT wireless technologies

Technology	Frequency	Data rate	Range	Mobility	Energy cons.
2G/3G	Cellular bands	10 Mb/s	Several km's	High	High
LTE Cat M1	Cellular bands	1–10 Mb/s	Several Km's	High	Medium
NB-IoT	Cellular bands	60 kb/s	Several Km's	Medium	Low
Bluetooth/BLE	2.4 GHz	1/2/3 Mb/s	<100 m	Low	Low
IEEE 802.15.4	2.4 GHz and Sub-GHz	40, 250 kb/s	<100 m	Low	Low
ZigBee	2.4 GHz and Sub-GHz	40, 250 kb/s	<100 m	Low	Low
WirelessHART	2.4 GHz	250 kb/s	<100 m	Low	Medium
ISA100.11a	2.4 GHz	250 kb/s	<100 m	Low	Medium
Z-Wave	Sub-GHz	40 kb/s	~30 m	Low	Low
IEEE 802.11	2.4 GHz, 5 GHz and Sub-GHz	0.1/54 Mb/s	<100 m	Low	Medium
LoRaWAN	Sub-GHz	<250 kb/s	~15 Km	Medium	Low
Sigfox	Sub-GHz	<1 kb/s	Several Km's	Medium	Low

Credit to https://blog.helium.com/802-15-4-wireless-for-internet-of-things-developers-1948fc313b2e

designer should consider these aspects to choose the best option according to the current cost of hardware devices, subscription fees, and the required number of deployed nodes.

3.5.1 Communications for Localization

A direct effect of Internet of Things applications is the emerging need to recover the location of such "things." Since many localization techniques are based on communications belonging to the previously described standards, it is worth introducing them briefly in this chapter. Localization services can be classified into *proximity services* and *positioning services* [94]. Proximity services are the simpler of the two categories and leverage communication protocols to determine the location of two devices relative to each other. One of the two is a transmitter, and the other determines if it is within the range and, in some cases, approximately how far away. Point of interest information (e.g., in a museum or in a shopping mall) and item finding are typical applications based on proximity services. Bluetooth LE beacons or NFC tags are often used in this case.

Positioning services aim to determine the physical location of "things" and involve more sophisticated infrastructure deployments. Real-time locating systems (RTLS) and indoor positioning systems (IPS) are two of the most popular types of positioning services. RTLS solutions are used for both asset tracking as well

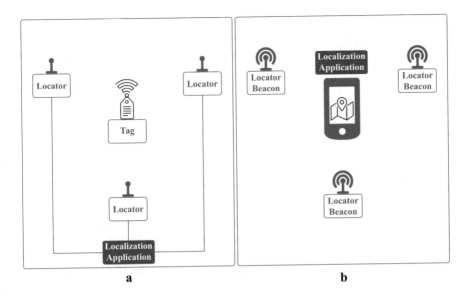

Fig. 3.41 Localization architectures: (**a**) tag-based, (**b**) beacon-based

as people tracking. Indoor positioning systems help visitors, such as shoppers in a mall, travelers in an airport, or workers in a large office building, navigate their way throughout a facility. As depicted in Fig. 3.41, there are two types of architectures, i.e., tag-based and beacon-based, depending on the purpose of the localization-based application. In the tag-based architecture, the "thing" transmits data to receivers, often referred to as "locators," in fixed and known locations throughout a facility. The locators estimate their distance from the "thing" and send such estimates back to a centralized server that acts as location engine. This configuration is used when the "thing" is a simple tag attached to an asset or a person that has to be localized by users far from it. Some tag-based solutions are using Bluetooth [94] or LoRaWAN [95]. The beacon-based architecture works oppositely. Instead of receivers, transmitters, commonly referred to as locator beacons, are deployed in fixed and known locations throughout a facility. The "thing" receives messages from locator beacons, estimates their distance, and performs localization. This configuration is used when the "thing" is a more sophisticated device such as a mobile phone, a vehicle or a robot which is directly interested in knowing its location. Bluetooth [94], Wi-Fi, and cellular network are used in this configuration. In recent years, positioning services have been also implemented by using passive RFID systems (usually involved in proximity services) by using arrays of either passive RFID tags or antennas always combined with complex tracking software [96].

Bluetooth 5.1, the last releases of cellular networks, and Wi-Fi introduce a direction finding feature that increases localization accuracy with respect to traditional multi-lateration techniques. Figure 3.42 shows the angle of arrival (AoA) method. The device to which direction is being determined, such as a tag in a

Fig. 3.42 Direction finding
by using Angle of Arrival
(AoA) and receiver antennas
array

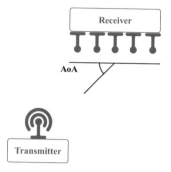

tag-based configuration, transmits a signal using a traditional antenna. The receiving devices, such as locators in that configuration, have multiple antennas arranged in an array. As the transmitted signal crosses the array, the receiving device sees a signal phase difference due to the difference in distance from each of the antenna in its array to the transmitting antenna.

3.6 Summary

This chapter has given the main concepts regarding the communication part of IoT applications. Network aspects have been discussed both in an abstract context and applied to specific well-known standards. The chapter has presented the main reference communication scenarios and architectures, as well as the primary standards behind them. To better describe network standards, the well-known ISO/OSI model has been adopted as reference, and the basic network terminology has been introduced. We presented several widespread standards such as Bluetooth, IEEE 802.15.4-based technologies, LoRaWAN, Sigfox, and, among cellular standards, NB-IoT and LTE-M. Then application-level technologies have been presented. Finally, we discussed the aspects to be considered in the design of the communication part of IoT applications, including localization aspects.

References

1. Y. Jadeja, K. Modi, Cloud computing – concepts, architecture and challenges, in *2012 International Conference on Computing, Electronics and Electrical Technologies (ICCEET)*, March 2012, pp. 877–880
2. F. Bonomi, R. Milito, J. Zhu, S. Addepalli, Fog computing and its role in the internet of things, in *Proceedings of the First Edition of the MCC Workshop on Mobile Cloud Computing*, ser. MCC '12 (ACM, New York, 2012), pp. 13–16. [Online]. Available: https://doi.org/10.1145/2342509.2342513

3. K. Tsilipanos, I. Neokosmidis, D. Varoutas, A system of systems framework for the reliability assessment of telecommunications networks. IEEE Syst. J. **7**(1), 114–124 (2013)
4. A. Tanenbaum, *Computer Networks*, 4th edn. (Prentice Hall Professional Technical Reference, Upper Saddle River, New Jersey, 2002)
5. W. Shang, Y. Yu, L. Zhang, R. Droms, *Challenges in IoT Networking via TCP/IP Architecture*, 2016. [Online]. Available: https://named-data.net/wp-content/uploads/2016/02/ndn-0038-1-challenges-iot.pdf
6. Wikipedia Contributors, *Standardization – Wikipedia, The Free Encyclopedia*, 2019. [Online; Accessed 27 March 2019]. [Online]. Available: https://en.wikipedia.org/w/index.php?oldid=884671676
7. 3GPP, *The 3rd Generation Partnership Project*. [Online]. Available: https://www.3gpp.org/
8. ITU, *International Telecommunication Union*. [Online]. Available: https://www.itu.int/
9. Wikipedia Contributors, *International Telecommunication Union – Wikipedia, The Free Encyclopedia*, 2019. [Online; Accessed 27 March 2019]. [Online]. Available: https://en.wikipedia.org/w/index.php?oldid=888401305
10. IEEE, *Institute of Electrical and Electronics Engineers*. [Online]. Available: https://www.ieee.org/
11. ISO, *International Organization for Standardization*. [Online]. Available: https://www.iso.org/
12. Wikipedia Contributors, *International Organization for Standardization – Wikipedia, The Free Encyclopedia*, 2019. [Online; Accessed 27 March 2019]. [Online]. Available: https://en.wikipedia.org/w/index.php?oldid=889433392
13. LoRa Alliance, *LoRa and LoRaWAN Specifications*. [Online]. Available: https://lora-alliance.org/
14. IETF, *Internet Engineering Task Force*. [Online]. Available: https://www.ietf.org/
15. C. DiBona, S. Ockman, *Open Sources: Voices from the Open Source Revolution* (O'Reilly Media, Inc., United States, 1999)
16. ZigBee Alliance, *ZigBee Specification*, 2012. [Online]. Available: https://www.zigbee.org/download/standards-zigbee-specification/
17. Bluetooth Special Interest Group, *Bluetooth Mesh Networking*, 2017. [Online]. Available: https://3pl46c46ctx02p7rzdsvsg21-wpengine.netdna-ssl.com/wp-content/uploads/2019/03/Mesh-Technology-Overview.pdf
18. Wikipedia Contributors, *Modbus – Wikipedia, the Free Encyclopedia*, 2019. [Online; Accessed 20 June 2019]. [Online]. Available: https://en.wikipedia.org/w/index.php?title=Modbus&oldid=897030889
19. V. Coskun, B. Ozdenizci, K. Ok, A survey on near field communication (NFC) technology. Wirel. Pers. Commun. **71**(3), 2259–2294 (2013). [Online]. Available: https://doi.org/10.1007/s11277-012-0935-5
20. Wikipedia Contributors, *Near-Field Communication – Wikipedia, The Free Encyclopedia*, 2019. [Online; Accessed 27 March 2019]. [Online]. Available: https://en.wikipedia.org/w/index.php?oldid=889005795
21. E. Ferro, F. Potorti, Bluetooth and Wi-Fi wireless protocols: a survey and a comparison. IEEE Wirel. Commun. **12**(1), 12–26 (2005)
22. S. Shahina, G. Shanmugapriya, A survey on Bluetooth technology and its features. Int. J. Inf. Technol. **1**(1), 28–37 (2015)
23. Wikipedia Contributors, *Bluetooth – Wikipedia, The Free Encyclopedia*, 2019. [Online; Accessed 27 March 2019]. [Online]. Available: https://en.wikipedia.org/w/index.php?oldid=889565151
24. K. Townsend, R. Davidson, C. Cufí, *Getting Started with Bluetooth Low Energy*. (O'Reilly, 2014). [Online]. Available: https://books.google.it/books?id=XjY4nwEACAAJ
25. Bluetooth Special Interest Group, *Bluetooth Core Specification – v5.1*, Jan 2019. [Online]. Available: https://www.bluetooth.com/specifications/adopted-specifications
26. D. E. Dilger, *Inside iOS 7: iBeacons Enhance apps' Location Awareness via Bluetooth LE*, 2013. [Online; Accessed 22 April 2019]. [Online]. Available: https://appleinsider.com/articles/13/06/19/inside-ios-7-ibeacons-enhance-apps-location-awareness-via-bluetooth-le

27. ISO/IEC/IEEE, *ISO/IEC/IEEE International Standard – Information technology–Telecommunications and Information Exchange Between Systems–Local and Metropolitan Area Networks–Specific Requirements–Part 15-4: Wireless Medium Access Control (MAC) and Physical Layer (PHY) Specifications for Low-Rate Wireless Personal Area Networks (WPANs), ISO/IEC/IEEE 8802-15-4:2018(E)*, April 2018, pp. 1–712

28. IEEE 802.15 WPAN Task Group 4 (TG4), http://www.ieee802.org/15/pub/TG4.html. Accessed 20 March 2019

29. Wikipedia Contributors, *Zigbee – Wikipedia, The Free Encyclopedia*, 2019. [Online; Accessed 27 March 2019]. [Online]. Available: https://en.wikipedia.org/w/index.php?oldid=889247689

30. P. Baronti, P. Pillai, V. W. Chook, S. Chessa, A. Gotta, Y. F. Hu, Wireless sensor networks: a survey on the state of the art and the 802.15. 4 and ZigBee standards. Comput. Commun. **30**(7), 1655–1695 (2007)

31. T. Kalaivani, A. Allirani, P. Priya, A survey on Zigbee based wireless sensor networks in agriculture, in *3rd International Conference on Trendz in Information Sciences & Computing (TISC2011)* (IEEE, 2011), pp. 85–89

32. Zigbee Alliance, *Zigbee IP and 920IP*. [Online]. Available: https://www.zigbee.org/zigbee-for-developers/network-specifications/zigbeeip/

33. M. Nixon, *A Comparison of WirelessHART and ISA100.11a*, 2012. [Online]. Available: https://www.emerson.com/documents/automation/white-paper-a-comparison-of-wirelesshart-isa100-11a-en-42598.pdf

34. E. Khorov, A. Lyakhov, A. Krotov, A. Guschin, A survey on IEEE 802.11 ah: an enabling networking technology for smart cities. Comput. Commun. **58**, 53–69 (2015)

35. Wikipedia Contributors, *IEEE 802.11 – Wikipedia, The Free Encyclopedia*, 2019. [Online; Accessed 27 March 2019]. [Online]. Available: https://en.wikipedia.org/w/index.php?oldid=889617201

36. ETSI, *European Telecommunications Standards Institute*. [Online]. Available: https://www.etsi.org/

37. R. S. Sinha, Y. Wei, S.-H. Hwang, A survey on LPWA technology: LoRa and NB-IoT. Ict Exp. **3**(1), 14–21 (2017)

38. J. Haxhibeqiri, E. De Poorter, I. Moerman, J. Hoebeke, A survey of Lorawan for IOT: from technology to application. Sensors **18**(11), 3995 (2018)

39. B. Reynders, Q. Wang, S. Pollin, A LoRaWAN Module for Ns-3: implementation and Evaluation, in *Proceedings of the 10th Workshop on Ns-3*, ser. WNS3 '18 (ACM, New York, 2018), pp. 61–68. [Online]. Available: https://doi.org/10.1145/3199902.3199913

40. Paul, *All What You Need to Know About Regulation on RF 868 MHz for LPWan*, 2017. [Online; Accessed 19 April 2019]. [Online]. Available: https://www.disk91.com/2017/technology/internet-of-things-technology/all-what-you-need-to-know-about-regulation-on-rf-868mhz-for-lpwan/

41. ERC, *70-03-ERC Recommendation 70-03 Relating to the Use of Short Range Devices (SRD)*, 2011. [Online]. Available: https://www.cept.org/Documents/srd/mg/933/Info_6_ERC_REC_70-03_August_2011

42. G. Dregvaite, R. Damasevicius, *Information and Software Technologies: Proceedings of the 22nd International Conference, ICIST 2016*, Druskininkai, Lithuania, 13–15 Oct 2016, vol. 639 (Springer, 2016)

43. Wikipedia Contributors, *Sigfox – Wikipedia, The Free Encyclopedia*, 2019. [Online; accessed 27 March 2019]. [Online]. Available: https://en.wikipedia.org/w/index.php?oldid=886299006

44. S.A. Sigfox, *Sigfox Web Site*, 2019. [Online; Accessed 31 May 2019]. [Online]. Available: https://www.sigfox.com

45. Wikipedia Contributors, *Z-Wave – Wikipedia, The Free Encyclopedia*, 2019. [Online; accessed 27 March 2019]. [Online]. Available: https://en.wikipedia.org/w/index.php?oldid=886885917

46. ISO/IEC, Communication Systems for Meters and Remote Reading of Meters – Part 4: Wireless Meter Readout (Radio Meter Reading for Operation in the 868 MHz to 870 MHz SRD Band), *ISO/IEC 13757-4*, Nov 2013

47. ISO/IEC, Information Technology – Home Electronic System (HES) Architecture – Part 3-7: Media and Media Dependent Layers – Radio Frequency for Network Based Control of HES Class 1, *ISO/IEC 14543-3-7*, Jan 2007, pp. 1–22
48. OMS Group, *Open Metering System*, Oct 2014. [Online; Accessed 21 June 2019]. [Online]. Available: https://oms-group.org/
49. Netbeheer Nederland, *P1 Companion Standard Dutch Smart Meter Requirements*, March 2014. [Online; Accessed 21 June 2019]. [Online]. Available: https://www.netbeheernederland.nl/_upload/Files/Slimme_meter_15_32ffe3cc38.pdf
50. DLMS User Association, *DLMS/COSEM: Architecture and Protocols*, Jun 2017. [Online; Accessed 21 June 2019]. [Online]. Available: https://www.dlms.com/files/Green-Book-Ed-83-Excerpt.pdf
51. Wikipedia Contributors, *Optical wireless communications – Wikipedia, The Free Encyclopedia*, 2018. [Online; Accessed 27 March 2019]. [Online]. Available: https://en.wikipedia.org/w/index.php?oldid=861097655
52. P.A. Mendez, R. James, Design of Underwater wireless optical/acoustic link for reduction of back-scattering of transmitted light. Int. J. Eng. Sci. **4**(5), 61–68 (2015)
53. IEEE, IEEE Draft Standard for Local and metropolitan area networks – Part 15.7: Short-Range Optical Wireless Communications, *IEEE P802.15.7/D3, August 2018*, pp. 1–412, Jan 2018
54. G. Mulligan, The 6LoWPAN architecture, in *Proceedings of the 4th Workshop on Embedded networked sensors* (ACM, 2007), pp. 78–82
55. Z. Shelby, C. Bormann, *6LoWPAN: The Wireless Embedded Internet*, vol. 43. (John Wiley & Sons, UK, 2011)
56. Wikipedia Contributors, *6LoWPAN – Wikipedia, The Free Encyclopedia*, 2019. [Online; accessed 27-March-2019]. [Online]. Available: https://en.wikipedia.org/w/index.php?oldid=887884134
57. G. Montenegro, C. Schumacher, N. Kushalnagar, *IPv6 over Low-Power Wireless Personal Area Networks (6LoWPANs): Overview, Assumptions, Problem Statement, and Goals*, RFC 4919, Aug 2007. [Online]. Available: https://rfc-editor.org/rfc/rfc4919.txt
58. G. Montenegro, J. Hui, D. Culler, N. Kushalnagar, *Transmission of IPv6 Packets over IEEE 802.15.4 Networks*, RFC 4944, Sep 2007. [Online]. Available: https://rfc-editor.org/rfc/rfc4944.txt
59. J. Nieminen, T. Savolainen, M. Isomaki, B. Patil, Z. Shelby, C. Gomez, *IPv6 over Bluetooth Low Energy*, RFC 7668, Oct 2015. [Online]. Available: https://rfc-editor.org/rfc/rfc7668.txt
60. J. Olsson, *6LoWPAN demystified*. Texas Instruments, Oct 2014. [Online]. Available: http://www.ti.com/lit/wp/swry013/swry013.pdf
61. Thread Group, *Thread Specification 1.1.1*, Feb 2017. [Online]. Available: https://www.threadgroup.org
62. E. Rescorla, N. Modadugu, *Datagram Transport Layer Security Version 1.2*, RFC 6347, Jan 2012. [Online]. Available: https://rfc-editor.org/rfc/rfc6347.txt
63. ZigBee Alliance, *Dotdot*, 2019. [Online]. Available: https://www.zigbee.org/zigbee-for-developers/dotdot/
64. Nest Lab, *OpenWeave*, 2019. [Online]. Available: https://openweave.io/
65. R. Kelsey, *Mesh Link Establishment*, Internet Engineering Task Force, Internet-Draft draft-kelsey-intarea-mesh-link-establishment-06, May 2014, work in Progress. [Online]. Available: https://datatracker.ietf.org/doc/html/draft-kelsey-intarea-mesh-link-establishment-06
66. Wikipedia Contributors, *Mobile telephony – Wikipedia, The Free Encyclopedia*, 2019. [Online; Accessed 27 March 2019]. [Online]. Available: https://en.wikipedia.org/w/index.php?oldid=887824246
67. Wikipedia Contributors, *2G – Wikipedia, The Free Encyclopedia*, 2019. [Online; Accessed 27 March 2019]. [Online]. Available: https://en.wikipedia.org/w/index.php?oldid=889521650
68. Wikipedia Contributors, *3G – Wikipedia, The Free Encyclopedia*, 2019. [Online; Accessed 27 March 2019]. [Online]. Available: https://en.wikipedia.org/w/index.php?oldid=888462727
69. Wikipedia Contributors, *4G – Wikipedia, The Free Encyclopedia*, 2019. [Online; Accessed 27 March 2019]. [Online]. Available: https://en.wikipedia.org/w/index.php?oldid=888700749

70. Wikipedia Contributors, *Narrowband IoT – Wikipedia, The Free Encyclopedia*, 2019. [Online; Accessed 27 March 2019]. [Online]. Available: https://en.wikipedia.org/w/index.php?oldid=887405493

71. D. Hanes, G. Salgueiro, P. Grossetete, R. Barton, J. Henry, *IoT Fundamentals: Networking Technologies, Protocols, and Use Cases for the Internet of Things*, 1st edn. (Cisco Press, USA, 2017)

72. Alok Sanghavi, *5G Infrastructure Needs Programmability*, 2019. [Online; Accessed 27 March 2019]. [Online]. Available: https://www.eenewseurope.com/news/5g-infrastructure-needs-programmability

73. Wikipedia Contributors, *Hypertext transfer protocol – Wikipedia, the free encyclopedia*, 2019. [Online; Accessed 3 June 2019]. [Online]. Available: https://en.wikipedia.org/w/index.php?title=Hypertext_Transfer_Protocol&oldid=900152978

74. A. Melnikov, I. Fette, *The WebSocket Protocol*, RFC 6455, Dec 2011. [Online]. Available: https://rfc-editor.org/rfc/rfc6455.txt

75. World Wide Web Consortium, *Web Interface Definition Language*, May 2019. [Online]. Available: https://heycam.github.io/webidl/

76. Wikipedia Contributors, *Web service – Wikipedia, The Free Encyclopedia*, 2019. [Online; Accessed 27 March 2019]. [Online]. Available: https://en.wikipedia.org/w/index.php?oldid=883348748

77. D. Box, D. Ehnebuske, G. Kakivaya, A. Layman, N. Mendelsohn, H. F. Nielsen, S. Thatte, D. Winer, *Simple Object Access Protocol (SOAP) 1.1*, 2000.

78. Wikipedia Contributors, *Representational State Transfer – Wikipedia, The Free Encyclopedia*, 2019. [Online; Accessed 27 March 2019]. [Online]. Available: https://en.wikipedia.org/w/index.php?oldid=889623686

79. F. Todd, *Using HTTP Methods for RESTful Services*. [Online; Accessed 22 April 2019]. [Online]. Available: https://www.restapitutorial.com/lessons/httpmethods.html

80. U. Hunkeler, H.L. Truong, A. Stanford-Clark, MQTT-A publish/subscribe protocol for Wireless Sensor Networks, in *2008 3rd International Conference on Communication Systems Software and Middleware and Workshops (COMSWARE'08)* (IEEE, 2008), pp. 791–798

81. Wikipedia Contributors, *MQTT – Wikipedia, The Free Encyclopedia*, 2019. [Online; Accessed 27 March 2019]. [Online]. Available: https://en.wikipedia.org/w/index.php?oldid=888839246

82. OASIS, *MQTT version 3.1. 1*, URL http://docs.oasis-open.org/mqtt/mqtt/v3, vol. 1, 2014

83. Wikipedia Contributors, *Advanced Message Queuing Protocol – Wikipedia, The Free Encyclopedia*, 2019. [Online; accessed 31-May-2019]. [Online]. Available: https://en.wikipedia.org/w/index.php?title=Advanced_Message_Queuing_Protocol

84. Z. Shelby, K. Hartke, C. Bormann, *The Constrained Application Protocol (CoAP)*, RFC 7252, Jun 2014. [Online]. Available: https://rfc-editor.org/rfc/rfc7252.txt

85. Wikipedia Contributors, *Constrained Application Protocol – Wikipedia, The Free Encyclopedia*, 2019. [Online; Accessed 27 March 2019]. [Online]. Available: https://en.wikipedia.org/w/index.php?oldid=888318701

86. M. Koster, A. Keranen, J. Jimenez, *Publish-Subscribe Broker for the Constrained Application Protocol (CoAP)*, Internet Engineering Task Force (IETF), Tech. Rep., 2019. [Online]. Available: https://www.ietf.org/id/draft-ietf-core-coap-pubsub-08.txt

87. Wikipedia Contributors, *XMPP – Wikipedia, The Free Encyclopedia*, 2019. [Online; Accessed 27 March 2019]. [Online]. Available: https://en.wikipedia.org/w/index.php?oldid=887947505

88. P. Saint-Andre, *Extensible Messaging and Presence Protocol (XMPP): Core*, RFC 6120, March 2011. [Online]. Available: https://rfc-editor.org/rfc/rfc6120.txt

89. P. Saint-Andre, K. Smith, R. Tronçon, R. Troncon, *XMPP: the definitive guide* (O'Reilly Media, Inc., United States, 2009)

90. Wikipedia Contributors, *OPC Unified Architecture – Wikipedia, The Free Encyclopedia*, 2019. [Online; Accessed 30 March 2019]. [Online]. Available: https://en.wikipedia.org/w/index.php?oldid=883275662

91. Unified Automation, *Introduction to OPC UA*, 2019. [Online; Accessed 20 June 2019]. [Online]. Available: https://documentation.unified-automation.com/uasdkhp/1.0.0/html/_12_opc_ua_overview.html

92. E. Fraccaroli, F. Stefanni, R. Rizzi, D. Quaglia, F. Fummi, Network synthesis for distributed embedded systems. IEEE Trans. Comput. **67**(9), 1315–1330 (2018)

93. S. Ulukus, A. Yener, E. Erkip, O. Simeone, M. Zorzi, P. Grover, K. Huang, Energy harvesting wireless communications: a review of recent advances. IEEE J. Sel. Areas Commun. **33**(3), 360–381 (2015)

94. Bluetooth Special Interest Group, *Enhancing Bluetooth Location Services with Direction Finding*, 2019. [Online]. Available: https://www.bluetooth.com/bluetooth-resources/paper-enhancing-bluetooth

95. LoRa Alliance Strategy Committee, *LoRaWAN Geolocation Whitepaper*, Jan 2019. [Online]. Available: https://lora-alliance.org/resource-hub/lora-alliance-geolocation-whitepaper

96. Y. Zhang, L. Xie, Y. Bu, Y. Wang, J. Wu, S. Lu, 3-dimensional localization via RFID tag array, in *2017 IEEE 14th International Conference on Mobile Ad Hoc and Sensor Systems (MASS)*, Oct 2017, pp. 353–361

Chapter 4
Architecting IoT Cloud

Farshad Firouzi and Bahar Farahani

> *It is better to have your head in the clouds, and know where you are... than to breathe the clearer atmosphere below them, and think that you are in paradise.*
>
> Henry David Thoreau

Contents

F. Firouzi (✉)
Department of ECE, Duke University, Durham, NC, USA

B. Farahani
Shahid Beheshti University, Tehran, Iran

© Springer Nature Switzerland AG 2020
F. Firouzi et al. (eds.), *Intelligent Internet of Things*,
https://doi.org/10.1007/978-3-030-30367-9_4

4.1   The IoT Cloud

The independent arenas of Cloud and IoT have been evolving quickly. Although these models are distinct from one another, their elements are often complimentary of one another. Integrating them provides benefits for specific application situations. IoT benefits from the limitless resources and capabilities of Cloud to tackle its technological constraints such as processing power, less storage, and communication. For example, Cloud offers IoT service management as well as the ability to deploy applications and services that use IoT things or generated IoT data. Cloud can also provide services in many real-life scenarios by acting as an intermediary between the things and applications. On the other hand, Cloud benefits from IoT's ability to address problems in a more dynamic and distributed approach. The primary motivations for Cloud and IoT integration can be summarized as follows [1]:

- *Communication* – Application and data sharing are two communication-oriented drivers for integration. Through a combined Cloud/IoT model, personalized, pervasive applications can be provided through IoT, while data collection and distribution can be automated for minimal cost. Cloud technology provides a cost-effective solution to manage, connect, and monitor anything from any location at any time through incorporated applications and customized portals. High-speed networks allow productive coordination, monitoring, communication, and control of remote IoT things as well as real-time data access. While the Cloud can remarkably improve IoT communication, it can also lead to some bottlenecks. In fact, while broadband capacity increased by a factor of only 10^4 over the last 20 years, data storage density has grown by a factor of 10^{18} and processor power has grown by a factor of 10^{15}. Therefore, limitations become apparent when trying to move huge amounts of raw data to the Cloud from the edge of the Internet.
- *Storage* – IoT incorporates a large number of data sources (i.e., IoT things/devices), generating a large amount of semi-structured or unstructured data. This data is characterized similarly to big data based on volume, data

variety, and velocity. An important Cloud/IoT integration motivator is the large-scale, long-life data storage made possible by the limitless, cost-effective, highly available storage capacity of the Cloud. Because Cloud is the most suitable solution to manage IoT data, it creates new opportunities for data aggregation, integration, and data sharing. After data has been stored in the Cloud, it can be treated through simple APIs, guarded by world-class security, and accessed and visualized remotely at any time.

- *Computation* – Due to processing and energy resource limitations, IoT devices cannot process complex data on-site. Collected IoT Data must be transferred to powerful nodes capable of aggregation and processing; however, scalability is difficult to attain without the right infrastructure. Cloud provides limitless processing and on-demand usage. This brings another important Cloud/IoT motivation to the forefront. With the help of Cloud, IoT processing requirements can be met (e.g., completing holistic data analysis, utilizing scalable, shared applications, and handling complex tasks).
- *Things as a Service* – The Internet of Things can be also be thought of as a network of networks where countless connections generate several opportunities. An integrated Cloud/IoT model allows the evolution of smart services and applications as the Cloud expands through things. This enables the Cloud to handle new, real-life situations and create a *Things as a Service* model. This is another key motivator for Cloud/IoT integration as new business models are created by this particular driver.

The most efficient way to design and implement an IoT Cloud solution is to break the problem into smaller pieces. Indeed, the solution can be better understood using a *layered architecture* that is separated into unique layers responsible for handling particular functions or roles. Figure 4.1 illustrates the overall architecture of the IoT Cloud.

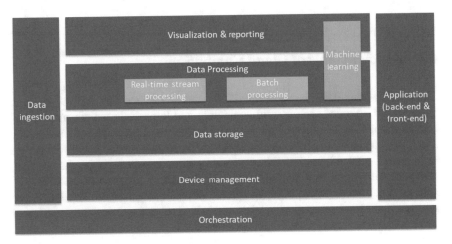

Fig. 4.1 Layered view of IoT Cloud architecture

- *Device Management Layer* – This layer is responsible for provisioning, registration, configuration, monitoring, control, and maintenance of connected IoT devices.
- *Data Ingestion Layer* – This layer is the initial stop for data coming in from different sources. Data is arranged and categorized here to enable data to move smoothly into the other layers.
- *Data Processing Layer* – This layer focuses on processing the data collected in the data ingestion layer. This is the first point where data analysis takes place as data is transmitted to different destinations.
- *Data Storage Layer* – Storage can be challenging depending on the size of data being collected. Utilizing a storage solution appropriate for large data size is important, and this layer focuses on storing vast amounts of IoT data as efficiently as possible. This layer includes different components such as data lake, data warehouse, and databases.
- *Application Layer* – IoT application is a software consisting of Presentation Tier/Layer (i.e., front-end), Business Logic Tier, Database Tier, as well as Application Integration Tier (i.e., APIs and other interfaces) that manages IoT devices, IoT data, users, and IoT services.
- *Data Visualization and Reporting Layer* – This layer is also known as a presentation tier and is likely the most important layer because here is where users can feel or see the value of the collected IoT data. It is important to grab the user's attention and make findings clearly understood.
- *Orchestration Layer* – IoT Cloud architecture often contains repeated processing operations inside encapsulated workflows. These workflows convert source data and transfer it among multiple sources or sinks. In addition, the management of IoT Cloud is a very complex task that directly results from the sheer number of virtual servers and application components. As the name of this layer suggests, the orchestration layer consists of a set of tools to orchestrate the Cloud and other layers.

4.2 Fundamentals of Cloud Computing

Cloud computing dates back to the 1950s and since then has evolved through different technologies such as grid computing and large-scale mainframes. Cloud computing has been defined by the National Institute of Standards and Technology (NIST) as, "a model for enabling ubiquitous, convenient, on-demand network access to a shared pool of configurable computing resources (e.g., networks, servers, storage, applications, and services) that can be rapidly provisioned and released with minimal management effort or service provider interaction." The NIST also suggests that Cloud computing consists of five basic characteristics, three models of service, and four deployment models [2].

4.2.1   Cloud Computing Key Characteristics

According to NIST, each Cloud should have the following basic characteristics:

- *On-Demand Self-Service* – Users are able to independently utilize computing capability including network storage and server time as needed without human intervention from individual Cloud service providers.
- *Broad Network Access* – Computing is accessible via a network and typical mechanisms that are useable by heterogeneous thick or thin client platforms such as laptops, mobile phones, workstations, or tablets.
- *Resource Pooling* – Computing resources such as processing, storage, network bandwidth, and memory are combined to serve many consumers via a multi-tenant model. Various virtual and physical components are allocated and reassigned based on client demand. Customers can sometimes at a higher level of abstraction specify the location (i.e., state, country, or datacenter), but they do not control or have exact knowledge of where provided resources are coming from.
- *Rapid Elasticity* – It is possible to provide and release (sometimes automatically) capabilities and quickly scale out on-demand. It may appear to clients that provided capabilities are limitless and always obtainable in any quantity.
- *Measured Service* – Cloud computing systems control and maximize the use of resources by utilizing metering abilities pertinent to a specific service type (i.e., active user accounts, processing, bandwidth, or storage). The consumption of resources can be tracked, controlled, and documented for reporting purposes, creating transparency for consumers and service providers.

4.2.2   Service Models

Providers of Cloud computing services utilize different models to offer services. According to the NIST, the three typical models include *Platform as a Service (PaaS)*, *Software as a Service (SaaS)*, and *Infrastructure as a Service (IaaS)*. Figure 4.2 illustrates the deployment options for IoT applications as well as the three service models of Cloud. Note that IoT users can deploy their applications in a private system (called *on-premises*) or on the Cloud as shown in Fig. 4.2.

- *Infrastructure as a Service (IaaS)* – This model gives organizations and companies computing resources such as networking, servers, data center space, and storage, using a pay-per-use structure.
- *Platform as a Service (PaaS)* – This model creates an environment that includes all things needed to facilitate complete life cycles for constructing and distributing Cloud applications without the complications or cost purchasing and managing all fundamental software, hardware, or hosting.

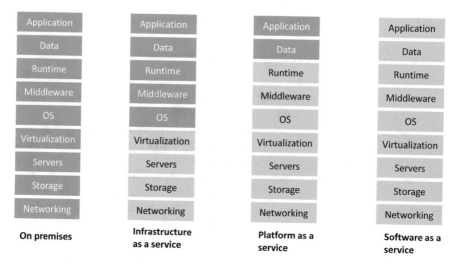

Fig. 4.2 Deployment options for IoT applications

- *Software as a Service (SaaS)* – In this model, Cloud Applicaiton (or other software as a service) are run on physically distant Cloud computers that are the property of and operated by others. These applications typically connect with user computers through an Internet browser.

4.2.3 Deployment Models

According to NIST, Cloud can be deployed in several ways such as:

- *On-Premise Private Cloud* – This is a Cloud infrastructure that is managed by a third-party vendor or internally for only one organization. A private Cloud can be hosted by the organization itself or externally.
- *Public Cloud* – Cloud services are provided through a network that is available to the public for use.
- *Hybrid Cloud* – A hybrid Cloud is made up of at least two (and possibly more) Clouds that are private or public and remain distinct from one another while being connected. This connection provides the benefits of more than one deployment model. Gartner suggests that a "hybrid Cloud is a Cloud computing capability made up of some aggregation of community, public, or private Cloud services available from various service providers."

4.3   Device Management Layer

The complexity of managing IoT-connected devices increases in relation to the number of devices added; therefore, an effective device management system is needed. As with other products, Internet-connected devices have a life cycle that includes the design and manufacturing phases as well as device installation, in-operation, and replacement or repair. For each part of the life cycle, the management solution must meet different requirements to handle the management needs of each phase appropriately. In general, a device management layer needs to address the following tasks:

- Provisioning

  - Registration
  - Configuration

- Monitoring and control
- Software updates and maintenance

4.3.1   Provisioning

The term "provisioning" can mean different things depending upon the industry it is used in. When it comes to provisioning IoT devices to a Cloud solution, provisioning is broken into two parts [3]:

- *Registration* – Registration is the initial phase of the provisioning process where the first connection between the Cloud and the device is made and IoT devices are enrolled in the Cloud.
- *Configuration* – The second phase of the provisioning process is setting up and configuring the device with an appropriate device configuration based on the particular requirements of the target IoT solution.

When both parts of the provisioning process have been accomplished, the device is considered completely provisioned. It is important to remember that there are some Cloud services that handle only the registration phase of provisioning by registering devices to the Cloud without configuration. A *Device Provisioning Service* is able to automatically handle both registration and configuration, which enables streamlined provisioning for a device.

Figure 4.3 illustrates a typical device provisioning scenario; however, it should be noted that the flow of the process can vary greatly depending upon the imple-mentation and application. Below are the general steps included in the provisioning process [3, 4]:

1. During the manufacturing process, the registration information, credentials, and identity (e.g., public and private key) of a device are integrated into the device's

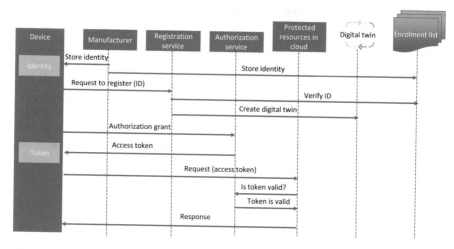

Fig. 4.3 A typical device provisioning and registration flow

storage. The manufacturer also inserts the device registration information on the Cloud enrollment list.

2. A device makes contact with the registration service and provides the credentials and registration information to confirm its identity.

3. The device's identity is confirmed by the registration service by validating the registration information in comparison to the enrollment list.

4. The device is then registered in the Cloud and a digital representation (i.e., shadow, manifesto, or device twin) of the actual device is created in the Cloud. The Cloud typically stores a common set of device characteristics and state information such as software details and hardware specifications. It should be noted that usually document-based NoSQL databases, such as MongoDB, are used to store heterogeneous device description in the Cloud.

5. Next, the registration service sends an authorization grant to the device.

6. Using the authorization grant, the device can contact the authorization service to be able to obtain an access token. This token is then utilized by the device when communicating with any resource in the Cloud.

7. The device connects with the Cloud and provides the access token for each request.

8. When a request is received from a device, the access token should be reviewed in the Cloud by authorization service to confirm the validity and to determine if the device has a suitable right and permission to access a resource/service. If it is accessible, the Cloud then processes the device's request.

It should be noted that when the device registration process is finished, users, applications, organizations, and groups can then be associated with the device.

Establishing relationships, or defining the right of access, between the owner, user, device, and organizations is an elemental part of the authorization and authentication mechanism. At the very least, connected devices can be accessible by three district user groups including maintenance and operations service providers, owners, and the original equipment manufacturer (OEM). The access must be thoughtfully designed and managed to protect against any possible cybersecurity problems.

4.3.2   Software Updates and Maintenance

In a holistic IoT solution, it is very important to be able to perform the over-the-air firmware and software updates. Indeed, the operations most frequently performed by administrators are software and firmware updates. To do so, it would be very helpful to automate these operations using workflow management, which executes a sequence of commands on the device. These tasks can be completed on-demand or they can be scheduled at a designated future time. Workflow engines are able to provide powerful, user-friendly dashboards that specify the details of workflows as well as scripting capabilities to program the engine.

Figure 4.4 depicts a general update process workflow. The process starts with a device management tool that notifies the digital twin of the device that a software update is available. Next, the digital twin sends a message requesting a firmware update to the device over a reliable and secure connection (i.e., WebSocket, HTTP, or MQTT). When the device receives the update request, it downloads the image of the new firmware from the pre-defined repository, verifies the image, extracts and applies the image, and finally reboots itself. Next, the device sends a message to the

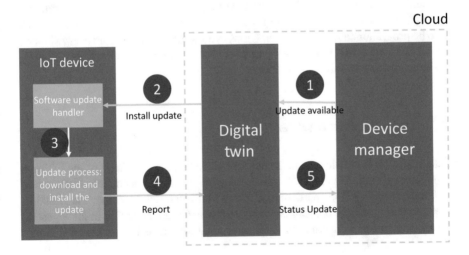

Fig. 4.4 A typical flow of firmware update

Cloud indicating that the software update was completed successfully. This message will cause a configuration synchronization task to begin so that the physical device's status and the device's digital twin can be in sync.

4.3.3 Monitoring and Control

Within a system comprised of thousands of remote devices, the secure and efficient operation of individual devices can impact the profit margin. Seemingly small problems can significantly impact customer satisfaction and hinder business success. Diagnostic and monitoring tools are the key to reducing the negative impact of device downtime caused by unpredicted operational issues or software problems. For example, problems or issues can be detected by tracking and analyzing storage, computing, input/output, and networking statistics at the processor task level.

The ability to download program dumps and logs is also vital to identifying and resolving software issues because it is not possible to travel to each device to debug them using a serial terminal. An application developer has to employ solid program logging and device management software to ensure that the necessary debugging information is available to be uploaded in the event of an error. In addition, the device management software must also generate practical problem insights for issues that occur across many devices.

Often, devices require additional configuration by the user, including attributes such as location, name, and specific application settings. For example, in IoT-based fleet management application, a device must monitor vehicle location and status and then send that information to the Cloud using a cellular connection. In this context, specific parameters must be written when the device is installed, including the distinctive truck or trailer ID (e.g., license plate number). Additional configuration settings such as the time interval for the sending position data can also be programmed into the device. Unless you want to preconfigure every shipped device, it is important to be able to remotely access and manage devices to provide on-demand support and configuration.

4.4 Data Ingestion Layer

Data ingestion is considered as the starting place and the first stage of the IoT data pipeline in Cloud. This layer is concerned with connecting to different sources of IoT data, bringing data (particularly, (semi-) unstructured data) into the Cloud as well as routing data to appropriate destinations in Cloud. Note that in the ingestion layer, data can be consumed in batches or streamed in real time. When it comes to real-time streaming, data is ingested as it arrives. On the other hand, in batch mode, the data is ingested in pieces at periodic time intervals. As the availability of IoT devices grows, the variety and volume of data sources are quickly expanding. Therefore, obtaining data can be a challenge. In general, major challenges facing data ingestion include [5, 6]:

- *Velocity and Volume of Data* – Data volume and the frequency of data generation are incredibly high in IoT.
- *Heterologous Data Sources* – When data sources have different formats and natures, it can be difficult to ingest quickly, process, and prioritize data.
- *Rapid Evolution* – Both data sources and ingestion technologies/frameworks change quickly over time.
- *Independent Data Changes* – Data can change independently from the ingestion application without any prior notice.
- *Semantic Data Changes* – Data semantics can evolve as the data is used in new business scenarios.
- *Detection and Capture* – Detecting and obtaining altered data can be challenging due to the unstructured or semi-structured nature of data as well as the low-latency requirements of specific business cases.

When designing a *Data Ingestion System*, it is important to consider the following:

- *Upgrade Capability* – The system must be able to upgrade in order to handle new data sources, applications, and technologies.
- *Data Integrity* – We have to ensure that the ingestion application is consistently obtaining correct and trustworthy data.
- *Dependability and Fault Tolerance* – The system must be fault tolerant of overcoming any failures.
- *Data Volume* – The ingestion layer should be able to handle the high volume of data. In general, the ability to store all data is preferable; however, in some instances it may be more appropriate to store aggregated (processed) data.
- *Scalability* – The system must rapidly consume data, be able to scale based on volume as well as the speed at which big data of IoT comes in from various sources including networks, machinery, sensors, human interaction, social media, and other media sites.
- *Heterogeneous Data Source/Format* – While data can take different forms, it is usually structured (i.e., tabular one), unstructured (i.e., video, audio, images), or semi-structured (i.e., CSS files, JSON files, etc.). The data ingestion layer should be capable of utilizing various data sources, data formats, technologies, and operating systems.

4.4.1 Data Ingestion Frameworks

4.4.1.1 Apache Flume

Apache Flume is a distributed system useful for capturing, aggregating, and routing massive quantities of streaming data into a data store such as Hadoop. The details of the Hadoop ecosystem will be discussed in Chap. 6. Flume includes several built-in channels, sources, and sinks (data destinations). Flume also offers some features necessary to be able to perform transformations (processing) of data during the ingestion process. Flume can also be scaled horizontally in order to provide high availability and scalability.

Apache Flume deployment includes starting one or multiple Flume agents. A Flume Agent is a JVM process comprised of three elements: Flume Source, Flume Channel, and Flume Sink (see Fig. 4.5) [7].

1. *Flume Source* – In Flume, an event is defined as a data unit consisting of a byte payload and a set of optional string attributes. Data source is responsible for ingesting the events triggered by an external source such as a Webserver or a connected sensor.
2. *Flume Channel* – Afterward, the event is stored in one or more channels by Flume Source. The stored data will stay in channel until it is consumed by Flume Sink.
3. *Flume Sink* – Finally, Flume Sink takes the event from channel storage and routes it to an external depository (e.g., HDFS in Hadoop).

Note that agents are capable of being chained, so multiple Flume Agents can be utilized. In such a case, Flume Sink of one agent sends the event on to the Flume Source of the next Flume Agent in the chain.

4.4.1.2 Apache Kafka

Apache Kafka is an open source, high-throughput, distributed message bus/queue/system to connect data consumers to data providers. In comparison to Flume, Kafka provides greater scalability and more durable messaging. Kafka is

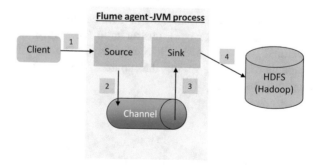

Fig. 4.5 Apache Flume Architecture

a publish-subscribe system in which messages persist and are categorized based on the topic. Message creators are known as publishers and the consumers of messages are known as subscribers. Consumers are able to subscribe to multiple topics and consume/receive all messages for that particular topic (see Fig. 4.6). There are five basic components required to move data in and out of Kafka as described below [8]:

- *Topics* – A topic is a category, channel, or a queue defined by the user under which messages are published. Within Kafka, topics are multi-subscriber, meaning they can have zero to many consumers subscribed to receive the published data.
- *Partitions* – Topics can be divided into partitions which allow users to scale and parallelize a particular topic by breaking its data across many different brokers (Kafka Server/Machine). In this approach, each partition can be allocated to an individual machine, enabling many consumers to read a topic simultaneously. Every message in Kafka is identifiable by a tuple made up of the message topic, partition, and offset inside the partition. An offset is a specific immutable order in which messages are arranged in a partition. In other words, an offset identifies the location of one record in the partition. Consumers are able to read messages beginning at a specific offset and can read from any chosen offset point, enabling consumers to join a topic/partition at any point. To support fault tolerance features, Kafka can replicate each partition across a configurable number of servers (see Fig. 4.7). In this case, a partition can function either as a leader or as a replica. All reads and writes for a partition must pass through the leader, and the leader handles all read and write requests for a partition. In addition, the leader is responsible for organizing replica updates with newer data. In the case of leader failure, a replica is able to function as the leader.
- *Producers* – Producers are responsible for posting a message to topics. A message is made up of a topic name referencing where the record will be sent, a partition number (optional), key (optional), and value (optional). If the partition is specified by the producer, the message will be routed to that specific partition. If a partition is not chosen by producer, but a key is indicated, the partition is picked by the broker based on the key's hash. In the case that no partition or

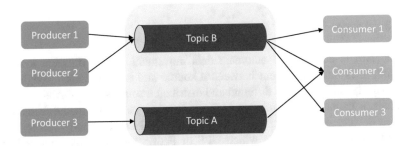

Fig. 4.6 The publish-subscribe mechanism in Kafka

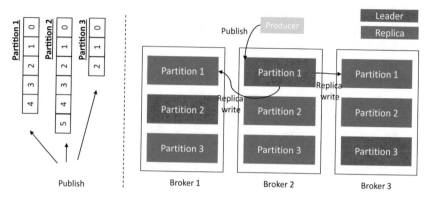

The concept of partitions of a topic   Leader and replica

Fig. 4.7 How topics and partitions work in Kafka

key is specified, a partition is chosen in a round-robin manner (i.e., partitions are selected based on a circular order). Kafka broker can also create a timestamp for each message when it wants to store the message in the partition. Producers can attach keys to messages, guaranteeing that all messages with the same key are sent to the same topic partition. Kafka ensures order in a partition but does not guarantee order across partitions within topics. Therefore, not utilizing a key will result in round-robin style distribution to all partitions, and it will not maintain the same order.

- *Consumers* – Consumers subscribe to a particular topic(s) and process messages that are posted in the corresponding topic(s).
- *Brokers* – A Kafka broker, Kafka node, or Kafka server all reference the same concept and the terms can be used interchangeably. Kafka brokers take in producer messages and store them on disk categorized by topics/partitions. The broker also enables consumers to obtain messages based on offset, topic, or partition.

4.4.1.3 Apache Nifi

Apache Nifi is designed to automate data movement among different systems. Nifi simplifies data movement between a source and a destination through real-time control, and it supports different and distributed sources with various formats, protocols, schemas, sizes, and speeds (i.e., clickstreams, social media feeds, geo-location devices, videos, log files, etc.). Apache Nifi can be configured to move data similar to the manner in which UPS or FedEx moves, tracks, and delivers parcels because this platform enables the user to track real-time data much like a package delivery. Nifi is also very useful for sensitive data flows which need robust compliance and security requirements. In contrast to Kafka and Flume, Nifi is capable of handling data objects of differing sizes. NiFi comes with a user-friendly drag-and-drop, web-based user interface which allows users to visualize the whole process and make necessary changes in real time. The core design concepts underpinning Nifi are similar to the basic idea of Flow-Based Programming (FBP). Below are the foundational Nifi concepts and components [9]:

- *FlowFile (Information Packet)* – It represents the data objects passing through the system. For each FlowFile, Nifi tracks a set of key/value pair characteristics of the object as well as its corresponding content (zero or more bytes).
- *FlowFile Processor (Processor)* – Processor handles a combination of data transformation, routing, and system mediation. Processors are able to access FlowFile characteristics as well its content. Processors are able to work on zero or more FlowFiles in parallel. NiFi processors can either commit the result/out or can roll back to their previous state (to address fault-tolerant issues). Nifi comes with a wide assortment of processors (at the time of this writing more than 260 processors) including connectors for Kafka and Flume that may be dragged, dropped, configured, and immediately put to work. There is also a possibility to design and implement custom processors for Apache NiFi.
- *Connection (Bounded Buffer)* – Connections serve as a link between processors. They function as queues and enable different processes to interact at various rates. Queues can be arranged dynamically and may include upper bounds on the load, allowing back pressure. Backpressure references scenarios where queues or buffers are at capacity (full) and unable to receive new data. In such a case, the backpressure mechanism ensures that new packets of data are not sent until the data bottleneck has been resolved or the buffer is no longer full.

4.4.1.4 Elastic Logstash

Elastic Logstash is an open-source data ingestion framework that takes in data from many sources concurrently, transforms the data, and sends the transformed data to the Elasticsearch database (a NoSQL Database). Logstash is capable of ingesting data of varying source, shape, and size from several sources such as logs, and databases, in a constant, streaming manner. Logstash has a large variety of output plugins, to support a range of different use cases. Additional Elasticsearch and Logstash details will be discussed later in this chapter.

4.5 Data Processing Layer

In this section, we will discuss the modern data processing and big data architectures created to manage huge amounts of data in order to extract the value of IoT data. Before reviewing data processing architectures, some fundamental terms are defined below [3, 5, 6, 10, 11]:

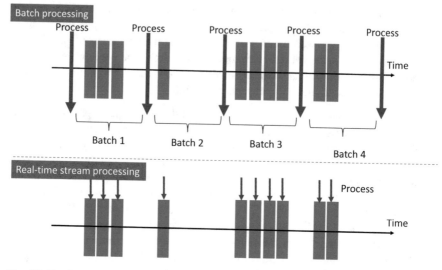

Fig. 4.8 Batch processing versus real-time stream processing

- *Streaming Data (Stream)* – Stream is a data that is continuously generated and transferred at a steady, high-speed rate.
- *Batch Processing* – A batch is made up of data points that have been gathered within a specified time period often known as a "window of data." Batch processing requires that a data set be collected and stored for a certain period of time. Then, the entire set is processed together at a designated future time. The parameters for processing can be determined in a number of ways including a scheduled time interval (e.g., new data is processed every 5 minutes) or through a specified condition (e.g., batch is processed when it contains five data elements or when it reaches 1 MB of data). An example of batch processing would include all of a financial firm's transactions submitted over the period of a week (see Fig. 4.8).
- *Stream Processing* – This type of processing enables real-time data processing and ascertains conditions in real time. Unlike batch processing, stream processing processes individual pieces of data as it comes in (arrives) rather than waiting to process at a specific interval. When it comes to performance, batch processing latency is minutes to hours while stream processing latency is only milliseconds to seconds. It should also be noted that although stream processing handles each new piece of data as an individual unit, many stream processing systems also allow "window" operations that enable processing to reference data that comes in during a specific time interval before and/or after current data (see Fig. 4.8).
- *Data at Rest* – This refers to data collected from a variety of sources but it is analyzed later. For example, a retail store owner may collect previous monthly sales data to make strategic business decisions. In this case the analysis of data occurs separately and after the data collection phase.

- *Data in Motion* – Data collection is similar to that of data at rest; however, data analysis happens in real time as the data-creating event occurs. For example, a connected wristband can constantly collect and record guest activity data in an event/conference/fair to customize the guest's visit with activity suggestions based on individual behavior, enabling personalized user experience in real time.

4.5.1   Data Processing Architectures

In this section we overview two well-known architectures for data processing, namely, Lambda architecture and Kappa architecture.

4.5.1.1   Lambda Architecture

Nathan Marz and James Warren first proposed Lambda architecture in "Big Data: Principles and best practices of scalable real-time data systems." Lambda was designed as a universal, fault-tolerant, scalable data processing architecture able to process massive amounts of data using stream processing and batch processing techniques. Figure 4.9 illustrates the three main elements found in Lambda architecture: the speed/real-time layer, batch layer, and serving layer [12].

- *Batch Layer* – Responsible for storing raw data and processing data in batch model to create batch views. The data scope in the batch layer can encompass hours to years.

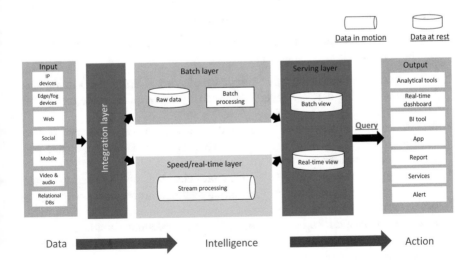

Fig. 4.9 Lambda architecture

- *Speed Layer* – Processes data in real time and computes real-time data views.
- *Serving Layer* – Stores the result of batch layer and speed layer (i.e., batch views and real-time views) and responds to ad hoc queries by returning previously computed batch and real-time views or building new views from the batch and speed layer outputs stored in the serving layer. This approach provides a more holistic and complete view by obtaining the best of two worlds (i.e., batch and real time). Batch views can be processed using more complex rules resulting in higher data quality and less skew. On the other hand, real-time views provide instant access to data.

The main benefits of Lambda architecture are as below:

- *Real Time* – The real-time layer can apply simple data processing and machine learning algorithms to provide real-time insights and alerts.
- *Improved Processing Without Data Loss* – The batch layer allows data processing to take place with great precision using complex algorithms without the loss of alerts, short-term information, or other insights generated by the real-time layer.
- *Reduced Storage Needs* – Because of the batch layer, the Lambda architecture minimizes the need for random write storage.
- *Tolerance to Human Errors and Hardware Crashes* – A well-implemented batch layer makes it difficult for hardware crashes or human errors to damage stored data because the system does not allow existing data to be deleted or updated. The real-time layer is more vulnerable to errors. Data can be lost or corrupted in this layer because the data stores are variable. However, if incoming writes are being transmitted to the batch storage area, the data results will eventually catch up when the next batch is processed. This means no data will be lost even if the real-time layer encounters an error. While results may be outdated if the real-time layer experiences failure, the batch layer data records will not be damaged and the results will sync again when the real-time layer is functioning correctly again.

While there are benefits to Lambda architecture, there are also shortcomings that should be considered:

- *Complexity* – Lambda architecture comprised of many layers, and thus maintaining proper syncing between layers can be costly and requires more thoughtful effort and handling.
- *Maintenance and Support* – Because this architecture is made up of two clearly defined, completely distributed layers (speed and batch), support and maintenance activities can be difficult.
- *Technology Mastery* – Many technology proficiencies must be used to create Lambda architecture. Finding and recruiting qualified professionals with expertise in these areas can be difficult.
- *Complex Implementation and Deployment* – Creating Lambda architecture using open-source technologies and then deploying it via the Cloud or on-premises servers can be complicated.

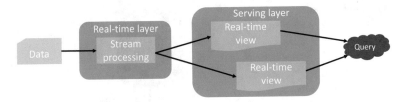

Fig. 4.10 Kappa architecture

4.5.1.2   Kappa Architecture

Kappa architecture was first designed by Jay Kreps. This architecture is focused on using stream processing, but it is not meant to replace Lambda architecture unless it truly fits your use case. In this architecture solution, all IoT data are routed through a real-time layer and results are moved to the serving layer where they can be queried (see Fig. 4.10). In other words, there is no batch layer in Kappa architecture [13].

The key idea behind Kappa architecture is to address real-time data processing by constantly reprocessing data through a single stream processing engine. This implies that the incoming data can be quickly replayed. For example, in the instance that there are code changes, a second stream process would then replay data through the real-time engine and then replace the previous views in the serving layer data.

Kappa architecture achieves simplicity compared to Lambda architecture by maintaining only one code instead of trying to manage two codes (i.e., one for batch layer and one for speed layer). Additionally, queries do not have to look in batch or speed views. Instead, queries look in only one serving location. Drawbacks to this architecture are centered on processing data in a stream. It is easier to handle duplicate events, cross-reference events, or manage order operations in batch processing. In addition, having access to the whole data set in the batch layer can lead to better optimizations, higher performance, and simple algorithms in Lambda architecture. Finally, in some sophisticated situations, streaming and batch algorithms can lead to different results.

4.5.2   Data Processing Frameworks

This section will overview the well-known frameworks designed for data processing.

4.5.2.1   Apache Storm

Apache Storm is a distributed, open-source stream processing framework known for its reliability and fault tolerance. Storm applications are created as a "topology" in

Fig. 4.11 Apache Storm

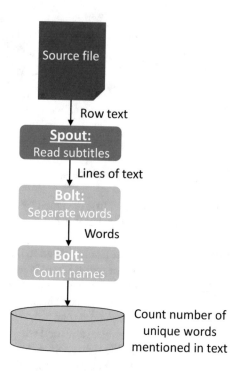

the form of a *directed acyclic graph* (DAG) that utilizes bolts and spouts. Storm's input stream is managed by a spout responsible for moving the data to a bolt that then transforms the data. A bolt can move the data to a storage area or move it to another bolt. Storm can be visualized as an interconnected chain of bolts that somehow transform data collected by the spout (see Fig. 4.11) [14].

4.5.2.2 Apache Flink

Apache Flink is also an open-source streaming framework providing extensive real-time data processing pipeline capability. It is highly scalable and able to address millions of events each second. Flink is designed based on the DataFlow model and processes data as it arrives. One of the great features of Flink is its fault-tolerance capability based on the checkpointing concept (i.e., saving internal states to external sources/storage and recovering the state of the system in case of any failure). Flink also provides an SQL API enabling individuals with a lack of programming knowledge to develop a Flink solution much easier and faster.

4.5.2.3 Apache Spark

Apache Spark is an efficient, in-memory engine used for data processing. It provides refined and articulate development APIs that enable data engineers and data scientists requiring quick iterative access to data, to execute SQL, machine learning, or streaming workloads. The details of Apache Spark will be discussed in Chap. 6.

4.6   Data Storage Layer: A Hybrid Architecture

Another challenge to realize holistic IoT Clouds is data storage. Relational databases were an appropriate place for data storage in the past; however, as enterprises strategically utilize big data and IoT applications, data persistence should not necessarily be relational. Ideally, we need to design and deploy a hybrid architecture that supports three types of repositories in any IoT environment (see Fig. 4.12) [15]:

- *Cold storage* – This repository is usually implemented by a data lake based on distributed file system technologies. Cold storage intends to store structured or unstructured data that is accessed infrequently.

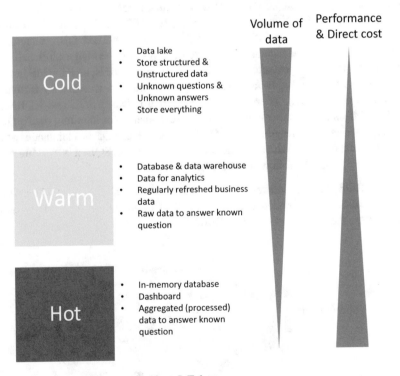

Fig. 4.12 A hybrid architecture to address IoT data

- *Warm storage* – This repository, which usually implemented by data warehouses or databases, intends to store structured data that is accessed moderately frequently.
- *Hot storage* – This repository is accessed very frequently and hence should be implemented by fast in-memory databases.

4.6.1 Database

Databases can be categorized into two different groups, namely, SQL databases and NoSQL databases.

- *What Is SQL?*

Structured Query Language, abbreviated as "SQL" and pronounced S-Q-L, is the standard language used when working with *Relational Database Management Systems* (RDBMS) including MySQL, PostgreSQL, Oracle, and MS SQL Server, etc. SQL can be used to simply insert, update, search, or delete database records, but it's also suitable for more complex database maintenance and optimization tasks.

Relational Database Management Systems (RDBMS) are the foundation of many modern databases such as MS Access, MS SQL Server, IBM DB2, Oracle, and MySQL. An RDBMS is a database management system (DBMS) established using a relational model first presented by E.F. Codd. In other words, an RDBMS is a type of DBMS with a row-based table structure which can be accessed by Structured Query Language (SQL). Therefore, the difference between SQL and RDBMS is that RDBMS is the actual database software responsible for handling data storage, updates, queries, and everything else. On the other hand, SQL is a language used to interact with the system. To better understand RDBMS and SQL, it is very important to get familiar with the following terminologies:

- *Table*: RDBMS uses database components known as tables to store data. Tables are a collection of related data comprised of columns and rows. Figure 4.13 illustrates the concept of tables in RMDBS using an example.

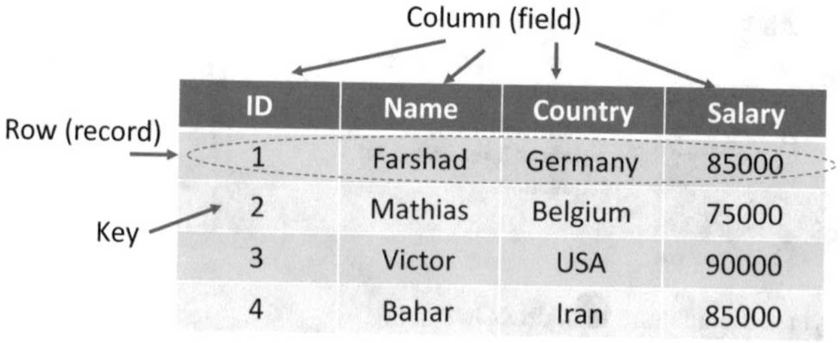

ID	Name	Country	Salary
1	Farshad	Germany	85000
2	Mathias	Belgium	75000
3	Victor	USA	90000
4	Bahar	Iran	85000

Fig. 4.13 A simple example of table in RDBMS

- *Record*: Records are also known as rows of data which are horizontal entities of a table. For each entry in the table, there is one row in the table.
- *Field*: Tables are made up of smaller entities known as fields; these are columns in a table that hold unique information regarding each table record.
- *Key*: A key can uniquely identify a row in the table. Keys also enable you to establish a relationship between tables.
- *Database normalization*: Database normalization is the means for efficiently organizing data within a database in order to (1) remove unnecessary data such as duplicate data across tables and (2) guarantee that data dependencies in database are properly enforced. These goals are important because they decrease the amount of space a database requires while ensuring logical data storage. Normalization guidelines are helpful in developing a solid database structure. In general, there are three major normal forms in a database:

  - *1st normal form* – There are no repeating groups of columns in a table.
  - *2nd normal form* – The table is in first normal form and all the columns are fully functional dependent on the table's primary key.
  - *3rd normal form* – A table in third normal form is a table in the second normal form that has no transitive dependencies. A transitive dependency can occur when: $P -> Q$ and $Q -> R$ is true, then $P -> R$ is a transitive dependency. Note that P, Q, and R are attributes (columns) of a table.

- *ACID compliance*: ACID compliance in transactions is one of the most important features of RDBMS. A transaction is a logical unit that typically consists of several low-level tasks which should be executed for data retrieval or updates in a database. A transaction must maintain atomicity, consistency, isolation, and durability (known as ACID) to guarantee the accuracy and completeness, as well as the integrity of data [16].

  - *Atomicity* – This property makes sure that all of a transaction's operations are executed or none are executed; partial completion of a transaction is not tolerated.
  - *Consistency* – The database must remain in a stable state before and after a transaction is completed. Transactions should not have a negative effect on data stored in the database.
  - *Durability* – The database must maintain all recent updates in the event that a system fails or restarts. If a transaction updates a portion of data and commits to it, then the database has to maintain the updated data. If a system fails before a committed transaction can write the data to the disk, then the data update must take place when the system becomes functional again.
  - *Isolation* – If multiple transactions are being executed at the same time and in parallel, then all transactions will be completed as if it was the only system transaction. No single transaction affects any other transaction.

- *What Is NoSQL?*

NoSQL which stands for "Not Only SQL" or "Not SQL" is a broad term covering a variety of technologies that address several key topics such as non-relational, distributed, open-source, and horizontally scalable. While the name may seem to imply such, NoSQL is not focused solely on the absence of SQL. For instance, several NoSQL query languages such as Hive's query language are heavily influenced by SQL. NoSQL does not mean that the database is schemaless. Schemaless means that the database does not have a fixed data structure (e.g., Table in RDBMS). For instance, PostgreSQL (RDBMS) has evolved to be able to serve as schemaless document storage as well. ACID-ity which is a main feature of SQL is not necessarily the focus either. Consider Hyperdex, a NoSQL database able to support ACID-transactions. When it comes to relationships, most NoSQL databases do not support joining in the same manner as a traditional database; however, some do. With some research, a distributed SQL database can also be found, and indeed more recently created databases are usually distributed in some way. Note that a join operation in SQL is used to establish a connection/relationship between two or more tables based on a set of given columns of the target tables. Based on the above facts, it is not entirely logical to narrowly define NoSQL [17].

A NoSQL database is meant to meet large, distributed data storage needs and is often used for big data and real-time use web applications such as Twitter, Google, or Facebook that constantly gather terabytes of user data. While traditional RDBMS utilizes SQL syntax to store or retrieve data, NoSQL encircles a diverse spectrum of database technologies that can house polymorphic, unstructured, semi-structured, or structured data. There are four types of unique NoSQL databases, each useful for different data needs (see Fig. 4.14):

- *Graph Database* – Uses graph theory and designed for data represented as a graph with an undetermined number of relationships among interconnected elements, for examples, Neo4j and Titan.
- *Key-Value Store* – A great starting place as it is one of the simplest NoSQL options, designed to store data comprised of an indexed key and value in a schemaless manner. Examples: DynamoDB, Cassandra, Redis, Azure Table Storage (ATS), and BerkeleyDB.
- *Column Store (Wide Column Stores)* – Rather than storing data in a row, these databases store data tables in vertical columns. While it may sound like a simple horizontal versus vertical adjustment, column stores actually offer high-quality performance and extremely scalable architecture, for examples, HyperTable, BigTable, and HBase.
- *Document Database* – This option extends the idea of key-value stores in that each document is given a unique key needed to retrieve a specific document, designed to store, manage, and retrieve semi-structured, document-oriented data, for example, MongoDB and CouchDB.

Document

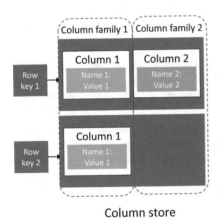

Column store

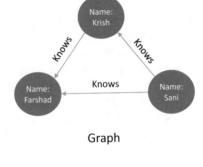

Graph

Fig. 4.14 Four different types of NoSQL database

- *When Should SQL/NoSQL Be Used?*

When choosing which database is most appropriate, it is most important to determine if a relational (SQL) or non-relational (NoSQL) data structure is needed. Both offer options; however, there are additional differences between SQL and NoSQL that must be considered. In general, SQL should be used in the following scenarios:

- When building custom dashboards
- Analyzes behavior-related or custom sessions
- When you need to store or extract database information quickly
- Preferable when using joins and complex queries
- When you need ACID transaction
- When you know the schema (structure of data) in advance and it does not change over time

On the other hand, NoSQL is used:

- When ACID support is not required
- When a traditional RDBMS model is insufficient
- When data being stored requires flexible schema
- When constraints or validation logic is not needed
- When logging data from distributed sources
- When storing temporary data or wish lists and session data

4.6.1.1 MongoDB

MongoDB is a document-oriented, cross-platform database that offers high performance and data availability while being easily scalable; MongoDB is able to support diverse schema, balance loads, handle replication, indexing, queries, and filing (i.e., to work as a file system) [16, 18].

- *Collection*: A group of MongoDB documents is called a collection. Documents in MongoDB are equivalent to tables in RDBMS. Documents housed in a collection may contain diverse fields; however, all documents in a particular collection are used for a similar or related purpose.
- *Document*: A document is known as a set of key-value pairs, often with dynamic schema, meaning documents within the same collection are not required to have identical fields or structures and common document fields may hold different data. Table 4.1 illustrates the relationship between RDBMS terminology and MongoDB. Figure 4.15 also shows the MongoDB document structure of a blog site using a comma-separated key-value pair.
- *Relationship*: In MongoDB, relationships illustrate how different documents are logically connected to each other. Unfortunately, MongoDB cannot support traditional primary key – foreign key relationships. However, in MongoDB, 1:1, 1:N, N:1, or N:N relationships can be modeled by two means, namely, embedded and referenced approaches. In the first approach, one document is embedded inside another document, and thus all the related data are located in a single document. On the other hand, in the referenced approach, documents will be maintained separately; however, documents have a reference field containing the address of the other documents.

Table 4.1 Relationship of RDBMS terminology with MongoDB

RDBMS	MongoDB
Database	Database
Table	Collection
Row	Document
Column	Field
Primary key	Primary key (MongoDB provides the default key _id itself)

```
{
    Person: {
        first_name: "Farshad",
        last_name: "Firouzi",
        address: [
            { "type": "home", "city": "Dusseldorf", "country": "Germany"},
            {"type": "home", "city": "Austin", "country": "USA"}
        ],
    }
}
```

Fig. 4.15 A sample MongoDB document

- *Index*: Indexes are used to execute queries efficiently. Without them, MongoDB has to scan every document in the collection to select documents matching the query. However, if an index exists for the stated query, MongoDB can then reduce the number of scanned documents.

Both RDBMS and MongoDB databases have their own uses. In this section we highlight the main advantages of MongoDB. However, based on their requirements of the target IoT application, any of them can be used.

- The main advantage of MongoDB is that it is schemaless. This means that documents are flexible and you do not need to enforce any schema/structure on them.
- MongoDB includes a built-in aggregation framework capable of performing ETL (extract, transform, and load) jobs. It also has a deep query ability with a document-based language almost as powerful as SQL.
- It is very straightforward to scale (scale-out) MongoDB horizontally.
- MongoDB is "object-oriented" and can easily represent any object structure in your domain.
- MongoDB enables organizations of all sizes to build data-driven applications faster. The reason is that MongoDB does not require complex object-relational mapping (ORM) layer to map objects in code to relational tables. Therefore, it is much simpler for developers to understand and map the data stored in the database to data in the application. Moreover, there is no need to spread the data across different relational tables and all the data can be found only in one single place.

4.6.1.2   Cassandra

Cassandra is a popular, open-source, wide-column, decentralized/distributed database used for storing high volumes of (semi)-structured data with high availability and no single point of failure. The Cassandra is a database-oriented by column. Rows contain what is often thought of as vertical data that would usually be contained in relational columns. Databases that are column-oriented are housed on disk column-wise. The most obvious benefit of this database is that queries can resolve quickly. For example, when exploring the average age of users, you could move to the location where age data is kept and read the required information rather than searching for age row-by-row as you would in a row-oriented database. Therefore, Cassandra allows you to ignore irrelevant data easily (see Fig. 4.16) [16, 19].

Cassandra's model is comprised of columns, column families, keys (column names), and keyspaces. Table 4.2 compares each element of the Cassandra model to its counterpart in an RDBMS model.

Query Language Cassandra Query Language (CQL) is similar to SQL, reducing barriers for users more familiar with relational databases. CQL also uses queries similar to SQL; however, CQL differs in that it does not support joins, group by, or foreign keys. Excluding these elements makes writing and retrieving Cassandra data more efficient.

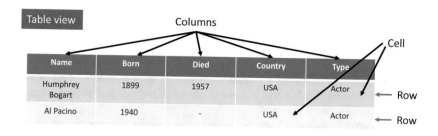

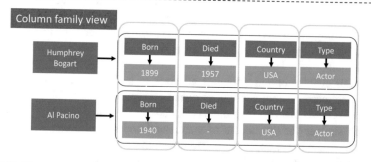

Fig. 4.16 How a table view can be converted to a column family model

Table 4.2 Relationship of RDBMS terminology with Cassandra

RDBMS	Cassandra
Database	Keyspace
Table	Column family
Primary key	Partition key
Column name	Column name/key
Column value	Column value

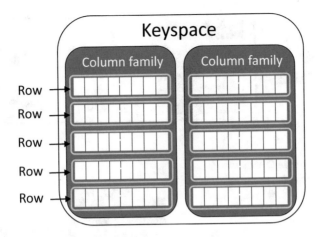

Fig. 4.17 Keyspace and column family in Cassandra

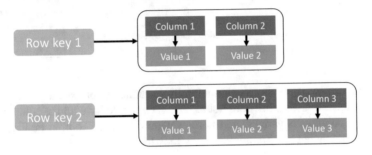

Fig. 4.18 Data model in Cassandra

Table 4.3 Main differences between the relational table and Cassandra column family

Relational table	Cassandra column family
The columns are fixed and cannot be changed over time. When a row is inserted in a table, all the pre-defined columns must have one value and none of them can be left empty. In case a column does not have value, it must be filled with *Null* value.	Columns are flexible and can be added to the column family whenever it is required
Only define columns and users fill in values	Can define columns or super column

Keyspace Cassandra Keyspace serves as the data container, with at least one column family located in each Keyspace. Figure 4.17 provides a schematic example of Keyspace.

Column Families Cassandra column families are collections of rows holding ordered columns that embody the structure of housed data (see Figs. 4.16 and 4.18). Table 4.3 details the points of differentiation between a column family and a relational database table [16, 19]:

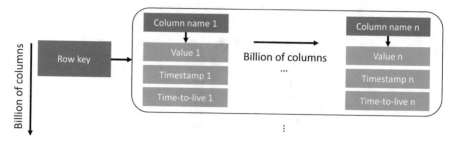

Fig. 4.19 Columns in Cassandra

Fig. 4.20 Super column in Cassandra

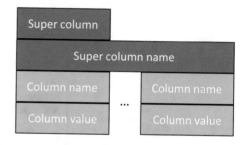

It's important to note that a column family's schema is not fixed as it would be in a relational table. Cassandra does not require each row to have all the columns. The number of columns is variable across rows within the same column family. Figure 4.18 illustrates what a column family may look like.

Column Columns serve as the foundational data structure with three possible values – key/column name, value, and timestamp. In Fig. 4.19, you will find the basic structure of a column.

Super Column Super Columns are special columns that serve as a key-value pair and store a map of sub-columns. A super column has a unique name and number of columns (see Fig. 4.20).

Cassandra has gained popularity based on the excellent technical features available and outlined below:

- *Elastic Scalability* – Highly scalable, allowing the addition of new hardware to accommodate higher customer volume and additional data.
- *Always On* – Architecture contains no single point of failure and is continuously available to house business-critical applications that cannot fail.
- *Linear-Scale Performance* – Increases throughput as the number of cluster nodes increases, maintaining a fast response time.
- *Adaptable Data Storage* – Houses unstructured, semi-structured, and structured data and can flex to accommodate changes in a data structure as needed.
- *Data Distribution* – Efficiently replicates data across many data nodes to distribute data where needed.

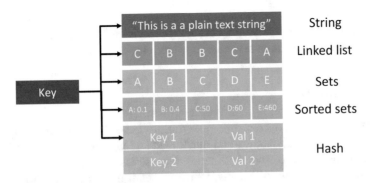

Fig. 4.21 An example of the Redis data model

- *Transaction Property Support* – Able to support atomicity, consistency, isolation, and durability (ACID) transactions.
- *Efficient Writing* – Runs on low-cost commodity hardware and writes astonishingly quickly while storing hundreds of terabytes of data without slowing down reading capability.

4.6.1.3   Redis

Redis, also known as *REmote DIctionary Server*, was created in 2009 as an in-memory, open-source key-value database. Redis is mainly used as a message broker or as a cache layer for other databases. Redis is known as a unique option in the key-value database arena because it can handle complex data types including (see Fig. 4.21) [20]:

- *Binary-safe strings:* A Redis string is binary safe, meaning that it can contain any kind of data (e.g., a JPEG image); however, its size should be maximum 512 megabytes.
- *Lists:* A list is a linked list of string elements meaning that it keeps the order of data insertion.
- *Sets:* A set is a collection of unique, unsorted string elements/members.
- *Sorted sets:* It is very similar to sets; they are a non-repeating collection of strings. However, in a sorted set, a numerical value (score) is assigned to each string element. The elements are always taken sorted by their score. Note that while members are unique, scores may be repeated. Sorted sets are a great means to index data to be able to add, remove, update, or retrieve elements.
- *Hashes:* Hashes are a map between string fields and string value.

As an in-memory, database, it is guaranteed that all data is in memory resulting in an incredible performance. Redis is also persistent on-disk database; however, writing to disk is optional. Another advantage offered by Redis is that all operations are atomic, guaranteeing that if clients access the database simultaneously, Redis will obtain the correct updated value. In addition, Redis is a multi-utility tool that can be used for caching, messaging queues (supports Publish/Subscribe), and short-life data such as application sessions, page hits, etc.

4.6.1.4 InfluxDB

InfluxDB is a big data, open-source NoSQL time-series database that supports high availability, massive scalability, and quick read and write functions. This NoSQL database is designed to store time-series data (series of regular or irregular data points across time) very efficiently, which makes it a great solution to store IoT sensor readouts. Regular data measurements take the form of fixed time intervals (i.e., heartbeat monitoring system data), while irregular data measurements are based on events such as sensor data, trading transaction data, etc. InfluxDB also provides query language similar to SQL that is easily tailored to search aggregated data [21].

In order to fully understand InfluxDB, it is important to define a few key concepts. The illustration in Table 4.4 demonstrates these vital concepts using a real-life example of car-counting sensors. This table shows the number of cars counted by two connected parking sensors mounted at Location 1 and Location 2 during a specific time interval.

- *Time* – Each InfluxDB database includes a time column that stores timestamps associated with corresponding data.
- *Field* – The next column (#Cars) in our example is field. Fields referred to as attributes as well. Fields are key-value pairs within the data structure responsible for recording real data values as well as metadata. Fields are comprised of field values and field keys. Field keys are strings and store metadata. The field value can be in the form of floats, strings, Booleans, or integers. As a time-series database, InfluxDB requires that each field value be associated with a particular timestamp. In our example, "#Cars" is the key and 10, 5, 5, 7, 9, and 2 are field values (see Table 4.4).
- *Tags* – The final two database columns in the sample (location and owner) are known as tags, which are comprised of tag values and tag keys. Tag values and keys are maintained as strings and represent metadata. In our example, "Owner' is tag key and "Farshad/Bahar" is tag value.

Table 4.4 An example of InfluxDB data model

Time	#Cars	Location	Owner
2019-05-31T00:00:00Z	10	1	Farshad
2015-05-31T00:06:00Z	5	2	Bahar
2015-05-31T05:54:00Z	5	1	Farshad
2015-05-31T06:00:00Z	7	2	Bahar
2015-05-31T06:06:00Z	9	1	Farshad
2015-05-31T06:12:00Z	2	2	Bahar

- *Measurement* – This can also be thought of as an SQL table because the measurement serves as a container for the time column, fields, and tags.
- *Retention Policy* – Retention policy describes the amount of time that an InfluxDB database stores data (duration) and how many data copies are kept within the cluster (replication factor). Data that is older than the given duration is automatically removed from the database. Generally, the minimum retention duration is 1 hour and the maximum duration is infinite.
- *Sharding* – Sharding refers to the horizontal partitioning of data with each partition known as a shard. InfluxDB also utilizes sharding to address the scalability problem. Each *shard* holds compressed, actual data of a particular series set. A shard can belong to only one shard group, and there may be many shards in a particular group. All points within a set series in a shard group will be stored in the same shard. The duration parameter specifies the amount of time a shard group covers. In other words, the shard duration determines the specific interval. For example, if shard duration is set to 1 week, each shard group will cover 1 week and include all data points with timestamps in that week timeframe.

4.6.1.5 Elasticsearch

Elasticsearch (ES) is a highly popular soft real-time search engine used by large-scale organizations including *The Guardian, GitHub, StackOverflow*, and *Wikipedia*. ES is considered as a document-oriented database created to hold, manage, and retrieve semi-structured or document-oriented data. Table 4.5 compares Elasticsearch with RDBMS [17, 22, 23].

There are a few basic concepts that one must grasp in order to understand the functionality and structure of Elasticsearch better.

Cluster Clusters are a collection of servers (nodes) that, when networked together, hold the complete data set and provide federated indexing and search capability for all connected servers.

Node A node is a solitary server that contains a portion of data and can contribute to the cluster's querying and indexing tasks. Note that each cluster is identified with a name and a node can be assigned to a particular cluster based on the cluster's name. Starting an instance of Elasticsearch from scratch results in the creation of a cluster with only one node. When the second instance of Elasticsearch starts (with the same "cluster name"), the cluster will contain two nodes. Additional instances

Table 4.5 Relationship of RDBMS terminology with Elasticsearch

RDBMS	Elasticsearch
Database	Index
Table	Mapping
Field	Field
Row	JSON object

of Elasticsearch can be started to form a cluster with any number of desired nodes. Each node within the cluster has knowledge of the other nodes within the same cluster because they communicate directly with one another via TCP. This is referred to as fully connected mesh topology. Every node within the cluster handles one or more roles, namely, a data node, a master node, a client node, or an ingest node [24]:

- *Master Node* – Responsible for constructing or eliminating indices and adding or removing nodes. Every time the cluster state changes, the master node notifies the other nodes within the cluster about the change. Each cluster contains only one master node at a time.
- *Data Node* – Each node contains a data shard (a partition of whole data) and handles data-related operations including creating, reading, deleting, updating, searching, and aggregating. A cluster may contain many data nodes. If one data node within the cluster should stop, the cluster continues operations and rearranges that node's data on the additional nodes.
- *Client Node* – It handles sending cluster-related requests to the master node as well as sending data-related requests to data nodes by serving as a "smart router." A client node does not contain data and is unable to become a master node.
- *Ingest Node* – Before actual indexing occurs, it handles preliminary processing of documents.

Index An index is a collection of documents with similar characteristics. An index is represented by a unique name which is used to refer to the index while executing operations such as index searches, updates, and deletions. A cluster may contain as many indexes as desired. In Elasticsearch, the index is comparable to the database schema in RDBMS and can be thought of as a set of tables with logical organization or grouping. In the same way, Type can be considered equivalent to Table and Document as equivalent to Row in RDBMS.

Document A document is simply a collection of fields organized in a specific JSON format. Each document belongs to a type, and it is stored in an index, associated with a unique identifier (UID).

Type/Mapping Type or mapping refers to a collection of documents that share a common set of fields found in the same index. For example, an index containing social networking application data can contain specific user profile data, another document containing messaging data, and yet another for social media comment data. We should note that Elasticsearch recently indicated that it would no longer be possible to include multiple types in an index, with the concept of types being eliminated in a later version.

Shards and Replicas An index is able to store massive amounts of data exceeding the hardware capabilities of the node. For example, an index containing 1 billion documents and requiring 1 TB of disk space may not fit on the node disk or become too slow to serve search requests from a node. To be able to address this large amount of data, Elasticsearch gives users the ability to horizontally subdivide an index into multiple, smaller horizontal pieces known as shards. Because of this, each shard contains all of the document properties, but contains fewer JSON objects than a full index. Horizontal separations allow shards to act as independent nodes that can be stored in any cluster. Sharding is vital because:

- You can horizontally split or scale content volume.
- You can distribute and parallelize operations across shards on multiple nodes, increasing performance and throughput.

Elasticsearch enables the user to create one or more copies of index shards known as replica shards or simply "replicas" (see Fig. 4.22). Replication is vital because:

- It provides increased availability in the case of shard or node failure; therefore, a replica should never be allocated to the same node as the original shard that it was created from.
- It supports scalability of search volume and throughput because searches can be performed on all replicas simultaneously.

The default in Elasticsearch allows each index to allocate five primary shards and one replica, meaning if you have two nodes in a cluster, the index will have five primary shards with five subsequent replicas (one complete replica), totaling ten shards for each index.

Utilizing Elasticsearch requires data to be stored in the JSON document format that is then queried for retrieval. The query domain-specific language used by Elasticsearch is Query DSL and requires that you query in JSON format. The user also has the ability to nest queries to perform very complex searching in stored documents. Figure 4.23 below demonstrates how an Elasticsearch query functions [25].

Aggregation Elasticsearch is also able to support aggregations, enabling the user to a group or extract statistics from data. You can think about aggregations by similarly equating it to the SQL "group by" and aggregate functions. Within Elasticsearch, users can execute searches that generate hits and simultaneously return aggregated results, all in a single response.

Logstash Generally, Elasticsearch is used with Logstash, an open-source, server-side data processing conduit that ingests data from a legion of sources simultane-

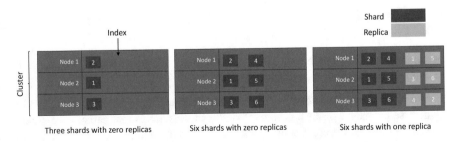

Fig. 4.22 A simple example of cluster, shards, and replica in Elasticsearch

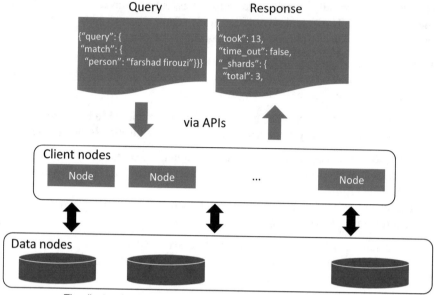

Fig. 4.23 How Elasticsearch query works

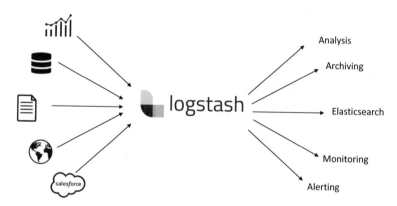

Fig. 4.24 How Logstash works

ously, converts it, and then transmits it to a destination, such as Elasticsearch. Main features of Logstash include (Fig. 4.24) [23]:

- *Data Ingestion* – Data can be scattered or siloed across multiple systems and in diverse formats; Logstash is able to handle various inputs that gather events/data from a mass of different sources such as logs, web applications, data stores, as well as AWS services in a streaming manner.

- *Data Filtering, Parsing, and Transformation* – As data moves from source to destination, Logstash is able to filter and parse each event and identify named fields to generate structure and then transform them into a common format to facilitate quicker analysis and added business value.
- *Data Transportation* – Uses various outputs to route data where needed, providing greater pliancy, and allows a deluge of downstream use cases.

Kibana Elasticsearch often includes Kibana, an open-source analytics and data visualization platform used to search, view, interact, and visually manipulate data via charts, maps, and tables that are housed in Elasticsearch indices. Although citizen data scientists may also perform basic data processing and analytics, Kibana is not considered as a holistic machine learning framework. Figure 4.25 illustrates how Logstash, Elasticsearch, and Kibana can create a pipeline to ingest, analyze, and visualize data.

4.6.1.6 Which Database Is Right for Your IoT Project?

Cassandra Greatest strengths include scalability without sacrificing reliability; Cassandra can be deployed across multiple servers without extensive extra work because it can replicate with minimal configuration. Cassandra is easy to set up and maintain regardless of data growth. It is also best used in industries where rapid database growth is needed. Cassandra does offer easier growth than MongoDB and, in general, is best for the following use cases:

- Sensor logs
- User preferences
- Geographic information
- Reporting systems
- Time-series data
- Write-heavy applications such as logging

MongoDB MongoDB works best for workloads containing highly unstructured data. If you are not able to anticipate the scale or type of data you will be using,

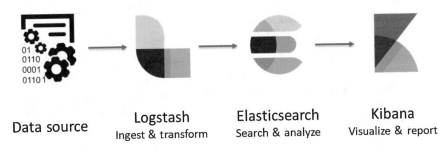

Data source	Logstash	Elasticsearch	Kibana
	Ingest & transform	Search & analyze	Visualize & report

Fig. 4.25 Logstash, Elasticsearch, and Kibana pipeline

MongoDB's flexible structure is preferable over Cassandra. If you have no clear schema defined, MongoDB is likely a solid choice. It can also be a good choice if you require scalability or caching for real-time analytics. MongoDB is not designed for transactional data such as accounting systems, etc.

Redis Although Cassandra was designed to handle huge amounts of big data, Redis is faster than Cassandra in retrieving and storing (key-value) data, especially when it comes to live streaming. Redis is best used when you have rapidly evolving data and you can estimate that the size of your final data can fit in memory. Redis is also great for analytics and real-time data communication.

Elasticsearch Primary use cases include:

- *Text Search* – Preferable when performing text searches where RDBMS cannot perform well due to poor configuration. Elasticsearch is customizable and extendable via plug-ins and allows you to create a high-quality search without extensive knowledge quickly.
- *Fuzzy Search* – This search allows for spelling errors such as finding "Levenshte" when searching for "Levenstein."
- *Instant Search* – Searching while the user is still typing via simple suggestions from existing tags, predicting based on search history, or creating a new search for each keystroke.
- *Content-Based Product/Document Recommendation* – Elasticsearch can function as a simple recommendation engine. In this case, Elasticsearch translates user content recommendation problems into a search query for implied interests of users. In addition, Elasticsearch includes document scoring by relevancy and document filtering by attribute.
- *Logging and Analysis* – Centrally stores logs from various sources for analysis. Kibana is useful in this case because it connects with Elasticsearch clusters and promptly creates visualizations.

4.6.1.7 CAP Theorem

CAP Theorem is a fundamental theorem which enables system architects to select the appropriate database platform for a data-driven solution (see Fig. 4.26). CAP Theorem states that a distributed database is only capable of meeting two of the following three conditions: consistency, availability, and partition tolerance (CAP).

Partition Tolerance Systems keep running, in spite of the number of messages delayed between nodes. A partition-tolerant system can sustain any level of network failure without reaching whole network failure. Data is replicated across multiple node combinations and networks to ensure the system stays up throughout intermittent outages.

High Consistency All nodes see data simultaneously. Executing a "read" operation returns the value of the newest "write" operation, causing all nodes to send back the

Fig. 4.26 CAP Theorem

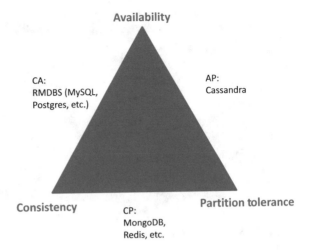

same data. A high-consistency system means a transaction is able to start and end with the system in a consistent state. Systems may move into an inconsistent state during the transaction, but the whole transaction is rolled back if an error is found during any part of the process.

High Availability Each request receives a success or failure response. Availability within a distributed system means that the system is operational 100% of the time – every client receives a response every time no matter the state of an individual system node.

4.6.2 Data Warehouse

A data warehouse (DWH) can also be referred to as an enterprise data warehouse (EDW), and it is the main method of data collection in use for the last 30 years. A DWH serves as a data integration point, responsible for organizing, arranging, and combining data from a variety of diverse relational databases. Note that in an enterprise there are several data silos. These silos are a grouping of organizational information that is separated from and inaccessible to other organizational departments. For example, it is common practice for each department (i.e., marketing, sales, accounting/finance) to have its own database. Therefore, a data warehouse is created to collapse the borders between these data silos and create a single point of truth (see Fig. 4.27). The DWH is also able to provide senior leadership with overarching insight regarding company performance via management reports, dashboards, or ad hoc analysis. A broad range of business data can be analyzed using a data warehouse, which becomes necessary when operational databases are unable to maintain performance while also meeting analytical requirements. Running a complicated database query means the database has to enter a short-term static state,

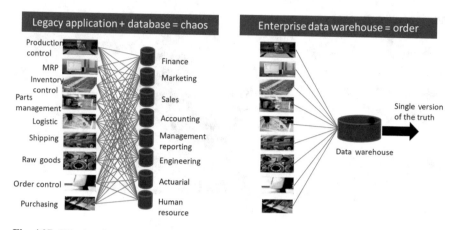

Fig. 4.27 Why is a data warehouse needed?

which is generally not supported by transactional databases. Therefore, a DWH is utilized to handle analytic needs, freeing traditional relational databases to focus on handling transactions. Additional characteristics of the data warehouse include the ability to analyze data from diverse sources, such as analyzing *customer relationship management* (CRM) data as well as Google Analytics data [26]. Note that DWH can only work with structured data. It is worth highlighting the main differences between data warehouses and relational databases.

- Data Optimization and Analytics – While relational databases and data warehouses are both relational data system, each of which is created with a different purpose in mind. Data warehouses are meant to house vast amounts of historical data and allow users to run quick and/or complicated queries involving all the data. On the other hand, relational databases are generally created to keep current daily transactions and empower quick access to clearly defined transactions for continuous business processes, referred to as *online transaction processing* (OLTP).
- Structure of Data – A second substantial difference between data warehouses and relational databases is the data normalization. While normalization is a usual practice in relational databases, data warehouses normally use demoralized data. The reason is that data normalization enables the database to occupy less disk space while minimizing transaction times. On the other hand, the fast response time of a query is not the main goal in data warehouses.
- Data Processing – Databases process an organization's daily transactions, so they usually do not include historical data. Current data is the most important component of a normalized, relational database. In contrast, data warehouses are utilized to meet analytical and business-reporting needs. Usually, data warehouses maintain historical data by combining transaction data copies from different sources over time.

4.6.3 Data Lake

The data lake has been created to address big data and the shortcomings of traditional databases and data warehouses (see Fig. 4.28). James Dixon, the founder and CTO of Pentaho, defined data lakes in the following way: "If you think of a traditional relational database as a store of bottled water – cleansed and packaged and structured for easy consumption – the data lake is a large body of water in a more natural state. The contents of the data lake stream in from a source to fill the lake, and various users of the lake can come to examine, dive in, or take samples."

In contrast to DWH which only holds cleaned transformed structured data, a data lake is a *data-centered architecture* that stores a vast amount of both structured and unstructured data in its raw format until it is needed. Indeed, a key advantage of data lakes is their capability to cost-effectively store data of unknown importance or value that would generally be removed because of the cost required to store the data securely. As a business' analytic abilities grow, the prospective data use cases are revealed, enabling historical data to be utilized in the training of machine learning models or answer future questions. Note that data lakes store the raw data, thereby, when they receive a business question, they need to query and transform data to be able to address the question. The data lake can be built using many technologies such as Amazon Simple Storage Service, Hadoop, NoSQL DBs, or various combinations to support a variety of formats such as images, logs, Excel, CSV, and sensor data. It has been discovered that as more data became available,

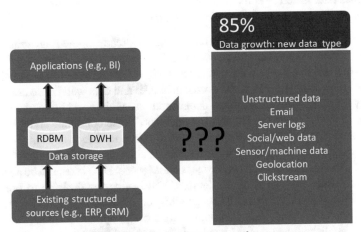

Fig. 4.28 The data lake has been designed to address big data

Table 4.6 The main differences between data lake and data warehouse

Data lake	Enterprise data warehouse (EDW)/RDBMS
Structured, semi-structured, unstructured data	Structured, processed data
Physical collection of un-curated raw data	Data of common meaning
System of insight: unknown data to make experimentation/data discovery	System of record: well-understood data to do operational reporting
Any type of data	A limited set of data types (i.e., relational)
All workloads – batch, interactive, streaming, machine learning	Optimized for interactive querying
Not suitable for transactions	Suitable for ACID transactions
Higher latency	Low latency (transactions)
Need skills to gain insights	Easily create reports (good support from BI tools)
Expensive for large data	Low-cost data storage
Schema-on-read (ELT)	Schema-on-write (ETL)

new applications could be created to serve business needs. Currently, data lakes support the following abilities:

- Cost-effective highly scalable storage of raw data
- Storage of diverse data types (structured, semi-structured, and unstructured data) in the same place
- Query and transform data when a business question arises
- Defines data structure at the time of use (i.e., schema on reading)
- Incorporates new methods of data processing

As previously discussed, corporations have already started integrating data lakes to address the requirements of their IoT projects. However, note that data warehouses and data lakes were created to serve different user groups and different purposes (see Table 4.6). In other words, data lake and data warehouse are complementary systems. Complementing a data warehouse with the addition of a data lake is an agile step forward for the most companies. These combined solutions create greater flexibility and improve speed when it comes to data processing, capturing streaming, semi-structured, or unstructured data. It also provides increased data warehouse bandwidth needed for *business intelligence*. Data lakes can be a powerful tool useful for individuals, data scientists, and businesses desiring to prepare or blend data or provide on-demand data profiling or to generate new insights from big data. On the other hand, data warehouses are a better option for those who require regularly published data that is already aggregated and processed. Finally note that data warehouses solely work with structural data based on ETL approach, whereas data lakes are low-cost storages to hold all types of data (structured and unstructured data) working based on ELT approach.

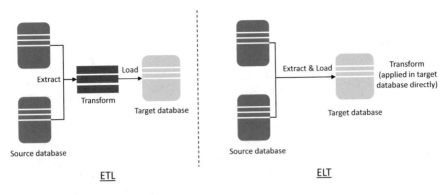

Fig. 4.29 A simple example of ETL and ELT

4.6.3.1 ETL (Extract, Transform, and Load) and ELT (Extract, Load, and Transform)

As discussed above, one of the key differences between data lakes and data warehouses is the way they process data. Let us examine the three stages – E, T, L (see Fig. 4.29):

- *Extraction* – Reading and collecting raw data from one or several data databases and routing it to a temporary repository
- *Transformation* – Converting, filtering, cleaning, processing, and aggregating the extracted data from the previous stage. Finally, structuring the output into a specific form to match the structure of the target database
- *Loading* – Writing the structured and converted data from the previous stage into the target database or data warehouse

ETL (Extract, Transform, and Load) occurs within a data warehouse while ELT (Extract, Load, and Transform) occurs within a data lake. Generally, ETL is a continuous process that occurs using a clearly defined workflow. First, ETL extracts data from similar or diverse data sources. Next, the data is cleaned, enhanced, transformed, and finally stored in a data warehouse or in a database. On the other hand, with the ELT approach, once data is extracted, loading begins immediately, and all data are transferred to a consolidated data storage area. However, transformations are performed in the target system. In other words, instead of transforming the data before it is written in the database, in ELT the process of transforming the data is completed by the target database. Therefore, ELT minimizes the processing on the source since the transforming is done in the target system.

4.6.3.2 Challenges of Data Lakes

While data lakes are a wonderful data management solution in our data-driven environment, they can come with challenges that are worthy of attention. Gartner has reminded, "the data lake will end up being a collection of disconnected data pools or information silos all in one place. Without descriptive metadata and a mechanism to maintain it, the data lake risks turning into a data swamp."

To address such risks, a governing layer must be added to the architecture to answer the questions below and ensure the four pillars of data governance are included. The data governance is a union of tooling and processes that usually increases the solution's *total cost of ownership* (TCO), but on the other hand, it increases the chance of *return on investment* (ROI) as well. The governance in data lakes must be created with the following questions and elements in mind [27]:

- *Data Catalog (What is the data and where is the data stored?)* – Data catalogs enable the user to compile and review metadata and index data so that they become searchable. Metadata is useful for auditing or for actively driving data transformation. Some data catalogs are also able to manually tag data to note that it includes *personally identifiable information* (PII) or utilize a machine learning algorithm to identify sensitive data.
- *Data Quality (Is the data accurate and useful for a specific purpose?)* – Achieving data quality can be done using *master data management* (MDM), a foundational process utilized to synchronize, categorize, centralize, organize, enrich, or localize master data based on a business' operational strategy. However, it is important to remember that MDM is originally established in relational data warehouses and databases, so it is compatible with structured data only. Best practices would indicate that MDM be applied selectively within data lakes.
- *Data Lineage (Where did the data come from and how has it been transformed?)* – Data lineage refers to the origin of data, what is done to it, and where it goes over time. Possessing the complete audit trail of data, including origin data, transformation information, and its analytic uses, is a necessity for meeting the regulatory requirements most organizations currently face. Data lineage information can also assist engineers in debugging or troubleshooting issues that arise while handling workloads.
- *Data Security (Is the data safe from unauthorized access?)* – When creating a data lake, it is important to address the following data security issues:
  - *Role-based access control* (at suitable granularity level)
  - *Network isolation* (e.g., security groups, firewall rules)
  - *End-to-end encryption* (e.g., SSL certificates)

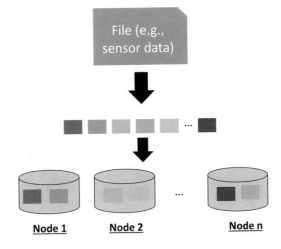

Fig. 4.30 Data that is split up into multiple blocks and stored across a set of connected nodes

4.6.3.3   Distributed File Systems

Data lake systems are mainly comprised of scalable data stores and distributed file systems that deploy data by partitioning larger files into smaller blocks or by partitioning tables that are appropriated across a pool of data servers (data nodes). In addition, data is usually replicated in several nodes to increase fault tolerance. In such a distributed system, a load balancer is utilized to increase scalability. This technique is demonstrated in Fig. 4.30.

Apache Hadoop's distributed file system and Amazon S3's Cloud storage are two well-known data lakes:

- *Hadoop Distributed File System (HDFS):* HDFS is a Java-based file system created to be deployed across several commodity servers easily and provides dependable and scalable data storage. The details of HDFS will be addressed in Chap. 6.
- *Amazon S3 Storage Service (Amazon S3):* Amazon S3 Storage Service stores data objects using an uncomplicated web service interface. It can store and fetch data in any amount. It is utilized as the main storage option for Cloud-native applications, as a bulk data depository or as a data lake for analytics. Amazon S3 is also useful for serverless computing, recovery, and backup.

4.6.3.4   Data Lake Tiers

Generally, it is highly recommended to engineer a data lake based on multiple tiers/layers (not less than two), with one being a quarantine zone. This is particularly important within tightly regulated industries that utilize highly sensitive data that must be manually verified by data stewards before transfer to another zone with wider user access. It is also highly suggested dividing the second tier into several

Table 4.7 Common file formats in data lake

File format	Properties	Use cases
JSON	Flexible and human readable	The consumer of data is an application which needs small data volume
CSV	Fixed schema, but human readable	The consumer is an analysis and the data volume is small
Orc	Columnar with schema	Efficient compression and improved performance for reading, writing, and processing data, e.g., suitable for read-heavy analytical loads
Parquet	Columnar with schema which supports complex nested data structures	Designed for projects in the Hadoop ecosystem. Suitable for read-heavy analytical loads
Avro	Row-based with schema	Suitable for write-heavy workloads

| Microservices (fine-grained) | Service-oriented architecture (coarse-grained) | Monolithic (one unit) |

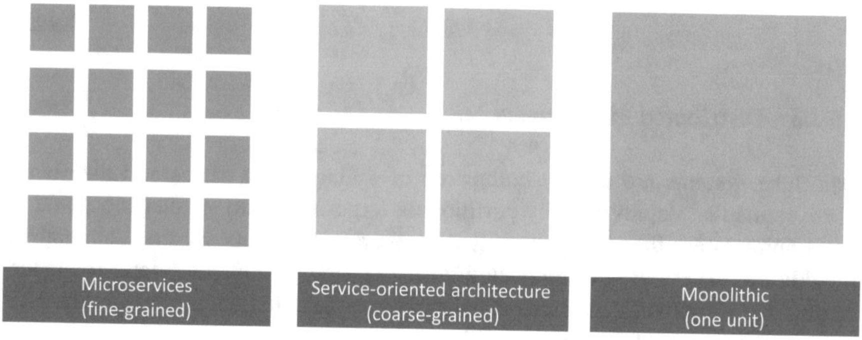

Fig. 4.31 Engineering server-side IoT applications via three different software architectures

zones based on the specific use case. For example, a tier could be separated into a warehouse staging zone, a machine learning training data zone, etc. Every tier/zone may include its own file formats and individual, granular access levels. The well-known file formats, their properties, and use cases are demonstrated in Table 4.7. Finally, note that transferring data between data lake tiers/zones involves a computing resource to complete the data transformation and process data between tiers/zones [27].

4.7 Application Layer

IoT applications located in the Cloud can be engineered based on three distinct architectures, namely, (i) monolithic architecture, (ii) service-oriented architecture (SOA), and (iii) microservice architecture. Figure 4.31 demonstrates the general differences between these three options [28].

When creating a server-side application, you can begin with a layered architecture. This architecture typically consists of the following layers/tiers [28]:

- *Presentation* – Handles all interface, manages and routes incoming requests to the business logic, and displays responses in a specific format (e.g., through web service APIs).
- *Application Business Logic* – Executes specific business rules associated with the request to prepare the response.
- *Database Access and Logic* – Implements the required logic to store and retrieve data from the database.
- *Application Integration* – Merging with other services (e.g., through REST API or messaging protocols such as MQTT).

Although applications usually have a modular nature, they are most often bundled and utilized as a monolith. There are several benefits of monolithic architecture for applications, including [28]:

- *Simple Development* – The implementation of monolithic applications is very straightforward compared to the other two options.
- *Ease of Testing* – The testing process is very straightforward. For example, the whole end-to-end testing can be realized by executing the application and testing the corresponding user interface (UI) via Selenium (a portable framework for testing web applications).
- *Ease of Deployment* – For deployment, you just need to copy the packaged application to a server.
- *Horizontal Scaling Simplicity* – The scalability is achieved by running several copies of the application behind a load balancer.

Many of the largest, highly successful applications in use today began as monoliths. While monolith architecture works well in early stage development, it also has drawbacks, namely:

- *Slow Change Speed* – Applications can be too large or complex to make changes quickly and accurately.
- *Inconvenient Updating and Deployment* – The entire application must be re-deployed for each update and uninterrupted deployment is difficult.
- *Manual Testing Needed* – Because the impact of changes in the application cannot be tracked and studied very well, extensive manual testing is usually required.
- *Scaling Compatibility* – Scaling is difficult if building-block modules have diverging resource needs.
- *Whole Application Reliability* – A bug in one module can bring the whole application down, and because all application instances (deployed in several servers) are identical, the bug can affect the whole application's availability.
- *New Technology Barriers* – Monolithic architecture cannot adapt to new technologies easily. The main reason is that any changes in the application frameworks or application programming language will impact the entire application.

Organizations including Netflix, eBay, and Amazon have resolved these issues by embracing a microservice architecture that breaks applications into small sets of interconnected services rather than bundling the application into a large, monolithic structure. Each microservice works as a miniature application with its own business logic and adapters. Note that at runtime, each instance of microservices can be deployed in a Virtual Machine or a Docker container. Some microservices offers an API used by other microservices or consumed by application users while other microservices might implement a web UI. The main benefits of microservice architecture include [28]:

- *Load Balancing* – Can better scale-out an application via advanced load balancing
- *Quicker Development* – Ability to develop faster, utilize siloed development teams, and decrease dependence
- *Versatile Deployment* – Can deploy each piece without impacting other systems
- *Simplicity* – Ease of development and maintenance
- *Content Caching* – Improves performance and reduces the load on the application
- *Troubleshooting* – Easier to recognize service failures

In the following use cases, it would be better to use monolithic architecture [28]:

- *Small Teams* – When you do not have enough developers experienced in and knowledgeable of different programming languages.
- *Building Minimum Viable Product (MVP) Versions* – When you just need to implement enough features very fast to satisfy your early requirements.
- *Lack of DevOps* – No multimillion-dollar investments to put toward software development (Dev) and IT operations (Ops) or extra time to spend on complex architecture.
- *Deep Development Experience* – You have extensive experience developing strong frameworks like Node.js, Go, etc.
- *No Bottlenecks* –There are no apparent issues with future application performance or scalability.

On the other hand, microservice architecture is more suitable in the following scenarios:

- *Flexible Deadlines* – Microservices require extensive research and planning.
- *Diverse Team Experience* – When your team members are experienced in and knowledgeable of different programming languages.
- *Have Scalability or Reliability Concerns* – When your server-side applications need to handle the load of millions of requests.
- *Co-Located Teams* – Development is spread across diverse departments located in different time zones or geographical locations.

- *Existing Monolithic App* – Your current monolithic application is experiencing issues that could be resolved by breaking the application into multiple microservices.

It is worth pointing out that software community skeptics consider microservices a simple brand remodeling of service-oriented architecture (SOA). Initial records regarding the use of distributed services as a means of software architecture can be dated back to the 1980s. However, SOA was not officially named until the mid-1990s when the Gartner Group labeled and adopted the new trend, popularizing it globally.

SOA was not able to address many issues created by monolithic architecture, and in many aspects, SOA is a monolith. Although it is comprised of multiple services, SOA is still primarily course-grained with high levels of interdependency. In SOA architecture, there are several software components that provide services to other components through a pre-defined communications protocol. In SOA, each module can function both as a provider of services and as a consumer of services. *Enterprise service bus* (ESB) is a form of integration architecture used within SOA that enables communication through a shared communication bus made up of various point-to-point consumer and provider connections. Additionally, all services in the SOA are able to share data storage. Figure 4.32 illustrates an example of SOA and ESB architecture [29].

Microservice architecture and SOA have several similarities as well as differences that make them unique. Both architectures contain services with a specific responsibility; therefore, service development can take place in several technology stacks, offering technological diversity to development teams. In both architectures, service development can be divided among multiple teams; however, all teams should understand the ESB integration in SOA which is not the case in microservice architecture [29].

Fig. 4.32 Service-oriented architecture and enterprise service bus (ESB)

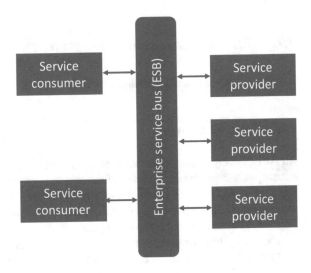

ESB can become a single point of failure in SOA, impacting the whole application. Because all services utilize the ESB to communicate, if one service slows, the ESB could become obstructed with requests for that particular service. On the contrary, microservices tolerate faults better. For example, a memory leak in one service only affects that particular service while alternate microservices continue managing requests.

Heterogeneous interoperability is also a large variation between the two architecture types. SOA supports the integration of many heterogeneous protocols using a messaging middleware. The microservice architecture seeks to simplify the architecture by reducing integration choices.

Another major difference is that unlike SOAs, microservices can both work and be deployed independently, making it easier to push out updated versions or scale a particular service. When it comes to deployment, SOA recreates and relaunches the whole application while microservice architecture allows individual services to be created and utilized separately. Therefore, microservice architecture empowers fast, continuous, and automatic deployments.

Data storage is shared among services in SOA; however, services can store data independently in microservices. Shared data storage has both benefits and shortcomings. For instance, shared data storage enables data to be recycled among all services but requires dependency and tight service connections.

Service granularity is another key difference between microservices and SOA. The diverse services in microservice architecture each does only one thing really well, while SOA service components can handle anything from small services to huge enterprise services.

Finally, SOA's approach to architecture focuses on "sharing as much as possible" while microservice architecture "shares as little as possible." SOA is most concerned with recycling business functionality while microservice architecture is based on the foundation of bounded context/domain.

4.7.1 Microservice Architecture Pattern

In this section we discuss the engineering details of microservice architecture.

4.7.1.1 API Gateway

An API gateway is an application/service that serves as the system entry point. API gateways abstracts the architecture of the internal system and then represents it via a set of APIs. API gateways can also be responsible for authentication, load balancing, monitoring, static response handling, caching, and request shaping or management. Figure 4.33 illustrates the basic concept of an API gateway [30].

The API gateway manages composition and requests routing and translation of protocols. Any client request has to pass through the API gateway where it is routed

Fig. 4.33 A typical
architecture of API gateway

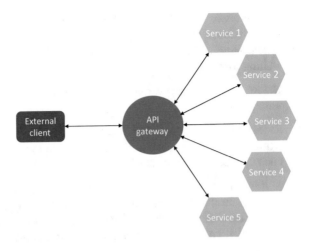

to the correct microservice. The API gateway generally manages requests by calling on (invoke) several microservices and combining results. Indeed, the API gateway provides a set of coarse-grained APIs for end users; however, behind the scenes, it actually manages each request by utilizing a variety of services (an average of six or seven fine-grained backend services). API gateway is also capable of translating between different web protocols like HTTP or WebSocket protocols.

One of the main drawbacks of API gateway is that it should be highly available. Therefore, it requires continuous development, deployment, and management, raising the risk that it could become a development bottleneck. The API gateway has to be updated in order to expose the endpoint for every microservice. The process for updating the gateway should be as light as possible to ensure developers are not left waiting. In light of both the positive and negative aspects of the API gateway, most state-of-the-art applications make use of the API gateway.

4.7.1.2   Service Invocation

The components of a monolithic application call on (invoke) each other using function calls or a language-level method while applications using a microservice architecture function as a distributed system using many machines and inter-process communication processes. Such mechanisms are grouped based on two specific features. The first feature is related to client interaction styles [30]:

- *One-to-One* – Requests from the client are processed by a single service instance.
- *One-to-Many* – Requests are processed by several service instances.

The second feature is focused on whether the communication is synchronous or asynchronous in nature:

- *Asynchronous* – While waiting for a response, the client does not block and any response is not immediately sent.
- *Synchronous* – Clients expect a timely service response and may block while waiting for the response.

Below you find the variety of synchronous and asynchronous one-on-one communications [30]:

- *Request and Response* – The client request is made and then the client waits for a timely service response. There are multiple protocol options, but two favored ones include REST and Thrift.

  - *Representational State Transfer (REST)* – An *inter-process communication* (IPC) mechanism that usually relies on HTTP. Resource is the main concept of REST. A resource is a concept that represents a business object (e.g., a product, customer) in the application or a combination of business objects. REST manipulates resources using verbs referred via URL. For example, GET requests to send back a resource representation, likely in the form of a JSON object or XML document, while a POST request actually creates a new resource that can be updated via a PUT request.
  - *Apache Thrift* – A framework used to write cross-language *remote procedure call* (RPC) servers and clients. Thrift comes with a C-style interface description language (IDL) defining APIs. Next, the compiler creates code for diverse programming languages such as Java, C++, PHP, Ruby, Python, and Node.js. Thrift is able to support different message formats such as binary, compact binary, and JSON. Note that binary is a more efficient choice because it takes up less space and is faster to decode than JSON.

- *Notification (a one-way request)* – Request is sent to service by the client without a reply being sent or expected.
- *Request/Async Response* – A request is sent to the server by the client and the server returns an asynchronous response; client waits without blocking and assumes the response will come later.

Noted below are the different kinds of one-to-many communications:

- *Publish/Subscribe* – Client publishes a message and it can be consumed by interested services.
- *Publish/Async Response* – A request message is published by the client and then the client waits for a designated timeframe for responses from services that are interested.

There are several messaging platforms that are open source and available including Apache Kafka, ActiveMQ, as well as RabbitMQ. Each system is designed around the idea of messages and channels. The messages are created by a header (i.e., metadata like security or identity information) and a body and then sent and received via channels. There are a variety of human-readable, text-based (JSON or

XML), or binary (Avro) message formats to choose from. Any number of consumers can receive channel messages and any number of producers can send messages.

4.7.1.3   Service Discovery

In Cloud-based applications consisting of several microservices, discovering the location of each microservice is a challenging task. Although infrastructure services (e.g., message brokers) often have a fixed static location/address identified through *operating system* (OS) environment variables, finding a service location can be difficult because application services use dynamically assigned locations/addresses. In addition, service instances change due to upgrades and auto-scaling. Therefore, a service discovery component is needed to locate each service. Service discovery can be driven either by the client or by the server [30]:

- *Client-Side Discovery* – The client governs network locations for service instances by querying a service registry containing available service instances; when a service instance starts, its network location is documented with the service registry and then removed when the instance ends. The instance registry is updated periodically with a heartbeat technique. Next, a client selects an available service instance and makes a request via a load-balancing algorithm.
- *Server-Side Discovery* – Client uses a load balancer to make a service request. In this approach, the balancer looks up the service registry and then sends a request to a free service instance. In the same manner as client-side discovery, in this approach, instances should also be registered in the service registry.

4.7.1.4   Service Registry

Service registries are a kind of databases that are able to keep track of service instance and network location information. Since service registry is an integral component of service discovery, it must be kept up to date and must be highly available. An example of a good service registry is Netflix Eureka. It includes a REST API used to query and register service instances. Another registry framework example would be Apache Zookeeper, originally a Hadoop sub-project that is now a separate top-tier project [30].

4.7.1.5   Deployment Strategy

Creating a monolithic application requires using one or many identical copies of one, large application. Generally, you need to prepare N servers (can be virtual or physical) and run M application instances on each of them. While deploying a monolithic application is not always an easy task, it is a less complicated process than that of a microservice application deployment, which is based on hundreds of services written in different frameworks and languages. Individual microservices are mini applications with their own monitoring, resource, and scaling needs. In addition, each service must have the appropriate memory, CPU, and I/O resources. Below you will find different approaches used to upload and run services [30].

- *Multiple Service Instances Per Host* – A more traditional application deployment approach, requiring the developer to provide one or more hosts (either virtual or physical) and then run several service instances on each host. Benefits of this approach include quick deployment and efficient resource usage because the server and OS are shared among multiple service instances. This approach provides little to no service instance isolation unless a service instance is designated as a separate process, which is a large drawback for this approach.
- *Single Service Instance Per Host* – Each service instance is run separately on its individual host. Two variations of this approach include:

  – *Service Instance Per Virtual Machine* – Each service is bundled as a *virtual machine* (VM) image and then initiated via VM. The details of VMs will be discussed in Sect. 4.10.

    - Benefits – Because the service instance is isolated, it uses only an isolated set amount of memory and CPU and does not rob resources from other services. Each microservice can also be monitored, maintained, managed, and scaled separately.
    - Drawbacks – Deployment new service versions can be slow, resources are used less efficiently, and each service instance requires an entire VM increasing the overhead of the system. Note that each VM has its own operating system (guest operating system), and thus its overhead is high (see Fig. 4.36).

  – *Service Instance Per Container* – Individual services run in separate containers. The details of the containers will be addressed in Sect. 4.10.

    - Benefits – Service instances in containers are isolated from each other. Similar to VM the underlying technology is encapsulated; container resource usage can be monitored. Each microservice can also be monitored, maintained, managed, and scaled separately. Since containers do not need to have an operating system, they can be built and started quicker than VM (see Fig. 4.36).
    - Drawbacks – Infrastructure lacks the maturity of VM infrastructure, and the level of container security is lower because host OS is shared among containers.

- *Serverless Deployment* – The terms of serverless and *Functions as a Service* (FaaS) are usually used interchangeably, although this distinction is still being defined by the tech community. The idea of FaaS is dividing the microservices into fine-grained functions and then deploying those functions to a third-party company which charges the application owner only based on the amount of time each function runs (i.e., serverless deployment). In other words, in serverless deployment, all you need to do is deploying your code. The vendor is responsible for handling all the operational requirements (e.g., scaling). On the other hand, in virtual machine or container-based deployment, you are responsible for monitoring and managing your system yourself. The details of serverless and FaaS will be covered in Sect. 4.10. As an example, AWS Lambda, one of the most well-known serverless deployment technologies, is compatible with Python, Java, and Node.js services. To deploy a service, one needs just to upload the corresponding microservices packaged as ZIP files. AWS Lambda handles microservice requests by automatically running a sufficient number of instances. However, this technology is not suitable for deploying long-running services. The reason is that all requests in AWS must finish in 300 seconds and all services must be stateless.

4.8   Data Visualization and Reporting Layer

The data visualization layer monitors project success and is the means by which users perceive data value.

4.8.1   Data Visualization Frameworks

- *Kibana*: As stated earlier, Kibana is a data visualization framework based on Elasticsearch. It enables users to understand data through a dashboard depicting a group of visualizations. Users can resize or rearrange data visualizations at will and save it to the dashboard so that it can be reloaded or shared. Kibana's main focus is enabling users to explore and analyze Elasticsearch's log data. It does not support other data sources, so if you are not utilizing Elasticsearch, Kibana is not a viable data visualization option.
- *Grafana*: Grafana is an all-purpose, open-source graph composer and dashboard that functions as a web application. Grafana provides built-in means for obtaining data from 30+ sources (i.e., Elasticsearch). It is best suited to visualize continuous time-series and streaming data such as metric reporting or sensor data.

4.8.2   Business Intelligence Frameworks

- *Tableau*: Tableau is a wonderful tool for data visualization because it is widely available and enables the user to manipulate big data. It has two derivatives including Tableau Server and the Cloud-based Tableau Online, specifically designed to handle big data from organizations. Utilizing Tableau does not require a coding skill. It comes with a user-friendly dashboard with drag and drops features. Almost all kinds of data can be connected and effectively analyzed – from a small amount of data like a spreadsheet to big data like Hadoop. Note that Tableau is considered as a business intelligence framework rather than a holistic machine learning platform. For example, users cannot develop cutting-edge predictive maintenance solutions in Tableau.

- *Microsoft Power BI*: Microsoft Power BI is a well-known business intelligence framework on the market, meant to analyze data and share and visualize subsequent data.
- *QlikView*: QlikView is a well-known option in the data analytics domain, and the QlikView tool is one of Tableau's closest competitors. It offers effective data visualization, business intelligence, and enterprise reporting options.

4.8.3 Advanced Data Analytical and Machine Learning Frameworks

- *Scikit-learn*: Scikit-learn, an open-source library of machine learning algorithms, is widely used and well documented. It seeks to provide commonly used algorithms for Python users. It is rapidly becoming the go-to platform for machine learning and is continuously evolving to provide greater efficiency, speed, and big data capabilities. Scikit-learn is commonly utilized with smaller data, but does provide a useful group of algorithms for clustering, out-of-core classification, decomposition, and regression.
- *TensorFlow*: TensorFlow, an open-source library of software useful for executing numerical calculations with data flow graphs, was recently created by Google. Many would argue that it is the best framework for deep learning and it has been utilized by top-tier organizations including Twitter, IBM, and Airbus because it is engineered based on a modular architecture resulting in significant flexibility. The most widely known TensorFlow use case is Google Translate which combines multiple functions such as forecasting; natural language processing; image, speech, and handwriting recognition; tagging; and text classification.
- *Caffe*: Caffe is another platform for deep learning that supports command-line interfaces as well as other interfaces like MATLAB, C++, Python, and C. It is widely recognized for its ability to model convolution neural networks (CNN) as well its speed and transposability. The greatest benefit of Caffe is that it has a large repository of networks that have been pre-trained and are ready for immediate use.
- *RapidMiner*: RapidMiner is predominantly concerned with speed in achieving data insights in complicated data science. The visualization interface includes ready-to-use workflows, machine learning elements, and data connectivity. RapidMiner can be integrated with several technologies such as Python and R. It is also capable of automating many tasks such as the selection of models, what-if gaming, data preparation, and predictive modeling.

- *Splunk*: At its beginning, Splunk was a log analysis platform. It has gained a solid foundation of loyal users and organizations that appreciate the ability to share graphs and dashboards as well as the way it enables data visualization and manipulation. It is well known for its analytic abilities as well as a web-based, user-friendly log review. These capabilities can be also be used to review big data stored in Hadoop.
- *H2O*: H2O is a distributed, in-memory, open-source platform for machine learning that supports linear scalability. It comes with several out-of-box machine learning and statistical algorithms (i.e., general linear models, deep learning, and gradient boosted machines, etc.).
- *Knime*: Knime is an enterprise-level, open-source analytics platform designed for use by data scientists. The visualization interface includes many nodes for a variety of uses from data extraction to data presentation, with a clear focus on statistical models.
- *MLlib*: MLlib is a component of Apache Spark and provides a scalable library for machine learning with several methods for regression, collaborative filtering, optimization primitives, classification, clustering, and dimensionality reduction. The details of the MLlib will be addressed in Chap. 6.
- *IBM Watson Studio*: IBM is one of the most widely recognized brands all around the globe. IBM Watson Studio is an appealing platform useful for creating and deploying deep learning and machine learning models. It also enables you to explore, refine, and transform data.

4.8.4 Load Balancing

Adding additional servers are necessary to enable the computing to scale efficiently and meet high request volumes. Load balancing is a means of effectively spreading incoming network traffic to a group of backend servers, sometimes referred to as a server pool or server farm (see Fig. 4.34). You can think of the load balancer much like the "traffic cop," of the server pool, sending requests to endpoint servers in a way that it keeps the workload evenly distributed and avoids lowering performance level. If one server goes down, the load balancer is able to send requests to the other servers, and if a new server is added to the pool, the load balancer automatically starts sending some incoming requests to the newly added server. Generally, a load balancer handles the following functions:

- *Request Distribution* – Spreads client requests or network load across multiple servers in an efficient manner
- *Performance Improvement* – Guarantees network dependability and high availability by only sending requests to online servers
- *Flexibility Increase* – Allows servers to be added or removed based on client demand

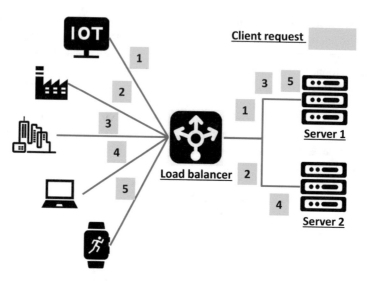

Fig. 4.34 A simple example of load balancing

There are many algorithms and techniques available to effectively balance the load of incoming requests across a server pool. The method chosen highly depends on the service/application, network status, as well as the condition of servers [31].

- *Round Robin* – Simply distributes requests to servers in a rotating sequential manner.
- *Weighted Round Robin* – In this version of Round Robin, each server has a static numerical weight/rank and servers with a higher rating receive a greater number of requests.
- *Chained Failover (Fixed Weighted)* – In this technique, a logical chain of servers is created and all requests go to the first server in the chain; when the first server cannot serve another request, the next server in the chain receives all requests, etc.
- *Least Connection* – Unfortunately, Round Robin methods do not consider current server loads when sending requests. To tackle this problem, in the Least Connection method, the most current request is sent to the server currently handling the smallest number of active sessions at the moment.
- *Weighted Least Connection* – Similar to the Weighted Round Robin method, each server has a numerical value used by the load balancer to send requests to servers appropriately. In this method, if two servers serve the same number of connections, the server with the higher weighting will receive the newest request.
- *Agent-Based Adaptive Load Balancing* – Every server in the pool utilizes an agent to report its current load level to the balancer. This information helps determine which server should receive which request. This method is often used in partnership with Weighted Least Connection or Weighted Round Robin.

- *Source IP Hash* – A unique hash key is created using an algorithm that combines destination and source IP addresses for the server and client. The key is then used to send the client to a designated server. Because the key can be recreated if a session is broken, the same client request will be sent to the same server it was allocated to before the session was broken. This can be particularly important if a client needs to connect to the same server after a disconnection.
- *Layer 7 Content Switching (URL Rewriting)* – This technique utilizes application layer information to route/send incoming requests in real time to the most appropriate server for processing.
- *Software Defined Networking (SDN) Adaptive* – In this approach, information from layers 4 and 7 and information from layers 2 and 3 are combined to determine how incoming requests should be apportioned and handled. This holistic approach enables the load balancer to consider network congestion, network infrastructure health, application status, and server status when making decisions.

4.9   Orchestration Layer

Administering Cloud resources across the complete life cycle requires a variety of processes and services to describe, deploy, select, monitor, and manage resources. Figure 4.35 illustrates what the term Cloud Resource Orchestration refers to. Generally, the orchestration of Cloud resources takes the shape of a multilayered model that includes [32]:

- *Description Layer* – This layer includes the models and languages used to represent distribution, configuration, tracking, and control of the Cloud resources.
- *Resource Provisioning Layer* – This layer is responsible for choosing, distributing, and administering software such as load balancers and database management servers and hardware (i.e., network, CPU, and storage). This layer considers the *service level agreement* (SLA) when providing services to users. An SLA is a required agreement between Cloud service providers and users that guarantees *quality of service* (QoS) in the areas of reliability, performance, availability, etc.
- *User Layer* – Consumers such as application developers or system administrators may utilize dashboards, *software development kit* (SDK), *integrated development environments* (IDEs), or *command-line interfaces* (CLIs) to interact with the services of other layers.

Fig. 4.35 Cloud Resource Orchestration

4.10 Virtualization

Virtualization is the foundational technology that brought the vision of Cloud computing to life. Virtualization includes the creation of a virtual (as opposed to an actual) form of something such as storage devices, network, and computing resources. Virtualization began in the 1960s, as a means of virtually portioning the mainframe's system resources across various applications. Since that time, the concept of virtualization has expanded. A few key benefits of virtualization include:

- *Cost-Efficiency* – Due to fear that an application may crash and take down another application running on the same machine, companies will often only run one application on a server. Current estimates suggest that most servers utilize only 10–15% of their total capacity. However, virtualization allows for a single physical server that would generally have only one purpose to function as a computing pool with many virtual servers, able to adapt to evolving workloads.
- *Energy Savings* – Powering servers that are utilizing only a small portion of their capacity can cost businesses significantly. Virtualization helps to reduce the number of actual servers required and lowers the amount of energy needed to cool and power them.
- *Efficient IT Operations* – Virtualization empowers IT professionals to work with more agility and efficiency, which lessens the time and effort needed to maintain resources. For example, IT professionals can easily complete archiving, recovery, or backups with greater speed and ease. Prior to the widespread use of virtualization, it could take weeks for a technical team to install and maintain software or devices on physical servers.

4.10.1 Main Categories of Virtualization

There are three main categories of virtualization, namely, network virtualization, storage virtualization, and computing server virtualization.

- *Network Virtualization:* Network virtualization unites all necessary physical networking components into one, software-based resource while separating the bandwidth into many, and independent logical channels that can be allocated to devices and servers in real time.
- *Storage Virtualization*: Storage virtualization is the collection of physical storage from many storage devices into what looks like one logical storage device manageable from a centralized computer. In short, storage virtualization requires adding a software layer that abstracts the details of the underlying physical storage devices from users while allowing all storage devices to be managed as a logical pool.
- *Server Virtualization (Hardware Virtualization)*: Currently, server virtualization is likely the most widely utilized kind of virtualization. Server virtualization

includes masking server resources so that they are not visible to users. This can be accomplished by a server administrator using a software application able to separate a physical server into many virtual environments. The three main approaches to server virtualization include system virtualization (virtual machine method), operating system (OS)-level virtualization (Containers), and function-level virtualization.

- *System Virtualization (Virtual Machine-Based Virtualization)*: System virtualization requires partitioning a server into multiple virtual servers or machines with separate operating systems (OS). Although sharing a single computer's resources, various virtual machines can run different operating systems and many applications. In short, system virtualization is the building of a virtual computer that functions as a physical computer with an OS. A *virtual machine* (VM) can also be referred to as a guest machine and functions like a software simulation of the hardware platform that generates a virtual environment for the guest OS. The software run on virtual machines is isolated from the elemental hardware. A *hypervisor*, also referred to as a *Virtual Machine Monitor* (VMM), is software that executes on host hardware and the host operating system to monitor and manage the running of virtual operating systems on the guest machines. For example, a computer operating with Microsoft Windows may host a computer that looks as though it is operating with the Ubuntu Linux.

- *Operating System Virtualization (Containerization or Container-Based Virtualization)*: In general, a container is a standard software unit that bundles code and its dependencies to distribute and run applications accessing a shared OS kernel without requiring VMs. Operating system virtualization is container-based virtualization that requires an OS structure includes a kernel which permits many separated user-space instances to exist. These instances are known as *Virtual Environments* (VEs), containers (Docker, Solaris), virtual private servers (OpenVZ), zones (Solaris), partitions, jails (chroot jail, FreeBSD), or virtual kernels (DragonFly BSD). They may appear to be actual computers from the perspective of the programs being executed in specific instances. Docker is the most well-known container technology. It requires that the application be packaged as an image. The image typically made up of the application and libraries needed to provide the designated service. After packaging the application, one or more containers are launched. In general, several containers run on each host. A cluster manager (e.g., Marathon or Kubernetes) can be useful in managing containers because the host is treated as resource pool, deciding where containers belong based on the resources needed and available on the host. It is very important to understand the main differences between container and VM (see Fig. 4.36). Virtual machines are software programs that mimic computing systems or physical hardware by running a hypervisor software able to replicate hardware resource functionality using the software. VMs encapsulate the OS as well as applications while containers are responsible for encapsulating applications and their resources

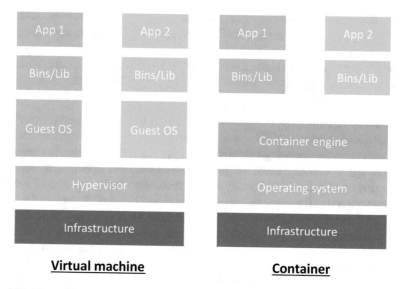

Virtual machine   **Container**

Fig. 4.36 The architecture of virtual machine and container

for deployment while sharing the host OS. Deploying IoT in Cloud requires a large network of servers, protocols, sensors, and applications with many endpoints including mobile, web, and firmware that need extensive integration among applications, data, and devices. These need to expand development timelines and increase the effort needed, even in an agile environment. Both microservices and containerization support efficient development by dividing IoT functions into more manageable, independent modules able to work in isolation without negatively impacting IoT performance. In an advanced IoT ecosystem, individual microservices are built with diverse programming languages such as Java, C, C++, Ruby, or Python. Containers increase service speed and ease of use by providing isolation within a single host and managing framework or library dependencies. Containers can also be turned on and off without adversely affecting subsequent containers located on the same host.

– *Function-Level Virtualization*: The term "serverless" does not mean that servers are not required, but that developers do not have to spend much time thinking about them anymore because the developer's focus can be shifted from managing servers to code. Serverless is also known as *Functions as a Service* (FaaS). As previously discussed, microservices are used to break monolithic applications down into smaller services that are more easily developed, managed, and scaled separately. FaaS takes this idea further by breaking microservices into even smaller pieces/functions (see Fig. 4.37). In FaaS and serverless computing, we run the code (function) only on-demand on a per-request basis (i.e., in response to events based on event-driven programming model). The current leader in this area is Amazon Lambda, while the products such as Azure Functions and PubNub Blocks are also

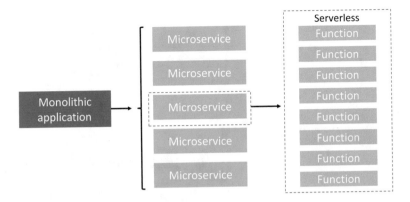

Fig. 4.37 The main concept of serverless and FaaS

working on this idea as well. Amazon Lambda allows you to run code for almost any backend service or application without any administration. You upload the code and Lambda handles everything needed to run or scale the code with high availability. The code can be automatically triggered by another AWS service or be called from any mobile or web application. It is expected that in a few short years, many more Cloud platforms will evolve to facilitate serverless architecture in some form. However, despite the rapid evolution of serverless architecture and the capability to program using highly popular languages, current serverless computing is limited and is not governed by a clear blueprint, technical standards, or options.

4.10.2 Behind the Scene of FaaS: OpenWhisk

In this section, we describe the detailed implementation of FaaS technology using one of the most well-known open-source FaaS platforms. Apache OpenWhisk is an open-source serverless platform driven mostly by IBM and Adobe. Figure 4.38 illustrates the high-level architecture as well as the programming model of Open-Whisk. Apache OpenWhisk is designed to function as an asynchronous and loosely coupled execution environment which can execute functions (event handler) in response to events at any scale. Events can come from any source such as datastores, message queues, mobile and web applications, sensors, chatbots, scheduled tasks (via Alarms), etc. In response to events, a stateless function (called action, event handler, or code snippet) is executed in serverless mode. Actions, which encapsulate application logic, can be invoked by the user-created APIs, *command-line interface* (CLI) of OpenWhisk, or by Triggers. Triggers are created to automate actions. In other words, triggers are endpoints and should be explicitly called by events. Rules create a binding between serverless functions (actions) and triggers. Put differently,

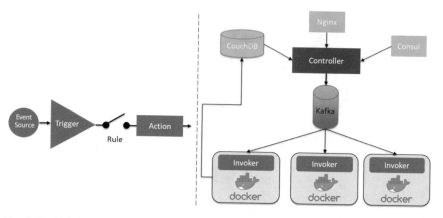

Fig. 4.38 High-level architecture as well as the programming model of OpenWhisk

a rule is a loosely coupled association between a trigger and an action. Behind the scene, OpenWhisk is implemented based on the following main building blocks:

- Nginx: Nginx is an open-source HTTP and reverse proxy server.
- Controller: Controller is an implementation of the actual REST API and Nginx forward all the HTTP requests to the controller which functions as the gatekeeper of the OpenWhisk. As its name suggests, the controller performs the authentication and authorization and finally decides the path that the request will eventually take. It should be noted that the controller is written in Scala programming language.
- Redis: Nginx can be optionally connected to Redis for caching.
- CouchDB: CouchDB is a document-oriented NoSQL database which stores the state of the system, credentials (for authentication and authorization), metadata, namespaces, and the definitions of actions, triggers, and rules.
- Consul: Consul is a tool that acts as a service registry and discovery in the system.
- Kafka: Kafka is a fault-tolerant distributed message broker which buffers and delivers the messages sent by the controller to invokers.
- Invoker: Invoker, which is written in Scala, is the component that is responsible for receiving invocation request from the controller via Kafka, spinning up a container, and executing the corresponding action. In OpenWhisk, containers are managed using Docker technology. Docker is the most well-known framework designed to make it easier to create, deploy, and run containers. When the execution of an action function is done, the invoker stores the corresponding results in CouchDB. It should be noted that inside each container, OpenWhisk puts a light HTTP server with two endpoints, namely, /init and /run. The /init endpoint is responsible for taking the action's code, preparing the code for execution, and the initialization of the container. For example, it might be necessary to compile the code to be ready for the execution. The /run endpoint triggers the execution of the action code. Note that the initialization of a container

comes at a runtime cost. To increase the speedup, in some cases, it might be possible to reuse containers. For example, if the same action is triggered for the second time, and the first action has already completed, OpenWhisk might be able to use the previous container without creating a new container. If a container is already available, we call it a *warm* container. On the other hand, when there is not an already initialized container available, the invoker needs to spin up a new one. This case is typically referred to as *cold start* and the newly created container is called *cold* container.

4.11   Scaling

Today's IoT Clouds are required to serve thousands or millions of simultaneous requests from devices or users and return accurate responses in a quick and dependable manner. Challenges around scalability are frequently faced by IT managers because it can be arduous to forecast the growth of applications, storage needs, and bandwidth consumption. Infrastructure scalability manages evolving application needs by adding or subtracting resources to accommodate demands as required. In many instances this can be achieved through *vertical scaling (scaling-up)* or through *horizontal scaling (scaling-out)*.

4.11.1   Vertical Scaling (Scale-Up)

Vertical scaling is used to obtain a certain level of performance through increasing the capacity of an existing machine (server) by addition of more power (CPU, RAM, storage) to it. For example, web servers or databases require extra resources such as storage, memory, computing power, or network in order to maintain an optimal performance level. When the Cloud is utilized to achieve this, applications are often migrated to a more powerful host or instance as the previous server is retired. Vertical scaling can also be accomplished in software through the addition of connections, threads, or by increasing cache size (applicable to database applications) as well. Note that vertical scaling is not a good option to scale dynamically, since it usually involves downtime for the system. MySQL is a good example of scale-up.

4.11.2   Horizontal Scaling (Scale-Out or Clustering)

Clustering, also known as horizontal scaling or scale-out, references the ability to connect two or more machines so that they appear to function as a single, virtual machine. Horizontal scaling is most often associated with distributed architectures and includes the addition of machines (nodes) to a system. This method of scaling requires that new machines be added to the system to share in handling the workload. When it comes to databases, horizontal scaling often means partitioning data (i.e., a single node only holds part of the data). Horizontal scaling is generally easier to achieve through the addition of new machines to the existing server pool. Horizontal scaling examples include MongoDB, Cassandra, and Google Cloud Spanner. Clustering is usually accomplished by using a load balancer at the front of the cluster to receive incoming requests and send them to nodes within the cluster. The primary advantages of clustering include:

- *Improved Performance* – Adding nodes as required and balancing the load across them enables quick and accurate responses to client requests.
- *Improved Reliability* – Clusters eliminate single points of failure because if a node fails the load balancer sends requests to other notes until that particular node is functioning properly again.
- *Reduced Cost* – Clustering is a cost-efficient method for achieving optimal performance and scale because it requires commodity hardware only.
- *Easy Maintenance* – Cluster nodes can be taken down for maintenance purposes or upgraded as required during business operating hours without interrupting performance because additional nodes within the cluster are able to handle incoming requests.

4.12 A Paradigm Shift from Cloud to Fog Computing

As the number of devices, data, and interactions continues to rise, Cloud architecture alone is unable to handle the deluge of information. The Cloud is able to provide computing access, storage, and easy, cost-effective connectivity; however, centralized resources can also generate delays and create performance issues if data or devices are distant from a public Cloud or data center. In this context, fog/edge computing has been created to tackle the aforementioned problems. Cisco's definitions for edge and fog computing are as below [33]:

- *Edge Computing (Known as "Edge")* – Moves the processing closer to the data source and does not require that data be sent to a remote Cloud or centralized system for processing. Because it removes distance and reduces time usually required to transfer data to a centralized system, the speed and performance quality of devices, applications, and data transport are improved.
- *Fog Computing* – Defines how edge computing works and supports operations for storage, computing, and network services among endpoint IoT devices and Cloud computing centers. In other words, while fog is the defining standard, edge is the foundational concept. Fog provides the required structure in edge computing, enabling enterprises to move computing from Clouds or centralized systems to edge, resulting in improved, scalable performance.

Despite the above definition from Cisco, many companies feel that fog and edge computing are basically the same because both are focused on utilizing local network computing capabilities to complete computing tasks normally completed by the Cloud. In this context, several companies suggest that the biggest difference between fog and edge computing is where the data processing occurs. Edge computing generally takes place on the actual endpoint IoT devices (IoT things) or on a gateway device that is physically close to sensors/actuators. However, fog computing pushes edge computing to data processing centers connected to the LAN or to actual LAN hardware that is more physically distant from actuators or sensors.

Key features of fog and edge computing include [34, 35]:

- *Heterogeneity* – There is a wide heterogeneity of IoT edge devices (e.g., vibration sensor, temperature sensor), there is a wide heterogeneity in communication technologies and protocols (e.g., OPC-UA, Modbus, CAN bus, BACnet, MQTT, REST, SICK SOPAS), and there is a wide heterogeneity in network technologies (e.g., Wi-Fi, LTE, 3G/4G, Bluetooth) as well. Fog acts as a multi-protocol building block that can be utilized in diverse environments for protocol translation, flexible integration, data delivery, and device management.
- *Interoperability* – Fog should be integrated into many solutions in order to support a broad range of different services such as data streaming.
- *Geographical Distribution* – Fog computing is deployed in a distributed manner to provide top-quality services for stationary and mobile end devices.
- *Edge Processing/Storage* – A wide range of applications can be executed on the edge node close to the data source in order to reduce response time and save the bandwidth between edge and Cloud. Edge can also be utilized as a short-time historical storage.
- *Quality of Service* – Fog computing emerged partly to address quality-of-service constraints of IoT endpoints. To name a few, real-time video streaming, gaming, and CCTV monitoring are among those applications that demand low-latency services.
- *Real-Time Interaction* – Fog can be used in real-time applications, including real-time traffic monitoring, require real-time processing speed, and capability as opposed to batch processing.
- *Large-scale Sensor Networks* – Fog computing is very useful to be utilized in large-scale sensor networks (e.g., in smart grid or for environmental monitoring applications) in which utilizing systems with distributed storage and computing resources are required.

As noted, Cloud computing in IoT networks offers several benefits including exceptional computing efficiency, enormous storage capability, and wide-area coverage. On the other hand, edge computing offers a device-centered process, increased mobility, high QoS, resource pooling at the edge, and the ability to manage data in real time. Table 4.8 presents the main differences between Cloud and edge computing.

Table 4.8 Main differences between Cloud computing and edge/fog computing

Characteristics	Cloud computing	Edge computing
Computing capacity	High	Low – medium
Server size and operating mode	Large, centralized servers	Smaller, distributed servers
Application suitability	High computational needs, the delay is acceptable	Low latency, requires a real-time operation, high QoS
Communication needs	High – devices require a constant Internet connection	Low – devices obtain cache contents via edge gateway
Deployment planning	Complicated planning	Possible ad hoc deployment with little to no planning

4.13 Summary

IoT and Cloud evolved separately as two distinct disciples over time; however, over the past few years, they have been integrated as complementary technologies. IoT Cloud paves the way for "device as a service" business model as well as for unlimited storage and processing power to be able to manage and process big data generated from millions of IoT devices. In this chapter, we overviewed the fundamentals of Cloud computing such as characteristics, services, and deployment techniques. Next, we presented a multilayer architecture for IoT Cloud including data ingestion, data storage, data processing, and data visualization. Finally, we detailed each of them covering the underlying technologies and their state-of-the-art frameworks.

References

1. A. Botta et al., Integration of Cloud computing and internet of things: A survey. Futur. Gener. Comput. Syst. **56**, 684–700 (2016)
2. *NIST: National Institute of Standards and Technology*. Available from: https://www.nist.gov/
3. *Microsoft Azure IoT*. Available from: https://azure.microsoft.com/en-us/services/iot-hub/
4. S.R. Sinha, Y. Park, *Building an Effective IoT Ecosystem for Your Business* (Springer, Cham, 2017)
5. *The Modern Documentation Service for Microsoft*. Available from: https://github.com/MicrosoftDocs
6. Data Ingestion, Processing and Architecture layers for Big Data and IoT. Available from: https://www.xenonstack.com/blog/ingestion-processing-big-data-iot-stream/
7. *Apache Flume*. Available from: https://flume.apache.org/
8. *Apache Kafka*. Available from: https://kafka.apache.org/
9. *Apache NiFi*. Available from: https://nifi.apache.org/docs.html
10. *Big Data Battle: Batch Processing Vs Stream Processing*. Available from: https://medium.com/@gowthamy/big-data-battle-batch-processing-vs-stream-processing-5d94600d8103
11. *Data in Motion Vs. Data At Rest*. Available from: https://www.inap.com/blog/data-in-motion-vs-data-at-rest/

12. N. Marz, J. Warren, *Big Data: Principles and Best Practices of Scalable Real-Time Data Systems* (Manning Publications Co, New York, 2015)
13. *From Lambda to Kappa: A Guide on Real-Time Big Data Architectures.* Available from: https://www.talend.com
14. J. Leibiusky, G. Eisbruch, D. Simonassi, *Getting Started with Storm* (O'Reilly Media, Inc, Sebastopol/Köln, 2012)
15. *Easy Access Big Data Insight.* Available from: http://www.datavirtualizationblog.com/easy-access-big-data-insights/
16. *Tutorialspoint.* Available from: https://www.tutorialspoint.com
17. *Elasticsearch as a NoSQL Database.* Available from: https://www.elastic.co/blog/found-elasticsearch-as-nosql
18. *MongoDB.* Available from: https://docs.mongodb.com/
19. *Cassandra.* Available from: http://cassandra.apache.org/doc/latest/
20. *Redis.* Available from: https://redis.io/documentation
21. *InfluxDB.* Available from: https://docs.influxdata.com/influxdb/v1.7/
22. *An Overview on Elasticsearch and Its Usage.* Available from: https://towardsdatascience.com/an-overview-on-elasticsearch-and-its-usage-e26df1d1d24a
23. *Elastic.* Available from: https://www.elastic.co/
24. *How Elasticsearch cluster works.* Available from: http://duydo.me/how-elasticsearch-cluster-works/
25. *What Is Elasticsearch and How Can It Be Useful?.* Available from: https://dzone.com/articles/what-is-elasticsearch-and-how-it-can-be-useful
26. *Data Warehouse Vs Data Lake.* Available from: https://www.sspaeti.com/blog/data-warehouse-vs-data-lake-etl-vs-elt/
27. *Essential Guide to Data Lakes.* Available from: https://www.matillion.com/
28. *Microservices Practitioner Articles.* Available from: https://articles.microservices.com/
29. *Microservices Vs SOA: How Are They Different?.* Available from: https://www.bmc.com/blogs/microservices-vs-soa-whats-difference/
30. *nginx.* Available from: https://www.nginx.com/
31. *Kemp Technologies.* Available from: https://kemptechnologies.com
32. R. Ranjan, B. Benatallah, *Programming Cloud Resource Orchestration Framework: Operations and Research Challenges.* arXiv preprint arXiv:1204.2204 (2012)
33. *Cisco Systems.* Available from: https://www.cisco.com
34. F. Bonomi et al., *Fog Computing and Its Role in the Internet of Things.* In *Proceedings of the first edition of the MCC workshop on Mobile Cloud computing*, 2012, ACM
35. B. Farahani et al., Towards fog-driven IoT eHealth: Promises and challenges of IoT in medicine and healthcare. Futur. Gener. Comput. Syst. **78**, 659–676 (2018)

Chapter 5
Machine Learning for IoT

Farshad Firouzi, Bahar Farahani, Fangming Ye, and Mojtaba Barzegari

Develop a passion for learning. If you do, you will never cease to grow.

Anthony J. D'Angelo

Contents

F. Firouzi (✉)
Department of ECE, Duke University, Durham, NC, USA

B. Farahani
Shahid Beheshti University, Tehran, Iran

F. Ye
Facebook, Menlo Park, USA

M. Barzegari
KU Leuven, Leuven, Belgium

© Springer Nature Switzerland AG 2020
F. Firouzi et al. (eds.), *Intelligent Internet of Things*,
https://doi.org/10.1007/978-3-030-30367-9_5

5.1 Fundamental of Machine Learning

Learning consists of such a broad range of processes that it is hard to define precisely. In the dictionary, learning is defined as "to gain knowledge, or skill in, or understanding of, by study, experience, or instruction" and "modification of a behavioral tendency by experience." In contrast to zoologists and psychologists specialized in defining learning from perspectives of biological behaviors, we focus on the learning process in machines. Many concepts in machine learning are brought from the efforts of psychologists to make more precise their theories of human learning through computational models. On the other side, during the development of machine learning, some concepts or technologies may also inspire certain aspects of biological learning [2–4].

As for the machine, we may impose the characteristic that a machine *learns* whenever the system changes its structure, programmed in such a manner that its expected future performance improves. The machine can only learn from the history of its inputs or in response to external information, where more information or changes shall render a more accurate response model. Some of these changes (e.g., augmenting a database by adding an event) fall easily within the province of other disciplines and are not essentially better understood for being called learning. For instance, when an image recognition machine can eventually differentiate cats from dogs after seeing several pictures of cats and dogs, we feel quite justified in that case to say that the machine has learned. A typical machine learning model or algorithm is similar to the following (see Fig. 5.1). A machine learning process consists of three main components: input, machine learning model, and output. Given the information, i.e., inputs, such as precipitation, humidity, and temperature, we would like to execute our task, i.e., output, predicting the weather to be either sunny or rainy. The core of this prediction process is a machine learning model, a.k.a how can we find an appropriate model to accurately map all combination of

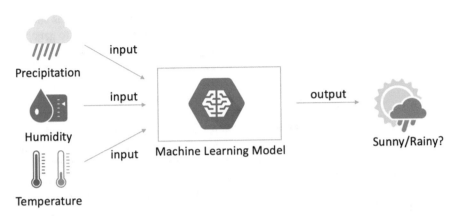

Fig. 5.1 An illustration of a machine learning model for predicting the weather

precipitation, humidity, and temperature to the actual weather. Fortunately, modern machine learning techniques provide a variety of choices, to name a few, neural networks, logistic regression, support vector machine, and deep learning.

5.1.1 Fundamental Terminologies

You have likely heard about artificial intelligence, data science, business intelligence, data mining, machine learning, data engineering, and deep learning; however, you might be unsure how these specialties are really different from one another. This section will focus on clarifying the specific focus of each area.

- *Artificial Intelligence (AI)*: According to the Merriam-Webster dictionary, intelligence is "the ability to learn or understand, or to deal with new or trying situations." The field of artificial intelligence is founded on the idea that machines or computer programs can have the capacity to reason, understand, learn, and think as a human being does. AI is focused on mimicking the intelligence of humans in computer systems or other machines through reasoning, self-correction, and learning.
- *Machine Learning (ML)*: The area covered by artificial intelligence is extensive, and machine learning is a subdivision of AI. In short, machine learning is a method utilized in achieving AI. Machine learning revolves around enabling computer systems to learn and make accurate forecasts based on data without requiring programming. This requires that an algorithm be given large amounts of data, enabling the machine to learn more through the processed information.
- *Deep Learning (DL)*: Deep learning, or deep neural network (DNN), is a subset of machine learning. The word "deep" is used because there are many steps required throughout the process of learning. Deep learning algorithms are

generally shaped by the human brain's data processing patterns. Data is subject to several nonlinear transformations through virtual neurons in order to generate a specific output. The output from one step becomes the input for another, and this process continues until a final output is achieved. The details of DL will be discussed in Sect. 5.6.

- *Data Science*: The term "Data Science" was born in the 1960s when it was used as an interchangeable name for computer science. Today, the phrase "data science" carries a very different meaning. Jeff Hammerbacher and D.J. Patil took the term in a new direction in 2008, when they became the first to refer to themselves as "data scientists" when describing their positions in Facebook and LinkedIn, respectively. Today, data science refers to a set of methods or techniques used to extract insights or information from data. While it intersects with AI, data science is not a subarea of AI or ML. It is a multidisciplinary field utilizing skills from a variety of areas, including visualization, statistics, and machine learning, to manipulate and analyze data, generate insights, or extract needed information from large amounts of data. In contrast, machine learning focuses on building programs and algorithms that learn independently and do not require human intervention to improve. For example, ML techniques are more appropriate than data science methods when it comes to realizing self-driving cars.

- *Data Mining*: Data mining became a widely used term in the database communities in the 1990s and is a subprocess of Knowledge Discovery in Databases (KDD), the process of gaining knowledge from information found in databases. Data mining is focused on recognizing patterns within a set of data and often requires analysis of massive amounts of historical data that was previously ignored or thought useless. These patterns are then used to predict future patterns, which is an important step in the KDD process. In contrast, data science is a broader field that includes various subareas from data visualization, big data analytics, and predictive modeling to data mining, statistics mathematics, and data visualization. The main differences between data science and data mining can be clarified with an example. If you wanted to review the previous 8 years' data in order to know how many sweets were sold during the festival seasons of three different cities, a data mining professional would review the historical data in legacy systems and use algorithms to extract patterns. On the other hand, if you need to know which of the sweets received the most positive reviews, the required data may not be located only in databases. This information could be spread across social media, customer surveys, or websites, requiring the skill of a data scientist.

- *Data Engineering*: The responsibilities and skills of data scientists and data engineers overlap significantly; however, the main point of difference is the specific focus of each. Data engineers focus primarily on creating data architecture or infrastructure. They develop, build, test, and maintain architectures like large processing systems or databases for data scientists. On the other hand, data scientists are concerned with utilizing statistical analysis and advanced mathematics to extract insights from data.

5.1.2 Review of Probability Theory

Probability plays a key role in machine learning because most learning algorithms rely on the probabilistic assumption of the data. Therefore, a basic understanding of probability theory is essential in learning and understanding machine learning techniques. Probability theory is the mathematical study of uncertainty. Below, we introduce some basic concepts in probability theory that will familiarize readers with the language used in machine learning [4].

5.1.2.1 Random Variable

A random variable is a set of possible values from a random experiment. Assume that we toss a coin, the outcome X of a coin toss can be either head (1) or tail (2). If the coin is fair, both outcomes $X = 1$ or $X = 2$ are equally likely to occur; hence we would see a probability of 0.5 in the outcome of such experiments. We could state that the probability of seeing heads when flipping a coin is ½. Note that a random variable denotes a whole set of outcomes, which means it can take on any of those values, randomly [5].

5.1.2.2 Distribution

Given a random variable, we can further characterize the probabilities associated with the random values it can take. If the random variable is discrete (i.e., it can have only a finite number of values), then this probability assignment is called a probability mass function (PMF). By definition, a PMF must be non-negative and must sum to one. Let us take coin flipping example again; if tails and heads are equally likely, then the random variable X takes values of $+1$ and -1 with probability 0.5 each. This can be described as

$$\begin{cases} \Pr(X = +1) = 0.5 \\ \Pr(X = -1) = 0.5 \end{cases}$$

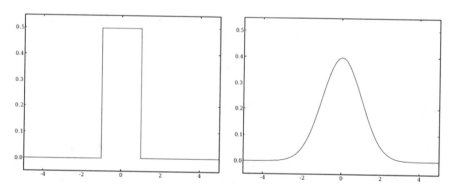

Fig. 5.2 The left is uniform distribution over $[-1,1]$; the right is Gaussian distribution with mean $= 0$ and variance $= 1$

In short, we can use an informal representation of the above equations:

$$p(x) := \Pr(X = x)$$

If we take continuous random variable into account, the distribution function can be rewritten as a probability density function (PDF). As in the case of a PMF, PDF must be non-negative and sum to one. We use a similar distribution format as PMF and illustrate with two distributions: the uniform distribution and Gaussian distribution, respectively (Fig. 5.2).

$$p(x) = \begin{cases} \frac{1}{b-a}, & if \ x \in [a, b] \\ 0, & otherwise \end{cases}$$

$$p(x) = \frac{1}{\sqrt{2\pi\sigma^2}} \exp\left(-\frac{(x-\mu)^2}{2\sigma^2}\right)$$

5.1.2.3 Mean, Variance, and Covariance

Mean is defined as the average of the numbers: a calculated "central" value of a set of numbers. Suppose we have an array of $[1, 4, 6]$, we can simply calculate the mean as $(2 + 7 + 9)/3 = 6$.

In statistics, we often need to know what the expected value of a random variable is. For example, we may ask a question on what the expected temperature is during a certain period of time. We also leverage the concept of "mean" to define expectations and related quantities of distributions.

We can define the mean of a random variable X as below:

$$\mathbb{E}[X] := \int x\, dp(x)$$

To make it more general, if $f : \mathbb{R} \to \mathbb{R}$ is a function, then $f(X)$ is also a random variable. Its mean can be calculated as below:

$$\mathbb{E}[f(X)] := \int f(x)\, dp(x)$$

If X is a discrete random variable, the integral in the above can be replaced by a summation:

$$\mathbb{E}[X] = \sum_x x p(x)$$

We can simply consider rolling a dice, which has equal probabilities of $1/6$. Therefore, the expected outcome of rolling dice is its mean $(1 + 2 + 3 + 4 + 5 + 6)/6 = 3.5$.

Variance is defined to measure how much on average $f(x)$ deviates from a probability distribution's expected value, as below:

$$\mathrm{Var}[X] = \mathbb{E}\left[(X - E[X])^2\right]$$

If we take rolling dice as the example, the variance of rolling a dice is $[(1-3.5)^2 + (2-3.5)^2 + (3-3.5)^2 + (4-3.5)^2 + (5-3.5)^2 + (6-3.5)^2]/6 = 2.91$

Variance only operates on one dimension; however, it would be possible to find and compute the correlation between two features using *covariance*. Covariance is a measure of how much two random variables vary together and defined as follows:

$$\mathrm{cov}\,(X, Y) = \mathbb{E}[(X - E[X])(Y - E[Y])]$$

If X and Y are discrete random variables, the corresponding covariance can be calculated using the following equation:

$$\mathrm{cov}\,(X, Y) = \frac{\sum_{i=1}^{n}(X_i - \overline{X})(Y_i - \overline{Y})}{(n - 1)}$$

in which $\overline{X}$ and $\overline{Y}$ illustrates the mean of variable X and variable Y, respectively. Variance and covariance are often represented together by a covariance matrix. In a covariance matrix, the diagonal elements are variance and off-diagonal elements are covariance. Note that a covariance matrix is a symmetric matrix. The following matrix demonstrates the covariance matrix of three variables:

$$\begin{bmatrix} \text{var}_1 & \text{var}_{1,2} & \text{var}_{1,3} \\ \text{var}_{1,2} & \text{var}_2 & \text{var}_{2,3} \\ \text{var}_{1,3} & \text{var}_{2,3} & \text{var}_3 \end{bmatrix}$$

5.1.3 Review of Linear Algebra

Linear algebra is a key foundation to the field of machine learning, and it mostly discusses vectors, matrices, and linear transformations. In this section, we briefly overview the fundamentals of linear algebra.

A *matrix* is a rectangular array of numbers organized in columns and rows. Numbers appear in a matrix are called entries or elements which can be addressed by their corresponding row number and column number. The number of rows and columns is called dimension or order of the matrix. For example, the order of the following matrix is 2_*3:

$$A = \begin{bmatrix} 1 & 3 & 3 \\ 2 & 5 & 9 \end{bmatrix}$$

The *transpose* of a matrix is a new matrix whose row and column indices are switched/flipped as shown by the following example. Note that the transpose of a matrix is usually represented by A^T or A':

$$A^T = \begin{bmatrix} 1 & 2 \\ 3 & 5 \\ 2 & 9 \end{bmatrix}$$

A *vector* is a matrix that has only one column as illustrated below:

$$v = [2 \ 0 \ 6]$$

A *square matrix* is a matrix where the number of its columns is equal to the number of its rows. A *symmetric matrix* is a matrix whose transpose is equal to itself:

$$A = A^T = \begin{bmatrix} 1 & 2 \\ 2 & 3 \end{bmatrix}$$

A diagonal matrix is a *square matrix* whose off-diagonal elements are equal to zero:

$$A = A^T = \begin{bmatrix} 1 & 0 \\ 0 & 3 \end{bmatrix}$$

A *scalar matrix* is a diagonal matrix whose elements along the diagonal are equal:

$$A = A^T = \begin{bmatrix} 2 & 0 \\ 0 & 2 \end{bmatrix}$$

An *identity matrix* (*unit matrix*) is a scalar matrix whose diagonal elements are equal to one. Note that the unit matrix is usually represented by I:

$$I = \begin{bmatrix} 1 & 0 \\ 0 & 1 \end{bmatrix}$$

A matrix is *orthogonal* when

$$A * A^T = I$$

A *shear* matrix (*transvection matrix*) is an identity matrix where one of its off-diagonal zero elements is replaced with one nonzero value (λ):

$$S = \begin{bmatrix} 1 & \lambda \\ 0 & 1 \end{bmatrix}$$

A *rotation matrix* is typically represented as follows:

$$R = \begin{bmatrix} \cos\theta & -\sin\theta \\ \sin\theta & \cos\theta \end{bmatrix}$$

A *transformation* from R^n to R^m is defined as a mapping function (T) that maps each vector (x) in R^n to a new vector ($T(x)$) in R^m. A transformation is linear when the following conditions are preserved:

$$T(V_1 + V_2) = T(V_1) + T(V_2)$$

$$T(\alpha V) = \alpha T(V)$$

In the above equation, V and V_2 are vectors, and α is a scalar value. Note that a linear transformation can be represented by a matrix. Figure 5.3 visually illustrates how scaling matrix, rotation matrix, shear matrix, and symmetric matrix can transform a vector. As shown in this figure, a symmetric matrix is actually a combination of rotation (R) and scaling (S) matrices. In other words, a symmetric matrix first rotates, then scales, and finally rotates back the vectors:

$$A = R(\theta) S R(-\theta)$$

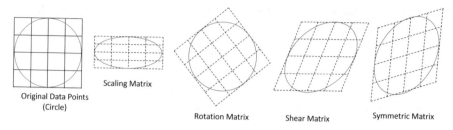

Fig. 5.3 A visual illustration of matrix transformation

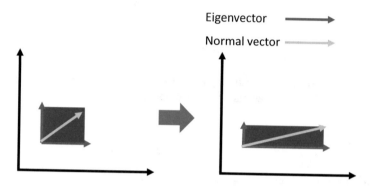

Fig. 5.4 An illustration of eigenvectors. In contrast to normal vectors, the directions of eigenvector do not change when a linear transformation is applied

All symmetric matrices (such as covariance matrix) can be decomposed into three matrices (i.e., rotate, scale, and rotate) as illustrated below:

$$\text{cov} = VDV^T$$

where V is an orthogonal matrix whose columns are the *eigenvectors* of the covariance matrix, and matrix D is a diagonal matrix whose elements are the corresponding *eigenvalues*. An eigenvector is a vector that changes by only a scalar factor and whose directions do not change when a linear transformation is applied (see Fig. 5.4). Eigenvector and eigenvalue can be defined formally by the following equation:

$$Av = \lambda v$$

In the above equation, A is a transformation matrix, v is a column vector that represents the eigenvectors of the matrix A, and finally λ is a scalar known as the eigenvalue.

5.1.4 *Supervised and Unsupervised Learning*

Based on the tasks to be solved and data available to the task, the machine learning model also varies. The most common way is to categorize machine learning models into *supervised learning* and *unsupervised learning*.

5.1.4.1 Supervised Learning

Supervised learning is the simplest model that readers can understand. The reason why this type of modeling is called supervised learning is that the supervised learning model is learned, or *trained* in the language of machine learning, from the training dataset, just like a teacher supervises a student through a learning process.

In a supervised learning model, input and output are clear to the reader, although the inner algorithm may not be so obvious. For instance, as described in Fig. 5.1, we would like to predict tomorrow's weather. The output is clearly defined – weather. It can be defined as sunny, cloudy, or rainy. Meanwhile, we collect a number of relative parameters, such as temperature, humidity, etc. These parameters are not directly reflecting the weather, but indirectly indicating the type of weather. Therefore, these parameters can be used for predicting the weather. We collected a dataset of historical records of weather and corresponding relative parameters, based on which we train the weather prediction model. In the training process, we already know the right answers. The algorithm iteratively makes predictions on the training data and is corrected by the *teacher*, the training process. The learning process stops when the algorithm achieves an acceptable level of accuracy. Thus, we can use the model to predict future weather with a certain level of confidence.

To further categorize supervised learning, those algorithms can be grouped into either (i) classification models or (ii) regression models (Fig. 5.5).

- *Classification*: Classification refers to a model in which the output is a category, such as weather (sunny, windy, rainy) or fruit (orange, apple, pear).

Classification
What will the weather be
like tomorrow? Cold or Hot?

Regression
What will be the
temperature tomorrow?

Fig. 5.5 Classification vs. regression

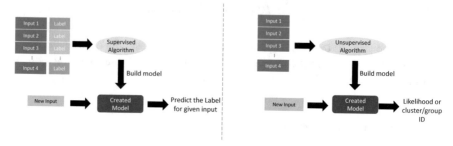

Fig. 5.6 Supervised vs. unsupervised machine learning

- *Regression*: Regression is a model in which output is a continuous value, such as weight (how many kilograms a person weighs) or price (how much a table cost).

5.1.4.2   Unsupervised Learning

In contrast to supervised learning, unsupervised learning is the process of inferencing from data without explicitly provided labels (see Fig. 5.6). Therefore, we are not clear about the output of the dataset. Tasks in unsupervised learning model include clustering, anomaly detection, latent variable learning, etc. Since no labels are provided, no obvious ground truth can be used for verifying the model, and it is difficult to compare or judge model performance in most unsupervised learning algorithms.

One of the commonly used unsupervised algorithms is clustering. This is the task of grouping a set of objects in such a way that objects in the same group share similar behavior to each other compared to those in the other groups. For instance, given a fruit basket, we can group red heart-shaped fist-sized fruits together as apple or orange-colored fruits together as orange.

5.1.5   Machine Learning in IoT

There is a vast amount of use cases of machine learning in IoT across vertical segments. In this section, we briefly review three of them to highlight the importance of machine learning (see Fig. 5.7).

- *Classification (Supervised)*: Classification is one of the most important machine learning techniques in IoT. For example, by combining machine learning with a readout of wearable health sensors, we can address several questions in the healthcare domain. Classification can be applied to ECG signals to detect and predict heart attacks in real time.

Fig. 5.7 A few use cases of machine learning in IoT

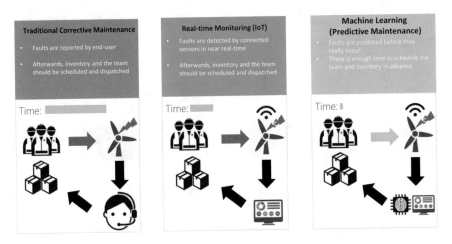

Fig. 5.8 A simple example of predictive maintenance in the energy industry

- *Regression (Supervised)*: Predictive maintenance (PdM) is a cutting-edge mainte-nance strategy, which has been adopted in several domains (e.g., manufacturing, energy and supply chain). The key idea of PdM is to identify which equipment needs maintenance and which component will fail in the future and to predict the remaining useful life (RUL) of machine parts (see Fig. 5.8). In this context, regression techniques can be utilized to predict RULs accurately.
- *Clustering (Unsupervised)*: An example of unsupervised learning in IoT is data processing in a factory producing car engines. Suppose that we want to design a machine to detect engines that require further adjustments. It is almost impossible to build a system to detect the defects visually, but this can be achieved by collecting several key parameters from each engine and then using a clustering algorithm to find groups/clusters. For example, if the parameters are temperature and the produced sound, the clustering algorithm (e.g., K-means clustering) will group the engines into different categories based on their similarity in producing sound at a specific range of temperature (Fig. 5.9). This will help engineers of that factory to detect the engines that belong to the problematic group quickly.

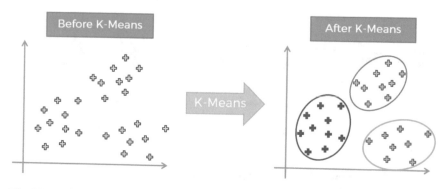

Fig. 5.9 Clustering of engine data

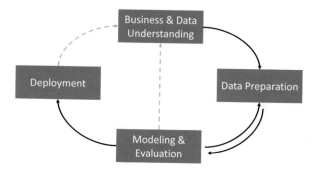

Fig. 5.10 Machine learning flow

5.1.6 Machine Learning Flow

5.1.6.1 Overall Flow of Machine Learning Projects

The most common methodology for machine learning projects consists of the following phases (see Fig. 5.10). It should be noted that in the data mining context, this methodology is known as cross-industry standard process (CRISP).

- *Business and Data Understanding*: In this phase, we need to define the scope of the project; understand the problem statement and pain points; study those factors which might be able to impact the project; construct, gather, and collect data from several sources (e.g., sensor readouts from IoT devices); and identify metrics and key performance indicators (KPIs) for measuring success.
- *Data Preparation*: In this phase, we need to prepare the data for machine learning algorithms. This phase includes (but not limited to) formatting data according to our machine learning algorithms, handling missing values, handling categorical variables, data normalization, and preparing training and test datasets.
- *Modeling and Evaluation*: In this phase, we build several machine learning models, evaluate the performance of each of which, and finally select the best

model which can address the problem. Note that this phase may provide feedback to the previous phases (i.e., business and data understanding phase as well as the data preparation phase).

- *Deployment*: In this phase, we deploy the selected model into production. Continuous monitoring and maintenance of models are also very important. Based on the feedback and performance of the deployed models, we might need to adjust our solution over time.

5.1.6.2   Data Preparation

The data preparation is also known as data preprocessing, data cleaning, and data cleansing. In general, the following steps are performed in the data preparation phase. Note that it is not mandatory to apply all of the following steps to your data. In reality, you need to decide case by case.

1. *Preparing Dataset*: The first step of the data preparation phase is constructing, collecting, and formatting data. As studied in previous chapters, in IoT projects, data comes in many forms including: (i) *Structured*: It concerns all data which can be stored in a table with rows and columns; examples of structured are CSV documents. (ii) *Semi-structured*: Semi-structured data cannot be stored directly in a table; however, with some process you can store them in tables; examples of semi-structured are XML and JSON documents. (iii) *Nanostructured*: It usually includes text and multimedia content (such as email messages, videos, photos, audio files, etc.). In the machine learning literature, we frequently see the word *dataset*, which simply refers to a collection of data. It is very common to use matrix and vector notations (in particular for structured and semi-structured data) to refer to data (see Fig. 5.11):

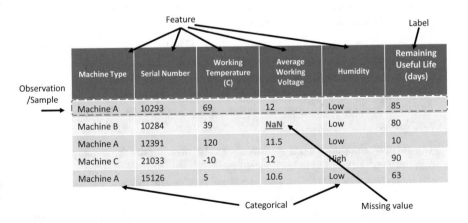

Fig. 5.11 A tabular view of a sample dataset

(a) Each row of the matrix corresponds to one single observation (it is also called a sample or a data point or a case).

(b) Each column represents a feature (also known as an attribute or a vector). A feature is an individual measurable property or attribute of a phenomenon being observed. In supervised techniques, features are also called independent variables.

(c) In supervised machine learning techniques, there is one column corresponding to "label" (also known as "response," "output," or "dependent variable").

2. *Handling Missing Values*: Possible variations include 'NaN', 'NA', 'None', '', '?' and others (see Fig. 5.11). Missing values can occur both in numerical features and categorical features:

(a) *Numerical Features*: Depending on the nature of the problem, you may consider using one of the following techniques to address missing values:

   (i) Ignore those rows with a missing value.
   (ii) Use the mean/median/mode value of the feature for those missing values.
   (iii) Use the previous/next values to replace missing values.
   (iv) Predict the missing values; for example, curve-fitting and regression algorithms can be used to find the missing values.

(b) *Categorical Features*: Note that a categorical feature is a variable that can take a limited number of possible labeled values. For example, a color feature with the values "Red" and "Blue." Dealing with categorical features is tricky. In general, you might be able to utilize one of the following techniques depending on the project's constraints:

   (i) Ignore those rows with a missing value.
   (ii) Replace the missing value with the most frequent category.
   (iii) Use the previous/next values to replace missing values.
   (iv) Predict the missing values.

3. *Handling Categorical Values*: Some machine learning techniques (e.g., decision tree) can directly use categorical features. However, to be able to use categorical features in other machine learning algorithms, typically, we need to use an encoding technique to convert them to numerical features. The most well-known encoding techniques are *integer encoding* and *one-hot encoding* (see Fig. 5.12).

(a) *Integer Encoding (Label Encoding)*: In this approach, a unique integer number is assigned to each category.

(b) *One-Hot Encoding*: Integer encoding can result in poor models because machine learning algorithms may consider some kind of order between categories (e.g., $0 < 1 < 2$). To tackle this issue, the one-hot encoding method can be applied to the feature. Let us explain the idea of one-hot encoding by a simple example. Suppose we have a feature with three

Machine Type	Machine Type	A	B	C
Machine A	1	1	0	0
Machine B	2	0	1	0
Machine A	1	1	0	0
Machine C	3	0	0	1
Machine A	1	1	0	0

Categorical Feature   **Integer Encoding**   **One-hot Encoding**

Fig. 5.12 A simple example of integer encoding and one-hot encoding

different categories (i.e., Machine A, Machine B, and Machine C). In One-hot encoding approach, we generate one boolean column for each category. Therefore, in our example, we have three columns and only one of these columns could have the value 1 for each observation (see Fig. 5.12).

4. *Normalizing Data*: In this step, we rescale all the features/attributes to have a common scale (usually into the range 0 to 1). For example, suppose we have a dataset containing two features, temperature and voltage. In this dataset, the temperature ranges from -30 to 120, whereas the voltage ranges from 0 to 12. Therefore, the temperature is about ten times larger than the voltage. In this case, we may consider rescaling those features into the range 0–1.

5. *Standardizing Data*: In this step, we rescale features so that they have a mean value of 0 and a standard deviation of 1. In this case, assume that our data has a Gaussian (bell curve) distribution.

6. *Splitting the Dataset into Training and Test Set*: In every machine learning project, there are two well-known techniques to split the dataset for training and evaluating a model, namely, *hold-out* and *cross-validation (k-fold cross-validation)*.

   (a) *Hold-out*: In this technique, the *original dataset* is divided into three groups: *training dataset, validation dataset*, and *test (hold-out) dataset* (see Fig. 5.13). The training dataset is the majority of data and typically contains 60–90% of the original set. The validation set is a subset of the training data, and it is used to evaluate the performance of the model during the training phase. In other words, the data of the validation test are not used for the training of the model, but instead, can be used to tune the hyperparameters of the model. The validation set is very useful to tackle some important problems in machine learning such as overfitting (it will be discussed in the next sections). The test dataset is just utilized to evaluate how well the trained

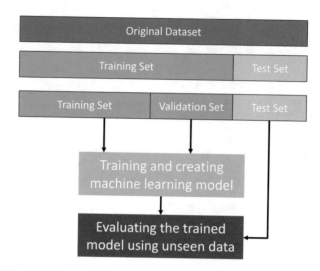

Fig. 5.13 Hold-out technique: splitting the dataset into training and test set

Fig. 5.14 Threefold
cross-validation

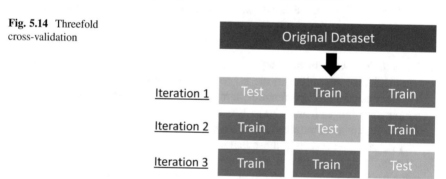

model can perform on unseen data. Figure 5.13 demonstrates the underlying concepts related to these datasets.

(b) *K-fold Cross-validation*: In this technique, the original dataset is repeatedly and randomly split into "k" equal-sized folds (also called groups, buckets, or sections). For each unique group, we take it as the test (hold-out) dataset and the other groups as the training set. This process is iteratively done for k times until each fold of k folds have been used as the test set (see Fig. 5.14).

5.2 Regression Analysis

Regression analysis is a subcategory of supervised machine learning. It is used to study the correlation between a dependent (usually called target or output) and an independent variable (called predictor or features or input). In this machine learning method, the impact of the independent variable(s) on the dependent variable(s) is analyzed. There are several well-known use cases for regression analysis:

- Predictive Analytics: Predictive analytics tries to model and predict future behaviors by analyzing historical data. This technique has a wide range of applications in IoT. As an example, predict maintenance predicts the time of machine failures based on sensor readouts (e.g., vibration data) mounted in machines in order to minimize downtime and to maximize productivity.
- Operation Efficiency: Regression analysis is used to optimize business processes and assets (e.g., machine, workstation, laborer on the shop floor). For example, on the production floor, IoT sensors mounted on machines can track inventory consumption in real time. Regression analysis can forecast future behavior and trigger automatic reordering or refill.
- Decision Support Systems: Regression helps make smarter and more accurate decisions based on the available data.
- Error Correction: Regression analysis can be utilized to correct wrong decisions, which are sometimes made based on some incorrect observations. For example, a technical manager on a shop floor may believe that increasing the temperature of a specific phase of the production line can increase the quality of the products. However, readouts of IoT sensor and regression analysis indicate that this assumption is not correct.
- New Insights: Regression analysis can reveal hidden patterns in data that are difficult to uncover by conventional approaches. For example, regression analysis techniques can be applied to IoT data of a production line to be able to find out the relationship between environmental/operating conditions and the quality of the products.

Formally, the regression task can be formulated as follows:

- There is a training set $((T = \{(x^{(1)}, y^{(1)}), \ldots, (x^{(T)}, y^{(T)})\})$ and we need to investigate the relationship (a mathematical equation) between the input (independent variable(s) or features) $x = (x_1, \ldots, x_D)$ and the output (dependent variable) $y^{(i)}$.
- Note that in a training set of regression analysis, the labels $y^{(i)}$ are *continuous*. This is one of the differences between regression tasks and classification tasks, in which the $y^{(i)}$ are *categorical*.

Some important terminologies related to regression analysis are:

- *Outliers*: An outlier (Fig. 5.15) is a data point in a dataset in which a very high or a very low value in comparison to other data points can be observed. An outlier can be deemed to be an extreme value in the dataset. The presence of outliers leads to less accurate results in regression; therefore, outliers are sometimes eliminated through a preprocessing step.
- *Multicollinearity*: Multicollinearity is the situation of having a high level of correlation among the inputs. This means that the predictors (independent variables)

Fig. 5.15 An example of an outlier data point

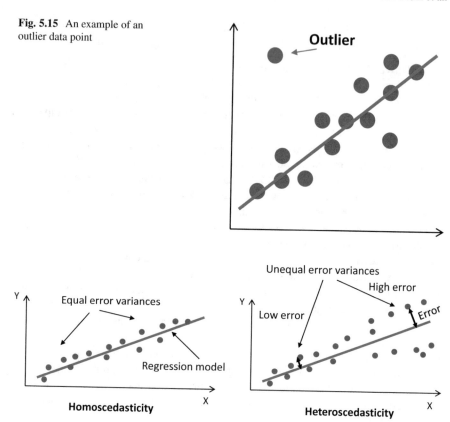

Fig. 5.16 Illustration of heteroscedasticity and homoscedasticity

are also correlated with each other. Multicollinearity leads to an increase in standard errors in regression analysis. This leads to inaccurate coefficients for some of the independent variables. In this case, this phenomenon makes some of the multicollinear variables mathematically insignificant (almost 0), while they are not.

- *Heteroscedasticity*: Heteroscedasticity occurs when the variance of the dependent variable (Y) depends on the independent variable (X). In other words, in this case, residuals of a regression model do not have a constant variance. This makes the analysis more complicated because regression analysis assumes that the variance across the independent variable is constant (called homoscedasticity). Figure 5.16 demonstrates these concepts visually. As shown in this figure (heteroscedasticity), when the value of X increases, the variance of Y also increases. On the other hand, when the case is homoscedasticity, the variance of Y is independent of the value of X.

Fig. 5.17 A linear regression
fit that minimizes the sum of
squared error of the difference
between the value of data
points and the fitted line

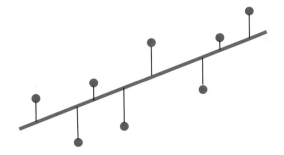

5.2.1 Linear Regression

A general regression model (Fig. 5.17) assumes a linear correlation between the
dependent and independent variables:

$$\hat{y} = h(x) = w_0 + w_1 x_1 + \cdots + w_D x_D = w_0 + \langle w, x \rangle = w_0 + x w^T$$

where $\hat{y}$ is the prediction made by the model, $h(x)$ is the linear model which
comprises a linear function with the coefficient of $w_0 \ldots w_D$, parameter w_0 is the
bias, and the $\langle w, x \rangle$ is the dot product.

An approach to extract the desired coefficient (i.e., find $w_0 \ldots w_D$) is to make $h(x)$
close enough to y for the provided training data. Therefore, we define a mathematical
term to calculate how close $h_w(x^{(i)})$ is to $y^{(i)}$, and this is called the cost function
(Eq. 5.1):

$$J(w) = \frac{1}{2} \sum_{i=1}^{m} \left(h_w \left(x^{(i)} \right) - y^{(i)} \right)^2 \tag{5.1}$$

Note that $x^{(i)} = (x_1, \ldots, x_D)$ represents the data point (entry) i in the training set
and $y^{(i)}$ is its corresponding output. There are several approaches to solve the above
equation. The most frequently used one is the *gradient descent algorithm*, in which
the cost function is minimized by moving in the opposite direction of the gradient
of $J(w)$ (the slope of the cost function). This method starts with an initial value of θ
and performs the following update iteratively:

$$w_j = w - \alpha \frac{\partial}{\partial w_j} J(w)$$

In summary, in a gradient descent algorithm, the following steps are followed:

1. Initialize the weights of the linear equation randomly.
2. Calculate the gradient of the cost function.
3. Update the weight proportional to the gradient ($W = W - \alpha G$), where $G = \frac{\partial}{\partial w_j} J(w)$.

Fig. 5.18 Graphical presentation of the way that a gradient algorithm works

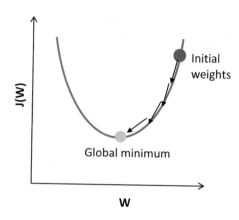

4. Repeat steps 2–3 until some termination criteria (such as the cost $J(W)$) stop reducing.

The update is performed for all of the training data, and the value of the gradient depends on the current values of the model parameters and the cost function. Parameter α is the learning rate, which controls the step size of each iteration. This value α should be selected carefully because a high learning rate leads to overshooting the minimum, and low rate results in reaching the minimum very slowly. A good approach for a proper selection of learning rate is starting with small values (such as 0.01 or 0.001) and redefining it based on the behavior of the gradient descent algorithm.

Figure 5.18 presents an intuition of the gradient descent algorithm. In this example, a blind man intended to reach the lowest altitude of rough terrain. One of the simplest approaches for him is to feel the slope of the ground and move in the direction that descends faster. If he keeps repeating this procedure, he will reach the lowest altitude point of the terrain. In comparison to the gradient descent algorithm, the slope is analogous to calculating the gradient; each step is similar to each iteration, and the cost function is to find lower altitudes.

Gradient descent algorithm can also be formulated stochastically. Stochastic gradient descent algorithm (SGD) computes the gradient by using a randomly chosen training data point in each update (instead of considering all data points of the training dataset). As a result, the algorithm runs faster, and yet, it moves in the same direction over many updates as traditional gradient descent.

5.2.2 *Regularization in Linear Regression*

One of the major challenges in regression analysis is overfitting/underfitting. Overfitting (high variance) occurs when the regression analysis algorithm works with a high level of efficiency on the training dataset but fails to perform correct

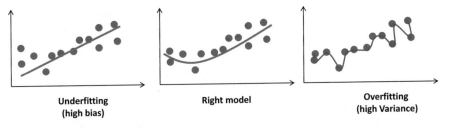

Fig. 5.19 An example of underfitting (high bias) and overfitting (high variance) regression

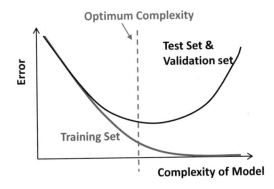

Fig. 5.20 Relation of the complexity of the model and the error of the training and test sets

predictions on the test dataset. This is also called the problem of *high variance*. On the contrary, when our algorithm works so poorly that it is unable to fit even training set well, it is said to be underfitting the data. It is also known as the problem of *high bias*.

Take Fig. 5.19 as an example. In this diagram, the straight line corresponds to a linear regression, which underfits the data and leads to large errors in the training set. A regression model of the polynomial kernel (the middle subfigure in Fig. 5.19) is the most suitable fit because it works well on both the training and test datasets. Note that the regression model of higher-order polynomial kernel (the right subfigure in Fig. 5.19) fits better on the training data, but it causes more error in the test dataset (see Fig. 5.20). In this figure, by moving from the left subfigure toward the right subfigure, the model tries to learn more details of the input data. Although higher-order kernel (more complex regression models) leads to higher accuracy/performance on training data, it may perform inaccurately on unseen inputs (test data), which means it loses its generality and gets worse. In other words, increasing the complexity of the model may decrease the training error, but it may eventually increase the test error, as explained above (Fig. 5.20).

Regularization is a technique of adding information to the learning algorithm to make the model more generalized. This, in turn, enhances the performance of the model on the unseen data (test data) as well. By using regularization, the learning algorithm is modified in a way that it acts more efficiently on unseen data. The modified regression model contains another term/component in its cost function, which it also should be minimized. Among regularization methods, L2 and L1 are the most frequently used methods, which modify the cost function by adding a generalization term as follows:

$$Cost\ function = Loss + Regularization\ term$$

Adding the regularization term results in smaller weights/coefficients, which leads to smaller overfitting. In other words, the regularization term punishes the cost function. The utilized regularization term is different in L1 and L2 methods. In L2, the cost function would be as follows:

$$Cost\ function = Loss + \lambda \sum \|w\|^2$$

where parameter λ is the regularization parameter, a hyperparameter that must be optimized for better performance. L2 regularization is also called weight decay because it forces the weights toward zero, however not exactly zero. A regression analysis method that performs L2 regularization is called ridge regularization. Ridge regularization is one of the well-known techniques to overcome overfitting.

Similarly, in L1, the cost function is written in this way:

$$Cost\ function = Loss + \lambda \sum \|w\|$$

in which the cost function penalizes the absolute value of the weights. However, unlike L2, the weights can be forced to be absolute zero. Since those input variables (features) with zero coefficients can be dropped from the regression model, the L1 regularization is useful for feature selection and reducing the complexity of the model. L1 regularization methods are also called lasso regularization. In other words, L1 regularization provides sparse solutions of the model by removing unimportant input variables. Getting sparse solutions could decrease the computational complexity of the model due to the presence of features with a coefficient of zero.

In case of highly correlated features, the ridge generalization distributes the coefficients among all of the features depending on the correlation, but lasso regularization chooses the features selectively and makes the coefficient of other features zero.

5.2.2.1 Geometric Interpretations of Regularization

One can define the L1 norm as the cumulative summation of absolute values of a vector's components. As an example, the L1 norm of the vector $[x1, x2]$ is $|x1| + |x2|$. With this definition in mind, we plot all the points whose L1 norms are equal to a constant value (c), as presented by a blue line in Fig. 5.21.

The geometry of L1 norm in Fig. 5.21 looks like a rotated square (an octahedron in higher dimensional space), in which the points on the tips are sparse (which

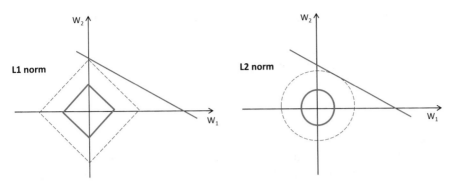

Fig. 5.21 Different forms of the constraint regions in lasso and ridge regression. *W1* and *W2* are the weights of regression features [*x1, x2*]

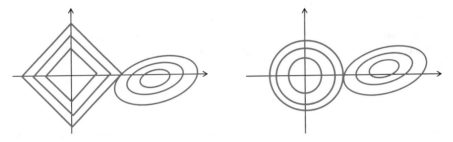

Fig. 5.22 Solutions of L1 and L2 norms and the effect of penalized parameters

means either *w1* or *w2* component is zero). Now if we make the box larger enough (i.e., increase the constant value (*c*)) to touch the red solution line, a sparse solution is achieved. It's worth mentioning that the L1 does not necessarily touch the solution by a tip, which means that the solution is not sparse in this case (i.e., we need to use both *x1* and *x2* in the regression model and none of them can be dropped). As the coefficients in ridge regression (L2 norm) are not set to zero, the shape of the L2 norm is different from the L1 norm. The shape of the L2 norm is a circle (Fig. 5.21), which is rotationally invariant and has no corner.

Let us examine the geometric interpretation of penalized linear regression by a simple regression example with two independent variables *x1* and *x2*. Recall that *w1*, *w2* are their corresponding coefficients/weights in the regression model. Suppose $y = f(w1, w2)$ is the original cost function (e.g., Eq. 5.1: mean square error in regression). We can plot its *contour* in the space X. Note that a contour plot is a visualization technique to represent a three-dimensional (*y, x1, x2*) surface by a two-dimensional (*x1, x2*) graph. A contour indicates the area at which the function (*y*) has fixed values. In our example, the counters are represented by the red diagram in Fig. 5.22. The minimum of the function ($y = f(x1, x2)$) is located in the center of red circles. In other words, the center of red circles is our solution (i.e., coefficients/weights of *x1, x2*), which minimizes the cost function. Now we add

L1 and L2 regularization parameters to cost function. Let us call the regularization function $g(\beta)$:

$$\text{For lasso regression} : g(W) = \lambda \left(|w_1| + |w_2| \right)$$

$$\text{For ridge regression} : g(W) = \lambda \left(w_1^2 + w_2^2 \right)$$

In the above equation, λ is the penalization parameter, and $w1$ and $w2$ are the coefficients of $x1$ and $x2$, respectively. $g(W)$ for lasso and ridge are depicted by the blue diagram in Fig. 5.22. In the cost function of lasso and ridge regression, we need to minimize $f(w1, w2) + g(w1, w2)$. This is equivalent to find those points that two contour plots (red and blue diagrams) meet each other. In other words, we should calculate the minimum of $f(W) + g(W)$, which is the intersection of two functions ($f(W)$ and $g(W)$). As shown in Fig. 5.22, in lasso regression, two contour diagrams can meet at a point where either $w1$ or $w2$ is zero. Therefore, the solution of lasso can be sparse. On the other hand, the contour plots in ridge regression do not have any tips, and thus it cannot result in any sparse solution.

5.2.2.2 Elastic Net Regularization

Zou and Hastie introduced the concept of the elastic net to overcome the weaknesses of L1 and L2 regularizations in 2005. *When the number of independent variables is more than the sample size ($p > n$), only one independent variable can be selected from any set of highly correlated independent variables using the lasso regression algorithm (up to n independent variables). In addition, if the number of independent variables is less than the sample size, the ridge regression method would have a better performance.*

Most of the times, highly correlated independent variables have similar regression coefficients. This situation is called the *grouping effect*. In real-world applications, the grouping effect can be beneficial for building the model. For example, in gene identification of diseases, the researchers are intended to find associated independent variables rather than only one from each set (which happens in lasso). Additionally, selecting a single independent variable from a set of highly correlated independent variables could result in a less robust model, which increases the precision error. This fact demonstrates why ridge regression performs more efficiently than lasso in this situation.

The elastic net algorithm is a combination of both L1 and L2 norms, in which some coefficients are shrunk (similar to ridge regression) and some are set to zero (like the lasso regression method). This method has two shrinkage parameters:

$$w^* = argmin \| y - xw \|_2^2 + \lambda_2 \| w \|_2^2 + \lambda_1 \| w \|_1$$

Fig. 5.23 Comparison of geometric interpretations of lasso (L1), ridge (L2), and elastic net

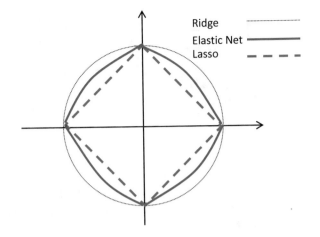

Elastic net is particularly useful when we have several correlated features. When two features have correlation, lasso selects one of them randomly, whereas the elastic net considers both. Therefore, it is also more stable in many cases. For example, imagine a problem in which there are two correlated features. In this case, lasso chooses one of them randomly, but the elastic net takes both into account. Like the ridge regression, the elastic net algorithm is more stable in comparison to other methods for most of the real-world problems. Figure 5.23 depicts the comparison of these algorithms [6, 7].

5.2.3 Bayesian Linear Regression

In linear regression, we have just one output value (y) for a given input. However, Bayesian has another point of view. In this approach, y is not a single value, but it is taken from a probability distribution. Recall that the linear regression approach models the relation between input data (features) and the output (target) by the following equation:

$$y = \beta^T X + \varepsilon$$

in which the response is produced by multiplying model parameters (i.e., weights: β) by the input (i.e., X) plus the model error (ε), which might be caused by random sampling noise or latent variables. In the ordinary least squares (OLS) approach, the model parameter (weights) can be determined by minimizing the sum of squared errors (Eq. 5.1). However, Bayesian linear regression uses a statistical approach based on probability distribution such as Gaussian distribution to model the mapping function between inputs (features) and output:

$$y \sim N\left(\beta^T X, \sigma^2\right)$$

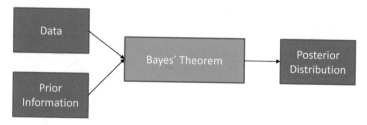

Fig. 5.24 Bayesian linear regression

In the Bayesian viewpoint, not only the output (y) has a distribution, but also all the model parameters (weights) have a distribution. In Bayesian regression, the following terms are defined:

Priors Priors are the initial or guess value of the model parameters that a domain expert can put into the model prior to training. If there is no knowledge about the parameters, non-informative priors, such as a normal distribution, could be used instead.

Maximum likelihood Maximum likelihood estimation is a method to determine the model parameter values in a way that the produced data (output of the model) is equal to the actual observed data. For example, for a Gaussian distribution curve, which has two parameters to be optimized (the mean, μ, and the standard deviation, σ), the maximum likelihood method can find the model parameters in a manner that the generated curve best fits the data.

Posterior Posterior indicates the output distribution of Bayesian linear regression based on the model parameters and priors. For a given dataset, one can estimate the posterior probability distribution by the Bayesian rule:

$$posterior = \frac{likelihood \times prior}{Normalization}$$

Bayesian regression algorithm enables us to compute the distribution of possible model parameters (posterior) based on the training dataset and the prior (see Fig. 5.24). Note that when we have infinite data, the posterior converges to the output of OLS linear regression. On the other hand, when we do not have enough data to train the model, the distribution of posterior spreads out.

5.3 Feature Selection

Feature selection (also called variable or attribute selection) is the process of selecting a subset of the input features to construct a high-performance machine learning model. In other words, feature selection enables us to implement a potential more accurate model while requiring fewer feature (input). This is accomplished by removing the *irrelevant* or *redundant* information from the input feature set. For example, in a supervised learning problem (either classification or regression), although there could be a large number of available features in the input dataset, only a subset of those features is relevant to the learning task. In this situation, incorporating all of the features may result in a risk of overfitting and high computational cost. Feature selection algorithms allow us to overcome this challenge.

An irrelevant feature is defined as a feature that contains no useful information regarding the problem (output variables) and is not capable of describing the relationship in data. Irrelevant features can also adversely impact the performance of the model. Note that there is a possibility to convert an irrelevant feature to a relevant one by combining it with some other features. For example, to approximate an XOR function by a machine learning algorithm, a single input is irrelevant, but as combined with the other input, their combination can be used to produce the output of the XOR function. This case is called *feature interaction*. In the case of feature interaction, which means that there are multiple interacting features, the impact of individual features on the output is not significant. However, they may show a correlation to the target variable when considered in combination.

Another important problem is the presence of highly correlated features. In this case, any individual feature may provide similar performance to the correlated feature subset. These correlated features are also called *redundant features*. Typically, not much additional information from this type of features can be provided to achieve a better machine learning model.

The main advantages and benefits of feature selection are listed below:

- *Overfitting reduction*: Less redundant data helps reduce noises and thus generates a more accurate output.
- *Accuracy improvement*: Less irrelevant data contributes to more accurate model.
- *Training time reduction*: Fewer data accelerates the algorithm training process.
- *Fewer attributes*: Feature selection results in a simpler model that requires less explanation.

5.3.1 Feature Selection Techniques

In general, there are two categories of feature selection techniques:

- *Univariate method*: Input variables are processed one by one to calculate their relationship with the output, and then the most powerful input variables (i.e., those inputs with the highest correlations with output) are selected. This approach works well in practice but may also fail because it does not take into account the intercorrelations among input variables and the impacts of inputs on each other.
- *Multivariate method*: The whole group of variables is processed together. Although this approach is more efficient than the previous one, it is more complex and requires more computational resources.

Feature selection techniques can also be classified as follows:

- *Filter methods*: Filter method is typically used as a preprocessing phase. Filter methods are mostly univariate and non-iterative. Filter tries to assess the predictive power of each feature. To do so, several statistical techniques can be utilized to be able to compute the score (power) of the feature that demonstrates its "level of relationship/correlation" with the output. Some famous examples are chi-squared, F score, information gain, ANOVA, regression, and Pearson correlation.

- *Wrapper methods*: Wrapper methods address the feature selection problem similar to a search problem. These methods are called wrapper because they wrap a machine learning (e.g., classification) inside the feature selection process. Wrapper methods can be implemented in several ways, including:

  - *Forward Selection*: Forward selection is an iterative method which starts by an empty set. Then, we need to execute the machine learning model for each feature to find the strongest one that results in the best performance. In the next iteration, the selected feature from the previous step is combined with all other features one by one to find the best pair of features leading to the highest performance. We keep these two features and move to the next iteration. In the next iteration, we try to find the best three features, and so on until the specified number of features are selected.

  - *Backward Elimination*: This is also an iterative approach. We start with all features, and in each iteration, we remove/delete one of the features that does not have a significant impact on the performance of the machine learning model. This process is iteratively performed until a stopping criterion is reached.

- *Embedded methods*: Embedded methods are implemented using those machine learning techniques that have built-in feature selection abilities. In other words, feature selection is integrated/embedded as part of the learning algorithm. Regularization methods, which we discussed before, are one of the most common approaches in this regard. These methods find the appropriate features by adding some constraints into the optimization and cost function of the machine learning. Lasso and elastic net regressions are examples of embedded techniques.

5.3.1.1 Chi-Square Test

The chi-square test (also called the chi-square test of independence, and chi sounds like "Hi" but with a "K") is used to study the significant correlation between two categorical variables. Note that you cannot use chi-square to compare continuous variables or a categorical with a continuous variable. Let us explain the fundamentals of this method with a simple example. Suppose we observed 100 people to see who is interested in IoT and who is interested in Arts. Therefore, we have one categorical feature (independent variable) which shows the gender (i.e.,

Fig. 5.25 An example of the chi-square test

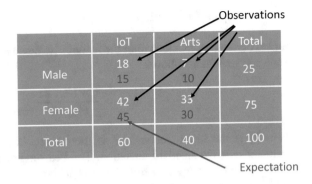

male, female) and one categorical dependent (target) variable that shows the interest of those individuals. As shown in Fig. 5.25, the observations can be summarized in a table called a *contingency table*. In our table, gender corresponds to rows of the table, interest corresponds to the columns of the table, and each cell corresponds to the frequency or the count of observations. Next, we need to define two hypotheses as listed below:

- *"Null" hypothesis (default hypothesis)*: Gender and interest (IoTs or Arts) are *independent* (i.e., there is no correlation between the feature and the dependent variable).
- *"Alternate" hypothesis*: Gender and interest are *not independent* (i.e., the feature and the target are correlated).

The next step is to calculate the expected value for each entry. To do so, we multiply each column total by each row total and divide by the overall total. As shown in Figs. 5.25 and 5.60 people are interested in IoT, and 25% of them are men. Therefore, we would expect 15 (25% of 60 persons) males to be the value (expected value) in the upper left cell. Next, $\chi^2 - statistic$ is computed based on the observed and expected variables as follows:

$$\chi^2 = \sum \frac{(observed - expected)^2}{Expected}$$

In our example, χ^2 is equal to 2. Finally, we need to test where the computed χ^2 lies on the χ^2 distribution curve to be able to accept or reject the hypothesis. Therefore, we look up the value 2 in the distribution curve (or in an χ^2 distribution table) to find the probability of this result. According to the distribution table, the corresponding probability for our example is 0.16. We also need a *significance level* which is usually 0.05 in chi-square test. The significance level is defined as the probability of rejecting the null hypothesis when it is true. For example, when the significance level is equal to 0.05, it indicates a risk of 5% in rejecting the null hypothesis. In our case, since 0.16 is bigger than 0.05, we retain our "Null" hypothesis meaning that there is no correlation between gender and interest.

5.3.1.2 Pearson Correlation

Pearson correlation methods filter the features based on their correlation coefficient, so one can write it as follows:

$$\rho_i = \frac{\text{cov}(X_i, Y)}{\sigma_{(X_i)}\sigma_Y}$$

where X_i is the input (feature), Y is the output, (X_i, Y) is the covariance, and parameter σ is the standard deviation. The Pearson correlation coefficient, which is also called a *sample correlation coefficient* or *sample Pearson correlation coefficient*, has a value in the range of $[-1,+1]$. The 0 coefficient indicates that there is no correlation between the two variables. A value greater than 0 implies that there is a positive correlation (i.e., when a variable goes up, the other variable also tends to increase) between variables, and a value less than 0 indicates a negative correlation.

5.3.1.3 Entropy

Entropy is a robust tool for correlation estimation and feature selection. Entropy is a parameter that measures the level of *impurity* in a group of examples (see Fig. 5.26). Entropy can be calculated by the following equation:

$$H(X) = Entropy = \sum_i p(x_i) \log_2 p(x_i)$$

In the above equation, $p(x_i)$ shows the probability of class i. Let us explain it with an example. Assume in our set (group), we have 16 red circles and 14 green triangles. Therefore, the probability of circle class is $16/(16 + 14)$, and the probability

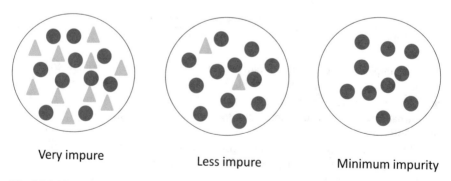

Very impure   Less impure   Minimum impurity

Fig. 5.26 Measuring the level of presence of different elements using the impurity metric

Fig. 5.27 Entropy vs.
probability (target class is 1)
for a two-class variable (X)

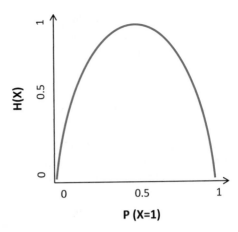

of triangle class is $14/(16 + 14)$. As a result, the corresponding entropy can be
calculated as follows:

$$Entropy = \left(\frac{16}{16+14}\right) \log_2 \left(\frac{16}{16+14}\right) + \left(\frac{14}{16+14}\right) \log_2 \left(\frac{14}{16+14}\right)$$

The entropy of a group that contains a single example class is zero. This means that
the group does not have any information. This also indicates that this group is not
a proper training set for the learning algorithm. In contrast, the entropy of a group
with 50% of either class is 1, which indicates that the group is a suitable training set
(i.e., the training set is balanced). Figure 5.27 illustrates the entropy vs. probability
of a two-class variable (e.g., red circle class and green triangle class). As you note,
the minimum of entropy is where the probability is equal to 0 or 1 (i.e., when we
have just red circles or green triangles). On the other hand, entropy rises to 1.0 at
a probability of 0.5 (maximum impurity) when the set is completely balanced (e.g.,
half of the examples are red circles and the other half examples are green triangles).

To be able to use entropy for feature selection, mutual information (MI) is
defined. MI shows how much information a variable has about another variable.
Larger mutual information (e.g., between the target (Y) and feature (X)) indicates
that the feature has more correlation with the target. Mutual information (MI) can
be calculated as

$$MI(Y, X) = H(Y) + H(X) - H(Y, X)$$

In the above equation, $H(Y,X)$ is a conditional entropy (entropy of a joint distribution):

$$H(Y, X) = -\sum_i p(y_i, x_i) \log_2 p(y_i, x_i)$$

Note that MI is zero, when there is not any correlation between X and Y, meaning that they are statistically independent variables. The maximum of MI happens when Y is completely dependent on X.

5.3.2 Feature Extraction

Feature extraction is slightly different from feature selection. While the latter selects a subset of the original input variables, the former generates some new variables (features) from the original ones. Principal component analysis (PCA), linear discriminant analysis (LDA), and spectral transformations (such as Fourier and wavelet transforms) are among the most well-known feature extraction techniques. Feature extraction improves the performance of the machine learning model by incorporating new relevant features.

5.4 Classification

Classification is another form of supervised machine learning, in which the output (target of the dependent parameter) labels are categorical. In other words, the output is classified into different groups. The goal of classification is to train and create a model (called classifier) based on the training dataset, which is then able to classify (i.e., predict the label or class) unseen data. Before presenting the details of different classification models and approaches, it is crucial to know about performance metrics, which is the first step for constructing classification models.

5.4.1 Measuring Performance for Classification Problems

Performance metrics are one of the key aspects of every machine learning projects, which determine how the performance of the algorithm is measured and is compared with other algorithms. Throughout this section, we explain each performance metric via a simple classification problem. We would like to predict if a given person has cancer (true/positive class) or not (false/negative class).

5.4.1.1 Confusion Matrix (Error Matrix)

The confusion matrix is a table to present the performance of a classification model. For a binary classifier, the confusion matrix is a 2×2 matrix similar to Fig. 5.28. In our example (prediction of cancer), the confusion matrix has two dimensions, namely, the actual dimension and predicted dimension. The actual dimension has

Fig. 5.28 Confusion matrix

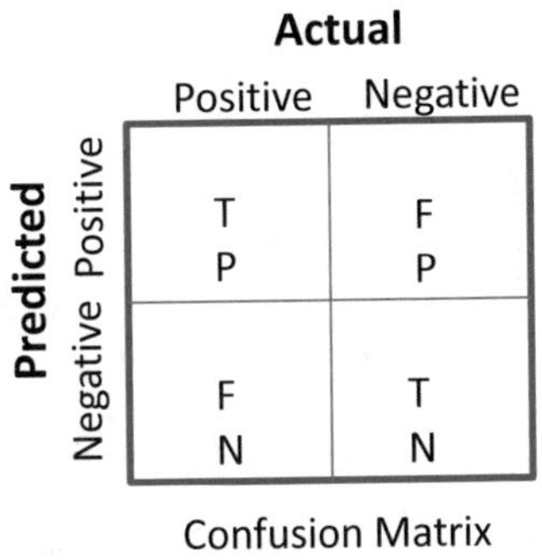

Actual

	Positive	Negative
Predicted Positive	T P	F P
Negative	F N	T N

Confusion Matrix

columns, and the predicted dimension has two rows corresponding to the number of available classes. Note that our problem has two classes (i.e., a person has cancer or not). The following quantities could be obtained through a confusion matrix:

- *True Positives (TP)*: TP show that the observation is positive and is correctly predicted to be positive.
- *True Negatives (TN)*: TN are negative cases, and they are correctly predicted to be negative.
- *False Positives (FP)*: FP indicate that the actual class was negative, but we incorrectly classified it as positive.
- *False Negatives (FN)*: FN show that the actual example was positive, but we incorrectly predicted it as negative.

Depending on the nature of the application, one of these four parameters can be minimized. For example, in our case (cancer prediction), missing a person with cancer is a big mistake because no further treatment or examination will be performed for him/her. As a result, we should minimize the false negative rate. Another example could be email spam detection system. In this case, true cases are spam emails. Now consider someone is waiting for an important email, but the system incorrectly marked the email as spam. This would be a huge mistake for the system. In this case, we need to keep the false positive rate as low as possible.

5.4.1.2 Performance Metrics

Some of the most important performance metrics that can be derived based on the confusion matrix are illustrated in Fig. 5.29.

Accuracy
Accuracy computes the ratio between the number of correct predictions (true positive and true negative) over all the predictions made by the model:

$$Accuracy = \frac{TP + TN}{TP + FP + FN + TN}$$

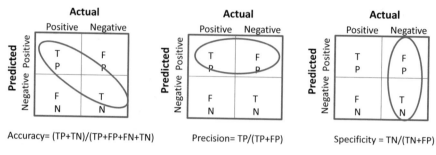

Fig. 5.29 Accuracy, precision, and specificity performance metrics

Accuracy is a good performance metric when the target classes are balanced. However, it should be avoided when the number of samples in each class is very different (i.e., imbalanced dataset). The reason is that in imbalanced datasets, the probability of instances belonging to a minority class is significantly low compared to a majority class. Therefore, the classifier tends to classify new observations mostly as the majority class. In our example, imagine that there are only five cases of cancer out of every 100 cases. In this situation, if the system predicts all 100 cases as noncancerous, the accuracy of the model is 95%, but apparently, the model is terrible at predicting cancer.

Precision (Positive Predictive Value)

Precision represents the proportion of true, relevant predictions (i.e., the percentage of your model results, which are relevant, or the ratio between the relevant instances and the total retrieved instances). Precision is formally defined as the ratio between the number of true positives and the number of true positives plus the number of false positives. In our example, the precision is defined as how many of the people detected as cancerous have cancer. In other words, precision indicates how much the model is precise. For example, if we predict just one cancerous patient, and the patient has cancer, the precision is 100%.

$$Precision = \frac{TP}{TP + FP}$$

Recall (Sensitivity)

Recall or sensitivity expresses the ability of the model to detect all the relevant cases (all the points of interest) within a dataset. Formally, recall is the number of true positives divided by the number of true positives plus the number of false negatives. In our cancer detection example, recall demonstrates how many of the cancerous patients are predicted as having cancer. If we mark every patient as cancerous, the recall is 100%.

$$Recall = \frac{TP}{TP + FN}$$

Choosing the proper metric between precision and recall depends purely on the problem statement and application. As a general rule, if the focus is on minimizing the false negatives, effort should be put on making the recall value close to 100%. On the other hand, if it is important to minimize false positives, the precision value should be close to 100%.

Specificity (True Negative Rate)
Specificity or true negative rate, as suggested by its name, measures the ability of the model to identify those cases that are negative correctly. In our example, specificity would be the proportion of the predicted healthy (i.e., negative: do not have cancer) people that are correctly marked as healthy. Therefore, the model will have 100% specificity when it correctly finds all patients without cancer.

$$\text{Specificity} = \frac{\text{TN}}{\text{TN} + \text{FP}}$$

F1 Score
F1 score is an optimal blend of recall and precision that can be calculated as follows:

$$\text{F1} = \frac{2 \times \text{Precision} \times \text{Recall}}{\text{Precision} + \text{Recall}}$$

This equation is a harmonic mean, and contrary to a simple average, it handles extreme values. For example, a classifier with a precision of 1.0 and recall of 0 would have an F1 score of 0, which would be 0.5 for a simple average. F1 score is also called the F score or F measure.

ROC Curve
A ROC curve (receiver operating characteristic curve) is a plot that demonstrates the performance of a classifier at different *thresholds*. In the ROC graph, true positive rate (TPR) is depicted as a function of false positive rate (FPR). As you will see later in this chapter, classifiers normally generate a probability as output for a given input (features). This shows the probability that a given input belongs to a specific class. To be able to classify the input, we need to choose a threshold. An output value above that threshold indicates positive class and a value below the threshold indicates negative class. Note that decreasing the classification threshold (decision threshold) results in predicting more cases as true, which in turn increases both FP and TP simultaneously. A typical ROC curve is plotted in Fig. 5.30.

AUC (Area Under the ROC Curve)
Area under the ROC curve (AUC) is a statistic parameter for model comparison. As its name implies, AUC measures the area underneath the entire ROC curve. This parameter aggregates the performance of the model across different classification thresholds (Fig. 5.31). This parameter enables us to identify which of the trained models predicts the classes best. In other words, it helps to rank and sort classification models.

Fig. 5.30 TP and FP rates at different classification thresholds

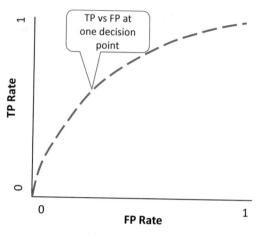

Fig. 5.31 Area under the ROC curve, known as AUC

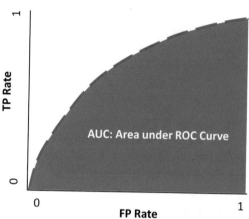

5.4.2 Over- and Undersampling

As noted in previous sections, classification metrics might be very confusing and misleading, specifically when the dataset is imbalanced. Over- and undersampling are widely used to overcome the challenge of imbalanced datasets, in which there is a majority of one class in comparison to other classes. Figure 5.32 illustrates an imbalanced dataset graphically. Undersampling, as its name suggests, selects only part of the majority class equal to the number of data points of the minority class for model creation. This results in a balance between probability distributions of classes.

Inversely, in oversampling, copies of the minority class are created in order to reach the number of examples in the majority class. The copies should be created in a way that does not affect the distribution of the minority class. Figure 5.32 demonstrates these concepts clearly.

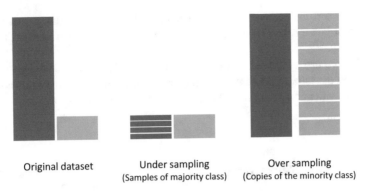

Original dataset   Under sampling   Over sampling
                     (Samples of majority class)   (Copies of the minority class)

Fig. 5.32 Undersampling and oversampling techniques to address the problem of imbalanced datasets

Fig. 5.33 A simple example of KNN classification

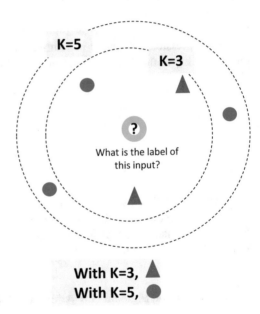

5.4.3 K-Nearest Neighbor (KNN)

k-nearest neighbors (KNN) is one of the simplest yet popular classification techniques that was first described in the early 1950s. KNN algorithm can be summarized as follows (see Fig. 5.33):

- First, we need to define K.
- Next, we calculate the distance between the given input (which should be classified) and all samples of the training set.

- Next, we sort the distances and select K-nearest neighbors of the given input. In other words, we select K samples of the training set which are closest to the given input.
- Finally, we use a simple majority to identify the label (class) of the given input based on the labels (classes) of its neighbors. In other words, the most common label/classification of its neighbors is selected as the label (class) of the given input.

5.4.4 Logistic Regression

Logistic regression is the go-to technique for binary classification in which the dependent variable (target) has just two classes. The main difference between linear regression and logistic regression is that in logistic regression the dependent variable is categorical and has as a binary value (two classes). On the other hand, in linear regression, the output is a numerical value. Logistic regression computes the probability of the default class. The mathematical form of logistic regression is given by

$$h_\theta(x) = g\left(\theta^T x\right)$$

$$g(z) = \frac{1}{1 + e^{-z}}$$

in which θ is coefficients/weights, x demonstrates the input (features), and $g(z)$ is a *sigmoid function* (also called *logistic function*) *which has an S shape* (see Fig. 5.35). To better understand the logistic regression, we need to study the fundamentals of logit and sigmoid functions.

5.4.4.1 Logit and Sigmoid (Logistic) Functions

Logit and sigmoid functions are widely used functions in machine learning applications. For a probability p, the corresponding odds (i.e., the ratio of the probability that an event will occur to the probability that the event will not take place) can be calculated by $\left(\frac{p}{1-p}\right)$. The logit function is the logarithm of the odds (Fig. 5.34):

$$logit(x) = log\left(\frac{x}{1 - x}\right).$$

The logit function leads to positive infinity and negative infinity as the value of p approaches 1 and 0, respectively. Due to the fact that the logit function maps the probability values to a full range of real numbers, it is frequently used in analytics.

Fig. 5.34 A plot of logit function. The logit function yields positive infinity and negative infinity as the value of x approaches 1 and 0, respectively

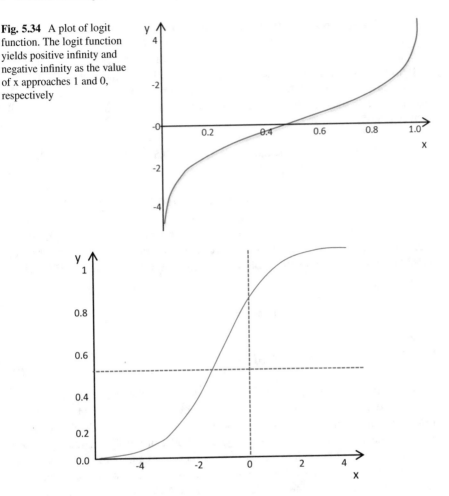

Fig. 5.35 A plot of sigmoid (logistic) function

For example, by taking advantage of logit function, one can transform a "yes-no" input to real-valued quantities. This is one of the fundamental concepts of logistic regression.

Sigmoid function (also called logistic function) is the inverse of logit function, so for a given probability p, $sigmoid(logit(p)) = p$. As a result, the sigmoid function maps a real value to the range of [0,1]. Larger inputs cause an output closer to 1. The sigmoid function is declared as $\sigma(x) = {}^{1}/_{1 + e^x}$ (see Fig. 5.35). One of the most significant applications of the sigmoid function is the situation that we want to map a real value into something similar to a probability, which is mostly used at the end stage of a classification algorithm. We will discuss the details of classification algorithms in the next section.

5.4.4.2 Decision Boundary (Decision Surface)

In order to map the returned score of the logistic function (which is a probability in the range of [0,1]) to a binary class, a threshold is defined, above which the classification output would be the class 1; otherwise the class would be 0. For example, when the threshold is 0.5, classes can be identified by the following rule (see Fig. 5.35):

$$Class = 1 : h_\theta(x) \geq 0.5,$$

$$Class = 0 : \ h_\theta(x) < 0.5$$

Now, let us explain the meaning of the decision boundary by an example. Recall that $h_\theta(x) = g(\theta^T x)$ in which $g()$ is a sigmoid (logistic) function and x represents the input (features). When the threshold value is equal to 0.5, according to Fig. 5.35, it implies that

$$Class = 1 : \theta^T x \geq 0,$$

$$Class = 0 : \ \theta^T x < 0$$

Now suppose that we have a training set similar to Fig. 5.36 and we want to classify the inputs into two classes (i.e., red circles and green triangles). We can draw several different hypotheses about $\theta^T x$. A very simple linear hypothesis might be $h_\theta(x) = g(\theta_0 + \theta_1 x_1 + \theta_2 x_2)$. As shown in the figure, this hypothesis represents a line (blue color) which divides the inputs into two different classes. This line is called the *decision boundary*. Formally, a decision boundary is a hyperplane/hypersurface that divides the underlying vector space into classes.

Note that a decision boundary does not need to be just linear. Adding more precision to the logistic regression model is possible by including higher-order

Fig. 5.36 Linear decision boundary

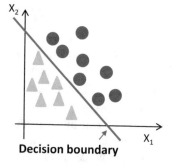

Fig. 5.37 Nonlinear decision boundary

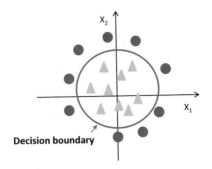

Decision boundary

polynomial terms and creating a nonlinear decision boundary. An example of a nonlinear decision boundary is depicted in Fig. 5.37 based on the following equation:

$$h_\theta(x) = g\left(\theta_0 + \theta_1 x_1 + \theta_2 x_2 + \theta_3 x_1^2 + \theta_4 x_2^2\right)$$

5.4.4.3 Cost Function in Logistic Regression

In the previous subsection, we learned what a linear/nonlinear decision boundary is. Now the question is how to compute the coefficients/weights of features (input). Instead of using mean squared errors (MSE) as the cost function (similar to linear regression), logistic regression uses a *cross-entropy* function, which is also called *log loss*. This cost function is divided into two cost functions for $y = 1$ (first class) and $y = 0$ (second class) separately [8]:

$$J(\theta) = \frac{1}{m} \sum_{i=1}^{m} Cost\left(h_\theta\left(x^{(i)}\right), y^{(i)}\right)$$

$$Cost(h_\theta(x), y) = -\log(h_\theta(x)) \ \ \text{if } y = 1$$

$$Cost(h_\theta(x), y) = -\log(1 - h_\theta(x)) \ \ \text{if } y = 0$$

which can be written as

$$J(\theta) = -\frac{1}{m} \sum_{i=1}^{m} \left[y^{(i)} \log\left(h_\theta\left(x^i\right)\right) + \left(1 - y^{(i)}\right) \log\left(1 - h_\theta\left(x^i\right)\right)\right]$$

The cost function for $y = 1$ and $y = 0$ is plotted in Fig. 5.38.

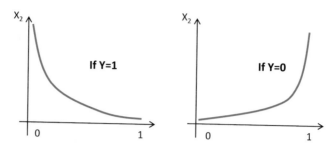

Fig. 5.38 Cost function of logistic regression

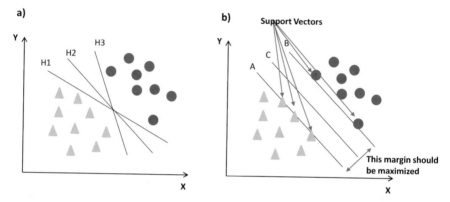

Fig. 5.39 A 2D support vector machine model. (**a**) There are many possible hyperplanes (e.g., H1, H2, and H3) that could be chosen to separate the data. (**b**) Optimal Hyperplane (has the maximum margin, i.e., the maximum distance between data points of both classes) using the SVM algorithm

5.4.5 Support Vector Machine

Support vector machine (SVM) is one of the popular classification methods that was created in the late 1990s. SVMs succeed in finding the optimal separation solution to classify between data points belonging to two classes. Figure 5.39a is an illustration of SVM in the 2D plot. Three separating hyperplanes (H1, H2, and H3) are plotted, which are called decision boundaries in classification. As you note, even for a simple classification problem, we can draw several hyperplanes to partition the underlying space and classify the inputs. The key question is which of these hyperplanes is the optimal one and how can it be computed. SVM enables us to address this issue.

The main objective of an SVM algorithm is to find a hyperplane with the maximum distance from data points of both classes (Fig. 5.39b). In other words, the goal is to find a hyperplane which has the largest margin (i.e., the one that creates a street with the largest width between classes). This specific hyperplane is called maximum-margin hyperplane. In this regard, support vectors play an important role. *Support vectors* are data points that are located closest to the hyperplane and

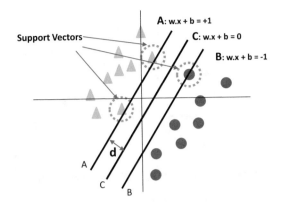

Fig. 5.40 Mathematical presentation of a support vector model

touch the boundary of the margin [9]. Therefore, support vectors determine the hyperplane position and orientation. For example, in Fig. 5.40 we have just three support vectors. Note that only these support vectors impact the location of the hyperplane, and the other data points are not important in the SVM algorithm.

Let us formulate the problem of linear SVM formally. The input of SVM is a training dataset of n points of the form

$$\left(\vec{x_1}, y_1\right), \left(\vec{x_2}, y_2\right), \ldots, \left(\vec{x_n}, y_n\right)$$

where each x_i represents an n-dimensional real vector (vector of features). Parameter y represents the output (target) and it is a two-class output (either 1 or -1). The goal is to maximize the margin. Any hyperplanes can be expressed as

$$\vec{w} . \vec{x} - b = 0$$

in which parameter $\vec{w}$ is a normal vector (weights), and $\vec{x}$ is the set of points. If the data points are linearly separable, we would be able to draw two hyperplanes (e.g., hyperplanes A and B in Fig. 5.40). Technically speaking, the margin is surrounded by these two hyperplanes, and the maximum-margin hyperplane (hyperplane C in Fig. 5.40) is located exactly in the middle of them. Given a normalized dataset, the two hyperplanes located at the border of the margin area are described as follows:

$$\vec{w} . \vec{x} - b = 1 \ (the \text{ class with label } 1)$$

$$\vec{w} . \vec{x} - b = -1 \ (the \text{ class with label } - 1)$$

To have better intuition and to understand the impact of $\vec{w}$ and b, see Fig. 5.41. From a geometrical point of view, the distance between these two hyperplanes is $\frac{2}{\|\vec{w}\|}$. As a result, minimizing the denominator ($\|\vec{w}\|$) results in maximizing the

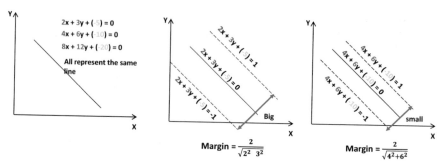

Fig. 5.41 An example of hyperplanes and the impact of coefficients/weights

Fig. 5.42 An example of not linearly separable data

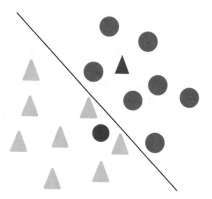

distance. To prevent data points from being in the margin area, we need to add the following constraints during the optimization:

$$\overrightarrow{w} \cdot \overrightarrow{x_i} - b \geq 1, \text{if } y_i = 1$$

$$\overrightarrow{w} \cdot \overrightarrow{x_i} - b < 1, \text{if } y_i = -1$$

which can be rewritten as (for each data point)

$$y_i \left(\overrightarrow{w} \cdot \overrightarrow{x_i} - b \right) \geq 1, \text{for all } 1 \leq i \leq n$$

To put it all together, we should consider minimizing $\| \overrightarrow{w} \|$ while considering the above constraints (when the classes are linearly separable). This is an optimization problem that can be solved by the Lagrangian multiplier method [9]. If the data are not linearly separable (Fig. 5.42), we cannot use the above constraints during the optimization. In this case, we need to find a hyperplane that penalizes data points on the wrong side. To do so, we can take advantage of the hinge loss function to penalize the points on the wrong side:

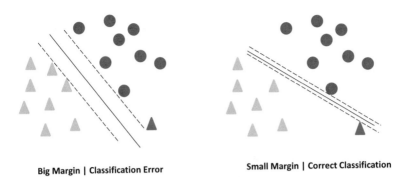

Big Margin | Classification Error   **Small Margin | Correct Classification**

Fig. 5.43 Tradeoff between classification error and margin

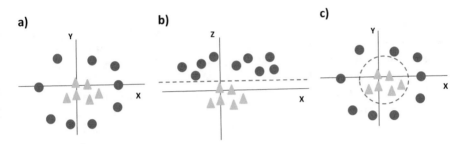

Fig. 5.44 Kernel trick in SVM. (**a**) Non-linearly separable data. (**b**) Data on higher dimension and a linear decision boundary. (**c**) Decision boundary in original dimensions

$$\max\left(0,\, 1 - y_i\left(\vec{w}.\vec{x_i} - b\right)\right)$$

This cost function is 0 if the actual and predicted values are on the same side, but for the wrong side points, the function's value (penalty) is increased proportionally to the distance of the wrong data points from the hyperplane. Now let us study the example of Fig. 5.43. As shown in this figure, we extract a hyperplane that has a big margin, but it cannot classify all the data points. On the other hand, we can have a hyperplane that has a very small margin (which is not good for generalization), but it can classify all the data points. In SVM, we can make a tradeoff between classification error and margin. To do so, we define a regularization parameter called C and the SVM cost function is defined as

$$\text{Cost} = \max\left(0,\, 1 - y_i\left(\vec{w}.\vec{x_i} - b\right)\right) + C * \left\|\vec{w}\right\|$$

In a case that data cannot be classified by a linear hyperplane (similar to Fig. 5.44a), we can apply a technique called *kernel trick*. The idea is to use a nonlinear function (kernel) to map the data points to a new high dimensional space, in which we can find a linear hyperplane using the abovementioned linear SVM technique [9, 10]. Let us apply the kernel trick to our example (Fig. 5.44a) to clarify

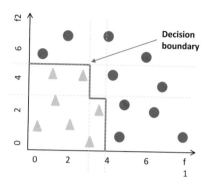 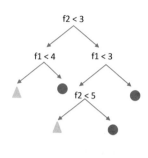

Fig. 5.45 A decision tree model

it better. In our example, we map our 2D data points to a 3D space. Suppose our mapping function is $x^2 + y^2$ which computes the value of data points in the z-axis. Now if we plot the data points using the newly computed values (x-z plane), we can realize that there is a linear hyperplane between two classes (Fig. 5.44b). The final step in the kernel trick is to move back from the higher dimensional space to the original space [9, 10]. In our example, transforming back this separation line (hyperplane) will create a circular boundary similar to Fig. 5.44c. Finally, we should note that similar to regression analysis, the SVM algorithm can also take advantage of regularization to generalize the solution and to avoid overfitting [9].

5.4.6 Decision Tree Classifier

A decision tree is a category of classification and regression algorithms. In these algorithms, a tree-like model of decisions is constructed. Classification of a new data point is accomplished by simply traversing down the tree. In decision tree algorithms, the domain (feature space) is divided into several regions, and each region is marked with a class label (or probability of a label) similar to Fig. 5.45. The common terms in decision trees are explained below:

- *Nodes*: In nodes, we check the value of a certain feature (attribute).
- *Edges*: Edges correspond to the output of the above test (i.e., test the value of a feature in node). An edge also connects a node to another one or to a leaf.
- *Leaves*: These are terminal nodes which show the actual prediction.

There are several implementations of decision trees such as C5.0, C4.5, and Iterative Dichotomiser 3 (ID3). C5.0 is the most well-known decision tree algorithm which has become the standard approach in the industry. A very simple technique to build the decision tree is the basic recursive divide-and-conquer algorithm consisting of the following steps:

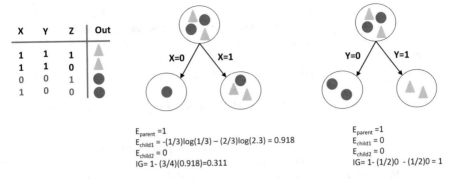

X	Y	Z	Out
1	1	1	△
1	1	0	△
0	0	1	●
1	0	0	●

$E_{parent} = 1$
$E_{child1} = -(1/3)\log(1/3) - (2/3)\log(2.3) = 0.918$
$E_{child2} = 0$
$IG = 1 - (3/4)(0.918) = 0.311$

$E_{parent} = 1$
$E_{child1} = 0$
$E_{child2} = 0$
$IG = 1 - (1/2)0 - (1/2)0 = 1$

Fig. 5.46 A simple example of selecting the split attribute based on information gain (IG)

- *Select* a feature (attribute) for the root node of the tree and then create an edge (i.e., a branch) for each value of the attribute.
- *Split* (divide) the data points (instances) into subsets (i.e., one for each branch created in the previous step).
- *Repeat* the above two steps recursively for each branch (each subset).
- *Stop* the above recursion for a branch when all its instances are in the same class (i.e., have the same label).

One important question that we need to address in the above algorithm is how to select the root in each iteration. In other words, how can we identify which feature to split upon? There are several methods to select the best attribute for splitting in each step. In general, a good attribute is the one that splits the successor nodes as pure as possible. It means that the corresponding branches contain mostly instances of one class. This can be explained by entropy. The dividing procedure should decrease the entropy because a good attribute (node) splits a set into subsets (regions) with the higher homogeneity (higher entropy means there is a mix of different classes in each region). To be able to implement this process, "information gain (IG)" is defined as follows:

$$Information\ Gain = Entropy\ (Parent\ node)$$

$$- [Average\ Enropy\ (children\ nodes)]$$

Information gains represent the decrease in entropy after a split on an attribute. It means that we need to select an attribute for splitting procedure that has the highest information gain. In other words, an attribute with the highest information gain leads to the most homogeneous branches (lower entropy or very pure subset). Let us explain the abovementioned process with an example. Assume we have three attributes (feature) and the target has two classes (see Fig. 5.46). If we split on X attribute, the IG will be 0.31. On the other hand, if we select Y as our splitting attribute, the IG will be 1. Therefore, we split based on Y.

The other challenge in the construction of decision trees is the stop criteria of splitting procedure, in other words, when we need to stop the splitting. One possible approach is to continue the splitting until each leaf node in the decision tree is completely pure (i.e., only the instances/examples of one class are in leaf). This approach might not be very efficient for all the applications, because it creates a large number of small regions in the feature space. From a technical perspective, this increases the overfitting. *Pruning* techniques are a great solution to tackle this issue.

- *Pre-pruning*: We stop growing the tree when the performance of the classifier is higher than a predefined threshold.
- *Post-Pruning*: First, we grow the complete decision tree. Although a complete tree can classify all the training data points correctly, it may suffer from overfitting. Next, a pruning algorithm is iteratively applied to simplify the tree by removing some of its nodes. This iterative algorithm decides to remove some of the nodes if the increase in entropy is below a predefined threshold.

5.4.7 Ensembles

An ensemble machine learning combines several weak models in order to create one single strong meta-model to be able to tack high bias (underfitting) and high variance (overfitting) issues. There are two broad categories of ensemble techniques, namely, *bootstrap aggregating (bagging)* and *boosting*.

5.4.7.1 Bootstrap Aggregating (Bagging)

Bootstrap aggregating, which is also called bagging, is an effective technique to reduce the model overfitting and to handle unstable datasets. It also improves the performance of training on a dataset with a limited number of training data. For example, bagging can be used to combine multiple decision trees as a forest model to obtain a stronger classifier.

This method generates n small training set from the original dataset. Each of them is called one sample. These samples are produced by random sampling of the input dataset with replacement, as illustrated in Fig. 5.47. Let us explain the concept of sampling with replacement with a simple example. Suppose we have three names and we need to sample two. The names are Farshad, Sani, and Krish. We put these three names in a hat and randomly choose one of them. Then we put the name back into the hat and select another name. The possibilities of two-sample names are (Farshad, Farshad), (Farshad, Sani), (Sani, Sani), and so on.

When all samples are constructed by random sampling with replacement technique, we independently learn and build n ensembles (i.e., individual learning models) corresponding to n samples. Each of these weak models is called the

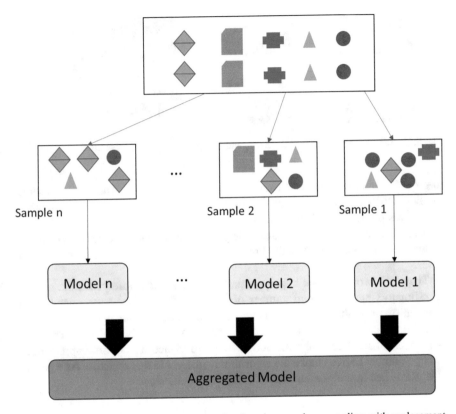

Fig. 5.47 An overview of bootstrap aggregating based on random sampling with replacement technique

estimator. Finally, the outputs of these weak models are combined by voting or simple averaging to provide a meta-model (see Fig. 5.47).

5.4.7.2 Random Forest

Random forest (RF) or random decision forest is one of the implementations of the bagging technique. Simplicity, flexibility, and great results have made RF one of the most widely used algorithms for machine learning. The term forest points to an ensemble of multiple decision trees (each of them is called *estimator*). This ensemble of decision trees (estimator) is merged in the random forest method to obtain predictions with higher accuracy.

That being said, RF adds additional randomness to bagging model by modifying the original decision tree. Recall that in the original decision tree, in each iteration, we need to select an attribute/feature to split upon. This attribute is selected among all available attributes. However, in the modified learning algorithm, when we want

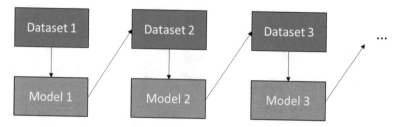

Fig. 5.48 Schematic presentation of the sequential steps in the boosting algorithm

to grow the tree, first we select a random subset of features. And then, we select the best attribute for that selected subset. The randomness of selecting a subset of features produces a wider diversity that contributes to better results. Full-grown decision trees are also at risk of overfitting. The randomness in selecting a subset of features in a random forest method prevents overfitting of the model in most of the cases.

There are two important hyperparameters in the random forest method: number of estimators and maximum number of features. The former corresponds to the number of trees the algorithm builds, and the latter is the maximum number of features the algorithm is allowed to evaluate in an individual tree. The main bottleneck of random forest method is its performance in real-time applications. A large number of trees make the algorithm inefficient in this scenario. As a general rule, RF is fast in training but makes predictions quite slow.

5.4.7.3 Boosting

Boosting is another method for producing ensembles, in which the ensembles are created *sequentially*. Recall that in bagging methods are *parallel* because we create and learn n base models in parallel. In contrast, the idea of boosting method is to use feedback from one base model to produce the next base model. In other words, we consider the training instances of the previously generated base models which are misclassified (see Fig. 5.48). In comparison to bagging, boosting shows a better performance for specific applications, but it increases the risk of overfitting the model.

Adaptive Boosting (AdaBoost)

AdaBoost is one of the well-known implementations of boosting. The core idea of AdaBoost is to sequentially modify a weak model to make it a better classifier. In fact, AdaBoost creates a better model by combining several weak classifiers linearly, and the final classification is performed using this combination, which can be expressed as

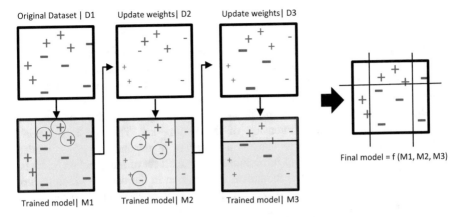

Fig. 5.49 The overall procedure of the AdaBoost algorithm

$$f(x) = \sum_{t=1}^{T} \alpha_t h_t(x)$$

in which the $h_t(x)$ are the basis classifiers. Here, the weak classifiers are single split decision trees (called decision stumps). In the beginning, all training data have equal weight. AdaBoost modifies the weights in a way that difficult to classify instances get more weight and adds new weak classifiers sequentially with the focus on more difficult instances. In this regard, each classifier (weak classifier) is trained by taking a random subset of the training set, and then, AdaBoost assigns higher weights to misclassified training items. As a result, this misclassified item will have a higher probability to appear in the next training subset for the next classifier.

A simple example of AdaBoost procedure is depicted in Fig. 5.49. This figure explains how AdaBoost updates the weights in each step to construct a final model by a linear combination of several weak classifiers. Bigger weights are illustrated by larger signs and smaller weights are shown by smaller signs.

Gradient Boosting
There is another type of boosting that is contrary to AdaBoost and works on the basis of training on the remaining errors (or pseudo-residuals) of stronger classifiers. This approach is known as gradient boosting, in which at each training iteration, a weak classifier is fitted on the computed pseudo-residuals. Then the effect of this weak classifier on the performance of the stronger one is computed based on a gradient descent optimization process.

5.5 Dimensionality Reduction

Computers cannot think in the way that human brains do, and developing neural networks is an attempt to address this issue. An artificial neural network, first developed in the 1950s, is a simulation of the neurons of the human brains in a manner that the computer can learn things in the humankind.

Dimensionality reduction is the process of reducing the number of machine learning features and the dimension of the feature set. The key idea behind dimensionality reduction is to convert a large set of dependent and correlated features (which have redundant information) to a small set of independent features. This enables us to remove the redundant information in the dataset which in turn improves the speed as well as the performance of machine learning algorithms.

One of the most well-known techniques for dimensionality reduction is principal component analysis (PCA). In PCA, the directions with the largest variances are considered as most "important" (i.e., the most principal) features. Therefore, to find the most important features, one needs to find the directions of the data that have the maximum amount of variance. PCA algorithm can be summarized in the following steps (see Fig. 5.50):

- The first step is to perform standardization (i.e., subtracting the mean and dividing by the standard deviation).
- The next step is to find the covariance matrix of the features.

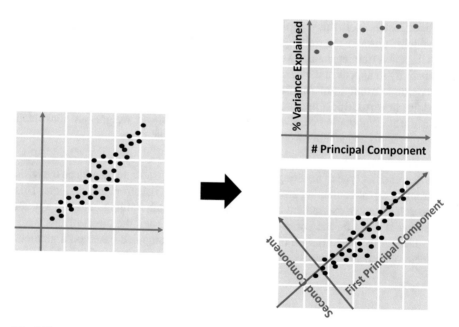

Fig. 5.50 An illustration of PCA

- Since the covariance matrix is a symmetric matrix, it can be decomposed to three matrices. After the composition, we can find its eigenvectors and eigenvalues. Principal components are indeed the eigenvectors of the covariance matrix.
- Next, we need to sort the eigenvectors by decreasing eigenvalues and select n important eigenvectors.
- The final step is to transform and project the original dataset using the eigenvectors onto a smaller subspace in which those eigenvectors form the axes of the new feature subspace.

5.6 Artificial Neural Networks

5.6.1 Neural Network Models

Computers cannot think in the way that human brains do, and developing neural networks is an attempt to address this issue. An artificial neural network, first developed in the 1950s, is a simulation of the neurons of the human brains in a manner that the computer can learn things in a humankind way. Generally speaking, a neural network is a class of machine learning techniques that mimic the behavior of neurons. Back in the 1950s, David Hubel and Torsten Wiesel, two famous neurophysiologists, performed experiments on cats and proposed their insights on the structure of the visual cortex, which was credited Nobel Prize in Physiology or Medicine in 1981. This work was the prototype of the neuron and was later developed to the entire neural network methodology. The finding of the visual cortex is that a single neuron only responds to stimuli in a restricted area (region) of the visual field that partially overlaps the region of close neurons, collectively covering the entire visual field. Neurons are different from each other. Some neurons are responsible to horizontal lines, while some are responsible to vertical lines, while some are responsible to larger areas, but some are responsible to small but complex patterns, which are a combination of low-level patterns. Therefore, these findings lead to the idea that layered neuron structure, where some layer neurons detect only simple patterns, and some layer neurons connected to previous layer neurons calculate previous layer neurons to detect more complex patterns. These studies gradually evolved into what we then called deep learning (DL) and Convolutional neural network (CNN) [11].

A neural network consists of neurons and weights. The neurons apply a function on the input values and pass the result to the output, and the weights carry this result to other neurons. Neurons are grouped into layers. Based on the way that the layers are arranged, they can be any of the *input*, *hidden*, or *output* layers, which are the main types of the layer in a typical neural network model. Figure 5.51 demonstrates a schematic view of a neural network model. In this model, every neuron is a computational unit that performs a function (called *activation function*) on the cumulative sum of its input values. The input of each neuron is the multiplication

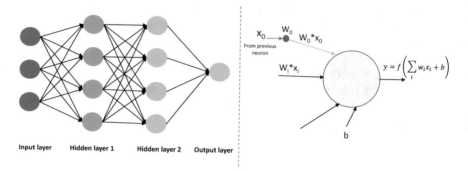

Fig. 5.51 Neural network architecture and neuron breakdown

of the output of previous layer neurons (x_i) by the corresponding *weights* that connect the neuron to the previous layer (w_i). Each neuron has also a bias (b), which contributes to the output of the neuron. As a result, the output of each neuron can be computed by $y = f\left(\sum_i w_i x_i + b\right)$, which is illustrated in Fig. 5.51.

5.6.2 Train a Neural Network Model

In the training process, the internal weights between the neurons of a neural network are updated in a cascading way. The training process can be summarized as follows [12]:

- *Step 1 (Model Initialization)*: In this step, each variable (e.g., weights) is a given value. Random initialization of the network is a common practice.
- *Step 2 (Forward Propagation)*: As the name suggests, in this step, input data is "forward propagated" through the network layer by layer (from the input layer to the output layer) to finally produce the output of the network. In other words, we perform an iterative process. In each iteration, neurons of each layer accept input, process it, and finally pass the corresponding output to the successive layer.
- *Step 3 (Backward Propagation)*: Once we computed the output and realized the error of the network (model), we backpropagate the errors from the output layer to the hidden layers and input layer to be able to update the weights accordingly.
- *Step 4*: We execute steps 2 and 3 iteratively until all the weights converge.

Let us explain the above three steps in more details. Before the neural network model training, the initial values of the weights should be selected properly. Zero is not a good choice because it causes the output of the first layer to be the same, which leads to a similar gradient during backpropagation. Instead, a commonly used approach is to initialize all weights randomly with small values. After the initialization of these parameters, the model can be trained with a *gradient descent algorithm*. To do so, a forward pass through the model generates an output value, which can be used to calculate the model error. We typically use a loss (cost)

Fig. 5.52 Visual chain rule, applied in the backpropagation algorithm

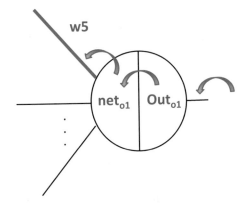

function to measure the error in relation to the correct output. The error is then used to update all the weights of the neural network model in the backpropagation phase using the *gradient descent* equation:

$$W^{[l]} = W^{[l]} - \alpha \frac{\partial E}{\partial W^{[l]}}$$

$$b^{[l]} = b^{[l]} - \alpha \frac{\partial E}{\partial b^{[l]}}$$

in which α is the learning rate. In other words, the new weight can be computed as (new weight = old weight—derivative rate $*$ learning rate). Recall that we used the same algorithm to find out the weights of the linear regression model parameters (see Fig. 5.18). The idea of gradient descent is to update the weight iteratively based on the derivative (or gradient or slope) of the loss function. Note that in the multilayered neural network, to be able to extract derivatives of cost/loss concerning an internal variable (weight), we need to use the *chain rule*. As an example, let us compute $\frac{\partial E_{total}}{\partial W_5}$ in Fig. 5.52. The corresponding chain rule can be written as follows:

$$\frac{\partial E_{total}}{\partial W_5} = \frac{\partial E_{total}}{\partial out_{o1}} \times \frac{\partial out_{o1}}{\partial net_{o1}} \times \frac{\partial net_{o1}}{\partial W_5}$$

It is also worth noting that in a neural network model, we might encounter several local optima (Fig. 5.53) because the training process of a neural network is a non-convex optimization problem. Similar to other machine learning methods, neural networks are also vulnerable to overfitting, which can be prevented by *generalization techniques* such as:

- Reducing the number of hidden layers and the number of neurons in the hidden layers
- Reducing the weight values by adding extra terms to the performance function (e.g., L2 regularization which we discussed before)

Fig. 5.53 Local optima in the training process of neural network models

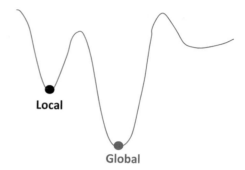

Local

Global

- Early stopping, which means the learning is stopped before the overfit occurs
- *Dropout* technique, in which randomly selected neurons are ignored (dropped out) during some iterations of the training process. This technique makes the training process noisy. As a result, neurons take the same level of responsibility and learn a sparse representation, which makes the model more robust.

Early stopping is an effective method for generalization. In this approach, the data is divided into three datasets: training set, validation set, and the test set. The training set is used for the backpropagation algorithm and updating the weights of the model. During the training process, the error of the validation set is monitored, which should normally show a decreasing trend. But as the network begins to overfit the data of the training set, the error of the validation set will increase. This behavior can be used to stop the training process, and the values of the weights and biases at the time that the validation set error was minimum can be chosen as the proper result of the training process. Finally, the test set is used to compare the efficiency of different models [12, 13].

5.6.3 Activation Function

The activation function is one of the key hyperparameters in a neural network. Sigmoid ($\sigma(z) = \frac{1}{1+\exp(-z)}$), tanh ($\tanh(z) = \frac{\exp(z)-\exp(-z)}{\exp(z)+\exp(-z)}$), and ReLU ($ReLU(z) = \max(0, z)$) are the most commonly used activation functions. Figure 5.54 represents the plots of these functions [14].

ReLU Basically, what ReLU does is keeping positive input as is while rectifying all the negative inputs as 0. Accordingly, one of the key advantages of ReLU compared to other activation functions is that it does not activate all neurons simultaneously by throwing out all the negative values. This makes it very computationally efficient, specially when there is a very big and deep neural network consisting of several layers with dozens of neurons. In practice, ReLU converges much faster than sigmoid and tanh activation functions.

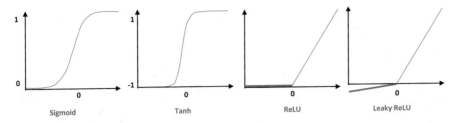

Fig. 5.54 Most popular activation functions

Sigmoid In practical applications, the sigmoid activation function is less used despite its popularity in the past because of two important problems:

1. Sigmoid function eliminates the gradient, which means when the input value is very big or very small, the output of the function is either 1 or 0. This will result in a *vanishing gradient problem* in the backpropagation algorithm. In this case, no signal is transmitted through the neurons and thus the neuron will not learn anything in the training phase.
2. The outputs of the sigmoid function are not zero centered, and in the backpropagation algorithm, this will create gradients that are either all positive or all negative. This is also not appropriate for the gradient updates of the weights.

Tanh Similarly, the tanh activation function has the vanishing gradient problem. However, in contrast to the second problem of the sigmoid function, its output is zero centered.

ReLU ReLU, an abbreviation of rectified linear unit, has gained great popularity in recent years. It has been reported that ReLU activation function greatly impacts the convergence rate of stochastic gradient descent algorithms, six times more than tanh and sigmoid functions [6–15]. The primary reason for this phenomenon is its linearity, which causes the gradient not to vanish. Besides, ReLU is less computationally expensive because it is simple thresholding at zero compared to tanh and sigmoid functions that depend on exponential and complex operations. Note that ReLU is sparsely activated because it is zero for all negative inputs. This sparsity (i.e., not all the neurons are active at the same time) might be a good thing because it can reduce the power of the neural network, resulting in less overfitting. The downside is that it can also lead to *dying ReLU* problem. A dead ReLU refers to the case where a ReLU always generates outputs with the same value (zero) which is not important for any inputs and next neuron layers, resulting in an incomplete learning process and very weak mode. ReLU indeed has another problem as well. The output of neurons with ReLU activation function can grow dramatically because the output of ReLU is a linear function of its input (for positive values). In other words, ReLU does not have any kind of boundary, and thus it cannot truncate the output. Compare it with sigmoid or tanh functions, in which the outputs are saturated.

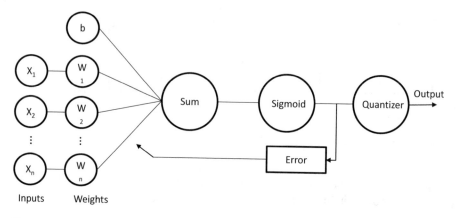

Fig. 5.55 Schematic presentation of a logistic regression model (a neural network with one neuron)

Leaky ReLU Leaky ReLU is a workaround for the "dying ReLU" problem. A leaky ReLU function is defined similar to a normal ReLU, but instead of the output of zero in the negative region of the x-axis, it has a slight slope. In other words, leaky ReLU provides a small, positive gradient when the input is below zero.

5.6.4 Softmax Function

Note that logistic regression can be seen as a simple neural network which has just one neuron with sigmoid function as its activation function (see Fig. 5.55). Also recall that logistic regression is indeed a two-class classifier which produces a value between 0 and 1.0. Consider an email classifier as an example. When the output of the logistic regression is 0.8, it suggests that the target email is spam with a chance of 80%, and with a chance of 20% is not spam. Softmax regression through a neural network layer extends this idea into a multiclass world. Softmax regression is a general form of the logistic regression (Fig. 5.55), which enables us to perform multiclass classification. This can improve the capability of conventional logistic regression because logistic regression is only suitable for binary classification tasks.

In softmax regression, we have a neural network with several outputs in a way that each output corresponds to one class. In softmax regression, we also need to replace the sigmoid function of the output layer with softmax function (see Fig. 5.56):

$$P\left(y = j \mid z^{(i)}\right) = \phi_{softmax}\left(z^{(i)}\right) = \frac{e^{z^{(i)}}}{\sum_{j=0}^{k} e^{z_k^{(i)}}}$$

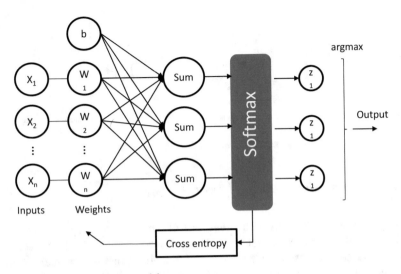

Fig. 5.56 A softmax regression model

in which the z is defined as

$$z = w_0 x_0 + w_1 x_1 + \cdots + w_m x_m = \sum_{i=0}^{m} w_i x_i = \boldsymbol{w}^T \boldsymbol{x}$$

Intuitively, the softmax function produces a probability for each output of the neural network. Each of them represents the probability of the belonging of given input (x) to a class. Note that for training the neural network in softmax regression, we also need to define a loss or cost function to be able to use it during the backpropagation step. In the softmax regression, generally we use the following cost function (based on entropy) which should be minimized during the training phase:

$$J(\boldsymbol{W}) = \frac{1}{n} \sum_{i=0}^{n} H(T_i, O_i)$$

in which parameter T_i represents the actual output (target) and O_i is the output of the softmax function. The cross-entropy function is defined as

$$H(T_i, O_i) = -\sum_{m} T_i \cdot \log(O_i)$$

In fact, the cost function is the average of all cross entropies of training samples.

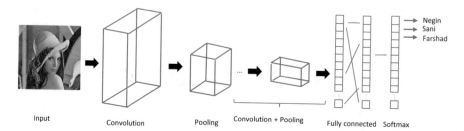

Fig. 5.57 Convolution neural network architecture

5.6.5 Convolution Neural Networks

Among all machine learning techniques, one of which called deep neural network (DNN) has been prevailing hot over the past several years. Both the academy and the industry witness their wide use and potential benefits in numerous areas. Convolution neural network (CNN or CovNet) is one of the main categories in a deep neural network. It is mainly used for visual image analysis tasks, such as image and video recognition and image classifications. Thanks to its state-of-the-art performance and revolutionary advances in the realm of the deep neural network, CNN can also be applied to natural language processing, drug discovery, strategy gaming, etc.

CNN follows the traditional neural network's architecture. First, it has an input layer, which is usually an image's pixels in the form of (255, 255, 255) RGB values. Next, it has hidden layers, where in traditional NN, the number of hidden layers is limited to 1. In CNN, the number of hidden layers becomes tens or even thousands in state-of-the-art systems. This is why CNN is called one of the deep neural networks. On the contrary, traditional NN is called shallow neural network. We will come with more details about what components are located in deep hidden layers soon. Then, the neural network has an output layer. This layer corresponds to the result, which is usually person identity in face recognition, age in age detection, or object type in general image recognition. Now, let us explore the mysterious hidden layers. In CNN, there are mainly three types of hidden layers (Fig. 5.57) [11]:

- Convolution (CONV) layer
- Pooling (POOL) layer
- Fully connected (FC) layer

5.6.5.1 Convolution Layer

The convolution layer is the main building block of a convolutional neural network that does most of the computational heavy lifting. Convolution, according to the definition, is the process of combining two or more functions/values to form a third

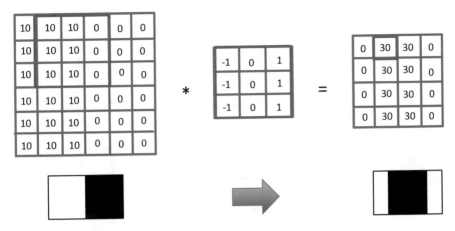

Fig. 5.58 Convolution computation

function/value. In the convolution layer, several pixels are convoluted to form a representative new value. The convolution layer takes two inputs, namely, input (image) matrix and a filter (kernel). As an example, consider a $6*6$ image matrix and a $3*3$ filter as shown in Fig. 5.58. Behind the scene, convolution is a dot product of the filter with a local zone of the image matrix. Note that we need to slide/move the filter across the whole matrix to be able to compute the output. The output matrix is so-called feature map. It should also be noted that convolutional layers are usually stacked on top of each other to detect more complex patterns and get in-depth information. To make an analogy, the brain combines low-level features such as basic shapes and curves and builds more complex shapes out of it. It first identifies low-level features and then learns to recognize and combine these features to learn more complicated patterns. These different levels of features come from different layers of the network.

Convolution and filters are not a new concept, and indeed it was used in image and signal processing for many years. Filters can be applied to extract features or perform operations such as edge detection or image blurring (image smoothing). In traditional image processing and signal processing, filters were mainly hand-engineered. However, CNN and deep learning technologies largely overcome the need for prior knowledge and human effort in feature design, since they can learn these filters/characteristics automatically [11, 16]. Let us take a closer look at edge detection filters. To detect different types of edge, we apply different filters; some are used to detect horizontal lines, and some are detecting vertical or diagonal lines (Figs. 5.59 and 5.60). To be more mathematical, we will study how the vertical filter works. We would like to utilize a filter to detect vertical edges from a 2D image. Take Fig. 5.58 as an example. The left side of this figure represents a 6×6 grayscale image. The left side of this $6*6$ image is white and its right side is gray. Therefore, we have a vertical line in the middle (between white and gray). The convolution operation computes a dot product of the filter with a local region of the

Fig. 5.59 Different types of filters

-1	-1	-1
0	0	0
1	1	1

Horizontal filter

-1	0	1
-1	0	1
-1	0	1

Vertical filter

-1	-1	1
-1	1	-1
1	-1	-1

45 degree filter

1	-1	-1
-1	1	-1
-1	-1	1

135 degree filter

Fig. 5.60 An example of filter (Left: original picture. Right: horizontal edges produced by horizontal filters)

image. Note that the local region should be exactly the same dimension as the filter. We iteratively apply the filter to the image, and in each iteration we slide/move/shift the filter just one pixel. As a result of this process, we can eventually generate a 4×4 matrix (on the right). As we can see, this matrix marks the vertical line in the middle while zeroing on the side.

5.6.5.2 Stride

The abovementioned sliding window technique has a name called stride. Stride is defined as the number of pixel shifts when sliding a filter over the input. For instance, if the stride is equal to 1, then we slide the filter by only 1 pixel at a time. When the

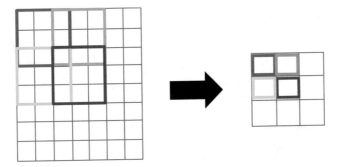

Fig. 5.61 Stride of 2 pixels with filter size of 3

Fig. 5.62 Zero padding

0	0	0	0	0	0	0	0
0	5	2	1	7	-2	1	0
0	3	3	3	5	3	8	0
0	1	1	-1	4	3	2	0
0	5	7	2	2	1	1	0

stride is equal to 2, it means that we need to slide the filter by 2 pixels at a time and so on. Figure 5.61 shows convolution would work with a stride of 2. This applies to both horizontal sliding and vertical sliding.

5.6.5.3 Padding

As you might note, when we slide the filter, those pixels that are located on edges are used (touched) less compared to the other pixels of the image. That implies that we are losing some information related to those edge pixels. In addition, the size of the output is shrinking in each step. *Padding* techniques are developed to address this issue. In this technique, we pad the image by placing zeros around the image to enable the filter to move (slide) on top and keep the size of the output equal to the size of the input (see Fig. 5.62).

One can use the following equation to compute the size of the output based on the size of the filter (f), stride (s), pad (p), and input size (n) [16]:

$$output\ size = \left(\frac{n + 2p - f}{s} + 1\right) \times \left(\frac{n + 2p - f}{s} + 1\right)$$

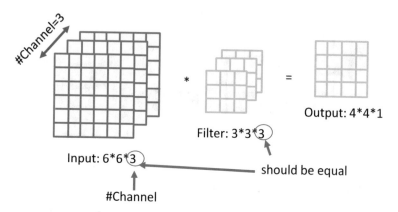

Fig. 5.63 Apply filter on three-channel matrix

Although we explained the concepts with a 2D matrix, in a real application, we usually do image recognition based on an RGB image (color image) which has three channels (i.e., R-red, G-green, B-blue). Thus, the input looks like a volume consisting of three parameters stacked on each other. Note that we can also have more than three channels. For example, astronomical images have extra channels for infrared and ultraviolet [11, 16]. The above procedure is explained by an example in Fig. 5.63.

Until now, we have used just one filter at a time. However, in real-life application, we need several filters to be able to identify different features. This explains the concept of building convolutional neural networks [11]. In this case, each filter provides its own output and then we combine (stack) them together to build an output volume (see Fig. 5.64). Given the growing dimension of inputs, the number of output can be recalculated as below [16]:

$$\text{Input}: (n \times n \times n_c) \quad \text{Filter}: (f \times f \times n_c)$$

$$\text{Output}: \left(\left[\frac{n + 2p - f}{s} + 1 \right] \times \left[\left(\frac{n + 2p - f}{s} + 1 \right) \right] \times n_c' \right)$$

5.6.5.4 Pooling Layers

Compared to the convolution layer, pooling layer is easier to understand. The task of pooling layer is to reduce the number of parameters and calculations in the network to be able to address the overfitting by gradually reducing the spatial size of the network. Two popular types of pooling layers exist: *max pooling* and *average pooling*. *Max pooling* (Fig. 5.65) *is to pick the max value from the pooling area.*

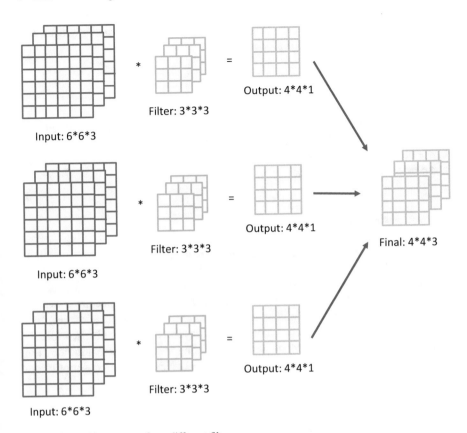

Fig. 5.64 Stacking outputs from different filters

Fig. 5.65 Max pooling

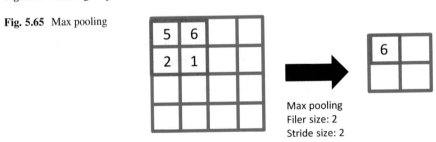

First, we define a filter (spatial neighborhood), and then as we slide it through the input, we select the largest item within the region covered by the filter.

Average pooling, as the name suggested, retains the average of the values encountered within the filter. Note that we need to select several hyperparameters including the filter size and the stride (it's common not to use any padding).

In contrast to the convolution layer, the pooling layer does not change the depth of the network and the depth dimension remains unchanged. The number of outputs after pooling is $\frac{N-F}{S} + 1$, where N is the dimension of input to pooling layer, F is the dimension of filter, and S is stride.

5.6.5.5 Fully Connected Layer

Similar to a traditional neural network, where all layers are fully connected, this layer also exists in CNN but stays right before the output layer; their activations can hence be calculated by a matrix multiplication followed by a bias offset. This is the last phase for a CNN network.

5.6.5.6 Well-Known CNN Architectures

LeNet, AlexNet, VGG, GoogLeNet, and ResNet are the most well-known publicly available CNN Architectures. Take the example of famous LeNet-5 architecture developed by Yann LeCun in 1998. This architecture consists of three convolution layers, two pooling layers, and one fully connected layer. The architecture is used for detecting hand-written digital recognition (MNIST) of 28 × 28 pixel images. Note that zero padding is used to construct the input layer as 32 × 32 pixels, while the rest of the convolutional layers do not have padding anymore. Activation function for each layer is tanh, except the output layer. The LeNet demo and description can be easily found on Yann LeCun's website.

5.7 Clustering

Clustering is a well-known unsupervised learning technique which allows us to group (cluster) data points in a way that data points in the same cluster are more similar to each other than those in different clusters. There are multiple ways of clustering. In this section, we overview two most popular techniques, namely, *K-means clustering* and *hierarchical clustering*.

5.7.1 K-Means Clustering

K-means extracts the clusters based on distance, which usually refers to *Euclidean distance* in most context. *K* in K-means represents the number of clusters in which we want our data to be divided into. There is a restriction in using K-means that the data shall be in continuous values rather than category values since K-means does not work on category data in nature. Moreover, it is advised to normalize the data before applying the K-means method. The reason still resorts to distance calculation. An example is that if we consider about clustering people based on both weights in

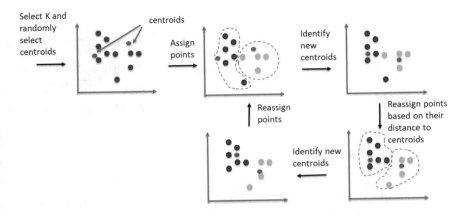

Fig. 5.66 An example of K-means clustering algorithm

kilograms and height in centimeters. Then, there is a weighting bias on height versus weight, which can be solved by a data preparation method called normalization.

Training the K-means algorithm is based on iterative refinement and centroid calculation that consists of the following steps (see Fig. 5.66):

1. Specify the number of clusters (K) and then randomly select a centroid for each cluster.
2. Assign/update each data point to the closest cluster (based on the distance of the data point and centroids).
3. Calculate/update centroid of the cluster. The centroid of each cluster is the average of all data points in the cluster.
4. Iteratively execute steps 2 and 3. Eventually, the algorithm is terminated when the change of objective function is below a threshold value.

Note that the accuracy of clustering in the presence of variations depends on the selection of an appropriate number of clusters k. Researchers usually use the elbow method for estimating k to make a tradeoff between the accuracy of the clustering method and the number of clusters.

5.7.2   Hierarchical Clustering

The method of hierarchical clustering produces a specific number of overlapping clusters of various sizes across a tree that creates a hierarchical system of classification. This clustering technique can be accomplished using a variety of methods, with the most widely used methods being *divisive approach* and *agglomerative approach*. An agglomerative method works from the bottom-up and it consists of the following steps:

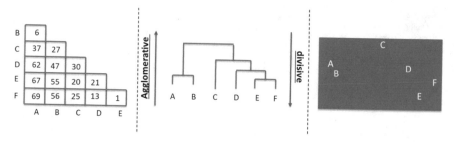

Fig. 5.67 A simple example of hierarchical clustering

- The algorithm starts by creating a background distance matrix often known as the Euclidean distance; a distance matrix illustrates the distance between items.
- Next, the algorithm treats each item as a single cluster.
- In an iterative approach, the algorithm merges two most similar clusters until all clusters are merged into one big cluster. This big cluster should contain all items.
- The output of the algorithm is a tree called dendrogram.

Figure 5.67 depicts what the agglomerative method of hierarchical clustering looks like. On the other hand, divisive clustering works from the top-down. All items begin in the same cluster (the root of the tree) and are divided into two separate clusters as the tree grows. The process of dividing is executed repeatedly until the designated number of clusters is obtained.

5.8 Summary

Machine learning is playing an important role in enabling IoT solutions to extract value and uncover insights from the generated data and to enhance the capabilities and intelligence of devices/applications. This chapter defined two broad categories of machine learning, namely, supervised and unsupervised techniques. Next, we presented the details of regression, classification, and clustering techniques. Afterward, the details of feature engineering including feature extraction and feature selection have been discussed. Finally, we presented the details of neural networks, deep learning, and convolutional neural networks.

References

1. C.M. Bishop, *Pattern Recognition and Machine Learning* (Springer, 2006)
2. H. Daumé III, A Course in Machine Learning. 2012.
3. R. Battiti, M. Brunato, *The LION Way – Machine Learning plus Intelligent Optimization* (LIONlab, University of Trento, Italy, 2017)

4. A. Smola, S.V.N. Vishwanathan, *Introduction to Machine Learning* (Cambridge University Press, Cambridge, 2008)
5. A. Smola, S. Vishwanathan, *Introduction to Machine Learning*, vol 32 (Cambridge University, Cambridge, 2008), p. 34
6. H. Zou, T. Hastie, Regularization and variable selection via the elastic net. J. R. Stat. Soc. **67**(2), 301–320 (2005)
7. O. Sosnovshchenko, O. Baiev, *Machine Learning with Swift: Artificial Intelligence for IOS* (Packt Publishing Ltd, 2018). https://www.amazon.com/Machine-Learning-Swift-Artificial-Intelligence/dp/1787121518.
8. Stanford Machine Learning. Available from: https://see.stanford.edu/Course/CS229.
9. I. Steinwart, A. Christmann, *Support Vector Machines* (Springer Science & Business Media, 2008). https://link.springer.com/book/10.1007/978-0-387-77242-4#about.
10. SVM (Support Vector Machine)—Theory. Available from: https://medium.com/machine-learning-101/chapter-2-svm-support-vector-machine-theory-f0812effc72.
11. H.H. Aghdam, E.J. Heravi, *Guide to Convolutional Neural Networks*, vol 10 (Springer, New York, NY, 2017), pp. 978–973
12. K. Huang, *Deep Learning: Fundamentals* (Springer, Theory and Applications, 2019). https://www.springer.com/gp/book/9783030060725#aboutBook.
13. A. Carvalhal, T. Ribeiro, Do artificial neural networks provide better forecasts? Evidence from Latin American stock indexes. Lat. Am. Bus. Rev. **8**(3), 92–110 (2008)
14. Basic Overview of Convolutional Neural Network (CNN). Available from: https://medium.com/@udemeudofia01/basic-overview-of-convolutional-neural-network-cnn-4fcc7dbb4f17.
15. A. Krizhevsky, I. Sutskever, G.E. Hinton, Imagenet classification with deep convolutional neural networks. in *Advances in Neural Information Processing Systems*, (2012), pp. 1097–1105.
16. Convolutional Neural Networks. Available from: https://medium.com/machine-learning-bites/deeplearning-series-convolutional-neural-networks-a9c2f2ee1524.

Chapter 6
Big Data

Natasha Balac

> *Errors using inadequare data are much less than those using no data at all.*
>
> Charles Babbage

Contents

N. Balac (✉)
University of California San Diego, San Diego, CA, USA
e-mail: nbalac@eng.ucsd.edu

© Springer Nature Switzerland AG 2020
F. Firouzi et al. (eds.), *Intelligent Internet of Things*,
https://doi.org/10.1007/978-3-030-30367-9_6

6.1 Introduction to Big Data

Big Data has changed the way we manage, analyze, and leverage data across all industry sectors. Big Data has the potential to examine and reveal trends, find unseen patterns, discover hidden correlations, reveal new information, extract insight, enhance decision making and automation, etc. Managing and analyzing data, in particular in the era of IoT, has always been one of the greatest challenges within organizations across industries. Finding an efficient and scalable approach to capturing, integrating, organizing, and analyzing information about IoT devices, products, and services can be a perplexing task for any organization regardless of the size or line of business. In the age of the Internet and digital transformation, the notion of Big Data reflects the changing world we live in. Everywhere around the world, more and more data is captured and recorded every day. Companies and organizations are becoming overwhelmed by the complexity and sheer volume of their data. While some data is still structured and stored in a traditional relational databases or data warehouses, a vast majority of the modern data sources are producing unstructured data including documents, conversations, pictures, videos, Tweets, posts, Snapchats, sensor readouts, click streams, and machine-to-machine data. Further, the availability and adoption of newer, more powerful mobile devices, coupled with ubiquitous access to global networks, is continuously driving the creation of new sources for data.

Although each data source can be independently managed and searched, the challenge today is how to make sense of the intersection of all these different types of data. When large amounts of data are coming from so many different forms, traditional data management techniques falter. While there has always been a challenging amount of data for existing IT infrastructure, the difference today is stark in terms of the sheer volume, speed, and complexity of the data. In parallel, the value of data is growing in terms of its volume and diversity. Data is emerging as the world's newest resource for competitive advantage, as it enables efficient, data-driven decision making. As the value of data continues to grow, new technologies are emerging to support the new requirements surrounding it.

6.1.1   Defining Big Data

There has been much hype about Big Data in the past several years for various reasons. Before investigating the major drivers of Big Data's popularity, we must first define the meaning of Big Data as a term. Dictionary.com defines Big Data as:

> ... data sets, typically consisting of billions or trillions of records, that are so vast and complex that they require new and powerful computational resources to process.

The *British Dictionary* defines Big Data as:

> Data held in such large amounts that it can be difficult to process.

Big Data can be defined and often is described in terms of its four major characteristics, as shown in Fig. 6.1: volume, velocity, variety, and veracity:

6.1.2   Volume

Volume indicates how much data has been collected. It is perhaps the most obvious characteristic of Big Data, as the current amount of data created is quite staggering. For example, in 1 minute on the Internet, Snapchat users share over half a million messages, YouTube viewers watch over 4 million videos, Twitter users post over four hundred thousand Tweets, and so on. All of these transactions can be archived for later consideration.

Fig. 6.1 4 Vs of Big Data

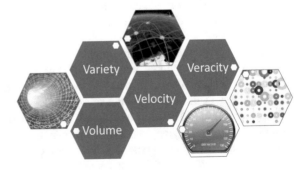

6.1.3 Velocity

Velocity refers to the speed at which data is being transacted. Streaming data can arrive in milliseconds and require a response within seconds or less. For example, Facebook's data warehouse not only stores hundreds of petabytes of data but needs to accommodate data coming in at the rate of more than 600 terabytes per data per day. Similarly, Google processes more than 3.5 billion searches per day, which translates to a velocity of over 40,000 search queries per second.

6.1.4 Variety

The huge variety of the types and structures of data has become one of the critical challenges of Big Data. Big Data requires systems to handle not only structured data but also semi-structured and unstructured data as well. As described above, most Big Data is in unstructured forms such as audio, images, video files, social media updates, log files, click data, machine and sensor data, etc. Unfortunately, most analytics techniques have traditionally been focused on analyzing only structured data, so new techniques and approaches need to be developed. This fact alone helps explain why there are such a large growing number of start-ups in Big Data.

6.1.5 Veracity

Perhaps the most nuanced of the four Vs is veracity. Veracity describes how accurate and trustworthy data is in predicting business value through Big Data analytics. Uncertainty in data is typically due to inconsistency, incompleteness, latency ambiguities, approximations, etc. Data must be able to be verified based on both accuracy and context. It is necessary to identify the right amount of high-quality data that can be analyzed in order to impact business or outcomes.

These four definitions suggest that Big Data typically requires resources (computation and data infrastructure, tools, techniques, expertise, etc.) beyond the current capabilities of many organizations [1]. Big Data solutions comprise a set of analytical tools that are geared toward the fast and meaningful processing of large data sets. This is a key aspect of Big Data analytics, as the goal is to derive meaning or insight from data that can be used for making data-driven business decisions. Big Data can be thought of as a process that is used when traditional data processing and handling techniques alone cannot uncover the insights and meaning of the underlying data. Often times, real-time processing is needed for a massive amount of different types and frequencies of data in order to reveal patterns, trends, and associations. This is especially true when relating to human behavior and interactions. The goal for Big Data is to enable organizations to gather, store,

Fig. 6.2 Big Data life cycle

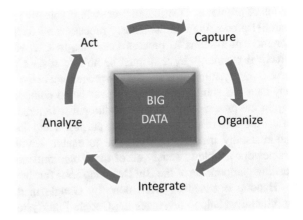

manage, and manipulate vast amounts of data at the right speed and at the right time in order to gain the most valuable insights.

The data growth in the past decade has been unprecedented. There are 2.5 quintillion bytes of data created each day, and this will only increase more rapidly with the growth of the Internet of Things (IoT): an estimated 50 billion IoT devices will be connected by 2020 [2]. With IoT's network of RFID tags, machines, appliances, smartphones, buildings, and many other devices with embedded technology that can be accessed over the Internet, estimates are projecting between 40 and 50 Zettabytes (or 2^{70} bytes equaling 1,180,591,620,717,411,303,424 bytes) of data will be created by 2020. Over 92% of the data in the world was generated in the last 2 years alone. A large contributor to this skyrocketing data generation are Internet applications like Snapchat, YouTube, Twitter, and Instagram [3].

In order to convert the vast amount of available data into insight, it is important to consider the functional requirements for Big Data. Figure 6.2 illustrates a set of iterative steps in the Big Data functional requirements life cycle. Data first needs to be captured, then organized, and integrated. The integration process includes data cleaning and preparation. Once integrated, data can be analyzed in order to solve the business problems at hand. Often times this analysis includes developing predictive models utilizing Data Science approaches. Finally, the organization can act based on the outcome of the Big Data analysis, frequently via the implementation and utilization of the predictive models.

6.2 Big Data Management and Computing Platforms

As the volume and velocity of data grows, so grows the need to manage and process it. Optimization, enabling the rapid formulation and testing of many diverse models and real-time operations, becomes essential, especially in the case of streaming data. Distributed and parallel processing approaches are well suited for these kinds of problems. Distributed processing typically segments large datasets, while parallel processing simultaneously processes all data or subsets of data. In order for modern systems to process Petabytes to Exabytes of data in a scalable and practical manner, a system must be able to sustain partial failure. Any Big Data processing system needs to continue processing in the face of failures without losing any data and should be able to recover failed components and then allow them to rejoin the process when ready. Additionally, failures during execution should not affect the final result for consistency purposes, and the addition of resources should automatically increase performance to enable seamless scalability. The Hadoop framework is able to satisfy all of these requirements and has become one of the leading components of the Big Data platforms for the last decade.

Hadoop is based on work done by Google in the early 2000s [4, 5]. More specifically, Hadoop leverages the Google File System's (GFS) MapReduce concepts, taking a fundamentally new approach to distributed computing [6–8]. Hadoop meets the requirement for a low-cost, scalable, flexible, and fault-tolerant large-scale system, while enabling the shared nothing architecture and the use of applications written in high-level programming languages. The overall goal of Hadoop is to execute computation where the data is stored, instead of moving large amount of data to computational resources. Additionally, data is replicated multiple times on the system for increased availability and reliability.

The Apache Hadoop software library [9] is an open source framework that enables distributed processing of large data sets across clusters of computers using simple programming models. It is designed to scale up from single servers to thousands of machines, each offering local computation and storage. Rather than relying on hardware to deliver high availability, the library itself is designed to detect and handle failures at the application layer. This enables highly available services on top of a commodity cluster where each machine may be prone to failures. Additional open source projects have been built around the original Hadoop implementation with the addition of Spark. Figure 6.3 depicts the basic Hadoop environment, which enables large scale data management and computing.

6.2.1 Big Data System Architecture Components

This section examines in depth the key concepts required when considering implementation of Big Data, both from the technical and an analytics perspective. As data systems became larger in recent years, performance became an acute concern. An approach to efficiently solving a wide range of problems without needing to change the underlying environment was needed. This type of approach would have to take advantage of parallel processing of data and additional CPUs availability. The goal of any analyst or data scientist, regardless of the field of study, is to work with as much data as is available, build as many models as possible, and improve model training time and accuracy by simply adding additional CPUs. The ultimate goal of any Big Data system is to enable development of the best data analytics

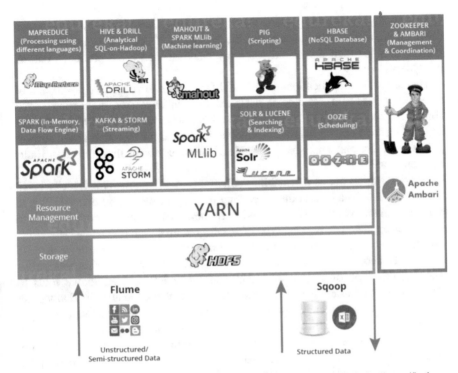

Fig. 6.3 Hadoop ecosystem. (A list of all of the Apache projects can be found at https://hadoop. apache.org/)

in a shortest amount of time. Finding the patterns hidden within a large, complex dataset requires processing, filtering, and analyzing massive amounts of data. Over time, several solutions have been proposed such as developing parallel data mining algorithms and parallel data source on parallel hardware.

From the technology perspective, Hadoop initially enabled the power of Big Data by providing large-scale computing sufficiently flexible and affordable storage, so that a wide variety of organizations could leverage Big Data technology. This is one of the reasons behind the synonymous use of the terms Big Data and Hadoop. However, the evolution of technologies has changed the high-tech landscape with a number of additional tools as discussed in the following sections.

6.2.2 Hadoop History

One of the pillars of the Big Data framework is Apache Hadoop. Hadoop is an Apache-managed software framework derived from MapReduce and Big Table. Hadoop is an Apache top-level project built and used by a global community of contributors and users. It is licensed under the Apache License Hadoop and was originally developed by Doug Cutting and Mike Cafarella in 2005 to support distribution for the Nutch search engine project at Yahoo [7]. Doug named the project after his then 2-year-old son's toy elephant, hence the product's current pachyderm logo.

Hadoop allows applications based on MapReduce to run on large clusters of commodity hardware. Hadoop is designed to parallelize data processing across computing nodes to speed computations and minimize latency. Two major components of Hadoop are the massively scalable distributed file system that can support petabytes of data and a massively scalable MapReduce engine that computes results in batch mode.

Hadoop started out as a scalability solution to high volume and velocity batch processing. The idea behind the processing paradigm called MapReduce [10] is to provide a simple yet powerful computing framework that enables computations to scale over big data easily. This type of system requires high efficiency. Therefore, instead of moving data to computation, computation is moved closer to data in Hadoop. Hadoop and MapReduce provide a shared and integrated foundation for enabling this type of computation and seamlessly integrate additional tools to support other necessary system functionalities. All of the modules in Hadoop are designed with a fundamental assumption that hardware failures, whether of individual machines or racks of machines, are common and thus should be automatically handled in software by the framework. The notion behind Hadoop's reliability requirements is based on the idea that if one computer fails once a year, then a 365-computer cluster will have a failure daily. If this number is scaled by an order of magnitude, the cluster could be expected to have a hardware failure hourly. It is essential for truly scalable systems to endure failure of any component.

Since data has been regarded as "new oil" or "new gold" in terms of an asset value, a new approach to storing and computing has emerged. As the cost of storing the data continues to decrease, organizations have developed a new approach of keeping all data. Since data is growing rapidly in size and complexity, the "schema on read style" has become the approach of choice. This approach enables all of the data to be ingested in a rough form and then projected into the schema on the fly, as it is pulled out of the stored location, thereby enabling experiments and new types of analysis.

6.2.3 The Apache Hadoop Framework Components

All of the Hadoop modules are designed around a fundamental assumption of hardware failures. The entire Apache Hadoop "platform" is now commonly considered to consist of a number of related projects including Apache Pig, Apache Hive, Apache HBase, and others. These components will be described in the following sections and are illustrated in Fig. 6.4.

Fig. 6.4 Apache Hadoop
basic components

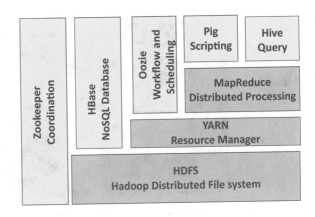

The Apache Hadoop framework is composed of the following modules:

- *Hadoop Common*: contains the libraries and utilities needed by other Hadoop modules
- *Hadoop Distributed File System (HDFS)*: a distributed file system that stores data on the commodity machines, providing very high aggregate bandwidth across the cluster
- *Hadoop YARN*: a resource-management platform responsible for managing compute resources in clusters and using them for scheduling of users' applications
- *Hadoop MapReduce*: a programming model for large-scale data processing

Although the MapReduce Java code is common, any programming language can be utilized to implement the "map" and "reduce" parts of the user's program. Apache Pig [12] and Apache Hive [13], among other related projects, expose higher level user interfaces like Pig Latin and a SQL variant, respectively. The Hadoop framework itself is mostly written in the Java programming language, with some native code in C and command line utilities written as shell scripts.

The two primary components at the core of Apache Hadoop version 1 are the Hadoop Distributed File System (HDFS) [14] and the MapReduce parallel processing framework (Fig. 6.5). These are both open source projects, inspired by technologies initially developed by Google. Hadoop's MapReduce and HDFS components originally derived from Google's MapReduce and Google File System (GFS) [11], respectively.

6.2.4 Hadoop Distributed File System

The Hadoop Distributed File System (HDFS) is a distributed, scalable, and portable file system written in Java. Each node in a Hadoop instance typically has a single NameNode, and a cluster of DataNodes forming the HDFS cluster [14]. Each DataNode provides blocks of data using a HDFS-specific block protocol (Fig. 6.6).

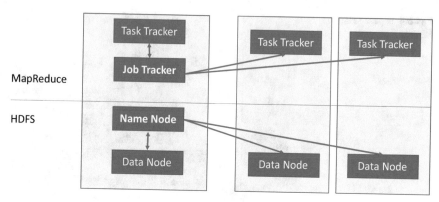

Fig. 6.5 Hadoop's distributed data storage and processing

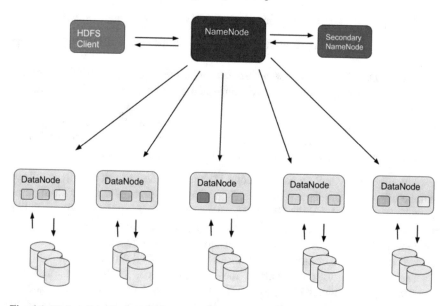

Fig. 6.6 Hadoop Distributed File System operations

The file system uses the TCP/IP layer for network communication. HDFS enables storing and manipulating large data sets by distributing them across multiple hosts. In addition, it enables reliability by replicating the data across multiple hosts, without requiring RAID (Redundant Array of Independent Disk) storage. Typically, data is duplicated on three nodes including two on the same rack, and one on a different physical rack. Data nodes can communicate to rebalance data, transfer copies of data and to keep the replication of data at the specified level [14]. HDFS provides the high-availability capabilities and automatic fail over in the event of failure.

The HDFS architecture includes a secondary NameNode; however its role is not to take over when the primary NameNode goes offline. The secondary NameNode connects with the primary NameNode on a regular basis and generates snapshots of the primary NameNode's directory information. This information is then saved by the system and can be used to restart a failed primary NameNode in order to create an up-to-date directory structure without having to repeat the entire set of file system actions [15].

In the original Hadoop release, NameNode was the single metadata storage and management information center. However, with the continually increasing number of files, this approach created challenges. HDFS Federation solves the bottleneck issue by allowing multiple name spaces served by distinct NameNodes and thereby enables data awareness between the job tracker and task tracker. The job tracker schedules map or reduces jobs, minimizing the amount of data movement. This can have a significant impact on job completion times, especially with data-intensive jobs.

6.2.4.1 Overview of Data Formats

There are a number of file and compression formats supported by the Hadoop framework, each with a corresponding set of application-specific strengths and weaknesses. HDFS enables several formats for storing data including HBase for data access functionality and Hive for data management and querying functionality. These file formats are designed for MapReduce or Spark computing engines for specific purposes, ranging from basic analytics to machine learning applications.

Choosing the most appropriate file format can have a significant impact on performance. It influences many aspects of the file system including read and write times, the ability to split files into smaller components, enabling partial reads and advanced compression support.

Hadoop enables the storage of text, binary, images, or other Hadoop-specific formats. It provides built-in support for a number of formats specifically optimized for Hadoop storage and processing. Some of the most common basic data formats include text, CVS files, and JSON records. More complex formats such as Apache Avro, Parquet, HBase, or Kudu can also be utilized. While text and CSV files are very common, they do not support block compression and therefore often come with a significant read performance cost. One common approach is to create a JSON document in order to add structure to text files and utilize structured data in HDFS. JSON is a common data format often used for asynchronous communication. JSON stands for JavaScript Object Notation records and is an open-standard file format that uses human-readable text to transmit data objects consisting of attribute-value pairs and array data types. JSON files store metadata and can "split" files; however it doesn't support block compression [16].

Several additional more sophisticated and specialized file formats are available in the Hadoop environment. One such format is Avro [17]. Avro is a data serialization standard for the compact binary data format used for storing persistent data on

HDFS. It provides numerous benefits and has evolved into the de facto standard. Avro's lightweight and fast data serialization and deserialization enables fast data ingestion. It stores metadata with the data itself and allows specification of an independent schema for reading the files. It can quickly navigate to the data collections in fast, random data access fashion. In addition, Avro files are "splittable," support block compression, and are accompanied by a wide and mature set of open source tools.

The RC (Record Columnar) file format was the first columnar file in Hadoop. It provides substantial compression and query performance benefits. However, it does not support schema evaluation. Optimized RC Files (ORC) represent the compressed version of RC files with additional improvements including enhanced compression and faster querying.

The Parquet file format is another column-oriented data serialization standard enabling compression, encodings, query performance benefits, and efficient data analytics. This format has gained popularity as it became the choice of format for Cloudera Impala. This optimization and usability contributed to its popularity in other ecosystems as well.

Apache HBase is a scalable and distributed NoSQL database on HDFS for storing key-value pairs. Keys are indexed, which typically enables fast access to the records.

The Apache Kudu file format is scalable and distributed table-based storage. Kudu provides indexing and columnar data organization to achieve a balance between ingestion speed and analytics performance. As in the case of HBase, Kudu's API enables modification of the data that is already stored in the system.

In general, three major factors should be considered when choosing the best format for the task at hand: write performance, partial read performance, and full read performance. These factors provide an indication of how fast the data can be written, how fast individual columns can be read, and how fast can data element be read from the data source. Columnar formats typically perform better in terms of read performance. CSV and other non-compressed formats typically demonstrate better write performance, but generally demonstrate slower reads due to lack of compression. Some additional key factors that should be considered while selecting the best file format include the type of the Hadoop distribution and associated formats. Additionally, querying and processing requirements should be considered alongside the processing tools. Further, extraction requirements merit attention, especially when extracting data from a Hadoop environment into an external database or other platforms. Finally, storage requirements are critical, as volume may become a significant factor, and compression may be required.

6.2.5 MapReduce

The MapReduce paradigm is the core of the Hadoop system. MapReduce is a distributed computing-based processing technique (Fig. 6.7). MapReduce was designed by Google in order to satisfy the need for efficient execution of a set of

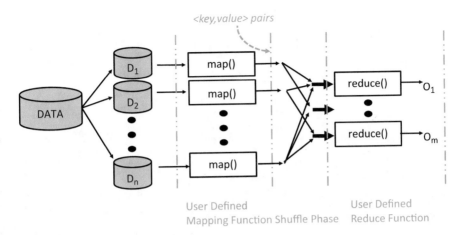

Fig. 6.7 The MapReduce process

functions on a large amount of data in batch mode. The "map" function distributes the programming tasks across a large number of commodity cluster nodes. It handles the placement of the tasks in a way that balances the load and manages recovery from failures. After the distributed computation is completed, another function called "reduce" aggregates all the elements back together in a "shuffle" and organizes the result. An example of MapReduce usage would be to determine a word count across thousands of newspaper articles.

MapReduce works with the underlying file system and typically consists of one JobTracker that receives the client's MapReduce job requests (Fig. 6.8). The JobTracker distributes processing to the available TaskTracker nodes in the cluster, while striving to keep the work as close to the data as possible. JobTracker is aware of which node contains the data and what neighboring processing is available. If for some reason processing cannot be executed on the same data hosting node, then priority is given to computing nodes in the same rack, thereby minimizing network traffic.

If a TaskTracker fails or times out, that part of the job is rescheduled. The TaskTracker on each node issues a separate Java Virtual Machine process to prevent the TaskTracker from failing. A heartbeat is sent from the TaskTracker to the JobTracker on a regular basis to check its status.

6.2.6 YARN

Apache Hadoop YARN is a sub-project of Hadoop. MapReduce underwent a complete retrofit in an early version (v0.23) and the MapReduce 2.0 was designed with YARN [18] as a key component. It separates the resource management and

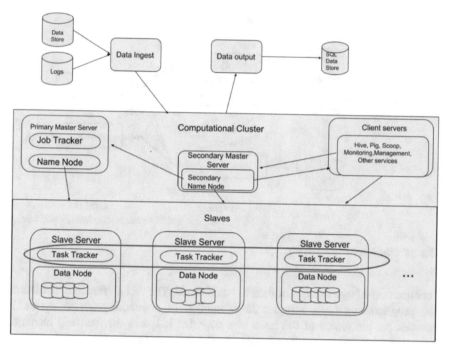

Fig. 6.8 MapReduce, detailed view

processing components enabling a broader array of interaction patterns for data stored in HDFS. The YARN-based architecture provides a more general processing platform that is not constrained by MapReduce limitations.

The fundamental concept underlying YARN is to split up the two major functionalities of the Job Tracker, resource management and job scheduling/monitoring, into separate functionalities. The Global Resource Manager (RM) and per-application Application Master (AM) are key components. The RM is the ultimate authority that arbitrates resources among all the applications in the system. The AM is a framework-specific library tasked with negotiating resources from the RM and working with the Node Manager(s) to execute and monitor the tasks [18].

YARN enhances the power of a Hadoop-based cluster in several key ways. First, it enables a higher level of scalability as the processing power in data centers continues to grow quickly. The YARN RM focuses solely on scheduling and therefore is able to manage extremely large clusters very efficiently. Furthermore, YARN enables compatibility with existing MapReduce applications without disruption to the existing processes. Additionally, YARN significantly improves cluster utilization by enabling workloads beyond that of MapReduce. The MapReduce RM is a pure scheduler that optimizes cluster utilization according to specified criteria such as capacity guarantees, fairness, and SLAs. In contrast, YARN enables additional programming models for real-time processing such as Spark, graph processing, machine learning, and iterative modeling. YARN's processing approach has the additional ability to evolve independently of the underlying RM layer in a much more agile manner.

6.3 An Introduction to Big Data Modeling and Manipulation

With the evolution of computing technology, it is now possible to manage immense volumes of data that previously could have only been handled at great expense by supercomputers. Prices of computing and storage systems continue to drop, and as a result, new techniques for distributed computing have become mainstream. The key inflection point for Big Data occurred when companies like Yahoo!, Google, and Facebook came to the realization that they there was an opportunity to monetize the massive amounts of data collected. New technologies were required to create large data stores, access those stores, and process huge amounts of data in near real-time. The resulting solutions have transformed the data management market. In particular, Hadoop, MapReduce, and Big Table proved to be the start of a new approach to data management. These technologies address one of the most fundamental problems: the need to process massive amounts of data efficiently, cost effectively, and in a timely fashion.

6.3.1 *Big Table*

Big Table was developed by Google as a distributed storage system intended to manage highly scalable structured data. Data is organized into tables with rows and columns. Unlike a traditional relational database model, Big Table is a sparse, distributed, persistent multidimensional sorted map. It is intended to store massive volumes of data across a scalable array of commodity servers.

6.3.2 *Pig*

Pig is the high-level programming component running on top of Hadoop's MapReduce component. Pig is a procedural language for creating MapReduce programs used with Hadoop [12]. Pig was originally developed at Yahoo Research in 2006 to enable ad hoc execution of MapReduce jobs on very large data sets. The initial language was called Pig Latin, enabling a variety of data manipulations in Hadoop. It is a SQL-like language that enables a multi-query approach on a nested relational data model where schema is optional. Apache Pig provides a rich set of built-in operators to support data operations including joining, filtering, sorting, ordering, nested data types, etc. on both structured and unstructured data. In addition, through the User Defined Functions (UDF), Pig can invoke code in many other languages including JRuby, Jython, and Java. This allows for the development of larger, more complex applications.

Pig is typically used in ETL applications for describing how a process will extract data from a source, transform it according to a rule set, and then load it into a data store. Pig can ingest data from files, streams, or other sources using the UDF. Once data is ingested, operations similar to the select command, various iterations, and other complex transformations can be performed. Once the processing is finalized, Pig stores the results of the transformations into the HDFS. Throughout the processing steps, Pig scripts are translated into a series of MapReduce jobs executed on the underlying Hadoop cluster.

6.3.3   Sqoop

Apache Sqoop is a tool designed for efficiently transferring bulk data between Hadoop and structured data stores such as relational databases [18]. Sqoop is a portmanteau that stands for SQL-to-Hadoop and is a simple command-line tool with the several valuable capabilities. Sqoop has the capability to import individual tables or entire databases to files in HDFS. It also generates Java classes to allow interaction with imported data. Additionally, Sqoop provides the ability to import from SQL databases directly into the Hive data warehouse within the Hadoop environment, thereby enabling computing on the data very rapidly.

6.3.4   Hive

Hive is the data warehouse software platform that enables a SQL-like language for facilitating, querying, and managing, large datasets residing in HDFS storage [13]. Often times referred to as the Hadoop data warehouse, Hive infrastructure sits on top of Hadoop and provides data query and analysis. HiveQL is the mechanism used to project structure onto the data and query the data using a SQL-like language [19]. HiveQL provides schema on read and transparently converts queries to MapReduce, Apache Tez, and Spark jobs. All three execution engines can run in Hadoop YARN. To accelerate queries, it provides indexes including bitmap indexes [13]. Additionally, Hive allows traditional and custom map and reduce mechanisms when HiveQL might be insufficient.

Initially developed by Facebook, Apache Hive is now used and developed industry wide. Hive supports analysis of large datasets stored in Hadoop's HDFS as well as compatible file systems such as the Amazon S3 filesystem.

6.3.5 HBase

Apache HBase is a column-oriented, distributed, and scalable database management system that runs on top of HDFS. HBase is a key component of the Hadoop stack, enabling fast, random access to large data sets. HBase is modeled after Google's BigTable, in order to handle massive data tables containing billions of rows and millions of columns.

It is well suited for sparse data sets, which are common in numerous Big Data use cases. Unlike relational database systems, HBase does not support a structured query language like SQL. HBase applications are written in Java similar to a typical MapReduce application. HBase also supports writing applications in Avro, REST, and Thrift.

6.3.6 Oozie

Apache Oozie is a scalable, reliable, and extensible system workflow scheduler system that manages and coordinates Apache Hadoop jobs while supporting MapReduce, Pig, Hive, Sqoop, etc. Oozie workflow coordinator jobs are Directed Acyclic Graphs (DAGs) of actions that are recurrent jobs triggered by time frequency and data availability [20]. Oozie is integrated with the Hadoop stack, typically with YARN, and supports numerous types of Hadoop jobs including Java MapReduce, Pig, Hive, Streaming MapReduce, Sqoop, general-purpose Java code, shell scripts, etc. Oozie itself is a Java Web application that combines multiple jobs sequentially into one logical unit of work. Oozie Bundle enables packaging of multiple coordinator and workflow jobs and management of the job's life cycle. It enables cluster administrators to develop complex data transformations with multiple component tasks, thereby providing greater job control recurrence.

6.3.7 Zookeeper

Apache ZooKeeper [21] provides operational services for a Hadoop cluster by enabling a distributed configuration service, a synchronization service, group services, and a naming registry for distributed systems [22]. Distributed applications use Zookeeper to store and mediate updates to important configuration information.

Due to the diversity of types of service implementations for applications, management can become rather challenging when the applications are deployed. ZooKeeper's purpose is to extract the essence of these different services into a simple interface via a centralized coordination service. The service itself is distributed and reliable supporting consensus, group management, and presence protocols. Application-specific utilization consists of a mixture of specific components of

ZooKeeper and application-specific conventions. ZooKeeper recipes [22] provide a simple service that can be used to build powerful abstractions.

6.3.8 Data Lakes and Warehouses

A data lake is defined as a centralized storage repository or system that contains a massive amount of structured, semi-structured, and unstructured raw data. The data structure and requirements are not defined until the data is needed to run different types of analytics. Often times a data lake is a single store of all raw enterprise data including copies of source system data, log files, clickstreams, social media, and output from IoT devices. It can also include transformed data used for delivering dashboards, reporting, visualization, new types of real-time analytics, and machine learning.

A data warehouse is a database optimized to analyze relational data coming from transactional systems and business applications. The data structure and schema are defined in advance to optimize for fast queries. The data from a warehouse is typically used for reporting and analysis. Data is cleaned, enriched, and transformed, so it can act as the "single source of truth" that users can trust [23]. Many organizations support both a data warehouse and a data lake, as they serve different needs and use cases. A data lake enables storage of non-relational data from mobile apps, IoT devices, and social media and does not require the schema to be defined when data is captured (referred to as "schema on read"). Massive amounts of data can be stored without careful schema design, thereby enabling flexibility in terms of the kinds of questions or data analytics that might need to be performed in the future. Data lakes enable different types of analytics including big data analytics, text mining, real-time stream data processing, and machine learning.

One great, early example of a successful data lake is the Big Data implementation at Mercy Hospital. The hospital leveraged technology to improve medical outcomes for patients by utilizing one of the first comprehensive, integrated electronic health record (EHR) systems to provide real-time, paperless access to patient information. Utilizing the EHR from Epic Systems, every patient's activity, including clinical and financial interactions, was captured. The hospital needed to address several typical challenges associated with Big Data implementations including scalability, data schema requirements, and response to large data queries. Mercy, in partnership with Hortonworks (one of the early Big Data providers at the time), has created the Mercy Data Library, a Hadoop-based data lake. This data lake enabled the integration, ingestion, and processing of large amounts of batch data extracts from relational systems, real-time data directly from Epic HER, and information from social media and even weather sources [24]. The combination of all of these data sets in a common platform enables the hospital to ask and answer questions at that were previously impossible due to scale, cost, or both.

One of the projects implemented on the Mercy system utilized thee advanced analytics techniques on a large amount of intensive care unit (ICU) patients' vitals data. When a patient is admitted to ICU, devices reading the patient's vitals send a new data record every second. Each ICU patient generates a large amount of data, which is often times very noisy. This data is traditionally summarized into 15 minute or longer intervals due to either the systems' inability to store and process the large amounts of more granular data or the cost-effectiveness of doing so. This approach limits the types of signal detection and analysis that can be performed on the data at scale. For example, determining which medicines bring down fever fastest would require a fine-grained measure of various streams (heart rate, breathing, movement, pain, etc.). Determining the efficacy of the medicine and enabling decision making in real time or near real time would require data time resolution of seconds or minutes. By implementing the data lake, researchers at Mercy Hospital were able to capture diagnostic data at high temporal rates, which in turn enabled real-time and near-real-time processing of the data.

This real-time data-on-demand model for researchers and clinicians is enabled by a combination of Sqoop, Storm, and HBase for more granular updates. In addition, Hive has provided a SQL-like approach to enabling the scalability of the Hadoop data lake.

6.4 An Introduction to Spark: An Innovative Paradigm in Big Data

One fundamental component to the Big Data Ecosystem not yet mentioned is Spark [25]. Although Hadoop captures the most attention for distributed data analytics, there are alternatives that provide advantages to the typical Hadoop platform. Apache Spark is an open source cluster computing framework originally developed in the AMPLab at the University of California, Berkeley, but was later donated to the Apache Software Foundation where it remains today [26].

Spark is a scalable data analytics platform that incorporates primitives for in-memory computing and typically demonstrates a significant speedup of the classic Hadoop's cluster compute and storage approach. Spark is implemented in the Scala programming language and provides a unique environment for large-scale data storage and processing. In contrast to Hadoop's two-stage, disk-based MapReduce paradigm, Spark's multi-stage, in-memory primitives approach provides performance up to 100 times faster for certain applications [25]. By allowing user programs to load data into a cluster's memory and query it repeatedly, Spark can enable a fast, efficient implementation of a variety of machine learning algorithms.

Similar to traditional Hadoop system, Spark requires a cluster manager and a distributed storage system. For cluster management, Spark supports the standalone native Spark cluster, Hadoop YARN or Apache Mesos. For distributed storage, Spark can interface with a wide variety of systems including HDFS, Cassandra, OpenStack Swift, Amazon S3, or custom solutions.

6.4.1 The Spark Ecosystem

While often compared to Hadoop and MapReduce, Spark is not a modified version of Hadoop. Hadoop is simply one of several ways of implementing Spark. In fact, Spark can run completely independently from Hadoop, powered by its own cluster management.

Spark typically leverages Hadoop in two ways by utilizing its storage and processing. As Spark has its own cluster management computation, it often times only uses Hadoop for storage.

Spark Core is the underlying general execution engine for the Spark platform. It enables in-memory computing and referencing datasets in external storage systems (Fig. 6.9).

Spark SQL is a component on top of Spark Core that introduces a new data abstraction called SchemaRDD, which provides support for structured and semi-structured data [27].

Spark Streaming leverages Spark Core's fast scheduling capability to perform streaming analytics. It ingests data in mini-batches and performs RDD (Resilient Distributed Datasets) transformations on those mini-batches of data [28].

Spark's machine learning library called MLlib is a distributed machine learning framework capable of taking advantage of the computational speedup related to the distributed memory-based Spark architecture. Spark MLlib is often an order of magnitude faster than Hadoop's original, but now retired, Mahout library [29].

GraphX is a distributed graph-processing framework layered on top of Spark. It provides an API for expressing graph computation. It also enables and optimizes user-defined graphs and processing by leveraging Pregel abstraction API [30].

Spark provides built-in APIs, supports, and is compatible with many languages and frameworks including Java, Scala, Python, R, Ruby, JavaScript, SparkSQL, Hive, Pig, H20, etc.

Spark can run standalone or on top of a cluster computing framework such as Hadoop. It handles batch, interactive, and real-time analysis within a single framework. It provides native integration with Java, Python, and Scala, thereby enabling programming at a higher level of abstraction.

One of the main advantages of Spark is that it is a general-purpose computing engine that seamlessly encompasses data streaming management, data queries, machine learning prediction, and real-time access to various analyses.

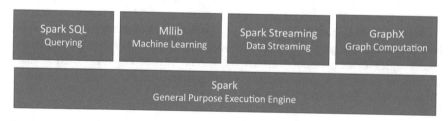

Fig. 6.9 Spark ecosystem

6.4.2 The Core Difference Between Spark and Hadoop

MapReduce can enable users to write parallel computations using a set of high-level operators without having to worry about work distribution and fault tolerance. However, it also demonstrates a number of limitations for complex computational tasks. While MapReduce is great at one-pass computation, it becomes rather inefficient for multi-step algorithms due to the lack of efficient data sharing. Every state between the map and reduce steps requires connection to the distributed file system and is slow due to replication and disk storage. In most current frameworks, the only way to reuse data between MapReduce jobs is to write data to an external storage system and then read from the same location at later time. Since, iterative and interactive applications require faster data sharing across parallel jobs, Hadoop's slow data sharing via MapReduce, due to replication, serialization, and disk I/O can cause serious computational delays. This is a major contributor to the often-presented statistic claiming that most of the Hadoop applications spend more than 90% of the time on the HDFS read and write operations [31].

Iterative operations on MapReduce typically have a need to reuse intermediate results across multiple computations in multi-stage applications. Figure 6.10 depicts how a traditional Hadoop framework enables iterative operations via MapReduce, showcasing data replication, disk I/O, and serialization overheads, causing the overall computational time consequences.

Interactive operations on MapReduce are typically performed when ad hoc queries are executed on the same subset of data. In this scenario, each query will perform the disk I/O, which can dominate application execution time. Figure 6.11 illustrates how the traditional Hadoop framework accomplishes the interactive queries on MapReduce.

As one might expect, large performance hits were experienced when complex workflows were executed on large amounts of data. The need for executing complex workflows without writing intermediate results to disk after every operation becomes a challenge. A new two-pronged approach was proposed to solve these

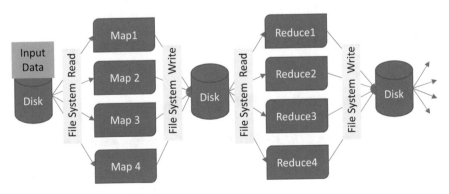

Fig. 6.10 MapReduce execution of iterative operations

Fig. 6.11 MapReduce
interactive execution
illustration

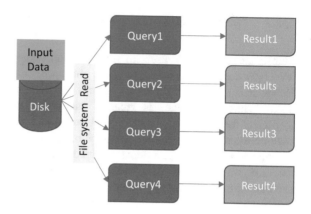

challenges. One aspect of the approach is to cache intermediate results in memory. The second is to allow users to specify persistence in memory and partition the dataset across nodes. In order to ensure fault tolerance, granular atomicity via partitions and transaction logging were used instead of replication.

The highest-level unit of computation in MapReduce is a job that loads data, applies a map function, shuffles, applies a reduce function, or writes data to persistent storage. In Spark, on the other hand, the highest-level unit of computation is an application that can be used for a single batch job, an interactive session with multiple jobs, or a server repeatedly fulfilling requests. A Spark job can consist of more than just a single map and reduce. Spark application processes can run on its behalf even when it's not running a job. Furthermore, multiple tasks can run within the same executor, resulting in orders of magnitude faster performance when compared to MapReduce.

Spark's goal was to generalize MapReduce to support new applications and enable more complex, often iterative or recursive, computations within same engine. Two main additions to Hadoop's approach were powerful enough to express these types of computation and overcome deficiencies of the previous models: fast data sharing and general Directed Acyclic Graphs (DAGs) for computation. These approaches will be presented in more detail in the following sections. This approach enables a much more efficient and much simpler approach, as end users can utilize libraries instead of specialized systems to run complex computational workflows.

6.4.3 Resilient Distributed Datasets in Spark

The Resilient Distributed Dataset (RDD) is Spark's fundamental data structure. It is an immutable foundational distributed collection of objects [32]. The original published paper proposed the concept of the RDD as a **resilient**, fault-tolerant data structure. The RDD lineage graph is able to recompute missing or damaged partitions due to node failures. RDDs are distributed with data residing on multiple nodes in a cluster. RDDs do not change once created and can only be transformed using transformations to new RDDs.

Each dataset in a RDD is divided into logical partitions, which are often computed on a variety of nodes of the cluster. By definition, RDD is a read-only, partitioned, fault-tolerant collection of records that can be processed in parallel manner [32]. RDDs can contain any type of Python, Java, Scala, or user-defined classes and objects. RDDs can be created in one of two ways: either by parallelizing an existing collection or referencing a dataset in an external storage system (HDFS, HBase, etc.). Spark makes use of the concept of RDD to achieve faster and more efficient MapReduce operations.

As shown in Fig. 6.12, iterative Operations on Spark RDDs store intermediate results in a distributed memory and thereby offer much faster computation. In cases when the distributed memory (RAM) is not sufficient to store intermediate results, Spark will default to storing those results on the disk.

The interactive operations on Spark RDD illustrated in Fig. 6.13 are typically utilized when different queries are executed on the same set of data repeatedly. This particular data can be kept in memory to realize significant improvements in execution times.

Each transformed RDD may be recomputed each time an action is executed on that RDD. However, RDDs can also persist in memory, in which case Spark will keep the elements on the cluster for considerably faster access the next time it is queried. There is also support for persisting RDDs on disk in Spark and replication across multiple nodes.

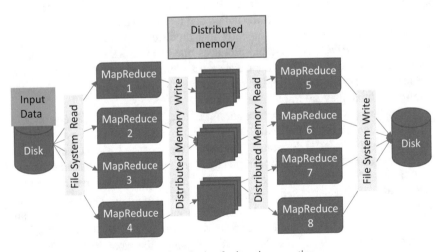

Fig. 6.12 Spark's approach to fast data sharing for iterative operation

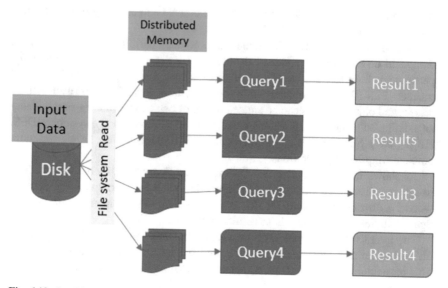

Fig. 6.13 Spark's approach to fast data sharing for queries

6.4.4 RDD Transformations and Actions

RDDs enable two main types of operations: *transformations* and *actions*. Transformation operations could be applied on RDDs and typically return another RDD, while action operations trigger computation and return values.

Spark transformation is a function that produces new RDDs from existing RDDs. It takes RDDs as input and produces one or more RDDs as output based on the transformation applied. Each time a transformation is applied, a new RDD is created. Note that the input RDDs cannot be changed due to the RDD design requirement that they are immutable by nature.

The process of applying transformations builds a **RDD** *lineage*. It keeps track of all of the parent RDDs of the final RDD(s). RDD lineage is also known as the **RDD operator graph** or **RDD dependency graph.** It represents a logical execution plan in the form of a Directed Acyclic Graph (**DAG**) of the entire set of parent RDDs.

RDD transformations are evaluated in a "lazy" manner, by performing the computation only when an action requires a result to be returned. Therefore, they are not executed immediately. Two of the most basic and often used transformations are the *map* and *filter*. A map function iterates over every line in a RDD and applies that function to every element of RDD, possibly enabling the flexibility that the input and the return type of RDD may differ from each other. For example, the input RDD type can be a string and after applying the map function the return RDD can be Boolean. **Filter** functions return a new RDD, containing only the elements that meet a predicate [32].

After the transformation, the resultant RDD is different from its parent RDD. It can be smaller if functions like count, filter, or sample are applied or larger when functions like Cartesian or union are applied. Alternatively, it could remain the same size when a map function is applied.

There are two types of transformations: *narrow* and *wide*. In **narrow transformations,** all the elements that are required to compute the records in single partition reside in the single partition of the parent RDD. A limited subset of partitions is used to calculate the result. Typical *narrow transformations* are considered to be the result of *map or filter functions. Alternatively,* in a wide transformation, all the elements that are required to compute the records in the single partition may reside in numerous partitions of the parent RDD. Wide transformations typically result from the application of join or intersection.

Transformations create RDDs from each other. However, in order to work with the actual dataset, actions need to be performed. Actions are Spark RDD operations that create non-RDD values. When an action is triggered and the result is calculated, the new RDD is not automatically formed as it was the case with transformations. The values of actions are stored to drivers or to the external storage system. In the example below, the first line defines a base RDD from an external file.

```
#Create an RDD from a file on HDFS
text = sc.textFile('hdfs://user1/mytext.txt')

#Transform the RDD of lines into an RDD of words
mywords=text.flatMap(lambda line: line.split())

#Transform the RDD of words into an RDD of key/value pairs
mykeyvals=mywords.map(lambda word:(word,1))
```

RDD transformation vs. action

```
#Map Transformation example counting length
lineLength = text_map(Lambda s: len(s))

#Reduce Action Example
totalLength = lineLength.reduce (Lambda a, b: a+b)

#More RDD manipulation examples

# The saveAsTextFile action writes the contents of
# an RDD to the disk

rdd.saveAsTextFile('hdfs://user1/myRDDoutput.txt')

# The count action returns the number of elements
# in an RDD

numElements=rdd.count();
numElements;
print(numElements)
```

This approach enables executions to be optimized while operations are automatically parallelized and distributed on the clusters. These operations are able to handle many machine learning algorithms that are iterative by nature, and the interactive ad hoc queries needed for many analytics applications in an efficient manner. This is enabled by reusing intermediate in-memory results across multiple data-intensive workloads, thereby avoiding movement of large amounts of data over the network.

6.4.5 Datasets and DataFrames in Spark

A *Dataset* is a distributed collection of data [33]. The Dataset is a new interface added in Spark 1.6 that provides the benefits of RDDs including the strong typing and powerful lambda functions, with the benefits of Spark SQL's optimized execution engine [33, 34]. A Dataset can be constructed from JVM objects and then manipulated using functional transformations mentioned in the RDD section including application of map, filter, etc. The Dataset API is available in Scala and Java. Python and R do not have the support for the Dataset API, but due to those languages' dynamic nature, many of the benefits of the Dataset API are already available and easily usable.

A *DataFrame* is a *dataset* organized into named columns. It is conceptually equivalent to a table in a relational database or a data frame in R or Python, but with richer built-in optimizations. DataFrames can be constructed from a wide array of sources such as structured data files, tables in Hive, external databases, or existing RDDs. The DataFrame API is available in Scala, Java, Python, and R, typically represented by a Dataset of Rows [33]. More details on DataFrames are presented below.

6.4.6 The Spark Processing Engine

While MapReduce is widely adopted for processing and generating large datasets with a parallel, distributed algorithm on a cluster, more and more iterative and interactive modes of operation have emerged that require faster data sharing across parallel jobs. Data sharing is slow in MapReduce due to replication, serialization, and disk IO. As noted earlier, studies have shown that most of the Hadoop applications spend more than 90% of the time doing HDFS read-write operations [25]. In contrast to Hadoop's two-stage disk-based MapReduce paradigm, Spark's iterative in-memory approach enables added computing flexibility and significant speedup. Additionally, Spark's ability to load data into memory and query it repeatedly enables scalable machine learning algorithm performance via the MLlib library [29]. The ability to perform sophisticated, advanced analytics is one of the main advantages of Spark, as it also supports SQL queries, streaming data, machine learning (ML), and graph algorithms.

While MapReduce can achieve complex tasks by defining and chaining various maps and reduce tasks, it is limited to one directional, sequential execution of the mappers and reducers. This limitation has been overcome by allowing task definition using DAGs in Spark. DAG in Apache Spark is an alternative to the MapReduce. It is a programming style used in distributed systems enabling multiple levels that form a tree structure without having to write the intermediate results to disk. A DAG is a finite directed graph with no directed cycles, consisting of finitely many vertices and edges. Each edge is directed from one vertex to another in a consistently directed sequence of edges that can never form a cycle [20].

The DAG concept has successfully been applied to Spark processing. A **DAG** is represented by a set of **vertices** and **edges,** where *vertices* represent the **RDDs** and *edges* represent the **operation to be applied on the RDD**. In order for action to be executed on the RDD, Spark creates the DAG of the tasks to be executed. The DAG is then submitted to the DAG scheduler that divides operators into stages of tasks in order to perform parallel computation.

The principal unit of Spark's computations is a *job*. It is typically a piece of code that reads some input from HDFS, performs computation on the data, and writes output data. Jobs are divided into stages. Stages are classified as a Map or Reduce stages and are divided based on computational boundaries. Most computations are executed over many stages. Each stage has some number of tasks, and typically one task is executed on one partition of data on one machine. The *Executor* is the process responsible for executing a task. The program/process responsible for running the job over the Spark engine is typically referred to as a *driver*. A driver runs on the *master* node, while the machine on which the executor runs is referred to as the *slave*.

6.4.7 Spark Components

Spark applications run as independent sets of processes on a cluster, coordinated by the SparkContext object located in the driver (see Fig. 6.14). The driver separates the process to be executed and creates the SparkContext in order to schedule jobs and negotiate with the cluster manager.

SparkContext can connect to several types of cluster managers (standalone Spark, Mesos, YARN, or Kubernets), in order to allocate resources across applications. Once connected, Spark acquires executors on nodes in the cluster. Executor processes run computations and store application data. Subsequently, it sends application code as defined by JAR or Python files passed to SparkContext, to the executors. Finally, SparkContext sends tasks to the executors to run.

Spark jobs contain a series of *operators* and run on a set of data. All the operators in a job are used to construct a DAG as shown in Fig. 6.15. The DAG is optimized by rearranging and combining operators where possible. For example, if the submitted Spark job contains a map operation followed by a filter operation, Spark's DAG optimizer will reorder the operators, as filtering reduces the number of records to before applying the map operation.

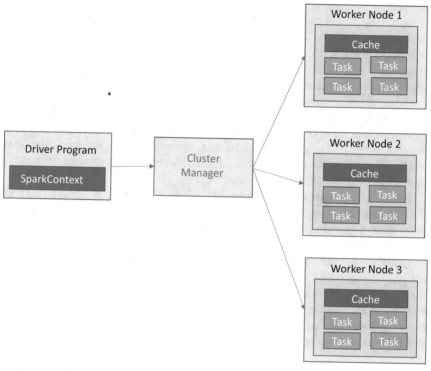

Fig. 6.14 Spark components

Fig. 6.15 DAG for a set of Spark jobs

The Spark system is divided into various *layers* with individual responsibilities in order to execute tasks efficiently. The layers are independent of each other. The primary layer is the *interpreter*, and Spark uses a Scala interpreter. As commands are entered in Spark console, Spark will create an operator graph. When an action is executed, the graph is submitted to a *DAG Scheduler*. The DAG scheduler divides the operator graph into a map and reduces stages. Each stage is comprised of tasks based on partitions of the input data. The DAG scheduler orders operators to optimize the graph, which is key to Spark's fast performance. The final result of a DAG scheduler is a set of stages that are passed to the *Task Scheduler*. The Task Scheduler launches tasks via a **cluster manager** (Spark Standalone, Yarn, Mesos, Kubernets). Nonetheless, it is unaware of any dependencies among stages as illustrated in Fig. 6.16. The *Worker* executes the tasks; however, it knows only about the code that is has received.

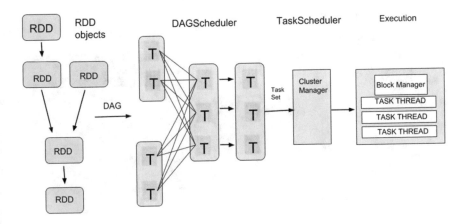

Fig. 6.16 Spark's internal job scheduling process

There are several advantages of DAG in Spark. In case of a lost RDD, Spark can recover the information using the DAG and, with multiple levels of execution, can execute a SQL query or ML operations with much more flexibility and efficiency than MapReduce.

6.4.8 Spark SQL

Spark SQL is a Spark module for structured data processing on very large data sets. Spark SQL provides Spark with additional information about the structure of data and computation and uses this additional information to perform optimizations. Spark SQL provides a fast execution engine by utilizing Spark as the underlying execution engine for low-latency, interactive queries. It also provides the ability for scale-out and failure recovery. The most common use of Spark SQL is to execute SQL queries. However, it is also Hive compatible via Hive Query Language (HQL). This allows it to read data from an existing Hive warehouse without a need to change queries or move data [26]. Spark enables querying of various data sources in addition to Hive tables including Parquet and JSON. In addition, Spark SQL enables combining SQL queries with the data manipulations and complex analytics supported by RDDs in Python, Java, and Scala [35].

Spark SQL provides three main capabilities for using structured and semi-structured data. First, it provides a DataFrame abstraction in Python, Java, and Scala, thereby simplifying the manipulation of structured datasets. Second, it can read and write data in a variety of structured formats including JSON, Hive Tables, and Parquet. Third, it enables data query using SQL. This can be accomplished both inside a Spark program and from external tools that connect Spark SQL to third-party tools via standard database connectors like JDBC and ODBC [36].

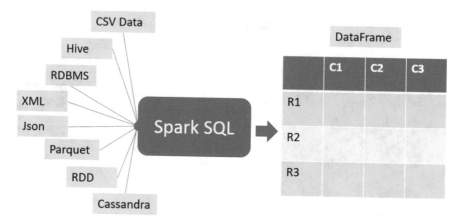

Fig. 6.17 Many ways to create a DataFrame in Spark

6.4.9 Spark DataFrames

A DataFrame in Spark represents a distributed collection of data organized into named columns [33]. A DataFrame is conceptually equivalent to a table in a relational database, a data frame in R or Python's Panda DataFrame, but with additional optimizations for the Spark engine. DataFrames support and can be constructed from a wide array of sources including structured data files, Hive tables, JSON, Parquet, external databases, HDFS, S3, etc. Additionally, through Spark SQL's external data sources API, DataFrames can be extended to support any third-party data formats or sources, including Avro, CSV, ElasticSearch, Cassandra, etc. DataFrames are evaluated lazily, just like RDDs, while operations are automatically parallelized and distributed on clusters. State-of-the-art optimization and code generation is woven throughout the Spark SQL *Catalyst optimizer* utilizing a tree transformation framework. DataFrames can be easily integrated with the rest of the Hadoop ecosystem tools and frameworks via Spark Core and provides an API for Python, Java, Scala, and R Programming (Fig. 6.17).

6.4.10 Creating a DataFrame

In order to start any Spark computation, a basic Spark session needs to be initialized using the sparkR.session() command [33]. Code presented below is adapted from the Spark http://spark.apache.org website.

```
From pyspark.sql import SparkSession
spark = SparkSession
    .builder
    .appName("Python Spark example")
    .config("myspark.config.option", "myvalue")
    .getOrCreate()
```

Reading data from a JSON file into a DataFrame is demonstrated in the code below:

```
# Read the data file in JSON format
df = spark.read.json("/user1/employees.json")
# Displays the content of the DataFrame
df.show()
```

6.4.10.1 Example of Reading DataFrame from the Parquet File

```
dfParquet=spark.read.parquet
                    ("/user1/employees.parquet")
display(dfParquet)
```

6.4.11 DataFrame Operations

DataFrames provide a domain-specific language for structured data manipulation in Scala, Java, Python, and R. DataFrames are Dataset of Rows in the Scala and Java APIs. These operations are also referred to as "untyped transformations" in contrast to "typed transformations" typically associated with strongly typed Scala or Java Datasets.

Basic examples of structured data processing using Datasets is demonstrated below:

```
# Print the schema in a tree format
df.printSchema()

# Select only the "name" column
df.select("name").show()

# Select employees with salary greater than 3000
df.filter(df['salary'] > 3000).show()

# Count people by salary
df.groupBy("salary").count().show()
```

The advantage of a SQL function on a SparkSession is that it enables applications to run SQL queries programmatically and returns the result as a DataFrame [33]. Temporary views in Spark SQL are session-scoped and will disappear if the session that creates it terminates. If there is a need for a temporary view to persist and be shared among all sessions until the Spark application terminates, a *global temporary view* should be created. A global temporary view is tied to a system preserved database global_temp and must use the qualified name to refer it, e.g., SELECT * FROM global_temp.employee.

```
#Temporary view utilized to query the data

df.createOrReplaceTempView("employees")
sqlDF = spark.sql("SELECT * FROM employees ")
sqlDF.show()
```

```
#DataFrame registered as a global temporary view
df.createGlobalTempView("employees")
spark.sql("SELECT * FROM
                   global_temp.employees").show()

#This will be available in the new Spark session
spark.newSession().sql("SELECT * FROM
     global_temp.employee").show()
```

Remember to only use SELECT * in cases of small data sets similar to the ones used in these illustrative examples; otherwise the WHERE clause should be utilized to prevent the possibly very large amount of queried data.

Some additional examples of the queries enabled by Spark SQL are shown below:

```
From pyspark.sql import functions as F

#Show all entries in the column named First name
df.select("firstName").show()

# Show all entries where salary >2000
df.select(df['salary'] > 2000).show()

# Show first name and 0 or 1 depending if they are
# older or younger than 25
df.select("firstName", F.when(df.age > 25, 1)
                   .otherwise(0)).show()
```

6.4.12 Spark MLlib

Spark MLlib is a library containing various machine learning (ML) functionalities optimized for the Spark computing framework. MLlib provides an extensive number of machine learning algorithms and utilities including classification, regression, clustering, association rules, sequential pattern mining, ensemble models, decomposition, topic modeling, and collaborative filtering [30]. In addition, MLlib supports various functionalities such as feature extraction, model evaluation, and validation. All of these methods are designed and optimized to scale across a Spark cluster. Spark's machine learning utilities enable construction of pipelines including tasks that range from data ingest and feature transformations, data standardization, normalization, summary statistics, dimensionality reduction, etc. to model building, hyper-parameter tuning, and evaluation. Finally, Spark enables machine learning persistence by saving and loading models and pipelines [37].

6.4.13   MLlib Capabilities

MLlib's capabilities enable utilization of the large number of major machine learning algorithms including regression (linear, generalized linear, logistic), classification algorithms (including decision trees, random forest, gradient-boosted tree, multilayer perceptron, support vector machine, naive Bayes, etc.), clustering (K-means, K-medoids, bisecting k-means,) latent Dirichlet allocation, Gaussian mixture model, and collaborative filtering. In addition, it supports feature extraction, transformations, dimensionality reduction, selection, and the designing, constructing, evaluating, and tuning of machine learning pipelines.

There are many advantages of MLlib's design including simplicity, scalability, and compatibility. Spark's APIs are simple by design and provide utilities that look and feel like typical data science tools such as R and Python. Machine learning methods can easily be executed with effective parameter tuning [38]. Additionally, MLlib provides seamless scalability by enabling the execution of the ML methods with minimal or no adjustment to the code on a large computing cluster. Spark is compatible with R, Python pandas, scikit-learn, and many other prevalent ML tools. Spark's DataFrames and MLlib provide common data science tool integration with existing workflows.

The goal of most machine learning experiments is to create an accurate model in order to predict on future, unseen data. In order to accomplish this goal, a training data set is used to "train" to fit the model, and a testing data set is used to evaluate and validate the model obtained on the training data set.

Utilizing the PySpark MLlib features, traditional approaches to machine learning can now be scaled to large and complex data sets. For example, we can use the traditionally utilized Iris data set to demonstrate the capabilities of the MLlib to develop predictive models on Spark.

```
from pyspark.ml import Pipeline
from pyspark.ml.classification import
                        DecisionTreeClassifier
from pyspark.ml.evaluation import
                MulticlassClassificationEvaluator
from pyspark.ml.feature import
                StringIndexer, VectorIndexer
from pyspark.ml.linalg import Vectors
from pyspark.ml.feature import VectorAssembler

#Get Spark Context from Spark Session
SpContext = SpSession.sparkContext

#Load the Iris.CSV file
df = spark.read.csv("Iris.data", inferSchema=True)
.toDF("sepLenght", "sepWidth",
                "petLenght", "petWidth", "class")


#Print the first 10 rows of the DataFrame
df.show(10)
```

Index the label by converting class to numeric using StringIndexer:

```
class_index = StringIndexer(inputCol="class",
                            outputCol="classIndex")
df = class_index.fit(df).transform(df)

# Split the data into train and test
(trainingData, testData) =
                      df.randomSplit([0.8, 0.2])

#train the Decision Tree Model
dt = DecisionTreeClassifier(labelCol="ClassIndex",
                            featuresCol="features")
model = dt.fit(trainingData)

predictions = model.transform(testData)

#Evaluate models' accuracy
evaluation = MulticlassClassificationEvaluator(
labelCol="labelIndex", predictionCol="prediction",
                            metricName="accuracy")
accuracy = evaluation.evaluate(predictions)

#Print models' error
print("Test Error = %g " % (1.0 - accuracy))

#Print Model summary
Print(model)
```

Spark enables solving multiple data problems on one platform, from analytics to graph analysis and machine learning. The Spark ecosystem also provides a utility for graph computations called GraphX in addition to streaming and real-time interactive query processing with Spark SQL and DataFrames [36].

6.4.14 Spark Streaming

Spark Streaming is a Spark component that enables processing of live streams of data and enables scalable, high-throughput, fault-tolerant data stream processing.

6.4.15 Intro to Batch and Stream Processing

Before looking into how specifics of how Spark Streaming works, the difference between batch and stream processing should be defined. Typically, batch processing collects a large volume of data elements into a group at once. The entire group is then processed simultaneously in a batch at a specified time. The time of batch computation can be quantified in a number of ways. The computation time can be determined on a prespecified scheduled time interval or on specific triggered condition including a number of elements of or amount of data collected. Batch data processing is a very efficient way to process large amounts of data collected over a period of time when there is no need for real-time analytics. Historically, this has been the most common data processing approach. Traditional databases and data warehouses, including Hadoop, are common examples of batch systems processing.

Stream processing typically utilizes continuous data and is a key component in enabling fast data processing. Streaming enables almost instantaneously data analysis of the data streaming from one device to another. This method of continuous computation happens as data flows through the system with no required time limitations on the output. Due to the near instant data flow, systems do not require large amounts of data to be stored.

The streaming approach processes each new individual piece of data upon arrival. In contrast to batch processing, there is no waiting until the next batch processing interval. The term *micro-batch* is frequently associated with streaming, when batches are small or processed at small intervals. Although processing may occur at high frequency, data is still processed a batch at a time in the micro-batch paradigm. Spark Streaming is an example of a system that supports micro-batch processing. Stream processing is highly beneficial if the events are frequent, especially over rapid time intervals, and there is a need for fast detection and response.

6.4.16 Spark Streaming

Spark Streaming is a Spark component that enables processing of live streams of data by providing an API for manipulating data streams similar to Spark Core's RDD API. It enables scalable, high-throughput, fault-tolerant data stream processing. Spark Streaming's API enables the same high degree fault tolerance, throughput, and scalability as Spark Core. Spark Streaming receives input data streams and divides them into batches called DStreams. DStreams can be created from a number of sources such as Kafka, Flume, and Kinesis or by applying operations on other DStreams (Fig. 6.18).

6.4.17 Spark Functionality

Spark Streaming receives input data streams and divides the data into batches. These batches are then processed by the Spark engine to generate the final stream of results in batches.

Discretized Stream or DStream is the core concept enabled by Spark Streaming. It represents a continuous stream of data. DStream is represented by a continuous series of RDDs. Operations applied to DStreams translate to operations on the underlying RDDs. Spark Streaming discretizes the data into small micro-batches. Spark Streaming receivers accept data in parallel and buffer it in the workers nodes'

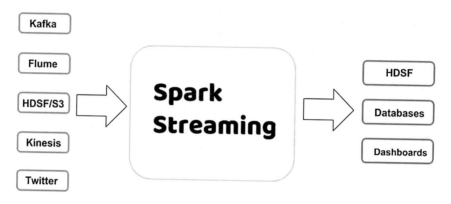

Fig. 6.18 Spark Streaming processing

Fig. 6.19 Spark Streaming functionality

memory. The Spark engine processes the batches, while optimizing latency, and outputs the results to external systems as shown in Fig. 6.19.

Spark Streaming maintains a state based on data coming in a stream often referred to as stateful computations. In addition, Spark Streaming allows window operations where a specified time frame could be used to perform operations on the data. The sliding time interval in the window is used for updating the window, utilizing the window length and sliding interval parameters. When the window slides over a source DStream, the underlying RDDs are combined and operated upon to produce the RDDs of the windowed DStream [28]. Spark tasks are assigned to the workers dynamically on the basis of data locality and available resources, therefore optimizing load balancing and fault recovery.

Spark Streaming's data stream can originate from the source data stream or the processed data stream generated by transforming the input stream. Internally, a DStream is represented by a continuous series of RDDs. Every input DStream is associated with a Receiver, which receives the data from a source and stores it in executor memory.

Analogous to Spark RDDs, Spark transformations enable DStream modifications. Input DStreams support many transformations that are applicable to RDDs, including map, filter, count, countbyvalue, reduce, union, etc. Spark Streaming enables two categories of built-in streaming sources: basic and advanced sources. *Basic sources* are typically directly available in the StreamingContext API, like file systems, and socket connections. *Advanced sources* typically include Kafka, Flume,

Kinesis, etc. and are available through extra utility classes. This requires linking against extra dependencies via linking utilities [28]. If the application requires multiple streams of data in parallel, multiple DStreams can be created. Multiple receivers, simultaneously receiving multiple data streams, can be created, often requiring allocation of multiple cores to process all receiver's data [28].

DStream's data output to external systems, including HDFS, databases or other file systems, utilizes output operations. Output operations trigger the actual execution of the DStream transformations as defined by one of many operations including print, saveAsTextFiles, saveAsObjectFiles, saveAsHadoopFiles, etc. DStreams similar to RDDs execute lazily by the output operations.

The example below illustrates a basic application of Spark Streaming: counting the number of words in text data received from a data server listening on a TCP socket, as adapted from [28].

```python
from pyspark import SparkContext
from pyspark.streaming import StreamingContext

# Create a local StreamingContext
# batch interval of set to 3 seconds
sc = SparkContext(appName= "NetworkWordCount")
ssc = StreamingContext(sc, 5)
#Create a DStream for the TCP data stream
#Specify local host and port number where the
 system will listen for streaming data
MyStream = ssc.socketTextStream("localhost", 9999)

# Split each DStream line into individual words
#Utilize flatMap to create new DStream of words
wordStream =
MyStream.flatMap(lambda line: MyStream.split(" "))

# Count each word per batch

wordPairs = wordStream.map(lambda word: (word, 1))
wordCounts = wordPairs.reduceByKey(_+_)

# Print the first ten elements of each RDD
wordCounts.pprint()

# In order to start the computation
ssc.start()
# Wait for the computation to terminate
ssc.awaitTermination()

# Run Netcat to enable application execution in Spark
# Open a socket on port 9999
$ nc -lk 9999
#Check that the port is open
$ nc localhost 9999
```

And type the text you would like to be counted – possible example:

```
Spark streaming is amazing!
This is a great example of a spark streaming application
Run
Execute
Run
```

```
#In another terminal run the
$ spark-submit mynetworkcount.py localhost 9999
```

The output on the screen will indicate the number of words counted.

This example illustrates the Spark Streaming process of ingesting data into a *Discretized Streaming framework. DStreams enable users to* capture the data and perform many different types of computations, as illustrated in this example by a simple word count of the incoming data set. DStreaming and RDDs are a crucial set of building blocks that enable construction of complex streaming applications with Spark and Spark Streaming.

6.5 Big Data Analytics: Building the Data Pipeline

Several different maturity levels can be considered with regard to Big Data Analytics. There are a number of organizations (DAMM, Gartner, IIA, HIMMS, TDWI, IBM, etc.) that have defined their own version of analytics maturity levels. However, they all agree on three general tiers. All organizations start with raw data and move first to cleaned, standardized, and organized data. They next progress to basic and advanced reporting. Finally, they may finally graduate to building predictive models. This process highlights the levels of sophistication in analytics moving from descriptive, to diagnostic, to predictive, and finally to prescriptive modeling. Descriptive analytics help understand what has happened in the past, while diagnostic analytics looks into reasons of why something might have happened. Predictive analytics techniques build machine learning models to predict what will happen. These models can then be fed into prescriptive models, which take the process directly to decision making and action by recommending what should be done under certain conditions.

6.5.1 Developing Predictive and Prescriptive Models

John Naisbitt famously said, "We are drowning in data, but starving for knowledge!" A great quote that is made more amazing when one considers that it was made in 1982. His observation rings ever more true today. While the scale of data has changed, the need for skills, tools, and techniques to find meaning in the mayhem of the Big Data world has not. It is costly to collect, store, and secure Big Data properly, and real return on investment (ROI) hinges on the ability to extract actionable information from the data. The field of Data Science is one angle from which to approach the data deluge. Data scientists endeavor to extract meaning and tell the story of the data in order to provide insight and guidance. Data scientists have established technologies that uncover relationships and patterns within large volumes of data that then can be leveraged to predict future behavior and events. For example, the development of predictive modeling techniques utilizing machine learning methods was driven by the necessity to address the data explosion. This technology learns from experience and predicts future outcomes in order to drive better business decisions. It extracts rules, regularities, patterns, and constraints from raw data, with the goal of discovering implicit, previously unknown and unexpected, valuable information from data.

6.5.2 The Cross Industry Standard Process for Data Mining (CRISP-DM)

The process of moving from raw data to effective models is an iterative and multi-phase one. As discussed in Chap. 5, the CRISP-DM standard, depicted in Fig. 6.20, identifies the six major phases of this data mining process. When approaching predictive model development, it is essential to deeply understand the application domain characteristics. This is the goal of phase one, the *Business Understanding* phase.

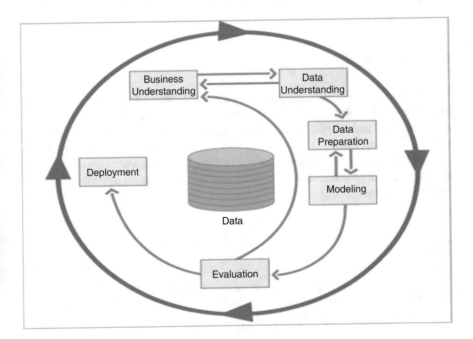

Fig. 6.20 CRISP-DM process model

Once the business problem and the overall project goals are fully understood, the project moves into the *Data Understanding* phase. Creating the proper dataset is the goal of this phase. It may involve bringing together data from different sources and of different types to be able to develop comprehensive models. The rate, quantity, and quality of the data are carefully considered. The execution of this phase may require reconsideration of the business understanding based on data availability, resource limitations, and such.

The *data preparation* phase is frequently the most time consuming and resource intensive phase of the process. The preprocessing and cleaning of the data undertaken in this phase can require considerable effort and should not be underestimated. Careful, advanced planning of data collection and storage can help minimize the effort expended in this phase.

The *modeling* phase can be initiated once the data has been sufficiently prepared. However, it is typical for data preparation efforts to continue and/or be revised based on the progress made and insights gained during the modeling process. The modeling phase involves applying one or more data science techniques to the data set in order to extract actionable insight.

Once models are developed (or "trained") in the modeling phase, the *evaluation* phase considers the value of the models in the context of the original business understanding. Frequently, multiple iterations through the process are required to arrive at a satisfactory data mining solution.

Finally, the *deployment* phase addresses the implementation of the models within the organization and completes the process. This may involve multiple personnel and expertise from a wide variety of groups in addition to the data science team.

6.6 Conclusion

Big Data is fundamentally changing the way organizations and businesses operate and compete. Big data and IoT also share a closely knitted future to offer data-driven analysis and insight. In this chapter, we explained how to build and maintain reliable, scalable, distributed systems with Apache Hadoop and Apache Spark. We also discussed how to utilize Hadoop and Spark for different types of big data analytics in IoT projects, including batch and real-time stream analysis as well as machine learning.

References

1. https://www.ibmbigdatahub.com/infographic/extracting-business-value-4-vs-big-data
2. https://spectrum.ieee.org/tech-talk/telecom/internet/popular-internet-of-things-forecast-of-50-billion-devices-by-2020-is-outdated
3. https://www.visualcapitalist.com/internet-minute-2018/

4. M. Cafarella, B. Lorica, D. Cutting, The next 10 years of Apache Hadoop. (O'Reilly Media, 2016). https://www.oreilly.com/ideas/the-next-10-years-of-apache-hadoop
5. G. Sanjay, G. Howard, L. Shun-Tak, The Google file system. SIGOPS Oper. Syst. Rev. **37**(5), 29–43 (2003). https://doi.org/10.1145/1165389.945450
6. J. Dean, S. Ghemawat, MapReduce: Simplified data processing on large clusters. Commun. ACM **51**(1), 107–113 (2008). https://doi.org/10.1145/1327452.1327492
7. M. Bhandarkar, in *2010 IEEE International Symposium on Parallel & Distributed Processing (IPDPS)*. MapReduce programming with apache Hadoop, (Atlanta, GA, 2010), pp. 1–1. https://doi.org/10.1109/IPDPS.2010.5470377
8. P. Merla, Y. Liang, in *2017 IEEE International Conference on Big Data*. Data analysis using Hadoop MapReduce environment, (Boston, MA, 2017), pp. 4783–4785. https://doi.org/10.1109/BigData.2017.8258541
9. https://hadoop.apache.org/
10. D. Jeffrey, S. Ghemawat, in *OSDI,* MapReduce: Simplified data processing on large clusters (2004)
11. S. Ghemawat, H. Gobioff, S. Leung, The Google file system, in *Proceedings of the nineteenth ACM symposium on Operating systems principles (SOSP '03)*, (ACM, New York, NY, USA, 2003), pp. 29–43. https://doi.org/10.1145/945445.945450
12. https://pig.apache.org/
13. https://hive.apache.org/
14. https://hadoop.apache.org/docs/r1.2.1/hdfs_design.html
15. K. Shvachko, H. Kuang, S. Radia, R. Chansler, in *2010 IEEE 26th Symposium on Mass Storage Systems and Technologies (MSST)*, The Hadoop Distributed File System, (Incline Village, NV, 2010), pp. 1–10. https://doi.org/10.1109/MSST.2010.5496972
16. https://www.json.org/
17. https://avro.apache.org/
18. http://sqoop.apache.org/
19. E. Capriolo, D. Wampler, J. Rutherglen, *Programming Hive: Data Warehouse and Query Language for Hadoop*, 1st edn. (O'Reilly Media, Sebastopol, CA, 2012). ISBN-13: 978-1449319335. ISBN-10: 1449319335
20. K. Thulasiraman, M.N.S. Swamy, *5.7 Acyclic Directed Graphs, Graphs: Theory and Algorithms* (Wiley, New York, 1992), p. 118. ISBN 978-0-471-51356-8
21. https://zookeeper.apache.org/
22. https://www.usenix.org/legacy/event/atc10/tech/full_papers/Hunt.pdf
23. H. Fang, in *2015 IEEE International Conference on Cyber Technology in Automation, Control, and Intelligent Systems (CYBER)*, Managing data lakes in big data era: What's a data lake and why has it became popular in data management ecosystem, (Shenyang, 2015), pp. 820–824. https://doi.org/10.1109/CYBER.2015.7288049
24. https://www.epic.com/
25. M. Zaharia, M. Chowdhury, M.J. Franklin, S. Shenker, I. Stoica, in *Proceedings of the 2nd USENIX conference on Hot topics in cloud computing* (HotCloud'10), Spark: Cluster computing with working sets. (USENIX Association, Berkeley, CA, USA, 2010), pp. 10–10
26. https://spark.apache.org/
27. https://spark.apache.org/sql/
28. https://spark.apache.org/streaming/
29. https://spark.apache.org/mllib/
30. https://spark.apache.org/grapx
31. V.J. Srinivas, P. Srikanth, K. Thumati, S.H. Nallamala, in *Proceedings International Journal of Computer Science Trends and Technology (IJCST)*, A review study of Apache Spark in Big Data processing, Vol. 4, Issue 3, May/Jun (2016)
32. M. Zaharia, M. Chowdhury, T. Das, A. Dave, J. Ma, M. McCauley, M.J. Franklin, S. Shenker, I. Stoica, in *Proceedings of the 9th USENIX conference on Networked Systems Design and Implementation* (NSDI'12), Resilient distributed datasets: A fault-tolerant abstraction for in-memory cluster computing. (USENIX Association, Berkeley, CA, USA, 2012), pp. 2–2

33. https://spark.apache.org/docs/latest/rdd-programming-guide.html
34. https://spark.apache.org/docs/2.2.0/sql-programming-guide.html
35. M. Zaharia, B. Chambers, *Spark: The Definitive Guide* (O'Reilly Media, Sebastopol, CA, 2018)
36. S. Kumar, in *2016 3rd International Conference on Computing for Sustainable Global Development (INDIACom)*, Evolution of Spark framework for simplifying big data analytics, (New Delhi, 2016), pp. 3597–3602
37. M. Assefi, E. Behravesh, G. Liu, A.P. Tafti, in *2017 IEEE International Conference on Big Data (Big Data)*, Big data machine learning using apache spark MLlib (Boston, MA, 2017), pp. 3492–3498. https://doi.org/10.1109/BigData.2017.8258338
38. D. Siegal, J. Guo, G. Agrawal, in *2016 IEEE International Conference on Cluster Computing (CLUSTER)*, Smart-MLlib: A high-performance machine-learning library, (Taipei, 2016), pp. 336–345. https://doi.org/10.1109/CLUSTER.2016.49

Chapter 7
Intelligent and Connected Cyber-Physical Systems: A Perspective from Connected Autonomous Vehicles

Wanli Chang, Simon Burton, Chung-Wei Lin, Qi Zhu, Lydia Gauerhof, and John McDermid

The need for connection and community is primal, as fundamental as the need for air, water, and food.

Dean Ornish

Contents

7.1   Introduction

Cyber-physical systems (CPS) integrate computation, communication, and physical processes. CPS range from small scale, such as wearable medical devices, to large scale, such as national power grids. The concept of CPS is frequently discussed

W. Chang (✉) · J. McDermid
University of York, York, UK
e-mail: wanli.chang@york.ac.uk

S. Burton · L. Gauerhof
Robert Bosch GmbH, Gerlingen, Germany

C.-W. Lin
National Taiwan University, Taipei, Taiwan

Q. Zhu
Northwestern University, Evanston, IL, USA

© Springer Nature Switzerland AG 2020
F. Firouzi et al. (eds.), *Intelligent Internet of Things*,
https://doi.org/10.1007/978-3-030-30367-9_7

357

together with IoT (Internet of Things). There are different opinions on how they relate to each other. We tend to think that IoT, while generally also involving both cyber and physical components, emphasizes more on the connectivity as well as the functionality enabled by such strong connectivity. CPS, on the other hand, stresses the interactions between the cyber and physical aspects. An example that belongs to CPS but not IoT is a vehicle without connection to the outside, neither other vehicles nor any infrastructure.

There is a trend observed that CPS are getting better and better connected, thus moving towards the regime of IoT. For instance, although the modern automobiles, a typical class of CPS, commonly have modules for connection, they cannot be classified as IoT, since such connection is not strong enough to enable significant functionality. However, the automotive industry is moving towards connected autonomous vehicles (CAVs), where the vehicle-to-vehicle (V2V) and vehicle-to-infrastructure (V2I) communication will completely change the way cars are driven, together with various kinds of sensors and algorithms. This will then become part of IoT.

Closed control loops are often found in CPS, and in each loop, there are three main operations involving sensors, controllers, and actuators:

- Sensing: The states of the physical plants to be controlled are measured with sensors.
- Computation: The control input is computed by the controllers based on the measured plant states.
- Actuation: The control input is applied to the plant through actuators.

The plant dynamics is then influenced by the control input, closing the loop. When the sensors, controllers, and actuators are distributed, a network is required for communication. A typical architecture of CPS is illustrated in Fig. 7.1.

Conventionally, there is a separate design paradigm for CPS. For instance, the development of control algorithms is based on idealistic models, and the details of the embedded implementation platforms are not considered. Large safety margins are commonly inserted. A new CPS design methodology taking account of all the layers as well as their interplays is being investigated, in order to achieve assurance of safety and security, robustness, and resource efficiency. These issues are highly important for the intelligent and connected CPS. For example, while many groups

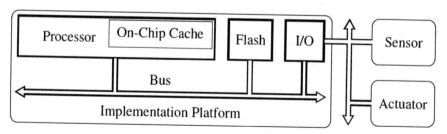

Fig. 7.1 Typical CPS architecture

around the world are able to develop full functionalities of CAVs, as evidenced by the prototypes and trials, mass production and deployment will not happen if any of the four issues mentioned above is not sufficiently addressed.

Some recent efforts contributing towards the new methodology are on resource-aware CPS design [1, 2]. That is, the cyber implementation resources are considered when developing control algorithms directly influencing the physical processes. This significantly improves the resource dimensioning and utilization and gains more trust in the system. There are mainly three types of resources: computation, communication, and memory. Note that the computational resources are often related to not only the processors, but also the operating systems. The communication resources vary with different protocols. The memory resources are commonly organized in a hierarchy.

The communication protocols are broadly classified into two groups – event-triggered (ET) and time-triggered (TT) networks. For instance, the Controller Area Network (CAN) is ET and has been widely used since its first official release in 1986. FlexRay, standardized in 2013, was designed to be faster and more reliable than CAN and can be found in many premium cars. Media access control in FlexRay is based on communication cycles of equal and predefined length in time. Each communication cycle is divided into a TT static and an optional ET dynamic segment. Messages can be sent with FlexRay over either the TT or ET segment using a bandwidth of 10 Mbit/s.

The TT static segment follows a timing division multiple access (TDMA) policy, where the entire segment is divided into multiple slots with the same predefined length in time. Each application involved in the TT communication is assigned a dedicated slot. This allows a predictable temporal behavior. Time-sensitive networking (TSN) is currently being developed to provide deterministic messaging on standard Ethernet. The key features are time synchronization and traffic scheduling. They are addressed by the 802.1AS and 802.1Qbv standards, respectively. All participating devices are synchronized to a global time and are aware of a network schedule that dictates when prioritized messages will be forwarded from each switch.

The memory resources often have a multi-level hierarchy. For example, the flash memory has a large size and stores all the application programs and data, but experiences high read/write latencies (hundreds of processor cycles). The cache (on-chip memory in Fig. 7.1) is faster with low read/write latencies (several processor cycles), but usually limited in size due to its high cost. The access times of cache and flash memory are denoted as t_c and t_m, respectively. When a processor executes an instruction, it checks the cache first. If this instruction is located in the cache, it is a cache hit and the access time is t_c. If this instruction is not in the cache, the memory block containing it is fetched from the flash memory and then written into cache. This is a cache miss and the access time is t_m. Afterwards, when the same instruction is called again by the processor, the access time is t_c if it is still in the cache without being replaced. This is a cache reuse.

A program usually has different execution paths resulting in different execution times. The worst-case execution time (WCET) is defined to be the maximum length of time a program takes to be executed. There are two general methods to reduce the WCET of a program – increasing the cache size and cache reuse. Many CPS applications are cost-sensitive, which makes it desirable to minimize the cache size. Therefore, the cache reuse should be maximized. When multiple applications share the same memory resources, the cache reuse depends on the execution schedule. For instance, a schedule that consecutively executes the same application increases the cache reuse.

Computational resources usually mean the available execution time, for a given processor with a certain operating frequency. When multiple applications share one processor, in general, the performance of an application can be improved if it is allowed to access the processor longer. Taking applications running periodic tasks as an example, a shorter period usually results in better performance. The downside is a higher processor utilization. On the condition that the performance requirement can be satisfied, reduction on the processor utilization of an application is desirable, as more applications can then be mapped to the processor, thereby saving the cost.

Due to the safety-critical nature of many CPS, time-triggered operating systems (TT OS) often run on the processor. For instance, OSEK/VDX (Open Systems and Their Corresponding Interfaces for Automotive Electronics/Vehicle Distributed Executive)-compatible OS are widely used in the automotive domain. OSEK/VDX OS only offer a limited set of predefined periods. In most cases, the optimal period is not directly realizable on the OS. The conventional way to handle it is to use the largest period offered by the OS that is smaller than the optimal one. This is a straightforward method, yet leads to a waste of computational resources. Sometimes, a mixture of periods may achieve a better trade-off between the performance and the processor utilization.

In the rest of this chapter, we will describe the essential technical background on CPS, organized into cyber components and physical components, emphasizing on the interactions between them. Resource-oriented efforts taking safety into account will be discussed to illustrate the new CPS design methodology, which could incorporate robustness and security. Two case studies on connected autonomous vehicles (CAVs) will be presented. One is on safety of machine learning (ML)-based perception for highly automated driving. The other is on robustness and security of connected vehicles.

7.2 Background

The technical background of CPS includes both the cyber and physical components, as well as their interactions. For the cyber components, we will focus on the memory architecture and analysis, real-time operating systems (RTOS), and scheduling. It is noted that there is a dedicated chapter in this book on communication. For the physical components, we will cover modelling of plant dynamics, safety requirements on control performances and physical constraints, control algorithms, and stability guarantee. The cross-layer interplay between the cyber and physical components will be discussed with memory-aware CPS design as an example.

7.2.1 Cyber Components

In the CPS architecture, there is often a hierarchy of memory resources, such as the flash memory and the on-chip memory in Fig. 7.1. As discussed in the introduction, cache reuse depends on the execution order of applications. Given a collection of applications (e.g., C_1, C_2, C_3) sharing the same processor and memory resources, it is conventional to run them in a round-robin fashion (C_1, C_2, C_3, C_1, C_2, C_3, $\cdots$). Since the code for different applications is usually different, the on-chip cache is frequently refreshed in this process. This results in poor cache reuse and long WCET. In order to address this issue, each application can be consecutively executed multiple times, which will increase the cache reuse and shorten the WCET.

An example execution order ($C_1(1)$, $C_1(2)$, $C_1(3)$, $C_2(1)$, $C_2(2)$, $C_2(3)$, $C_3(1)$, $C_3(2)$, $C_3(3)$, $\cdots$) is illustrated in Fig. 7.2, where $C_i(j)$ denotes the jth execution of the application C_i. Before the first execution $C_i(1)$, the cache is either empty (i.e., cold cache) or filled with instructions from other applications, which are not used by C_i (equivalent to cold cache). The WCET of $C_i(1)$ can be computed by a number of existing standard techniques. Before the second execution $C_i(2)$, the instructions in the cache are from the same application C_i and thus can be reused. This results in more cache hits and hence shorter WCET. Depending on which execution path the program takes, the amount of WCET reduction varies. Therefore, a technique is required to compute the guaranteed WCET reduction of $C_i(2)$ and $C_i(3)$ relative to $C_i(1)$, independent of the path taken.

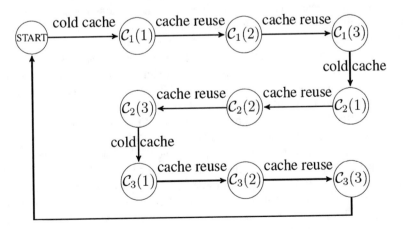

Fig. 7.2 An example memory-aware execution order with three applications

This technique starts from a control flow graph (CFG) and then sets up the equations to compute the reaching cache states (RCS) and live cache states (LCS) of each node, based on which the fixed-point computation is performed. Afterwards, the guaranteed WCET reduction can be calculated [3, 4]. We will begin our discussion with some basics.

In the two-level memory hierarchy shown in Fig. 7.1, there are N_c cache lines, denoted as $CL = \{c_0, c_1, \ldots, c_{N_c-1}\}$ and the flash main memory has N_m blocks, denoted as $M = \{m_0, m_1, \ldots, m_{N_m-1}\}$. Each memory block is mapped to a fixed cache line. A basic block is a straight-line sequence of code with only one entry point and one exit point. This restriction makes a basic block highly amenable for program analysis. There are three key terms in memory analysis that are described as follows:

- Cache states: A cache state cs is described as a vector of N_c elements. Each element $cs[i]$, where $i \in \{0, 1, \ldots, N_c - 1\}$, represents the memory block in the cache line c_i. When the cache line c_i holds the memory block m_j, where $j \in \{0, 1, \ldots, N_m - 1\}$, $cs[i] = m_j$. If c_i is empty, it is denoted as $cs[i] = \perp$. If the memory block is unknown, it is denoted as $cs[i] = \top$. CS is the set of all possible cache states.

- Reaching cache states: RCS of a basic block b_k, denoted as FCS_{b_k}, is the set of all possible cache states when b_k is reached via any incoming path.

- Live cache states: LCS of a basic block b_k, denoted as LCS_{b_k}, is the set of all possible first memory references to cache lines at b_k via any outgoing path.

Since the focus is on WCET reduction between two consecutive executions of C_i, e.g., $C_i(1)$ and $C_i(2)$, it is necessary to compute the RCS of the exit point in $C_i(1)$ and the LCS of the entry point in $C_i(2)$. By comparing all possible pairs of cache states, the guaranteed number of cache hits and thus WCET reduction can be calculated. Conceptually, the program RCS is the set of all possible cache states after the program finishes execution by any execution path, and the program LCS is the set of all cache states, where each cache state contains memory blocks that may be firstly referenced after the program starts execution, for any execution path to follow. Both the RCS and LCS could contain multiple cache states. Each pair with one cache state cs from the program RCS and one cache state cs' from the program LCS represents one possible execution path between the two consecutive executions. For any cache line c_i in a pair, if $cs[i]$ is equal to $cs'[i]$ and they are not equal to $\top$, then there is certainly a hit and thus WCET reduction.

As discussed in the introduction, CPS often run TT OS due to the safety-critical nature. We will take OSEK/VDX OS, which is a class of RTOS widely used in the automotive industry, as an example. In general, OSEK/VDX OS supports preemptive fixed-priority scheduling. That is, priorities are assigned to applications and at any point in time, the task with the highest priority among all active ones is executed. Tasks can be triggered by events (e.g., interrupts, alarms, etc.) or by time. In the TT scheme, each application gets released and is allowed to access the processor periodically. There are various periods of release times and each application is assigned one. Different applications may have different periods. Every time an application is released, its program gets the chance to be executed.

A timetable containing all the periodic release times within the alleged hyperperiod (i.e., the minimum common multiple of all periods) needs to be configured. An example with a set of three periods 2 ms, 5 ms, and 10 ms is illustrated in Table 7.1. The hyperperiod is equal to 10 ms and the timetable repeats itself every 10 ms by resetting the timer. Independent of the triggering mode (i.e., be it ET or TT), the assigned priority will still determine the execution order of applications. In the TT scheme, a higher priority is typically assigned to the application released within a shorter period, since this generally results in a more efficient use of the processor.

An example with two applications C_1 and C_2 sharing one processor is illustrated in Fig. 7.3. C_1 has a period of 2 ms and C_2 has a period of 5 ms. The execution time of C_1 is assumed to be 0.7 ms and the execution time of C_2 is assumed to be 2 ms. C_1 has a higher priority than C_2. Within a hyperperiod of 10 ms, C_1 is released at 0 ms, 2 ms, 4 ms, 6 ms, 8 ms, and 10 ms. C_2 is released at 0 ms, 5 ms, and 10 ms. It can be seen that C_2 is executed only when C_1 does not require to access the processor. For instance, at 0 ms, both C_1 and C_2 are released and require access to the processor. C_1 is permitted to be executed while C_2 has to wait. At 0.7 ms, C_1 completes its execution and C_2 gets the access to the processor.

Denoting E_i^{wc} to be the WCET of an application C_i, if the period is h, the processor utilization for C_i is

$$L_i = \frac{E_i^{wc}}{h}.$$

The upper bound on the utilization of any processor is 1. Considering a single processor p,

Table 7.1 An example OSEK/VDX OS timetable of application release

Time	Release
0 ms	Applications with periods of 2 ms/5 ms/10 ms
2 ms	Applications with the period of 2 ms
4 ms	Applications with the period of 2 ms
5 ms	Applications with the period of 5 ms
6 ms	Applications with the period of 2 ms
8 ms	Applications with the period of 2 ms
10 ms	Repeat actions at 0 ms

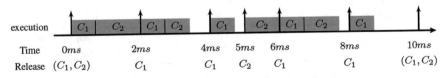

Fig. 7.3 Release and execution time of two applications sharing one processor

$$\sum_{\{i|C_i \ runs \ on \ p\}} L_i \le 1.$$

Clearly, increasing the period of an application decreases its processor utilization and thus potentially enables more applications to be integrated on the processor.

It is assumed that the set of available periods restricted by OSEK/VDX is ϕ. As discussed in the introduction, a mixture of periods may achieve a better trade-off between performance and processor utilization [5]. An example is shown in Fig. 7.4. Switching between two periods can only occur at the common multiplier of them. For instance, switching between 2 ms and 5 ms is possible at the time instant of 10 ms, 20 ms, and so on. Therefore, following this rule, possible sequences of periods are $\{2ms, 2ms, 2ms, 2ms, 2ms, 5ms, 5ms, repeat\}$, $\{5ms, 5ms, 10ms, repeat\}$, and so on.

Scheduling is one of the core tasks of an OS. Different scheduling algorithms have different properties, and the choice of a particular algorithm may favor one class of processes over another. In choosing which algorithm to use in a particular situation, we must consider the properties of the various algorithms. Many criteria have been suggested for comparing scheduling algorithms. Which characteristics are used for comparison can make a substantial difference in which algorithm is judged the best. The criteria include the following:

- Processor utilization: We want to keep the processor as busy as possible. Conceptually, the processor utilization can range from 0% to 100%. In a real system, it should range from 40% (for a lightly loaded system) to 90% (for a heavily loaded system).
- Throughput: If the processor is busy executing tasks, then work is being done. One measure of work is the number of tasks that are completed per time unit, called throughput. For long tasks, this rate may be one task per hour; for short tasks, it may be ten tasks per second.

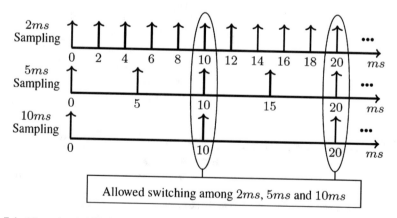

Fig. 7.4 Allowed switching instants among multiple periods

- Turnaround time: From the point of view of a particular task, the important criterion is how long it takes to execute that task. The interval from the time of submission of a task to the time of completion is the turnaround time.
- Waiting time: The processor-scheduling algorithm does not affect the amount of time during which a task executes. It affects only the amount of time that a task spends waiting.

It is desirable to maximize the processor utilization and throughput and to minimize the turnaround time, waiting time, and response time. In most cases, we optimize the average measure. However, under some circumstances, we prefer to optimize the minimum or maximum values rather than the average.

There are many different scheduling algorithms. The simplest one is the first-come, first-served (FCFS) scheduling algorithm. With this scheme, the task that requests the processor first is allocated the processor first. The average waiting time under the FCFS policy is often quite long. Note also that the FCFS scheduling algorithm is nonpreemptive. Once the processor has been allocated to a task, that task keeps the processor until it releases the processor, often by terminating. The FCFS algorithm is thus particularly troublesome for time-sharing systems, where it is important that each task get a share of the processor at regular intervals. It would be disastrous to allow one task to keep the processor for an extended period.

A different approach to processor scheduling is the shortest-job-first (SJF) scheduling algorithm. This algorithm associates with each task the length of the task's remaining execution time. When the processor is available, it is assigned to the task that has the shortest remaining execution time. The SJF scheduling algorithm is provably optimal, in that it gives the minimum average waiting time for a given set of tasks. Moving a short task before a long one decreases the waiting time of the short task more than it increases the waiting time of the long task. Consequently, the average waiting time decreases. The real difficulty with the SJF algorithm is knowing the length of the remaining execution time, which often has to be approximated. The SJF algorithm can be either preemptive or nonpreemptive. If the newly arrived task is shorter than what is left of the currently executing task, a preemptive SJF algorithm will preempt the currently executing task, whereas a nonpreemptive SJF algorithm will allow the currently running task to finish.

The SJF algorithm is a special case of the general priority-scheduling algorithm. A priority is associated with each task, and the processor is allocated to the task with the highest priority. Equal-priority tasks are scheduled in FCFS order. Priorities are generally indicated by some fixed range of numbers, such as 0–7 or 0–4095. However, there is no general agreement on whether 0 is the highest or lowest priority. Priorities can be defined either internally or externally. Internally defined priorities use some measurable quantity or quantities to compute the priority of a task. For example, time limits, memory requirements, and the number of open files have been used in computing priorities. External priorities are set by criteria outside the operating system, such as the importance of the task. Priority scheduling can be either preemptive or nonpreemptive. A major problem with priority scheduling algorithms is indefinite blocking, or starvation. A priority scheduling algorithm can leave some low-priority tasks waiting indefinitely. A solution to this problem is ageing. Ageing involves gradually increasing the priority of tasks that wait in the system for a long time.

There are many other scheduling algorithms that will not be discussed in detail here, such as the round-robin scheduling and multilevel scheduling. For RTOS, latency becomes the most critical factor. Common scheduling algorithms include the rate-monotonic scheduling, earliest-deadline-first (EDF) scheduling, proportional share scheduling, etc. Note that in the above we focus on scheduling of a single processor. For the multi-processor system, the scheduling becomes more complicated and a number of issues need to be considered, such as processor affinity and load balancing.

7.2.2 Physical Components

CPS often involve control applications. A control application is responsible for controlling a plant or dynamical system. For linear single-input single-output (SISO) control applications, the dynamic behavior is modelled by a set of differential equations:

$$\dot{x}(t) = Ax(t) + Bu(t),$$

$$y(t) = Cx(t),$$

where $x(t) \in \mathbb{R}^l$ is the system state, $\dot{x}(t)$ is the derivative of $x(t)$ with respect to time, $y(t)$ is the system output, and $u(t)$ is the control input. The number of system states is l. The system (or state) matrix is A. The input matrix is B. The output matrix is C. These matrices A, B, and C are physical properties of the plant. System poles are eigenvalues of A. In a state-feedback control algorithm, $u(t)$ is computed utilizing $x(t)$ (feedback signals) and then applied to the plant, which is expected to achieve certain desired behavior.

In most applications, the controller is implemented in a digital fashion on a computer. This implies that the system states must be sampled when measured by the sensors. Assuming the sampling period to be h, the sampled system state is denoted as

$$x[k] = x(t_k), t_k = kh, k = 0, 1, 2, 3, \cdots.$$

Similarly, the sampled system output is

$$y[k] = y(t_k).$$

The control input taking discrete values is denoted as $u[k]$, which is passed through a zero-order hold (ZOH) and applied to the plant. The output of the ZOH is given by

$$u(t) = u[k], t_k \le t < t_{k+1}.$$

Now the discretized dynamics can be derived:

$$x[k+1] = A_d x[k] + B_d u[k],$$

$$y[k] = Cx[k],$$

where

$$A_d = e^{Ah}, B_d = \int_0^h \left(e^{A\tau'} d\tau' \right) B.$$

Settling time is a widely used metric to quantify the control performance, especially for real-time control applications [6, 7]. The time it takes for the system output $y[k]$ to reach and stay in a closed region around the reference value r (e.g., $0.98r$ to $1.02r$) is the settling time of a control loop and denoted as t_s. Shorter settling time implies better control performance. In order to ensure safety of the CPS, there is often a requirement on the settling time. That is, t_s must be shorter than or equal to certain bound t_s^0.

Besides the control performance, there are system constraints related to the CPS safety. For instance, in almost every real-world system, due to the physical constraint of the actuator, there is some maximum available control input signal, and the controller needs to be designed such that the maximum value of $u[k]$ does not exceed this limit U_{max}, i.e., $u[k] \le U_{max}$. This is the constraint of the input saturation. Another constraint is on the peak overshoot, which is defined as

$$y_{max} - r \le \phi_0 r,$$

where y_{max} is the maximum system output and ϕ_0 is the overshoot threshold. The constraint on the steady-state error has been discussed when defining the settling time. The system output $y[k]$ has to reach and stay in a closed region around r, i.e., the system has to settle. If the region is $[0.98r, 1.02r]$, then the steady-state error tolerance is $\phi_e = 2\%$.

In the state-feedback control algorithm, the control input $u[k]$ is computed based on the system state $x[k]$. There can be both linear and nonlinear controllers, depending on the relationship between $u[k]$ and $x[k]$. The general structure of a linear controller is as follows:

$$u[k] = Kx[k] + Fr,$$

where K is the feedback gain and F is the feedforward gain. Clearly, the relationship between $u[k]$ and $x[k]$ is linear. With this controller, the system dynamics becomes

$$x[k+1] = (A_d + B_d K) x[k] + B_d F r,$$

i.e., closed-loop dynamics.

Different locations of closed-loop system poles, i.e., eigenvalues of $(A_d + B_d K)$, result in different system behavior. In pole-placement, poles are placed in desired locations (eigenvalues are set) often to fulfil various high-level goals, such as control performance maximization and system constraints satisfaction. The desired poles p can be decided with empirical or optimization techniques. This method is feasible since there is freedom to choose the feedback gain K. Once pole locations are decided, the following characteristics equation of z can be constructed with these poles as roots:

$$z^n + \gamma_1 z^{n-1} + \gamma_2 z^{n-2} + \cdots + \gamma_n = 0.$$

Then the following is defined:

$$\gamma_c(A_d) = A_d^n + \gamma_1 A_d^{n-1} + \gamma_2 A_d^{n-2} + \cdots + \gamma_n I.$$

According to Ackermann's formula, the feedback gain used to stabilize the closed-loop system is calculated as

$$K = [0 \cdots 0 \ 1] CO^{-1} \gamma_c(A_d),$$

where

$$CO = \left[B_d \ A_d B_d \cdots A_d^{n-1} B_d \right]$$

is the square controllability matrix. The static feedforward gain F used to make the system output $y[k]$ track the reference r is computed by

$$F = \frac{1}{C_d (I - A_d - B_d K)^{-1} B_d}.$$

All eigenvalues of $(A_d + B_d K)$ must have absolute values of less than unity in order to ensure system stability. This is illustrated with a double integrator example as follows:

$$A = \begin{bmatrix} 0 & 1 \\ 0 & 0 \end{bmatrix}, B = [0 \ 1]^T, C = [1 \ 0].$$

Fig. 7.5 Different system output responses for stable and unstable poles

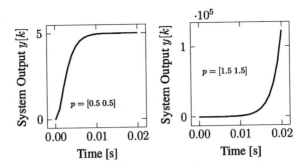

The initial state is $\begin{bmatrix} 0 & 0 \end{bmatrix}$ and the reference value is 5. The sampling period is set as $h = 0.001$s. The system output responses for two sets of poles $p = \begin{bmatrix} 0.5 & 0.5 \end{bmatrix}$ and $p = \begin{bmatrix} 1.5 & 1.5 \end{bmatrix}$ are shown in Fig. 7.5.

If the system is controllable, i.e., CO has full rank, there is no restriction on pole locations. Controllability of a discrete system is defined as the ability to transfer the system from any initial state $x[0] = x_0$ to any desired final state $x[k_f] = x_f$. The controllability condition is equivalent to the non-singularity of the controllability matrix CO. If CO does not have full rank, some of the poles cannot be modified with a choice of K and thus are uncontrollable. Note that CO is not invertible in this case. If the uncontrollable poles are stable (with absolute values of less than unity), then the system is stabilizable. Restricted pole-placement can be used for stabilizable systems in the way that only controllable poles are placed in the desired locations and uncontrollable poles remain untouched. Therefore, in the CPS design, the system is required to be at least stabilizable, if not controllable.

7.2.3   Cyber and Physical Interactions

We will use an example on memory-aware CPS design to illustrate the cross-layer interplay between the cyber and physical components. First of all, the link between the WCET of the control programs and the control timing parameters needs to be established. As discussed in the introduction, the overall control loop performs three operations: measure, compute, and actuate. The general timing model of a control loop is illustrated in Fig. 7.6. The compute operation executes the control program, which takes E time units. As mentioned before, the sampling period is denoted by h. The time interval between the measure and the corresponding actuate operations in the same sampling period is the sensor-to-actuator delay τ^{sa}, which is equal to the WCET of the control program E^{wc}.

Two example sampling orders are used to show the derivation of control timing parameters from the WCET results. As illustrated in Fig. 7.7, S1 is the conventional memory-oblivious scheme and summarized as follows:

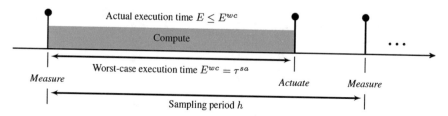

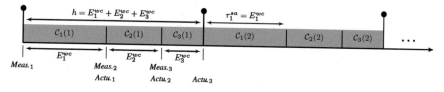

Fig. 7.6 The general timing model of a control loop

Fig. 7.7 The conventional memory-oblivious sampling order S1

$$S1 : C_1(1) \longrightarrow C_2(1) \longrightarrow C_3(1) \longrightarrow C_1(2) \longrightarrow C_2(2) \longrightarrow$$
$$C_3(2) \longrightarrow C_1(3) \longrightarrow C_2(3) \longrightarrow C_3(3) \longrightarrow \cdots .$$

There is no cache reuse in S1 between consecutive executions, considering that different control applications typically have different instructions to execute. In other words, when $C_i(j)$ starts execution, all instructions of C_i need to be brought into the cache from the flash memory. Therefore,

$$E_i^{wc}(1) = E_i^{wc}(2) = \cdots = E_i^{wc},$$

where $E_i^{wc}(j)$ is the WCET of the jth execution for C_i. The WCET of the application C_i is denoted by E_i^{wc}, since all executions of the same application have equal WCET. This can be computed with standard WCET analysis techniques, as discussed before. Clearly, all control applications run with a uniform sampling period of

$$h = \sum_{i=1,2,3} E_i^{wc}.$$

Moreover, for the sensor-to-actuator delay,

$$\tau_i^{sa} = E_i^{wc}.$$

As illustrated in Fig. 7.8, S2 is an example memory-aware sampling order and summarized as

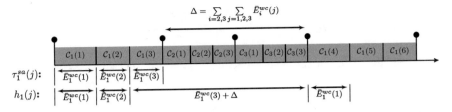

Fig. 7.8 An example memory-aware sampling order

$$S2 : C_1(1) \longrightarrow C_1(2) \longrightarrow C_1(3) \longrightarrow C_2(1) \longrightarrow C_2(2) \longrightarrow C_2(3) \longrightarrow$$
$$C_3(1) \longrightarrow C_3(2) \longrightarrow C_3(3) \longrightarrow \cdots .$$

The effective WCET taking into account the cache reuse between consecutive executions is denoted as $\overline{E}_i^{wc}(j)$. From the above discussion,

$$\forall i \in \{1, 2, 3\},$$

$$\overline{E}_i^{wc}(1) = E_i^{wc},$$

since there is no cache reuse from the previous program for the first execution of every application $C_i(1)$. $\overline{E}_i^{wc}(2)$ and $\overline{E}_i^{wc}(3)$ are shorter than $\overline{E}_i^{wc}(1)$ due to cache reuse. The amounts of cache reuse are the same for $C_i(2)$ and $C_i(3)$ in the worst case. Denoting the guaranteed WCET reduction as $\overline{E}_i^g$,

$$\forall i \in \{1, 2, 3\},$$

$$\overline{E}_i^{wc}(2) = \overline{E}_i^{wc}(3) = \overline{E}_i^{wc}(1) - \overline{E}_i^g .$$

From these varying WCETs, the sampling periods of all three applications can be calculated. Taking C_1 as an example, there are three sampling periods $h_1(1)$, $h_1(2)$, and $h_1(3)$, which repeat themselves periodically:

$$h_1(1) = \overline{E}_i^{wc}(1), h_1(2) = \overline{E}_i^{wc}(2), h_1(3) = \overline{E}_i^{wc}(3) + \Delta,$$

where Δ is computed as

$$\Delta = \sum_{i=2,3} \sum_{j=1,2,3} \overline{E}_i^{wc}(j).$$

Similar derivation can be done for C_2 and C_3. The average sampling period of an application h_{avg} is

$$h_{avg} = \frac{\sum_{i=1,2,3} \sum_{j=1,2,3} \overline{E}_i^{wc}(j)}{3} < h.$$

Moreover, the corresponding sensor-to-actuator delay $\tau_i^{sa}(j)$ also varies with cache reuse as

$$\forall i \in \{1, 2, 3\},$$

$$\tau_i^{sa}(1) = h_i(1) = \overline{E}_i^{wc}(1), \; \tau_i^{sa}(2) = h_i(2) = \overline{E}_i^{wc}(2), \; \tau_i^{sa}(3) = \overline{E}_i^{wc}(3).$$

As all control timing parameters have been derived, it can be seen that the sampling period $h_i(j)$ of a control application is non-uniform for the memory-aware scheme. The average sampling period of S2 is shorter than the uniform sampling period of S1, due to the WCET reduction resulting from cache reuse. The sensor-to-actuator delay $\tau_i^{sa}(j)$ varies. The next task is to develop a controller design method to exploit the shortened non-uniform sampling periods and achieve better control performance.

For an application C_i with l system states under the conventional memory-oblivious sampling scheme S1, the constant sampling period h is larger than the constant sensor-to-actuator delay τ_i^{sa}. Therefore, the discrete-time system is

$$x[k+1] = A_d x[k] + B_1\left(\tau_i^{sa}\right) u[k-1] + B_0\left(\tau_i^{sa}\right) u[k],$$

where

$$B_0\left(\tau_i^{sa}\right) = \int_0^{h-\tau_i^{sa}} e^{At} dt \times B, \; B_1\left(\tau_i^{sa}\right) = \int_{h-\tau_i^{sa}}^{h} e^{At} dt \times B.$$

It is assumed that $u[-1] = 0$ for $k = 0$. Clearly, the system dynamics depends on both $u[k]$ and $u[k-1]$. Thus, a new system state is defined as

$$z[k] = \left[x[k] \; u[k-1]\right]^T,$$

and the transformed system becomes

$$z[k+1] = A_{S1} z[k] + B_{S1} u[k],$$

$$y[k] = C_{S1} z[k],$$

where

$$A_{S1} = \begin{bmatrix} A_d & B_1 \left(\tau_i^{sa} \right) \\ 0 & 0 \end{bmatrix}, B_{S1} = \begin{bmatrix} B_0 \left(\tau_i^{sa} \right) & I \end{bmatrix}^T, C_{S1} = \begin{bmatrix} C & 0 \end{bmatrix}.$$

A_{S1} is a square matrix.

Next, the following input signal is applied:

$$u\,[k] = K_{S1} z\,[k] + F_{S1} r.$$

The closed-loop system is then

$$z\,[k+1] = (A_{S1} + B_{S1} K_{S1})\,z\,[k] + B_{S1} F_{S1} r.$$

In order to find the poles resulting in the best control performance with the pole-placement technique, a constrained optimization problem is formulated. Decision variables are the controllable closed-loop system poles, i.e., the controllable eigenvalues of $(A_{S1} + B_{S1}K_{S1})$. The optimization objective is the control performance. One constraint is that the closed-loop system is stable, i.e., the decision variables have absolute values of less than unity. Another constraint is the input saturation. Constraints on the overshoot and steady-state accuracy are also considered. A heuristic can be developed to solve this challenging non-convex optimization problem. After the poles are placed, the feedback gain K_{S1} is then calculated and then the feedforward gain F_{S1} is computed. As long as (A_{S1}, B_{S1}) is stabilizable, i.e., uncontrollable poles have absolute values of less than unity, the above design is feasible.

For the memory-aware scheme, the sampling is non-uniform. The number of consecutive executions for any application C_i, where $i \in \{1, 2, \ldots, n\}$, in one period is denoted by m_i. Then, the periodically repeating sampling order is denoted by $(m_1, m_2, \ldots, m_n)$. For the ease of understanding, a simple sampling order $(2, 2, 2)$ of three control applications is considered. Generalization to any periodic sampling order is straightforward.

As shown in Fig. 7.9, there are two sampling periods $h_i(1)$ and $h_i(2)$, which are repeated periodically. The two switching systems are

$$x\,[k+1] = A_1 x\,[k] + B_1 u\,[k],$$

$$x\,[k] = A_2 x\,[k-1] + B_2^1 u\,[k-1] + B_2^2 u\,[k],$$

where B_2^1 and B_2^2 depend on the second sampling period $h_i(2)$ and the sensor-to-actuator delay of the second execution $\tau_i^{sa}(2)$. The system output is

$$y\,[k] = C x\,[k].$$

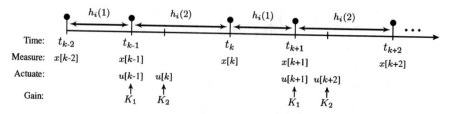

Fig. 7.9 Periodically switched sampling periods for C_i in the schedule $(2, 2, 2)$

It should be noted that $x[k]$ is influenced by both $u[k-1]$ and $u[k]$, since $\tau_i^{sa}(2)$ is smaller than $h_i(2)$, i.e., $u[k]$ is applied before the sensing of $x[k]$.

Introducing a new state $z[k] = \begin{bmatrix} x[k] \ u[k] \end{bmatrix}^T$, the system becomes

$$z[k+1] = A_1^{hol} z[k] + B_1^{hol} u[k+1],$$

$$z[k] = A_2^{hol} z[k-1] + B_2^{hol} u[k],$$

where

$$A_1^{hol} = \begin{bmatrix} A_1 & B_1 \\ 0 & 0 \end{bmatrix}, B_1^{hol} = \begin{bmatrix} 0 & I \end{bmatrix}^T, A_2^{hol} = \begin{bmatrix} A_2 & B_2^1 \\ 0 & 0 \end{bmatrix}, B_2^{hol} = \begin{bmatrix} B_2^2 & I \end{bmatrix}^T.$$

Both A_1^{hol} and A_2^{hol} are square matrices. The system output is

$$y[k] = \begin{bmatrix} C & 0 \end{bmatrix} z[k].$$

There are two control inputs that need to be designed within one period:

$$u[k+1] = K_1 z[k] + F_1 r,$$

$$u[k] = K_2 x[k-1] + F_2 r.$$

The closed-loop system dynamics is then

$$z[k+1] = \left(A_1^{hol} + B_1^{hol} K_1 \right) z[k] + B_1^{hol} F_1 r,$$

$$z[k] = \left(A_2^{hol} + B_2^{hol} K_2 \right) z[k-1] + B_2^{hol} F_2 r.$$

The number of poles to place, i.e., the number of eigenvalues in $\left(A_1^{hol} + B_1^{hol} K_1\right)$ and $\left(A_2^{hol} + B_2^{hol} K_2\right)$, is $2l + 2$. This is a constrained non-convex optimization problem. The objective to maximize is the control performance of C_i. Decision variables are the poles and thus the number of dimensions in the decision space is $2l + 2$. Heuristics can be used to solve this pole-placement optimization problem. Once poles are placed, K_1 and K_2 can be computed. Then F_1 and F_2 can be calculated. With this method, both feedback gains are designed together taking all the information into account. The maximum control performance can be obtained if the optimization technique returns the optimal poles. However, when C_i is consecutively executed m_i times in a sampling order, the number of dimensions in the decision space becomes $m_i(l + 1)$, which compromises the scalability.

The above memory-aware CPS design is able to achieve better control performance with the same given memory resources or equivalently satisfies the control performance requirement with fewer memory resources. This can be generalized to other types of resources as well. Note that in this design process, the system safety is always guaranteed, in the sense that the control performance requirements are always met. Other aspects such as security and robustness can be also addressed by this new cross-layer CPS design methodology.

For security properties, it is generally difficult to incorporate cryptographic algorithms and message authentication code into the CPS, since they consume substantial computational power and communication bandwidth. New approaches can be developed that design the controllers together with cryptographic algorithms with a thorough timing analysis of the complete system. A trade-off analysis between the degree of security, control performance, and platform schedulability can be performed.

For robustness properties, new techniques can be developed that the control algorithms are able to tolerate the faults on the implementation side, such as from sensors, actuators, processors, memory, bus, etc. There are mainly two directions. First, the statistically expected performance of the CPS can be improved, taking all sorts of possible faults into account. Second, certain performance of the CPS, e.g., the stability of a plant under control, is guaranteed for a limited set of faults.

7.3 Case Studies

7.3.1 Assuring the Safety of Machine Learning-Based Perception for Highly Automated Driving

This case study describes how to assure the safety of machine learning approaches using their application to the perception functions of highly automated driving as an example. An assurance case strategy is described based on an understanding of the causes of functional insufficiencies in machine learning. A number of techniques for demonstrating the performance of the functions are discussed, and it is shown how these can contribute to a structured assurance case. Finally, we reflect on the current state of the art in this area and the overall potential for providing a convincing argument for using machine learning technique in safety-critical applications.

7.3.1.1 Introduction

The transition from *hands-on* (Levels 1–2 of [8]) driver assistance to *hands-off* highly automated driving (HAD) (Levels 3–5) requires a number of changes to system safety concepts. For example, a higher level of component availability is required as the system cannot be simply deactivated upon detection of a component hardware fault. The conditions for being acceptably safe with respect to functional safety for passenger vehicles are set by ISO 26262 [9]. Adherence to this standard remains a necessary prerequisite in order to ensure a reliable and fault-tolerant implementation of the system with respect to random hardware and systematic failures. For highly automated driving, demonstrating the sufficiency of the system to meet its overall safety goals becomes more challenging due to the inherent complexity and unpredictability of the operational design domain. This continually evolving environment is in itself observed via channels that are imperfect due to the technical limitations of the sensors. Thus, the understanding and decision-making components of the system are presented with noisy, incomplete, and partly inconsistent data about the current situation. Based on this partial understanding of the environment, the system must make the decisions required to implement a driving strategy capable of safely navigating the vehicle to its ultimate destination. The dominating challenge facing the safety assurance of highly automated driving systems is the derivation and validation of adequate system safety goals and the demonstration of their fulfilment under all feasible situations. This needs to be achieved despite the complexity and uncertainty inherent in the domain, sensing and understanding/decision algorithms. The issue of the insufficiency of the system to meet the safety goals, due to inherent performance limitations in sensors or actuators or the inadequacy of the intended function itself, is not directly addressed by ISO 26262. The "Safety of the Intended Functionality" (SOTIF) approach described in the draft standard ISO 21448 [10] aims to address these issues. However, the standard was developed with driver assistance (Levels 1 and 2 of SAE model) systems in mind. It is therefore unclear whether or not the approaches defined by the standard scale to the level of complexity of HAD systems (Levels 3 and upwards).

The performance limitations in the perception task are typically counteracted by using multiple sensing channels and finely tuned heuristics. Recent advances in machine learning algorithms and the availability of increased computing power have led to the promise of such algorithms being able to solve the perception tasks required by highly automated driving functions operating in unrestricted environments. Algorithms such as deep neural networks [11] can make sense of unstructured data using efficient computations in real time. By providing enough labelled images as training data, the algorithms learn to identify and classify objects such as vehicles and pedestrians with accuracy rates that can surpass human abilities. Neither of the above-mentioned standards address the application of machine learning techniques to automated driving tasks. As a result, assurance methods must be developed and the ability of the system to meet its safety goals must be systematically argued based on "first principles" where adherence to a standard is only one part of the overall argument. An assurance case [12] provides a convincing and valid argument that a set of claims regarding the safety of a system is justified for a given function based on a set of assumptions over its operational context.

The rest of this case study is structured as follows. First, safety requirements allocated to machine learning functions are described from a systems engineering perspective and an example function, camera-based pedestrian recognition, is introduced. Next, the potential causes of functional insufficiencies in machine learning functions are discussed, for which mitigation measures will form a key component of the safety assurance argument. A number of sources of evidence of the performance of the machine learning functions are then described. Finally, we reflect on the current state of the art in this area and the overall potential for providing a convincing argument for using machine learning technique in safety-critical applications.

7.3.1.2 Safety Requirements on the Machine Learning Function

The challenges involved in providing a convincing system-level assurance case will depend on the functional scope of the machine learning application as well as whether it is trained and validated during development, or whether it continues to learn in the field. It is expected that, in practice, the initial applications of machine learning in series development of highly automated driving will be based on pre-trained functions, implementing well-specified detection tasks which can be supported by plausibility checks based on alternative channels within the system context. One such example application, which shall be referenced in the rest of this case study, is the application of Convolutional Neural Networks (CNNs) [11] to detect objects such as pedestrians based on camera images as part of a collision avoidance system for self-driving vehicles. CNNs are a class of feedforward neural networks (NN) that consist of a large number of connected neurons – computational units that calculate a weighted sum of their inputs and apply a nonlinear activation function on this sum. The weights are determined by minimizing a loss function of the network over a given set of training data (labelled images) and back-propagating the respective error terms through the network. In this manner, CNNs allow a classification annotated with a confidence level for each class and a localization of an object within a given image (e.g., frames of a video).

Performance requirements must be defined and allocated to the machine learning function in order to ensure that, at a system level, the safety goals are met. The derivation of performance (Safety) requirements within the system context is one of the key contributions to ensuring overall system safety and requires deep domain and system knowledge. Deriving a suitable set of requirements for open context systems is a non-trivial task in itself, systematic approaches to systems engineering are therefore indispensable. Incorrect functioning of the pedestrian detection function can cause hazards such as "unnecessary emergency breaking or steering" and "too late or no emergency braking when necessary." These hazards potentially violate the safety goal "do not harm pedestrians" of the automated driving system. Thus, we consider the pedestrian detection function as safety relevant.

Figure 7.10 summarizes the system-level context of a CNN-based object detection function. The function takes camera images as an input and operates in parallel to traditional computer vision algorithms as well as alternative sensing channels such as Radar. In this case study, pedestrian detection is divided into two subtasks: (1) classification and (2) localization of the pedestrian within the image. The specification of each task is derived from the driving context (e.g., ego speed, distance to object) and system boundaries (e.g., braking distance). For example, for the first subtask, the specification is derived from the need to detect persons of a minimum height from a particular distance travelling with a maximum relative velocity which results in a minimum amount of pixels inhabited by the object within a single image frame from the camera. The following requirements need to be defined in detail for each pedestrian class:

- Pedestrian of minimum height (A1 pixels) and of minimum width (A2 pixels) are classified.
- Pedestrians are detected if B % of the person is concealed.
- There are less than C1 false positives per 1000 frames.
- There are less than C2 false negatives per 1000 frames.
- There are less than D1 misclassified detections.
- Vertical deviation less than E1 pixels to ground truth.
- Horizontal deviation less than E2 pixels to ground truth.

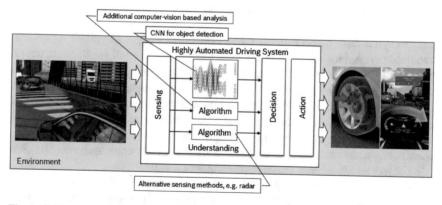

Fig. 7.10 System context of CNN-based object detection

Such requirements provide a clear measure of performance for the machine learning environment, but also imply a number of assumptions on the system context. These assumptions might include that the braking distance and speed are sufficient to react when detecting persons, for example, 100 m ahead on the planned trajectory of the vehicle and that other system measures (e.g., alternative sensing channels) are available to decrease the overall false-negative and false-positive rate to a sufficiently safe level, etc. The following list contains typical assumptions that are relevant for the assurance case:

- Assumptions on the operational profile of the system's environment. For example, the types and occurrence distribution of objects in the environment.
- Assumptions on attributes of inputs to the machine learning function. For example, the camera resolution is sufficient to be able to detect persons from a distance of 100 m with the required accuracy.
- Assumptions on the performance potential of machine learning. For example, the chosen CNN approach has the intrinsic potential, given the right parameterization and set of learning data to fulfil the allocated performance requirements.

Arguing safety when applying machine learning functions therefore not only requires a demonstration that the performance requirements are met, especially when evaluated on a specific data set that might be not representative for all driving situations. Moreover, it is necessary to argue that requirements are appropriate given the role of the function in the overall automated driving system and environmental context and that all relevant assumptions have been explicitly defined and validated.

7.3.1.3   Causes of Functional Insufficiencies in Machine Learning

The inherent uncertainty associated with machine learning techniques coupled with the open context environment lead to different causes of performance issues compared to traditional, algorithmic, and control-theoretic approaches to vehicle control. In order to argue the claim that functional insufficiencies within the machine learning function (here: camera-based object detection, supervised training) are minimized, it is important to understand the causes of such insufficiencies. As interest in machine learning safety has grown, a number of authors [13, 14, 15] have investigated different causes of performance limitations in machine learning functions. Some examples applicable to the use case described here are described below:

- *Scalable oversight and distributional shift.* One of the key differences in machine learning techniques compared to algorithmic approaches is the lack of a detailed specification of the target function. Instead, the functional specification can be seen to be encoded within the set of training data. Therefore, if the training data does not reflect the target operating context, then there is a strong likelihood that the learned function will exhibit insufficiencies. Critical or ambiguous situations, within which the system must react in a predictably safe manner, may occur rarely or may be so dangerous that they are not well represented in the training data. Consider, for example, the situation where a small child enters the road ahead between two parked vehicles. This leads to the effect that critical situations remain undertrained in the final function (scalable oversight). In addition, the system should continue to perform accurately even if the operational environment differs from the training environment (distributional shift) [14]. This effectively can be formulated as the robustness of the system to react in a shift of distribution between its training and operational environment. Distributional shift will be inevitable in most open context systems, as the environment constantly changes and can adapt to the behavior of actors within the system. For example, car drivers will adjust their behavior within an environment in which autonomous vehicles are present, vehicle and pedestrian appearances change over time, etc.

- *Robustness of the trained function.* Machine learning techniques are typically chosen for their ability to approximate target functions based on a finite set of training data. This has advantages over procedural techniques where the function to be implemented may be too complex to specify or implement algorithmically due to an open context environment or due to the unstructured nature of the input data. In other words, when presented with new data, the function will predict a correct answer based on already observed input/output pairs. An often cited problem associated with neural networks, is the possibility of adversarial perturbations [16, 17, 18]. An adversarial perturbation is an input sample that is similar (at least to the human eye) to other samples but that leads to a completely different categorization with a high confidence value. It has been shown that such examples can be automatically generated and used to "trick" the network. Although it is still unclear to what extent adversarial perturbations could occur naturally or whether they would be exploited for malicious purposes, from a safety validation perspective, they are useful for demonstrating that features can be learnt by the network and assigned an incorrect relevance. Therefore, methods are required to minimize the probability of such behavior especially in critical driving situations. One of the factors that is often attributed to this class of problems is that the set of possible functions is exponentially larger than those that can be represented through machine learning techniques. Therefore, the likelihood that a machine learning technique would select an appropriate approximation appears at first glance very unlikely. The authors in [19] argue, however, that deep learning is nevertheless effective because the function to be approximated is rooted within the physical universe and physics favors certain classes of exceptionally simple probability distributions that deep learning is uniquely suited to model. The challenge, therefore, is how to ensure that the machine learning algorithms focus on those physical properties of the inputs relevant to the target function without becoming distracted by irrelevant features. In other words, act within the same hierarchical dimensions as the target function [19].

- *Differences between the training and execution platforms.* As discussed above, machine learning functions can be sensitive to subtle changes in the input data. When using machine learning to represent a function that is embedded as part of a wider system as described here (see Fig. 7.10), the input to the neural network will have typically been processed by a number of elements already [15], such as image filters and buffering mechanisms. These elements may vary between the training and operation environments leading to the trained function becoming dependent on hidden features of the training environment not relevant in the target system. In addition, typical reliability issues in the target hardware (e.g., random hardware failures) may not manifest themselves directly in obviously erroneous outputs due to the data driven approach, where deviations of individual parameters or calculations may have subtle but relevant effects on the overall decision made by the neural network.

7.3.1.4 Sources of Evidence and Structuring the Assurance Case

The development of methods for demonstrating the performance of machine learning functions to the level of integrity required by safety-critical systems is currently an emerging field of research. It is expected that, analogous to traditional algorithmic-based software approaches, a diverse set of complementary evidence based on constructive measures, formal analysis, and test methods will be required to make a robust assurance case. In this section, we discuss different categories of potential evidence that can be used to support such an assurance case.

The choice of training data has a direct impact on accuracy of a machine learning function. Criteria are therefore required in order to determine whether or not the training data has the potential to lead to a sufficient level of performance, including:

- *Training data volume:* A sufficient amount of training data is used to provide a statistically relevant distribution of scenarios and to ensure a stabilization of a strong coverage of weightings in the neural network.
- *Coverage of known, critical scenarios:* Domain experience based on well-understood physical properties of the system and environment as well as previous validation exercises ensures the identification of classes of scenarios that should exhibit similar behavior in the function.
- *Minimization of unknown, critical scenarios*: Some critical attributes of the input space may not be known during system design [20]. A combination of systematic identification of equivalence classes in the training data and statistical coverage during training and validation will therefore be essential to minimize the residual risk of insufficiencies due to inadequate training data.

Key components of demonstrating the correctness of traditional safety-critical software are introspective techniques that include manual code review, static analysis, code coverage and formal verification. These techniques allow for an argument to be formulated on the detailed algorithmic design and implementation but cannot be easily transferred to the machine learning paradigms. Other arguments must therefore be found that make use of knowledge of the internal behavior of the neural networks.

- *Saliency maps:* Based on the back propagation of results in the neural network, saliency maps [21] highlight those portions of an image that have greatest influence on classification results and can be used to provide a manual plausibility check of results as well as to determine potential causes of failed tests.
- *Explanations:* Another line of research tries to generate natural language explanations referring in human understandable terms to the discriminating contents of an input image to explain which features were relevant for the classification [22].

Due to the inherent restrictions of the applicability of white-box approaches to the verification of the trained function, a strong emphasis will remain on testing as a means to estimate the achieved performance of the trained function. Standard approaches to testing machine learning functions involve reserving a proportion of the data collected for training purposes to performing validation tests. These tests naturally suffer from the same inadequacies as described above for the training data. Several additional test approaches are therefore being developed.

- *Synthetic data generation and search-based testing:* Based on advances in computer graphics realism as well as the possibility to generate data with specific properties, the use of synthetically generated data may also play a role [23] in the assurance case. Synthetic data can be used to generate huge numbers of test cases, in particular to cover critical or rare situations, otherwise not adequately represented in naturally occurring data. The use of synthetic data also allows test cases to be automatically generated together with the corresponding ground truth. This allows for search-based optimization approaches to be applied to automatically generate (physically feasible) images which produce incorrect classifications. However, the use of synthetic data also implies the introduction of the additional assumption in the assurance case that the synthetic data would lead to test results that are indeed representative of the operational environment.
- *White-box coverage tests:* At present, there is no clear consensus on which stopping criteria to apply when testing machine learning functions. Due to the fact that deep neural networks operate in a highly dimensional feature space, choosing test cases based on a set of domain-specific equivalence classes is less likely to be effective, as there is a high chance that these do not match the feature dimensions learnt by the neural network. White-box criteria have been proposed based on the concept of neuron coverage to determine the completeness and effectiveness of the test data. This involves calculating the ratio of activated neurons (activation values above a given threshold) to the total number of neurons for a given set of input data [24, 25]. These approaches have also been combined with search-based testing techniques to create variations of test data that achieve coverage. These techniques are only applicable in combination with functional criteria, and it is as yet unclear how effective such white-box techniques are at discovering performance issues in the neural networks.

The objective of applying techniques such as those described above should not only be to demonstrate that a given performance requirement has been met but also to understand under which sets of conditions the function does not meet its expectations. This information can be used to design redundancy concepts and run-time measures. For example:

- *Run-time plausibility checks:* Plausibility checks on the outputs of the neural network could involve tracking results over time (e.g., objects detected in one frame should appear in contiguous frames, until out of view) or by comparing against alternative sensor inputs (e.g., radar or LIDAR reflections). Such plausibility checks may mitigate against inaccuracies that occur spontaneously for individual frames.
- *Run-time monitoring of assumptions:* If certain assumptions regarding the operational distribution are determined to be critical, then they could also be monitored during run-time. Discrepancies between the distribution of objects detected at run-time and the assumptions could indicate either errors in the trained function or that the system is operating within a context for which it was not adequately trained. If such a situation is detected, appropriate actions for mitigating the effect of the discrepancy can be initiated.

7.3.1.5 Summary

The strategy for arguing the safety of an automated driving system that makes use of machine learning in its perception functions is summarized in Fig. 7.11. A systematic analysis of the operating domain is required to identify classes of relevant scenarios and environmental characteristics that could impact the performance of the function. This includes explicitly stating assumptions made when selecting the training data (e.g., typical size and appearance of pedestrians). Based on a system-level understanding of the functional and performance requirements, a set of specific requirements must be derived and allocated to the machine learning function based on an understanding of the inherent performance potential given

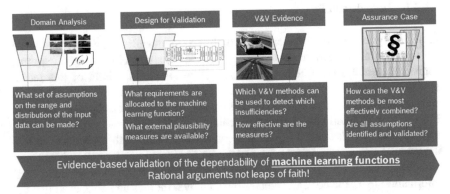

Fig. 7.11 Summary of assurance approach for machine learning in automated driving

the particular learning algorithm and its system context (e.g., camera resolution and therefore distance at which pedestrians of a given size can be detected). The allocation of performance requirements should also take into account the ability to perform plausibility checks based on domain knowledge (e.g., maximum speed of pedestrians) and alternative sensing channels (such as radar). The verification of the function itself should consider attributes of the training data as well as diverse analysis and test methods. Due to inherent robustness issues resulting from the high dimensional input space and unpredictable manner in which features are learnt, black-box, statistical testing evidence is not seen as a convincing argument on its own. The more diverse the set of evidence that can be collected, the greater the chance that all causes of potential deficiencies in the function are covered. Ideally, verification and validation activities will provide insight into residual performance limitations of the trained function to allow for system-level measures to be argued that compensate these allowing for an overall safe system design.

Some of the approaches described in this case study have yet to be proven in a series production context, and therefore their ability to predict the performance of machine learning for safety-critical highly automated driving applications is as yet unclear. Furthermore, it is expected that the safety arguments will need to be application specific. For this reason it is crucial that a robust safety argument based on a diverse set of complementary evidence is made. Such safety argumentation approaches also need to be supported by further research to validate the effectiveness of techniques at reducing and detecting various classes of insufficiencies. An industry consensus must also be developed in order to identify strategies and methodologies that would form the basis of future internationally recognized development standards.

7.3.2 Assuring the Security and Robustness of Connected Vehicle Applications

7.3.2.1 Introduction

In recent years, automotive makers, high-tech companies, startups, and even governments are developing autonomous vehicles aggressively. At the same time, connected vehicles (or Internet of Vehicles) are another spotlight of this recent automotive technology revolution. In most cases, connectivity and autonomy can and should work together and realize a good application. For example, as shown in Fig. 7.12, at an intelligent intersection controlled by its intersection manager, the approaching vehicles and the manager can communicate with each other so that the vehicles can go through the intersection in a safe and much more efficient way without traffic lights. Here, "connectivity" is necessary to provide environmental information (e.g., the coming vehicles blocked by buildings and obstacles) that the sensors on a single vehicle cannot sense, while "autonomy" is necessary to provide

Fig. 7.12 An intelligent intersection with connected autonomous vehicles can achieve safe and efficient traffic control without traffic lights

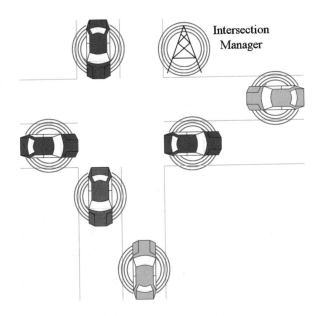

Intersection Manager

precise control (e.g., entering the intersection after 5 seconds) that human drivers cannot achieve.

However, there are some fundamental concerns of connected autonomous vehicles as follows:

- Robustness of connectivity. The connectivity is based on wireless communication which may suffer message corruption and loss due to the open and uncontrolled communication environment.
- Robustness of autonomy. The autonomy is based on many machine learning mechanisms whose models do not provide guarantees, especially when the corresponding training data is missing. This has also been discussed in the first case study.
- Security of connectivity. The wireless communication is also vulnerable to security attacks such as spoofing and jamming.
- Security of autonomy. Those machine learning mechanisms are also vulnerable to malicious security attacks such as evasion attacks (e.g., misleading classifiers) and poisoning attacks (e.g., providing misleading training data). Without human drivers, those attacks can lead to catastrophic incidents.
- System integration. There are many subsystems in a connected autonomous vehicle. However, the resource, such as computational capability and communication bandwidth, on a connected autonomous vehicle is usually limited, and there are tight and hard real-time deadlines. Any solution or subsystem for robustness or security must be compatible with existing systems without violating system requirements.

It should be emphasized that a non-malicious fault or error (the first and second concerns) and a security attack (the third and fourth concerns) are different. The source of a non-malicious fault or error can usually be measured by probability. In contrast, a security attack is malicious, and it is possible to attack the weakest part of a system.

In this case study, we will introduce the security challenges in connected vehicle applications and discuss three important topics: key management system, intrusion detection system, and system integration.

7.3.2.2 Security Challenges in Connected Vehicle Applications

As news reported that some cases that a vehicle can be successfully hacked and then patched, security is a rising concern for automotive systems. However, there are still many open questions, especially for connected autonomous vehicles (the security of a single vehicle has been well studied [26, 27, 28]). This is because, inevitably, autonomous vehicles will rely on decisions of vehicles themselves, and connected vehicles will make decisions based on external information, so totally separating internal and external networks cannot match the need. Furthermore, there is still a gap for connected autonomous vehicles to be realized, and many applications are still under development, which needs us to address security in advance. Last but not least, traditional automotive design does not have security in mind, making system integration more difficult and thus unresolved at this point.

There are many different kinds of security attacks. In this case study, we abstract them to *outsider* attacks and *insider* attacks. An outsider attack is from an entity which has not been authenticated, while an insider attack is from an entity which has been authenticated but compromised. Some examples of insider attacks include a tempered sensor, a discovered hardware or software implementation flaw, a leaked security key, or a legitimate but malicious user. For outsider protection, a key management system such as a Public Key Infrastructure (PKI) is the target system in this chapter. For insider protection, an intrusion detection system is the target system, as they focus on the information legitimacy which cannot be verified by a key management system.

7.3.2.3 Key Management System

A key management system is fundamental to protection against outsider attacks. The proof-of-concept of Secure Credential Management System (SCMS) [29] has been proposed by the United States Department of Transportation (USDOT) in recent years to establish trust between connected vehicles and vehicular infrastructures and then support security and privacy for vehicular networks. Based on traditional PKIs, the goal of SCMS is to provide scalability to support millions of vehicles and trade-offs between security, privacy, and efficiency. A simplified SCMS architecture design is shown in Fig. 7.13. There are several Certificate Authorities

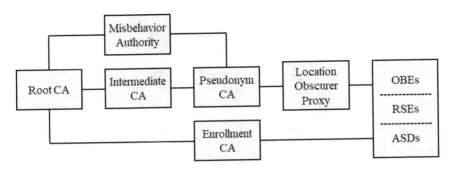

Fig. 7.13 A simplified SCMS architecture design [29]

(CAs) establishing a chain of trust and issuing certificates with different lifetime validity to form the basis of secure communication between connected vehicles and vehicular infrastructure. Onboard Equipment (OBE), Road Side Entity (RSE), and Aftermarket Safety Device (ASD) must be registered in the system and must obtain certificates. Then message exchange can be protected by public and private keys. As a proof-of-concept system, there are still some detailed design issues which should be addressed. Please refer to one previous publication for those issues [30].

It should be mentioned that a key management system should be integrated with the underlying communication protocols. There are two main approaches to realize the connectivity. One is based on Dedicated Short-Range Communications (DSRC), and the other one is based on cellular networks. The DSRC has a longer history, and many field tests have been done by automotive makers. The DSRC provides security services at the middle layers (network layer, transport layer, and message sublayer) [31]. Message authentication is supported by using the Elliptic Curve Digital Signature Algorithm (ECDSA), which is an asymmetric cryptographic algorithm. When a vehicle intends to send a message, it signs the message with its private key and sends the message with its signature and certificate digest. A vehicle receiving the message then uses the public key corresponding to the private key to verify the message. The generation time of a message and the location of a vehicle are optionally included in a signed message to protect against replay attacks. Besides DSRC, many automotive makers and high-tech companies have also formed the 5G Automotive Association (5GAA). It is necessary to further investigate its security services.

7.3.2.4 Intrusion Detection System

Besides a key management system which can detect abnormal security key usage (probably by outsider attackers), an intrusion detection system is needed to monitor systems, networks, and/or information between vehicles and infrastructures. We expect an architecture where cloud servers and edge servers can also install intrusion detection systems, as shown in Fig. 7.14.

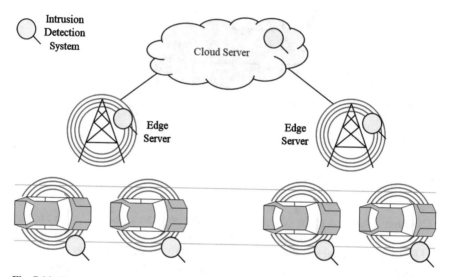

Fig. 7.14 The cloud server and the edge servers can also install intrusion detection systems

Some initial solutions to intrusion detection for Cooperative Adaptive Cruise Control (CACC) have been proposed [32]. CACC is more advanced than Cruise Control (CC) and Adaptive Cruise Control (ACC). In a platoon scenario, each vehicle collects the information of positions of vehicles (or gaps between vehicles), velocities of vehicles, and accelerations of vehicles. Usually, radars and LIDARs are used for sensing velocities and gaps, and accelerations are provided by connectivity. CACC controls the vehicle's behavior based on the collected information and achieves better vehicle-following than CC and ACC due to the additional acceleration information. In the previous work [32], it is assumed that fake information (no matter what the source is) of position, velocity, and acceleration affects the platoon, and three types of intrusion detection systems based on the rules of physics, principal component analysis, and hidden Markov model, respectively, are proposed to detect the fake information. It is observed that there is trade-off between detection capability and computational efficiency, and thus different intrusion detection systems have different appropriate locations, such as vehicles themselves, edge servers (roadside units), and cloud servers.

There are some limitations for a single intrusion detection system. As mentioned above, the limited computational resource may restrict the detection capability, and different intrusion detection systems have different strengths against different types of intruders (the analysis result can be a probabilistic estimation of an intruder). One potential solution is a consensus algorithm to combine the analysis results from different intrusion detection systems and achieve a stronger "cooperative" intrusion detection system. It should be mentioned that a consensus algorithm has a more general usage, and it has been well studied in the domain of distributed systems [33, 34, 35]. For connected autonomous vehicles, the challenges come from the unstable connection topology (as vehicles are moving) and strict timing constraints. However, connected autonomous vehicles have the feature that it is easier to identify neighbors, which can support secure consensus algorithms.

7.3.2.5 System Integration

With a key management system, we can sign and verify messages and much increase the difficulty for outsider to perform attacks. However, the resource on a vehicle is usually limited, and there are tight and hard real-time deadlines, implying that signing and verifying all messages are probably not applicable. One previous piece of work explored the design space of receiving group assignment within a single vehicle [36]. The receiving group assignment affects not only the number of Message Authentication Codes (MACs) transmitted on the network but also the level of security threats between nodes in the same receiving group. This concept can be generalized to partial signing and verification without violating security and other system constraints. This is doable as most messages in automotive systems are periodic, so a few missing, corrupted, or fake messages may not lead to dangerous states. In some cases, this can be taken care of by a proper controller design, following the discussion in the beginning of this chapter.

In a more formal way, the problem of system integration can be formulated by the Contact-Based Design (CBD) methodology [37, 38] and the Platform-Based Design (PBD) paradigm [39]. The behavior of different subsystems and components, including security mechanisms, can be defined by property specification languages, e.g., Linear Temporal Logic (LTL) and its extensions. Then, formal verification with assume-guarantee contracts can be applied to guarantee the correctness of system design. Given that formal verification may have limited efficiency and scalability, one alternative is the run-time verification which serves a similar role to an intrusion detection system. On the other hand, the previous work applied the PBD paradigm to security-aware system design [36]. As shown in Fig. 7.15, the application space and the architecture space are abstracted (otherwise, they are too complicated and too specific to be synthesized or analyzed) to models which keep relevant features but remove irrelevant details. Similarly, security mechanisms also need to be abstracted and fed into the mapping process to see if they can be integrated into the application and the architecture. Based on the mapping process, we can select appropriate security mechanisms, satisfy different kinds of design constraints, and guarantee the correctness of system design.

7.3.2.6 Summary

Security is a rising concern for automotive systems. To address this concern, security protection is needed at different layers including external networks (between vehicles and infrastructures), gateways, in-vehicular networks, and components. In this case study, we focused on the emerging connected autonomous vehicles and

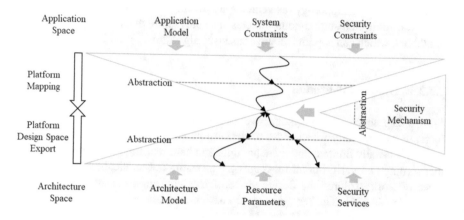

Fig. 7.15 Security-aware system design by the PBD paradigm [36]

abstract attacks to outsider attacks and insider attacks. To protect against them, we introduced the key management system and the intrusion detection systems and pointed out the need of system integration due to the limited resource and strict system constraints. We admit that they are just part of the big picture of automotive security, but we believe that they can provide important insights to secure connected autonomous vehicles and even other IoT systems.

7.4 Concluding Remarks

CPS are fast developing, demanding a new design methodology that unifies all the layers, aiming for safety, security, robustness, and resource efficiency. This chapter gives the technical background on the cyber components, physical components, and their interactions. A perspective from CAVs is taken with two case studies on assuring the safety of machine-learning-based perception and assuring the security and robustness, respectively.

Research and development along the direction of CPS require multidisciplinary expertise. There are still many challenges that need to be addressed. For instance, in the perception function of autonomous systems, machine learning algorithms are dominating in the performance. However, they have obvious drawbacks. First, they are vulnerable to adversarial attacks. Second, they are hard to analyze and provide guarantee.

References

1. W. Chang, S. Chakraborty, Resource-aware automotive control systems design: a cyber-physical systems approach. Found. Trends® Electron. Des. Autom. 10(4), 249–369 (2016)
2. W. Chang, D. Roy, L. Zhang, S. Chakraborty, Model-based design of resource-efficient automotive control software, ICCAD, 2016
3. W. Chang, D. Goswami, S. Chakraborty, L. Ju, C. Xue, S. Andalam, Memory-aware embedded control systems design. IEEE Trans. Comput. Aided Des. Integr. Circuits and Syst. 36(4), 586–599 (2017)
4. W. Chang, D. Roy, X. Hu, S. Chakraborty, Cache-aware task scheduling for maximizing control performance, DATE, 2018
5. W. Chang, D. Goswami, S. Chakraborty, A. Hamann, OS-aware automotive controller design using non-uniform sampling. ACM Trans. Cyber-Phys. Sys. 2(4), 26 (2018)
6. W. Chang, A. Proebstl, D. Goswami, M. Zamani, S. Chakraborty, Battery- and aging-aware embedded control systems for electric vehicles, RTSS, 2014
7. W. Chang, A. Proebstl, D. Goswami, M. Zamani, S. Chakraborty, Reliable CPS design for mitigating semiconductor and battery aging in electric vehicles, CPSNA, 2015
8. SAE, J3016, Taxonomy and definitions for terms related to on-road motor vehicle automated driving systems, 2013
9. ISO, ISO 26262, *Road Vehicles – Functional Safety* (ISO, Geneva, Switzerland, 2011)
10. ISO, ISO/PRF PAS 21448, *Road Vehicles – Safety of the Intended Functionality* (ISO, Geneva, Switzerland, 2018)
11. I. Goodfellow, Y. Bengio, A. Courville, *Deep Learning*, vol 1 (MIT Press, Cambridge, 2016)
12. IEEE, *IEEE Standard Adoption of ISO/IEC 15026-1 – Systems and Software Engineering – Systems and Software Assurance* (IEEE, New York, 2014)
13. K. Varshney, Engineering safety in machine learning, ITA, 2016
14. D. Amodei, C. Olah, J. Steinhardt, P. Christiano, J. Schulman, D. Mane, Concrete problems in AI safety, arXiv preprint, 2016
15. D. Sculley, G. Holt, D. Golovin, E. Davydov, T. Phillips, D. Ebner, V. Chaudhary, M. Young, J. Crespo, D. Dennison, Hidden technical debt in machine learning systems, *Advances in Neural Information Processing Systems*, 2503–2511, 2015
16. A. Nguyen, J. Yosinski, J. Clune, Deep neural networks are easily fooled: High confidence predictions for unrecognizable images, CVPR, 2015
17. A. Kurakin, I. Goodfellow, S. Bengio, Adversarial examples in the physical world, arXiv preprint, 2016
18. J. Metzen, T. Genewein, B. Bischoff, *On Detecting Adversarial Perturbations* (ICLR, 2017). [Online]. Available: https://openreview.net/forum?id=SJzCSf9xg
19. H. Lin, M. Tegmark, D. Rolnick, Why does deep and cheap learning work so well? J. Stat. Phys. 168(6), 1223–1247 (2017)
20. J. Attenberg, P. Ipeirotis, F. Provost, Beat the machine: challenging humans to find a predictive model's unknown unknowns. J. Data Inf. Qual. 6, 1–17 (2015)
21. K. Simonyan, A. Vedaldi, A. Zisserman, Deep inside convolutional networks: visualising image classification models and saliency maps, arXiv preprint, 2013
22. L. Hendriks, Z. Akata, M. Rohrbach, J. Donahue, B. Schiele, T. Darrel, Generating visual explanations, ECCV, 2016
23. S. Richter, V. Vineet, S. Roth, V. Koltun, Playing for data: Ground truth from computer games, ECCV, 2016
24. K. Pei, Y. Cao, J. Yang, S. Jana, Deepxplore: Automated whitebox testing of deep learning systems, SOSP, 2017
25. Y. Sun, X. Huang, D. Kroening, Testing deep neural networks, arXiv preprint, 2018
26. S. Checkoway, D. McCoy, B. Kantor, D. Anderson, H. Shacham, S. Savage, K. Koscher, A. Czeskis, F. Roesner, T. Kohno, Comprehensive experimental analyses of automotive attack surfaces, USENIX, 2011

27. P. Kleberger, T. Olovsson, E. Jonsson, Security aspects of the in-vehicle network in the connected car, IV, 2011
28. K. Koscher, A. Czeskis, F. Roesner, S. Patel, T. Kohno, S. Checkoway, D. McCoy, B. Kantor, D. Anderson, H. Shacham, S. Savage, Experimental security analysis of a modern automobile, SP, 2010 .
29. United States Department of Transportation, Security credential management system (SCMS)
30. H. Liang, M. Jagielski, B. Zheng, C.-W. Lin, E. Kang, S. Shiraishi, C. Nita-Rotaru, Q. Zhu, Network and system level security in connected vehicle applications, ICCAD, 2018
31. J. Kenney, Dedicated short-range communications (DSRC) standards in the United States. Proc. IEEE **99**(7), 1162–1182 (2011)
32. M. Jagielski, N. Jones, C.-W. Lin, C. Nita-Rotaru, S. Shiraishi, Threat detection for collaborative adaptive cruise control in connected cars, WiSec, 2018
33. R. Carli, F. Fagnani, P. Frasca, S. Zampieri, Gossip consensus algorithms via quantized communication. Automatica **46**(1), 70–80 (2010)
34. L. Lamport, Paxos made simple. ACM SIGACT News (Distributed Computing Column) **32**(4), 51–58 (2001)
35. R. Olfati-Saber, R. Murray, Consensus problems in networks of agents with switching topology and time-delays. IEEE Trans. Autom. Control **49**(9), 1520–1533 (2004)
36. C.-W. Lin, B. Zheng, Q. Zhu, A. Sangiovanni-Vincentelli, Security-aware design methodology and optimization for automotive systems. ACM Trans. Des. Autom. Electron. Sys. (TODAES) **21**(1)., 18:), 1–26 (2015)
37. A. Benveniste, B. Caillaud, D. Nickovic, R. Passerone, J. Raclet, P. Reinkemeier, A. Sangiovanni-Vincentelli, W. Damm, T. Henzinger, K. Larsen, Contracts for system design, research report RR-8147, INRIA, 2012
38. B. Meyer, Applying "design by contract". IEEE Comput. **25**(10), 40–51 (1992)
39. L. Carloni, F. Bernardinis, C. Pinello, A. Sangiovanni-Vincentelli, M. Sgroi, Platform-based design for embedded systems, *Embedded Systems Handbook*, pp. 1281–1304, 2005

Chapter 8
Distributed Ledger Technology

Xing Liu, Bahar Farahani, and Farshad Firouzi

The chains of habit are too weak to be felt until they are too strong to be broken.

Samuel Johnson

Contents

X. Liu (✉)
Kwantlen Polytechnic University, Surrey, BC, Canada
e-mail: xing.liu@kpu.ca

B. Farahani
Shahid Beheshti University, Tehran, Iran

F. Firouzi
Department of ECE, Duke University, Durham, NC, USA

© Springer Nature Switzerland AG 2020
F. Firouzi et al. (eds.), *Intelligent Internet of Things*,
https://doi.org/10.1007/978-3-030-30367-9_8

393

8.1   Introduction to Distributed Ledger Technology and IoT

8.1.1   What Is a Distributed Ledger?

Distributed ledger technology is a general term that is used to describe technologies for the storage, distribution, and exchange of data between users over private or public distributed computer networks. Essentially, a distributed ledger is a database that is spread and stored over multiple computers located at physically different locations. Each of such computers is frequently referred to as a node. A distributed ledger can also be considered as a common datasheet stored on multiple distributed nodes.

Figure 8.1 shows a centralized, a decentralized, and a distributed system. Distributed ledger technology is based on distributed systems.

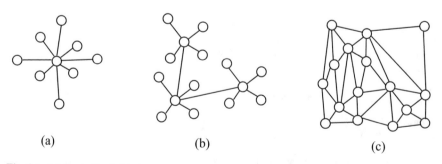

Fig. 8.1 (**a**) Centralized, (**b**) decentralized, and (**c**) distributed system

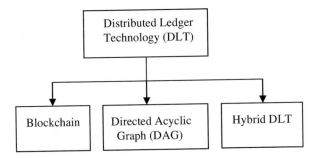

Fig. 8.2 Three different implementations of distributed ledger technology

Distributed ledger technologies can also be classified into three categories based on the way the technology is implemented. These include blockchain, DAG, and hybrid DLT, as shown in Fig. 8.2.

8.1.2 Blockchain

Blockchain is the underlying technology of Bitcoin. It has become widely known since 2008 because of the publication of a paper titled *Bitcoin: A Peer to Peer Electronic Cash System* [1] which was authored by perhaps a group of people pseudonymously named Satoshi Nakamoto.

Although it was not until 2008 that blockchain became well known to the world, research about digital documents can be dated back to 1991 when Haber and Stornetta published their research paper titled *How to Time-Stamp a Digital Document* [2]. The paper discussed how to use a computer server to timestamp and link digital documents as a chain with pointers attached to the data in each document. Any change in the data would render the pointers invalid. This guaranteed that data stored could not be tampered after the server had signed the documents. The paper also coined key concepts and terms such as timestamping, hashing, signature, data linkage, and distributed trust. These concepts are the cornerstone concepts and terms of modern blockchains.

The idea of digital currency can be traced even further back than blockchain. In 1982, David Chaum published his paper titled *Blind Signatures* [3]. This paper led to the creation of perhaps the first digital currency in the world called *eCash* which was released in 1993. In 1996, Douglas Jackson created a gold-backed digital currency called *e-gold* [4]. In 1998, a digital currency called *b-money* was proposed by a computer scientist named Wei Dai [5]. After that, Adam Back came up with his "hashcash" in 2002 [6] which made the first implementation of Proof-of-Work. About 5 years before 2008 when the Bitcoin paper was published, a decentralized currency system powered by Proof-of-Work (PoW) was created by the team led by Emir Sin Gün who published their paper in 2003 [7].

Since the publication of Satoshi Nakamoto's Bitcoin paper, an overwhelming amount of attention has been paid to Bitcoin and other similar digital currencies, together with the technology used in the currency transfer systems. The enthusiasm was due to the claimed advantages provided by Bitcoin: there is no need for a middle man in the transaction process so trustless parties can do business with each other; there is no central point of failure so system reliability is greatly enhanced; there is no double-spending so fraud can be avoided; transaction history is traceable so transactions can be verified; financial benefits are provided to participants and they are rewarded for their contribution to the operation of the Bitcoin system, for example, by mining blocks. In recent years, numerous Bitcoin-mining data centers have been built around the world to make profits. Due to its enhanced security, privacy, speed, and reduced cost, some organizations have started accepting Bitcoin as a means of payment, such as tuition fees. Up to date, people have been primarily focusing on the financial aspects of Bitcoin.

However, since 2014, greater attention has been paid to the technology which Bitcoin is based on: blockchain. People have realized that blockchain can be separated from Bitcoin and the technology can be beneficial to other areas where data security and privacy are of prime importance. Quick surveys of the literature have revealed that blockchain applications can be found in virtually every aspect of our lives nowadays. For this reason, the business value of blockchain has been predicted to be multibillion dollars by 2030.

8.1.3 Types of Blockchain

Blockchains are of different types. First of all, they can be *permissionless* or *permissioned*.

8.1.3.1 Permissionless Blockchains

Permissionless blockchains are *public*, open source, and based on the Proof-of-Work consensus algorithm. Anybody can participate in a permissionless blockchain without obtaining approval beforehand. The person can simply download the required software program and start running it on his or her own computer. The person can send transactions to the blockchain and these transactions will be included in the blockchain only if they are valid. Transactions are transparent so that everybody on the blockchain can view them, although the transactions are anonymous. In addition, anybody can participate in validating transactions before they are added to the blockchain. Well-known examples of public blockchains are Bitcoin and Ethereum.

8.1.3.2 Permissioned Blockchains

Blockchains can be permissioned as well. Permissioned blockchains can be further divided into two types: *federated* and *private*.

- *Federated* – Federated blockchains are usually operated by a special group of participants. They do not allow the arbitrary participant to validate transactions. Only preselected participants are involved in the process and are responsible for the validity of transactions. In federated blockchains, the public can be granted the right to read the transactions, but only the selected participants can write the transactions. Examples of federated blockchains are R3 which is for banking and EWF which is designed to be used in the energy sector.
- *Private*: A private blockchain is typically centralized to one organization and only this organization can validate transactions. The public can be allowed to read the transactions, or only some selected parties can read the transactions. Multichain is an example of private blockchains.

The difference between a federated blockchain and a private blockchain is the number of organizations that operate the blockchain. A private blockchain is operated by one organization. However, a federated blockchain is operated by multiple organizations, although federated blockchains can still be considered to be private blockchains.

Private and public blockchains differ in the execution of consensus algorithms, maintenance of the common ledger, and the authorization to join to the blockchain network. Table 8.1 shows the differences between the various types of blockchains.

Blockchains use different *consensus algorithms* which make blockchains different from traditional distributed database technologies. Consensus algorithms are essentially about decision-making in a group and how the decisions can be made for the benefit of the majority of group members. Well-known consensus algorithms are Proof-of-Work (PoW) and Proof-of-Stake (PoS).

8.1.4 Directed Acyclic Graph (DAG)

Similar to a blockchain, a DAG can also store data transactions. In a DAG, a transaction is represented by a node and is linked to one or several other transactions.

Table 8.1 Comparison of different types of blockchains

	Public	Private	Federated
Permission type	Permissionless	Permissioned	Permissioned
Reading	Anybody	Restricted	Restricted
Writing	Anybody	Restricted	Restricted
Validation	Anybody	Limited to one	Limited to several

However, the links are *directed* because they point from earlier transactions to newer transactions, in a way called *topological ordering*. A DAG does not allow loops. That means, a node is not allowed to traverse back to itself by following the directed links. In this sense, a DAG is acyclic. Essentially, from the standing point of computer science, a DAG is a graph with transactions being the nodes of the graph and the edges which have directions.

Unlike blockchains, DAGs do not have blocks. There is no mining in DAGs. Transactions provide validation for each other but a transaction cannot validate itself. A new transaction is required to validate one or more previous transactions when it joins the DAG. Every new transaction refers to its parent transactions, signs their hashes, and includes the hashes in the new transaction.

8.1.5 Hybrid DLTs Based on Blockchains and DAGs

Blockchains and DAGs can be combined to create hybrid DLTs. An example of hybrid DLTs is Bexam [8]. Bexam is a platform that leverages blockchain and DAG technologies with greatly improved speed and scalability.

Bexam uses a flexible chain structure together with a node hierarchy so that it has the security of a blockchain and the speed of a DAG with approximately 0.2 seconds per block and about 40 million transactions per second. It is highly scalable because it combines the concepts of DAG. Bexam uses a new consensus algorithm called Proof-of-Rounds (PoR) and a KYC (Know Your Customer) verification process to identify and prevent malicious actions. The electric power and computing resource requirements of Bexam are also very low.

In addition, a token technology is used in Bexam for its transactions. It is convenient to integrate Bexam into existing enterprise infrastructures.

8.1.6 Internet of Things (IoT)

Internet of Things is part of the Industry 4.0 revolution. IoT is a research and industrial focus in recent years.

Essentially, IoT is all about having *smart things* which are equipped with sensors and actuators collaborating over the Internet. IoT systems currently in use have a centralized architecture where data is stored in the cloud. Cloud is a central place with databases and services.

The IoT ecosystem is very complex. The complexity is due to the vast number of devices connected, the types of wired and wireless communication networks involved, and the varieties of software programs used. This complexity makes the IoT ecosystem vulnerable and susceptible to attacks.

The current IoT ecosystem was developed largely based on available Internet technologies in the past and did not have a systematically designed secure structure in the first place. Therefore, people are concerned with the security of current IoT systems.

There is a strong belief that blockchain is the solution for IoT security due to its intrinsic advantages such as distributed data storage and immutability. Blockchain may be able to improve the overall security of the IoT ecosystem.

8.2   Benefits of DLTs

8.2.1   Blockchain Benefits

A blockchain can be considered as a system that stores an identical copy of a spreadsheet called ledger on multiple distributed computers. The system frequently has no central authority. Transactions submitted by participants are validated and recorded in the ledger which is accessible to the community of participants. The transactions are cryptographically signed and are then assembled into blocks. The blocks are linked one after another and are added to the blockchain by consensus mechanisms. Transactions are immune to changes after they are published. Blocks are replicated across all computers of the distributed system so that they all hold the same ledger.

The way a blockchain is created and maintained leads to numerous benefits in comparison to traditional databases. Prominent features of blockchains such as decentralization and consensus are the intrinsic attributes of blockchain that give rise to benefits which are essentially out of the box:

1. First, blockchains *allow trustless participants to interact with each other*. No trusted third parties are required to serve as intermediaries and validate transactions. It is the consensus algorithms that validate the transactions. A blockchain can maintain itself by handling conflicts automatically and creating forks if necessary, so that the ledger is always in good standing.
2. Blockchains also make data storage more *reliable*. This is again due to the elimination of third-party agents which reduces the risk of unauthorized access and unwanted modification of stored data. The ledger is not stored in a single location or managed by any single company. The cryptographic linking between blocks ensures data unchangeability. Blockchains are robust. Transactions are processed by multiple participating nodes; therefore, no single node is critical to the entire database. No central point can be exploited so the system is much better against hacking and fraud. This equips blockchains with high fault tolerance capabilities. Blockchains are immune to malicious modifications. It is impossible to change it back once data has been written into the blockchain. No change of history is allowed either. Third parties cannot make any changes to the system as well.
3. Transactions stored in blockchains are *permanent*. Therefore, they are *verifiable* and *auditable*. Such verifications can be applied not only to data but also to

interactions and message exchanges among participants. Every transaction is tagged with a signature and a timestamp so that data ownership is auditable and traceable. Transactions can be traced back to its origin. Transaction history can be used to verify data authenticity and prevent fraud. This eliminates backdoor transactions and possible disputes and prevents data tampering.

4. *Transparency* is another benefit of blockchain technology. In a public blockchain, transactions, once they are made, are accurate and consistent among participants. All changes to a public blockchain are accessible to all participants. Users have full visibility of transaction information in the system. Anyone can verify the correctness of the system. Using a single public ledger avoids the complications of multiple ledgers. Transparency increases trust among participants too. This is particularly important in scenarios such as fair disbursement of funds or benefits. Everyone can maintain a copy of the ledger and verify its correctness. .This provides resiliency and trust among participants.

5. Decentralization in blockchains leads to high *availability*. A blockchain is based on a large number of nodes working in a peer-to-peer manner. Data is replicated and updated on all nodes. Being inaccessible to a single node will not cause the system to stop functioning. Therefore, a system based on blockchains is highly available.

6. Blockchains have enhanced *security* and *integrity* over traditional database systems. Transactions will not be recorded before they are agreed upon by participants. Approved transactions are encrypted and linked as a chain. Together with the distributed copies, a blockchain is very difficult for hackers to break.

7. Blockchains can lead to *reduced transaction costs*. Transactions can be completed in a peer-to-peer or business-to-business manner. No third-party intermediaries are required. This avoids the cost induced by using a third party such as a bank. Cost that can be saved includes overhead, governance, auditing, and other fees.

8. Blockchains make transactions *faster*. In a blockchain, transaction time can be even reduced to just a few minutes. However, current interbank transfers and final settlement could take days. Transaction time is extremely important for industries such as transportation and energy. Time reduction could potentially save billions of dollars. The situation is similar in the financial industry. Blockchains can save time because they eliminate verification, reconciliation, and clearance which are usually lengthy processes. The reason is that a single version of data already agreed upon by participating financial institutions is available on the shared ledger of the blockchain.

9. The other benefit of blockchains is *smart contracts*. A blockchain such as Ethereum not only stores data, but also provides a programming logic called smart contracts. Smart contracts can execute business logic. They are programs that execute agreements and manage the transfers of digital assets between participants under specified conditions in a blockchain. They can be considered a digital version of traditional contracts written in a programming language. Because smart contracts are deployed and executed on blockchains, they are

therefore secured as well. The execution is also transparent, immutable, and decentralized. Essentially, smart contracts make program execution secured.

8.2.2   DAG Benefits

From computer science point of view, a DAG is a graph with directed edges and no cycles. It is a treelike data structure that is suitable for storing, organizing, and finding transactions.

As a distributed ledger technology, DAG has advantages over other technologies. For example, DAG does not have blocks and miners. Validation is done by the transactions themselves. New transactions validate old transactions in a distributed manner when they are added to the DAG. This greatly increases the speed of DAG – hundreds of thousands of transactions can be processed in a DAG in a second.

Distributed validation between transactions leads to much-improved scalability as well. The newer transactions are added to DAG, the more transactions that are available for validation, the faster the validation is done. In theory, DAG has infinite scalability. Because DAG does not have blocks and miners, there is no mining fee associated with DAG. This makes DAG an appropriate technology for the Internet of Things which has a large number of transactions between sensors and devices, and it is not logical and realistic to charge fees.

DAG is also easily made quantum-proof. That is, DAG is safe to use even when quantum computers become available in the future, because DAG does not rely on cryptography which could be potentially broken by quantum computers. Algorithms have been implemented in DAG to make DAG quantum-resistant. An example of DAG is IOTA technology which is already believed to be resistant against quantum computing attacks. However, DAG should have a substantial amount of traffic before it can start working. Greatly reduced traffic will make DAG vulnerable to attacks. Solutions based on coordinators have been suggested to get a DAG system up and running. The effectiveness of the suggestion of using coordinators is still under debate.

It should be noted that there are a number of differences between DAG and blockchains. First, DAG is blockless. In DAG, transactions validate each other and transactions are not assembled into blocks. On the other hand, blockchains assemble transactions into blocks. Secondly, DAG is more scalable. In fact, DAG is infinitely scalable in theory. This means that the performance of DAG will not deteriorate as new transactions are added to the graph. On the contrary, blockchains will experience slowdown when the blockchain gets longer. DAG does not require mining as well. Therefore, DAG uses much less electric power. However, blockchains based on PoW use a lot of electric power.

Another difference is that DAG does not charge fees whereas blockchains do. Furthermore, DAG is much faster because it does not require mining and validation is done in parallel and not in a chained manner. Finally, DAG is quantum-proof. Blockchains are susceptible to quantum attacks because they are based on cryptography and consensus algorithms which are breakable by quantum computers. Table 8.2 lists the differences between DAG and blockchains.

8.3 How Blockchain Works

The functionality of a blockchain in Bitcoin is to facilitate money transactions and the recording of the transactions. A slightly simplified microscopic operation of a blockchain can be illustrated by looking at the process of how a transaction is started and settled, as shown in the flowchart in Fig. 8.3.

In general, a successful transaction has to go through a sequence of steps, as shown in the following:

1. Step 1: Form a transaction. In this step, sender information, receiver information, sender's public key, amount of fund to transfer, receiver's public key, and timing information are required.
2. Step 2: Form a block. In this step, the previous block hash, the current block containing the transaction in Step 1, and other transactions are included.
3. Step 3: The block is broadcasted to the entire network.
4. Step 4: Nodes on the network validate the block.
5. Step 5: The block is added to the blockchain.
6. Step 6: Fund transfer is completed.

Table 8.2 Comparing DAG and blockchain

	DAG	Blockchain
Using block	No	Yes
Scalability	Good	Poor
Mining	No	Yes (for PoW)
Fees	No	Yes
Speed	Fast	Slow
Quantum-proof	Yes	No

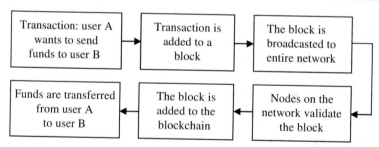

Fig. 8.3 Operation of a blockchain

8.3.1 Transaction, Block, Ledger, and Blockchain

There has been some confusion about the concept of the blockchain and the ledger that stores the data for a blockchain. To be precise, a *blockchain* refers to the system of nodes that make a blockchain operational, whereas the term *ledger* is the database stored in the nodes. However, the term *ledger* is used for accurate descriptions. A ledger is a data structure that is replicated and shared among distributed nodes of the blockchain network. A ledger can be considered as *a chain of blocks*. Each block in the chain carries a list of transactions and other data. A transaction has a transaction ID, an input which contains the type of the asset to be transferred and the amount and signed with the sender's public key, and an output which includes the type of asset to be received and the amount and signed with the receiver's public key. Figure 8.4 shows an example of a transaction which is simplified for the convenience of description.

After being validated, transactions are assembled into blocks. A block consists of three parts: the block header, the hash of the block header, and the transactions inside the block. The block header is made in a special way. It contains the hash of the header of the previous block, a timestamp when this block is created, and a Merkle root hash which is derived from the hashes of the transactions of this block. A Merkle root hash is the hash of all the hashes of all the transactions that are part of a block. The block also contains two other important parameters, namely, nonce (which stands for *number used only once*) and difficulty target. These two parameters are what make mining (in a Power-of-Work blockchain) tick. The details of the mining process will be discussed in the next section. Figure 8.5 shows the details of a block (the shaded area is the block header).

	Input	Output
Transaction ID	Type, Amount, Sender Key	Type, Amount, Receiver Key

Fig. 8.4 A transaction

Fig. 8.5 A block

Version
Hash of Header of Previous Block
Timestamp
Difficulty Target
Nonce
Merkle Root Hash
Hash of Header of This Block
Transactions

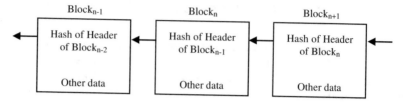

Fig. 8.6 A ledger

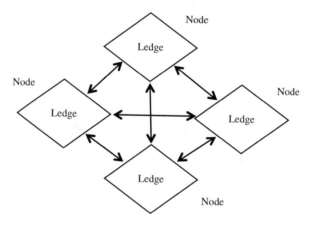

Fig. 8.7 A blockchain as a distributed system of nodes

The very first block of a blockchain is called *genesis block* which is common to all nodes in the blockchain and has no parent. The hash of each block is obtained cryptographically and is the block's identity. Each block contains the hash of the previous block; this way a chain of blocks is established. This chain of blocks is the ledger which is frequently referred to as blockchain. Figure 8.6 depicts a ledger.

It is the ledger shown in Fig. 8.6 that is stored in multiple networked distributed computers. These computers are called *nodes* and they form the blockchain. So, it is the ledger that is stored in the nodes of a blockchain. The nodes communicate via wired or wireless communication networks in a peer-to-peer manner. A blockchain is shown in Fig. 8.7.

8.3.2 Transaction Validation and Block Mining

A user who wants to interact with a blockchain must do it through a node, although several users can share the same node. Through this node, a user can sign and initiate transactions. Every transaction is signed with the user's private key and can be accessed through the user's public key, which essentially serves as the "address" of the transaction. Transactions are broadcasted by a user's node to its immediate neighboring nodes.

The neighboring nodes validate each transaction and propagate it further along possible pathways. All nodes of the blockchain will have this valid transaction after some time. The neighboring nodes will block and discard transactions that are not invalid.

After a given time period, a node will have received a number of valid transactions. The node will then have the transactions organized in order, have them validated and packed into a timestamped candidate block, find a nonce value to create a hash which satisfies the difficulty level set by the blockchain, and have the candidate block broadcasted to all other nodes in the blockchain for verification.

The nodes in the blockchain all participate in verifying the validity of the candidate block. They make sure that the format of the block is correct. They make sure that each transaction in the block is valid and is signed by the suitable parties. They make sure that all hashes in the new block were computed correctly. They also make sure that the candidate block references to the hash of an appropriate previous block in the ledger. If the result of the verification process turns out to be positive, every node will add the block to its own copy of the ledger. If the candidate block is not valid, then it will be discarded. This process will repeat indefinitely as long as the computer network is not down for any reason.

A critically important question is how a node should decide if a transaction is valid. First of all, a node needs to ensure that the signatures (hashes) of the sender and the receiver are valid. That is, the sender and receiver are both legitimate registered participants of the blockchain and they do have valid "accounts" in the blockchain. The amount to be sent should also be valid in terms of the type of assets and minimum allowed value based on the kind of applications. A node also must validate if the sender has sufficient unspent funds. Figure 8.8 shows how a transaction is validated.

However, the above validation process assumes that every node can be trusted, which is usually not the case for a public blockchain. A public blockchain usually consists of a group of non-trusting participants. Therefore, a set of rules are required for the nodes to agree on the validity of the transactions. Because the transactions are assembled into blocks, blocks need to be validated after the transaction validation is carried out. In blockchains, consensus algorithms are employed to validate blocks. The opinion of the majority of the nodes on the blockchain will decide the validity of the blocks.

The problem is that a bad user can create multiple participant identities via one specific node and can therefore potentially control the entire blockchain. In order to avoid such a problem, what Bitcoin does is making the finding of a new valid block very *computationally expensive* so that a bad node is not able to beat other nodes collectively on the blockchain because of a single node's limited computing power. This is the *consensus mechanism* called Power-of-Work. Based on this mechanism, malicious blocks from a bad node are unlikely to be accepted because it is up to the majority of the nodes on the blockchain to approve the validity of a candidate block.

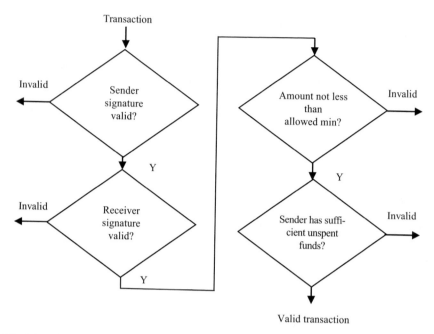

Fig. 8.8 Validate a transaction

In a Power-of-Work blockchain, any node can make efforts by conducting mining to find and recommend a new block as the next valid block for the blockchain.

During mining, a node strives to find a suitable random number called nonce ("number only used once") which is embedded in the block's header (see Fig. 8.9). A valid nonce is a value that makes the hash (e.g., SHA-256 for Bitcoin mining) of a block header have the required number of leading zeros, as set by the difficulty parameter of the blockchain. The number of leading zeroes is called the *difficulty* which is set by the blockchain and can be adjusted over time. In other words, difficulty is a measure of how difficult it is to find a suitable hash based on the given difficulty target. Note that in blockchain, usually the network automatically adjusts the difficulty level for mining over time. The validity of a new block can be easily verified by other nodes because they only need to validate the hash using the nonce value already found by the node that is recommending the new block. This takes only a very short amount of time because it involves only the calculation of one hashing algorithm. Other nodes will adapt and add the validated recommended block to its own copy of the ledger. In general, nodes will validate and adopt broadcasted recommended blocks rather than trying to mine its own. It is better to validate and adopt an existing recommended block and start mining the next recommended block because this at least gives the node the chance of winning a reward for successfully mining the next valid block. The rationale of this strategy is because of the way a blockchain resolves conflicts: only the block which gives the longest chain will be adopted.

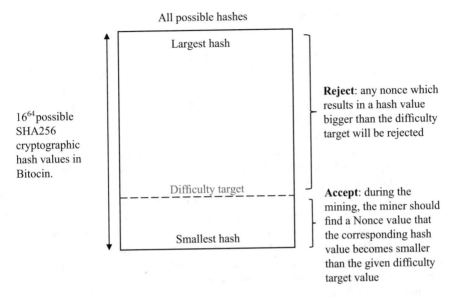

Fig. 8.9 How mining works. Miners search for a valid hash that satisfy the given difficulty target. In Bitcoin nonce range contains 4 billion possible values

Fig. 8.10 Forking in blockchain

Fig. 8.11 Longer chain is adopted

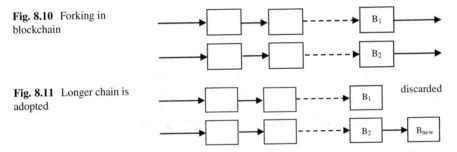

Therefore, in general, the nodes will validate and adopt the first recommended block broadcasted over the blockchain, and then they will all start mining the next recommended block. If there are two nodes that publish a valid candidate block at the same time, conflict occurs because both candidate blocks will be added to the ledgers on different nodes. It is very likely that these blocks contain different transactions; therefore, the last blocks in the ledgers are not the same. If this happens, a fork is created. The strategy to resolve this conflict is to wait for a new block to be added. Then all nodes will adopt the ledger that has the longest chain because it carries the greatest amount of work in a Proof-of-Work-based blockchain. This way consensus is reached if a block should be in the ledger. The conflict (fork) scenario is shown in Fig. 8.10 where blocks B_1 and B_2 are both valid and have been added to the ledger, but they may contain different transactions inside. Figure 8.11 shows how a conflict is resolved – the longer chain is adopted as a valid ledger.

8.3.3 Smart Contracts

The second generation of blockchains such as Ethereum [9] uses smart contracts. Smart contracts are executable computer code stored on a blockchain. Similar to transactions, smart contract code is transparent and can be examined by all participants of the blockchain. Smart contracts are accessed via their addresses and users can activate them by sending them transactions. In an Ethereum blockchain, smart contracts run on every node of the blockchain which has a virtual machine running. The smart contracts execute under the virtual machine.

To a large extent, smart contracts are very similar to stored procedures in database management systems. Functions in smart contracts are defined based on business rules. Participants can initiate transactions to call functions in smart contracts along with the required data. Smart contracts are deterministic and the same input to a smart contract always generates the same output. Smart contracts can call each other as well. The code of a smart contract can be examined by participants so that they can predict the outcome before they commit the contract. The outcomes of executing smart contracts can be verified by participants too. Smart contracts help prevent possible contract disputes.

8.3.4 Consensus Algorithms

When designing or adopting blockchains, several factors need to be considered. The first factor to consider is who can access the blockchain. This depends on whether the blockchain will be openly accessible by the public or not. If it is, public blockchains must be used. The decision also affects the selection of consensus mechanisms. A private blockchain needs to be adopted if only specific participants can access the blockchain. Private blockchains can employ special consensus algorithms so that time-consuming minings are not required.

As stated previously, a blockchain consists of multiple identical ledgers stored in distributed computers. A blockchain may not be controlled by any central authority. It is therefore obvious that malicious users will be very much attempted to take advantage of the blockchain. This situation of the blockchain is very similar to the famous Byzantine Generals' Problem. From this sense, blockchains require *Byzantine Fault Tolerance* (BFT). Without BFT, bad users will be able to break the blockchains by sending malicious transactions. If damages do occur, they will not be repaired because no central authority is available to carry out corrective actions. For this reason, consensus algorithms are used to provide BFT.

So far, large numbers of consensus algorithms have been developed for blockchains. With consensus algorithms, mutually distrustful participants can work together. Each of the consensus algorithms has its own strengths and weaknesses and is suited for specific applications. Although the number of proposed consensus algorithms is large, only several of them are widely known. Selected consensus algorithms are discussed below.

8.3.4.1 Proof-of-Work (PoW)

Proof-of-Work, or PoW [10], is the most well-known consensus algorithm which was used by Bitcoin. The purpose of this algorithm is to validate transactions and add validated new blocks to the blockchain. Due to its public and distributed nature, a blockchain needs a mechanism to prevent malicious transactions and attacks. This is the responsibility of participating nodes called miners through a process called *mining*. Essentially, PoW presents a complex mathematical puzzle for the nodes to solve. Very strong computing power is needed to solve this puzzle in a timely manner. However, proving the correctness of a solution for the puzzle is easy. In the meantime, miners receive rewards for solving the complex puzzle. In summary, in a PoW-based blockchain, miners strive to validate transactions, solve the puzzle, propose candidate blocks, and receive rewards. The amount of work done by a miner determines its chance of successfully mining a single block and receiving a reward.

PoW not only provides a solution for the Byzantine Generals' Problem, but also provides defense against denial-of-service (DoS) attacks because it takes a tremendous amount of computational power for an attack to be successful. In a PoW-based blockchain, the available fund/credit in the attacker's wallet does not increase its ability to publish new blocks. What really matters is a node's computational power to solve a puzzle and generate new blocks. PoW strongly discourages DoS attacks on a blockchain because it is highly unlikely that an attacker has the ability to acquire enough hardware and energy resources to overpower the rest of the nodes on a blockchain as a whole.

However, PoW has weaknesses and users should be aware of them. First, a PoW-based blockchain is vulnerable to the so-called 51% attack, in which case the attacker has the majority of the mining power for whatever reason. With 51% or more of the mining power, the attacker is able to control the operations of the blockchain and prevent other mining nodes from creating new blocks. By doing this, only the attacker will get the rewards. With 51% of the computing power, the attacker can even reverse transactions.

The second weakness of PoW is huge power consumption due to the need for solving complex puzzles. It has been observed that the Bitcoin blockchain is currently using more power than the whole country of Ireland and will use more power than the whole country of Denmark by 2020.

8.3.4.2 Proof-of-Stake (PoS)

Proof-of-Stake, or PoS [11], was designed to overcome the weaknesses of PoW. The basic rationale of PoS is that a node who owns more stakes in the blockchain will more likely want it to succeed. To be able to be admitted to the blockchain, a node needs to have a specific amount of assets stored in its wallet. Furthermore, a node needs to deposit some assets as stake in order to qualify as a miner. Although every node is entitled to validate and mine new blocks based on their asset possession, actual miners are randomly chosen by the blockchain based on the assets stored in their wallets. The blockchain will examine all nodes with their stakes and choose some of them as miners based on the ratio of their stakes with respect to the overall system stakes. That is, if a node owns 10% of the total stakes, then it has 10% of the chance to be selected as a miner. A node with only 1% of the total stakes will only be selected 1% of the time. The next new block will be voted for by all users with stakes. However, in PoS-based blockchains, although the nodes with more initial stakes can potentially accumulate more and more digital assets, the blockchain is designed in such a way that it is extremely difficult for several nodes to acquire the majority of assets within the blockchain. This way, no nodes will be able to dominantly manipulate the blockchain as they wish.

Comparing to PoW-based blockchains, PoS-based blockchains do not need powerful computing hardware. A functional computer with a stable Internet connection is all that is needed to work as a node. PoS-based blockchains are much more energy efficient than PoW-based blockchains because they do not use much electric power in their operations. Not having mining operations also enables PoS-based blockchains to run much faster than PoW-based blockchain. A PoS-based blockchain has very little chance of having a 51% attack because of its design.

The main disadvantage of PoS-based blockchains is that it is impossible to achieve full decentralization. The reason is that in a PoS-based blockchain, only limited numbers of nodes are participating in creating new blocks.

8.3.4.3 Delegated Proof-of-Stake (DPoS)

Another well-known consensus algorithm is the Delegated Proof-of-Stake (DPoS) [12] invented by Daniel Larimer. In DPoS, there are three groups of entities: stakeholders, witnesses, and delegates. The responsibilities of stakeholders are the election of witnesses. The responsibilities of witnesses are the creation and addition of blocks to the blockchain. The responsibilities of delegates are maintaining the blockchain and suggesting changes to the blockchain.

Witnesses are elected by the stakeholders. Each stakeholder has one vote for one witness. Witnesses with the highest number of votes are elected. Stakeholders vote to increase the number of witnesses until at least 50% of the stakeholders consider the blockchain has achieved sufficient decentralization.

Elected witnesses take turns to produce new blocks in given timeframes. However, the quality of their work is monitored by stakeholders via a reputation scoring system. Poorly performing witnesses will lose scores or their titles. Stakeholders will continuously vote for the witnesses. Part of the witnesses is replaced at regular intervals as well.

Delegates are also elected by stakeholders. However, their responsibility is to maintain the blockchain. For example, delegates can suggest block size changes, paid incentive, and transaction fee changes. The stakeholders will decide if the proposed changes should be implemented. Delegates may receive rewards as well.

Energy saving and decentralization promotion are the two main advantages of DPoS. DPoS needs less energy than PoW because witnesses generate blocks based on specific time schedules, rather than competing with each other to add blocks. The computing hardware requirement is no longer as demanding as PoW as well. In addition, greater decentralization is achieved in DPoS because its consensus mechanism allows stakeholders to choose suitable witnesses to validate transactions.

The main disadvantage of the DPoS consensus mechanism is that it can never achieve full decentralization, although decentralization can be increased by having more witnesses validate blocks, due to scalability constraints.

8.3.4.4 Practical Byzantine Fault Tolerance (PBFT)

The practical Byzantine Fault Tolerance (PBFT) consensus algorithm [13] is another popular consensus algorithm used in blockchains. PBFT enables a blockchain to tolerate Byzantine faults, i.e., defend against attacks from malicious nodes. The algorithm is designed to work in asynchronous systems. PBFT has low overhead time and low latency.

In PBFT, all nodes of a blockchain are organized into a sequence. A specific node is designated as the leader node. Other nodes are designated as backup nodes. When a node sends out a message, the rest of the nodes will exchange information with each other to validate the message in case it is tampered during transmission. It is expected that the good nodes will reach an agreement on the state of the blockchain through majority.

Each round (called *view*) of the PBFT works as follows:

1. A client sends a request to the leader node.
2. The leader node broadcasts the request to backup nodes.
3. The backup nodes execute the request and send a response to the client.
4. The client waits to receive $f + 1$ node responses with the same result which will be used as the result of the operation, where f represents the maximum number of potentially faulty nodes.

To secure its role, the leader node may be changed in a round-robin fashion during every view. The leader node can even be replaced if it does not broadcast a request after a given time interval. The majority of good nodes also have the power to identify a faulty leader node and replace it with the next leader.

To give more details, here is how PBFT works in Fabric:

1. One of the nodes is elected as a leader.
2. Transaction requests are submitted to the leader.
3. The leader organizes the transactions into an ordered list and broadcasts this list to all other nodes in the blockchain for validation.
4. Every validating node executes the ordered transactions one by one. Then it calculates the hash code for the new block which is based on the received transactions. Then this validating node broadcasts the hash code to other validating nodes and starts counting the responses from them.
5. If a validating node realizes that two-thirds of all validation peers have the same hash code, it will add the new block to its own copy of the ledger.

The PBFT model works only if the number of malicious nodes in a blockchain does not exceed one-third of the total nodes in the system in a given time window. The more nodes are there in the blockchain, the more unlikely for the malicious nodes to reach one-third of the total nodes.

The PBFT algorithm has two main advantages compared to other consensus algorithms. The first advantage is that it can finalize transactions and blocks without needing confirmations as what is done in PoW. The second advantage of the PBFT model is that it uses significantly reduced energy, again as compared to PoW.

There are two limitations to the PBFT consensus algorithm. First, it works well only for blockchains of small sizes due to its communication model among nodes. Second, it is susceptible to Sybil attacks. Due to the first limitation, the size of the blockchain cannot be increased significantly just to mitigate Sybil attacks. Luckily, possible solutions have been identified to solve this problem. For example, PBFT can be interlaced with PoW to overcome both limitations.

8.3.4.5 IOTA

A totally different technology in the cryptocurrency family is *IOTA* [14]. IOTA is an open-source distributed ledger with great potential for applications in the Internet of Things.

IOTA works on the platform called Tangle. Tangle hashes use Winternitz signatures [15] which is a hash-based cryptography, unlike blockchains that use elliptic curve cryptography or ECC. Winternitz signatures are much faster than ECC. The actual hash function used by Tangle is Kerl [16] which is a version of SHA-3. Kerl works based on ternary operations, which is more secure than other crypto technologies used in blockchains. Currently, many crypto algorithms can be broken by superfast quantum computers. However, it is very difficult for a quantum computer to break ternary operations used by Kerl. The chances of Tangle suffering from a quantum attack are roughly 1 million times less than the blockchain.

8.4 Directed Acyclic Graph (DAG)

8.4.1 What Is a DAG

As discussed previously, DLT, or "distributed ledger technology," has its set of records (the ledger) held by multiple distributed nodes. For instance, the cryptocurrency Bitcoin has a blockchain which is a DLT with its ledger (transactions) stored in multiple computers. Each new transaction added to the ledger is copied to other computers. This ensures that multiple copies of the ledger are available.

DAG is a type of ledger. A DAG is a graph with directed edges and no cycles. A DAG has its nodes sorted in a special order, which is called *topological sorting*. In a DAG, each transaction is linked to at least one other transaction. The edges are directed from earlier transactions to recent transactions. Loops are not allowed in DAGs, which means that a transaction cannot travel back to itself if it follows along the directed edges. Figure 8.12 shows a DAG.

8.4.2 How IOTA Tangle Works

Tangle is IOTA's DAG that operates in a special way. In Tangle, each new transaction must validate at least two previous transactions before it can be added to the DAG. With Tangle, all nodes on the IOTA network can issue and validate transactions at the same time. In Tangle, data are attached to transactions. However, Tangle does not assemble transactions into blocks. Therefore, Tangle is blockless.

Tangle does not require mining to reach consensus. This avoids powerful mining computers and extensive use of electric energy. No mining also means no fees are needed to reward miners. Users do not need to pay transaction fees as well.

Tangle is highly scalable because of its use of DAG as its ledger and simultaneous transaction processing. Increased transactions in a DAG do not slow down the IOTA network. In fact, performance will improve as the number of transactions increases due to the characteristic of simultaneous validation. IOTA with Tangle has a higher speed than blockchains.

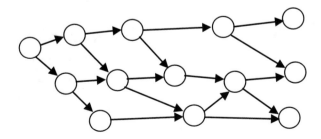

Fig. 8.12 A directed acyclic graph (DAG)

When Tangle gets started, it uses a coordinator to prevent malicious activities because it does not have enough transaction nodes to conduct validation. This coordinator will be obsolete after the IOTA network becomes more established. The use of an initial coordinator creates the possibility of a central point of failure.

8.5   DAG Versus Blockchain

Similar to blockchains, DAGs store transactions on a distributed ledger. However, the ledger is quite different in a blockchain than it is in a DAG. In a blockchain, the distributed ledger is a chain of blocks which are built using transactions. Blocks are validated and chained up in chronological order. Chained blocks are not modifiable. A blockchain is very similar to a linked list concept in computer science. On the contrary, a DAG is a collection of transactions linked in special ways. There are no blocks in a DAG. A DAG can be compared to a tree in computer science. Figure 8.13 compares the structures of blockchain and DAG.

Consensus is achieved differently in blockchains and DAGs. In blockchains, consensus is achieved by validating transactions block by block via mining. On the other hand, DAGs have transactions validate their immediate predecessors.

Based on the above discussions, blockchains have better immutability than DAGs, whereas DAGs are better at handling a large number of transactions. Blockchains do not scale well, but DAGs do. DAGs are vulnerable to attacks if the volume of transactions is too low.

8.6   Blockchain and Internet of Things

8.6.1   Internet of Things

Internet of Things (IoT) [17] is a natural extension of the human being's efforts of connecting the world through computer networks. So far computers around the world have been connected for sharing information. The World Wide Web (or the

Fig. 8.13 (**a, b**) Comparison of blockchain and DAG structures

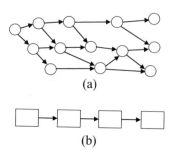

(a)

(b)

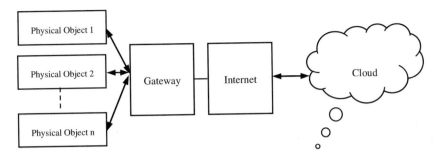

Fig. 8.14 A typical centralized architecture of the Internet of Things

Web) is an indicator of this usage. The Web has been used to share and exchange digital textual or visual information in the form of electronic documents. As the digital revolution continues, the natural next step is to connect all the physical objects in the world, have them not only exchange data, but also interact with each other, in order to make our lives more convenient, more efficient, and safer. These efforts led to the birth of the Internet of Things, or IoT for short.

Currently, most IoT solutions are designed based on a centralized architecture (see Fig. 8.14).

A current IoT system consists of the physical objects/devices, the gateway, the Internet, and the cloud. Physical objects can send data through the gateway and the Internet to the cloud and the data can get stored and analyzed there. Physical objects can also receive commands from other physical objects through the cloud to perform specified actions. Commands can be issued from a central manager from the cloud directly as well. The IoT stack and standard protocols create the layers of an architecture that provides services to IoT physical objects.

In the history of people's efforts in trying to connect physical objects in the real world, large peer-to-peer (P2P) wireless sensor networks (WSNs) were conceived and were once the focus of research. The missing pieces in these researches in terms of fundamental architecture design are privacy and security, which should really have been considered at the beginning of system design. WSNs originally were not designed to operate at a global scope as well.

8.6.2 Weaknesses of Internet of Things

Core recent developments on IoT moved towards the above cloud-based centralized architecture. However, the centralized IoT architecture has numerous *weaknesses*. In this centralized architecture, all information is sent from the physical objects to the cloud where data is processed using analytics tools. Responses are sent back from the cloud to the IoT physical objects if necessary. This type of centralized structure has *poor scalability*. The problem will become even worse when billions of new physical objects are to be added to IoT networks in the near future.

The second weakness of the centralized IoT architecture is a *single point of failure*, because every physical object is potentially a vulnerable point and can compromise the security of the entire IoT network. Failure of a single physical object can potentially bring down the entire IoT network as well.

The third weakness of the centralized IoT architecture is to do with *maintenance*. Updating software in the current IoT network is extremely difficult due to the fact that software updates need to be distributed to a huge number of physical objects which can be physically located anywhere.

The fourth weakness is related to *security and privacy*. Data spoofing and corruption can occur anywhere on the IoT network, ranging from the physical objects, the communication networks over which IoT data travel through, and the cloud storage where IoT data are gathered, stored, and processed. Unauthorized access to personal data in the cloud can happen which has always been the concern of the general public.

The fifth weakness is that IoT systems frequently use *resource-constrained computing devices* such as microcontrollers. These microcontrollers lack the computing power and storage capacity to support advanced and computation-intensive algorithms which can assist in protecting data security and privacy.

The sixth weakness is that current IoT systems have *no immutable records* of the history of interactions among physical objects. Because of this weakness, it is very difficult to track down the causes if problems do occur.

Another weakness of IoT is that the current centralized structure has only one copy of the data stored in the cloud. If this copy of data is tampered, there is no way to know what has been changed. There is no way to prevent the tampering from happening as well.

Because of these weaknesses, IoT faces the challenge of people *lacking trust* in technology, primarily due to their concerns on privacy and security. Their perception of the scale and complexity of IoT systems makes the situation worse because it is beyond their comfort zone. Granting device access and control to technological service providers is frequently a difficult decision and is a sensitive matter for IoT system owners as well.

IoT devices such as connected actuators are often required to perform actions according to the commands they receive from the cloud or other IoT devices. If such commands are hijacked, the consequence could be disastrous. A small example would be that the door of a house is wrongly opened for a burglar. Improper actions of devices could also lead to fires and flood in buildings and offices.

Overall, current IoT systems are subject to physical object identity-based attacks, manipulation-based attacks, cryptanalytic attacks, and service-based attacks.

8.6.3 Blockchains and IoT

Blockchain technologies have exactly what is needed to fix the weaknesses of centralized IoT. It is easy to perceive that the decentralized structure, the way that data is created and stored, and the consensus mechanism used will help overcome most of the weaknesses of the current IoT systems.

Depending on the use cases, blockchain technologies can be applied to each level of the IoT systems. Blockchains can be used to store and manage device IDs, encode and verify data packets on the communication networks, and secure data in the cloud and data stored in the distributed devices.

Blockchain technologies can be applied at a small and local scale such as smart homes and smart buildings, or to larger scales such as in smart cities, or even at a global scale for cross-continent IoT systems.

Blockchain technologies will help reduce IoT operational costs and prevent threats and attacks. Blockchains are unique and attractive because they have the following features: transactional privacy, security, data immutability, auditability, integrity, system transparency, and fault tolerance.

Wired and wireless communication technologies have reached new high levels. The technologies are still evolving, witnessed by the growing interest of adopting 5G technologies in IoT. It can be predicted that the requirements of data transmission speed by IoT will be up to users' expectations.

For this reason, privacy, security, and transparency and trust should be at the center of future IoT system designs. They should be considered right at the beginning when an IoT system is conceived.

In summary, as an emerging technology, IoT is promising and has a great future. Current IoT systems use resource-constrained devices which are ideal targets for cyberattacks. They have poor scalability and have the problem of a single point of failure. Maintenance is difficult. IoT data are not immutable. Privacy and security are critical concerns of IoT.

Blockchains can mitigate IoT risks and issues by using a large number of individual nodes that exchange data on a peer-to-peer (p2p) basis. Data records are immune to tampering and corruption. The consensus mechanism of blockchain can prevent malicious nodes from joining the IoT network, rejecting the data they send, and ensuring data integrity.

Among the various blockchains, practical Byzantine Fault Tolerance (PBFT)-based blockchains appear to be especially suitable for IoT, due to their abilities to defend against attacks from malicious nodes, work in asynchronous systems, and have low overhead time and low latency.

The other promising blockchain IoT platform is IOTA. It was designed specifically for the Internet of Things. IOTA is blockless and does not use computation-intensive mining algorithms. Instead, users verify the transactions of other users. The main advantage of IOTA is greater scalability.

With the support of blockchain technology, IoT systems will have the characteristics of decentralized resource management, robustness against threats and attacks, fault tolerance, and improved trust.

8.6.4 How to Combine Blockchains and IoT

According to the use cases and goals, blockchains can be combined with IoT in different ways. Figure 8.15 shows a block diagram for the current cloud-based centralized IoT system.

Blockchains can be applied to IoT systems in two ways, depending on the purposes of the application. The most comprehensive implementation uses a blockchain to record all data and interactions between physical objects [18], as shown in Fig. 8.16.

In Fig. 8.16, all data and interactions go through the blockchain. In this architecture, data and interactions are validated and their records are immutable. This is useful if both data and device interactions are important for the application. The drawback of this architecture is increased latency, increased bandwidth requirement for the communication network, and increased data flow on the network.

The other choice of combining blockchain and IoT is storing only IoT data in the blockchain, as shown in Fig. 8.17.

Fig. 8.15 Data and interactions of IoT physical objects are stored in a central server in centralized IoT

Fig. 8.16 Data and interactions of IoT physical objects are stored in the blockchain

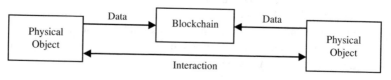

Fig. 8.17 Only data of IoT physical objects are stored in the blockchain

In Fig. 8.17, interactions between physical objects are not stored in the blockchain. In this architecture, only IoT data are validated and only data records are immutable. This is useful if IoT data are important for the application, but interactions between physical objects are not critical. This architecture has less latency, reduced bandwidth requirement for the communication network, and reduced data flow on the network.

In addition, smart contracts can be employed in the blockchain to set up specific requirements and agreements that govern data flow and usage, as well as monitor, allow, or disallow interactions to occur.

8.7 Prominent Enterprise DLT Platforms

Although there are tens of DLT platforms introduced in the literature, so far at the enterprise level, three of the platforms are most prominent. They are Hyperledger Fabric, Ethereum, and IOTA (See Table 8.3).

Enterprises have special requirements for DLTs. Ideally, enterprises require decentralized data storage. Data stored should be immutable and permanent. Security and privacy of data are of prime importance. Enterprise DLTs should support smart contracts and tokens.

8.7.1 Hyperledger Fabric

Hyperledger refers to several technologies. Hyperledger Fabric, developed by IBM, is one of them. Hyperledger Fabric was designed for B2B applications. IBM's client server-based architectures are used in Hyperledger Fabric to provide decentralized

Table 8.3 Comparing Ethereum, Hyperledger, and IOTA

	Data storage	Security and privacy of data	Support for tokens	Support for smart contracts	Immutability and persistency
Ethereum	Truly decentralized	Offered	Offered	Offered	Supported
Hyperledger Fabric	In members of a private consortium	Offered	No Support	Supported	Supported
IOTA	Not decentralized, but maintained by several central components	Offered	Not Offered	Not Supported	Not Supported

data storage. Private transactions are used to provide data security and privacy. However, Hyperledger Fabric does not support tokens.

Hyperledger Fabric supports smart contracts through chaincode. The operation of Hyperledger Fabric depends on a number of central participants.

Overall, Hyperledger Fabric is suitable for use cases where data is to be exchanged between a closed group of companies. It is not suitable for fully distributed applications.

8.7.2 Ethereum

Ethereum is a truly decentralized DLT which can run as a public blockchain, private blockchain, or consortium blockchain. Ethereum provides truly decentralized data storage through its architectural design. Data in Ethereum is less secure and private because it originally focused on the public chain. Tokenization is supported so that real assets can be digitally represented. Companies can build digital business models using Ethereum.

Smart contracts are seamlessly integrated into Ethereum and can be programmed using its built-in programming language called Solidity. Smart contracts are executed in the Ethereum Virtual Machine (EVM). Distributed apps (DApps) can be developed and run under the EVM. DApps can be deployed on Ethereum without additional infrastructure. Data immutability is guaranteed by Ethereum's architecture.

8.7.3 IOTA

IOTA can be used as a data layer on top of IoT to facilitate transactions between machines. The nodes in an IOTA network can both generate and confirm transactions. IOTA is a "feeless" DLT. Currently, IOTA is not truly decentralized because it depends on central maintaining elements. Data security and privacy are achieved via transaction validation, data encryption, and subscriber authorization. IOTA does not support tokenization. IOTA does not support smart contracts as well. High speed and high scalability are two main advantages of IOTA.

The transaction referencing structure in Tangle provides data immunity. In theory, data can be traced back to the very first transaction(s) in Tangle, although the snapshotting mechanism makes this impossible.

8.8 Applications of Blockchain

Blockchain has the potential to be applied to virtually every aspect of our life. Figure 8.18 shows several of the application domains.

8.8.1 Financial Services

Financial service is no doubt the most prominent application domain of blockchain technology due to its relationship with Bitcoin. Unlike traditional financial services, blockchain enables transactions to occur in a peer-to-peer manner without involving third parties. This eliminates intermediary financial services such as banks and saves costly service fees. Blockchain also records transaction history and such records cannot be tampered. This will help with verification and in avoiding disputes.

Blockchain will also greatly speed up transaction processing and can reduce the time needed for processing to seconds, even if the transactions are cross-border, in which case processing delays can be up to several days. Blockchain-based financial services are also available to customers around the clock.

Stock trading platforms based on blockchains allow investors to purchase and sell stocks almost instantly in a secure manner. Funds created from selling stocks can be made available right after the transactions so that investors can reinvest the funds without wait times.

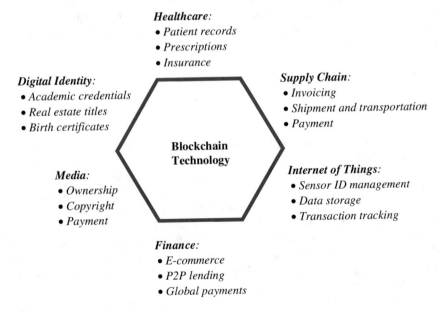

Fig. 8.18 Blockchain applications

The way business is conducted will change because of blockchain. For example, in legal practice, blockchains can be used to store wills and other inheritance records and wills will not be able to be tampered. Smart contracts can be used to set up inheritance criteria as well.

A more effective insurance industry will be in place because of blockchains. It will be very easy to verify asset owners and therefore avoid fraudulent claims. The whole insurance industry will be more effective and reliable.

Copyrighted digital contents can be better protected when blockchains are used. Ownership rights can be made transparent. Content creators will be able to receive royalties speedily.

8.8.2 Healthcare

Healthcare practices will change greatly because of blockchains. Healthcare is a complex business because it involves doctors, nurses, staff, medical service providers, insurance companies, testing labs, and pharmacies. Parties involved in healthcare are distributed in geo-location. However, they can all initiate transactions and the results of these transactions are supposed to be stored in one system and are used in an integrated manner. Currently, such transactions are stored in different systems which are inconvenient, time-consuming, and error-prone to use by stakeholders. Blockchains will change this completely. All transactions will be stored in the same ledger. This means patient medical history, test results, benefits and eligibility, insurance coverage, medication, and allergies are all available on the same blockchain. Management of healthcare systems will be more efficient because demands of medication, equipment, and other consumables can be managed by using blockchains as well.

8.8.3 Energy

The *energy industry* is another major blockchain application domain [19]. Applications can be made in electric grid, or in the oil and gas industry. When applied to the electric grid, blockchains can be used in wholesale or in peer-to-peer electricity distribution systems. When used in wholesale electricity distribution systems, blockchains can directly connect end users to the electric grid. Users can trade energy via the electric grid. No retailers will be necessary any longer so that electricity costs will be lowered. In fact, it has been envisioned that interconnected electric grids will be made available and will allow participants to buy and sell renewable energy at greatly reduced prices. Table 8.4 shows the applications of blockchain in the energy industry.

Blockchains can be specifically implemented in electricity data systems which manage fuel prices, market prices, and marginal costs. Such data can be recorded,

Table 8.4 Blockchain applications in the energy industry

In oil and gas industry	In electric grid
Gas and commodity trading	Wholesale electricity distribution
Supply and data tracking	Peer-to-peer electricity distribution
Consortiums	Electricity data management

stored, and tracked by blockchains. With blockchains, clerical errors can be avoided. Data will not be misreported or unreported. Blockchains will also allow the public to view transactions and their prices and monitor money movements.

Because privacy and trade secrets are particularly important for oil and gas companies, they are more interested in private permissioned blockchains and consortium blockchains. With these blockchains, companies can limit data access to selected parties. Oil and gas companies are looking into using blockchains for commodity trading and supply and data tracking.

The potential benefits of blockchains for oil and gas companies are increased data security, reduced time delays, and reduced data management costs.

8.8.4 Identity Management

Another application area where blockchains are well suited is *digital identity management*. This applies to both the identities of properties and human beings. When blockchains are applied to properties, the use case is called *smart property*. Tangible properties such as cars, bikes, houses, appliances, and jewelry can have their digital identities embedded when they are manufactured. These identities, together with their owners, will then be stored in blockchains. Ownership transfers can be recorded and traced through blockchains. The authenticity of the properties can be verified. Blockchains can manage intangible properties such as patents and company stock shares too.

It has been envisioned that blockchains will be used in managing the identities of human beings. Human beings can receive their identities when they are born. The same identities can be used for governmental registration, health records, motor vehicle licensing, life insurance, schooling, and employment. This will greatly simplify the currently used information system structures which are independent from each other and difficult to synchronize. Current information systems are susceptible to frauds and mishandling, whereas blockchain-based systems are secure.

One of the serious problems with current information systems is that, after users submit their personal data, they do not know exactly where the data is located, for what purpose the data will be used, who will see the data, and who will use the data. These are not only problems for the people who submitted the data, but also problems for organizations who own the databases. Online companies are able to abuse user's personal data or sell the data to advertisers. Blockchains create a so-called single point of trust and protect our privacy. Data will be encrypted and people will have control of their own data.

8.8.5 Supply Chain Management

Supply chain management is another best-suited application for blockchains. Blockchains can be used to record the entire process of transferring physical goods from the producers to the consumers. Details of farms or greenhouses, delivery trucks, warehouses, supermarkets, and retail stores, as well as the movements of physical goods between them, can all be recorded in a blockchain, along with the temperature, humidity, time, etc.

Blockchains make supply chain management more efficient. Managers can use blockchains to help with planning and avoid overstocking. Goods can be located in real time. Causes of problems can be traced back to its origin.

8.8.6 Other Applications

Blockchains will bring revolutions to *voting and elections*. Digital voting will be more secure than ever before. Votes will become transparent and immutable. Voters will be able to find out if their votes have been counted. Election data will not be able to be tampered. Millions of dollars will be saved in running elections.

Another industry that will benefit from blockchains is *real estate*. Property titles and their transfers can be stored in blockchains. The records are transparent to the public and are permanent.

Student records and certificates can be stored in blockchains. Verification of credentials can be made instant and will be reliable.

Driver licenses, violations, and accident records can be stored in blockchains as well. The whole process of vehicle ownership and policy verification will be efficient without errors.

Blockchains can be used just for *data backup*. Data backed up are immune to tampering. However, current cloud-based data storage systems are not immune to hackers.

8.9 Other Aspects of DLTs

8.9.1 Scalability and Other Practical Considerations

When it comes to adopting DLTs in an organization, practical considerations become important. These considerations are to do with common-sense parameters: memory size and speed. For DLTs and blockchains, these parameters are transaction size, block size, and transactions per second (TPS). Before each DLT technology is examined, some information about VisaNet (the credit card processing system) is useful. It is reported that VisaNet can handle an average of 150 million transactions per day [20]. This is equivalent to about 1736 transactions per second on average.

8.9.1.1 Bitcoin

Bitcoin generates 1 block every 10 minutes. The size of the block is 1 megabyte (1,048,576 bytes). This block is broadcasted to the Bitcoin network which had 10,198 nodes on January 17, 2019. The Karlsruhe Institute of Technology reported that, on January 17, 2019, it took 13,989.42 milliseconds or approximately 14 seconds to propagate the block to 99% of the nodes on the Bitcoin network [21]. This means that the block propagation time of Bitcoin is about 14 seconds.

The average Bitcoin transaction size is 380.04 bytes on January 17, 2019 [21], although on May 12, 2019, it was 352.23 bytes per transaction on average [22]. Among the 380.04 bytes, 346 bytes are overheads for the transaction, and only 34 bytes are real data for the transaction.

Therefore, the average number of transactions per block in Bitcoin is 1,048,576/380.04 = 2759.12. This gives 2759.12/(60 × 10) = 4.548 transactions per second. This is far less than the 1736 transactions per second of VisaNet. In summary, Bitcoin has the following parameters as shown in Table 8.5.

8.9.1.2 Hyperledger Fabric

The performance of Hyperledger Fabric [23] is a function of the number of endorsing peers, number of channels, endorsement policy, ordering service configuration

Table 8.5 Bitcoin performance data

Block generation time	Block size	Transactions per block	Block propagation time	Transaction size	Actual data in transaction	Overhead of transaction	Transactions per second (TPS)
10 minutes	1 MB	2759	14 seconds (99% of nodes)	380.04 bytes	34 bytes	346 bytes	4.548

(i.e., block size and frequency), number of organizations, and ledger database used. It is also to do with execution complexity of chaincode or smart contracts, transaction sizes, use of mutual TLS security in network traffic, number of vCPUs, memory allocation, disk type and speed, and network speed. Furthermore, it is to do with data centers, CPU speed, and crypto acceleration. An experiment conducted by [24] tested Hyperledger Fabric 1.3.0 in a single Kubernetes cluster running on the IBM Container Service. The worker nodes were configured as 4vCPU and 16Gb memory with SSDs. A two-organization cluster executed on a single channel and 2, 4, and 8 endorsers were used respectively. The corresponding throughput and average latency are shown in Table 8.6. In this table, TPS stands for "transactions per second." The table indicates that, with 2 endorsers, the tested system carried out 785.58 transactions per second, and it took 715 milliseconds for 95% of the nodes to commit a transaction. Whereas when 8 endorsers were employed, the tested system was able to finish 1265.5 transactions per second, and it took only 686 milliseconds for a transaction to be committed by 95% of the nodes in the system.

It should be noted that Samsung SDS revealed that it had developed an accelerator software to speed up Hyperledger Fabric transactions to 3500 TPS, with experiments succeeded in achieving 20,000 TPS [25].

8.9.1.3 Ethereum

Unlike Bitcoin, Ethereum does not have a fixed block size. Instead, Ethereum has a gas limit for each block which determines how many transactions can fit in a block. The block generation time of Ethereum has achieved about 13 seconds [26]. The TPS of Ethereum is about 15 transactions per second [27]. The history of the Ethereum block size can be found in [28]. Ethereum performance data is shown in Table 8.7.

8.9.1.4 IOTA

IOTA does not have blocks. A transaction in IOTA consists of 2673 trytes [29]. Using the IOTA converter [30], 2673 trytes can be converted to 1589 bytes. IOTA can execute 500–800 transactions per second on average and it will be even faster

Table 8.6 Hyperledger Fabric test results

Number of endorsers		2		4		8	
TPS	95% (ms)	785.58	715	948.2	667	1265.5	686

Table 8.7 Ethereum performance data

Block generation time	Block size	Transactions per second (TPS)
Around 13 seconds	Variable	15

Table 8.8 Ethereum
performance data

Transaction size	Transactions per second (TPS)
1.598 bytes	500–800 and above

when more users have participated [31]. IOTA performance data can be found in Table 8.8.

8.9.1.5 Scalability of DLTs

Gartner defines scalability as *the measure of a system's ability to increase or decrease in performance and cost in response to changes in application and system processing demands.* In other words, if a system is scalable, it should be able to grow in size and performance if user demand increases.

Bitcoin does not scale well. In its original design, the block generation time is fixed at 10 minutes, the block size is fixed at 1 megabyte, and the TPS is fixed at 4.5. If more and more users participate, the wait time for a transaction to go through is not acceptable. Technologies such as Segwit (Segregated Witness) have been developed to mitigate the scalability problem of Bitcoin.

IOTA is scalable. This is due to the fact that IOTA does not store transactions in blocks which have limited size. In IOTA transactions approve other transactions. Therefore, the more transactions IOTA has, the more transactions it can approve simultaneously. IOTA performance will increase with the increase of users.

8.9.2 Token and Token Economics

Technically speaking, a token in a blockchain represents a programmable currency unit embedded in a blockchain and is part of smart contract logic. In simple non-technical terms, a token is a kind of private digital currency. A more comprehensive definition is given by Mougayar [32], where a token is defined as *a unit of value that an organization creates to self-govern its business model, and empower its users to interact with its products, while facilitating the distribution and sharing of rewards and benefits to all of its stakeholders.*

Tokens can be used to grant rights to use a product, or the rights to vote. Tokens can also be used as a unit for exchanging values in a blockchain ecosystem. Tokens can be incentives earned by doing useful work and can be spent when using a service or product. Tokens can serve as a payment method. Tokens can be distributed in ICOs (Initial Coin Offerings).

Essentially, tokens help build self-sustainable mini-economies in distributed autonomous organizations (DAOs) based on blockchains. This is interestingly termed as "tokenomics" or "cryptoeconomics."

8.10 Vulnerabilities of Blockchain

Although blockchain technology provides numerous advantages and application potentials, it is not perfect. It is important to be aware of its weaknesses. Potential attacks can occur on several aspects of blockchain technology. Systems based on blockchain technology can even be used to commit crimes.

The first vulnerability of blockchain technology is originated from its *consensus mechanism*, which is susceptible to a 51% attack [33]. Specifically, in a Power-of-Work-based blockchain network, if the computational power of a single miner node exceeds 50% of the total power of the entire blockchain network, then the entire blockchain could potentially be controlled by that attacker. In a Power-of-Stake-based blockchain network, the 51% attack can also occur if the number of stakes owned by a single node is more than 50% of that of the total blockchain network. The attackers of a 51% attack are able to reverse transactions, conduct double-spending, exclude transactions, reorder transactions, cause problems for operations for normal transaction confirmation, and stop the mining operations of other mining nodes.

Sybil attack is also a vulnerability of blockchain which takes advantage of the fact that public blockchain networks have no centrally trusted nodes and every transaction is sent to a number of other nodes for processing. A Sybil attack is initiated by assigning a number of identifiers to the same node. During a Sybil attack, the attacker is able to outvote honest nodes and takes control of the network. Therefore, the consequence of a Sybil attack is equivalent to a 51% attack.

Private keys are another source of vulnerabilities in a blockchain network. A private key is the identity of a user. It is used to sign transactions and verify asset owners. Private keys are also used in transaction validation and candidate block verification. However, a legitimate user's private key can get lost. If this happens, there is no way to recover the private key. The legitimate user will not be able to access his/her account on the blockchain network anymore and will therefore lose the assets he or she owns. If a private key is stolen by a criminal, the legitimate user's blockchain account can get tampered. Whatever damage the criminal does is difficult to track, repair, and recover because there are no centralized third-party trusted institutions to seek assistance from.

Although it is commonly known that, by introducing consensus algorithms, a blockchain network can prevent the *double-spending attack*, as claimed by the Bitcoin paper [1], it is still possible for double-spending to occur in a blockchain network. It is misleading to believe that double-spending is fully eliminated by the consensus mechanism during validation. Among all blockchains, the Power-of-Work-based blockchain network is especially vulnerable, as the attacker can exploit the time interval between the initiation and confirmation of two transactions to quickly launch a double-spending attack. Double-spending refers to the fact that a malicious user spends the same cryptocurrency for multiple transactions. Knowing it takes time to mine a block and reach consensus, the attacker could launch a race attack involving two consecutive transactions. Before the second transaction is invalidated, it is possible that the attacker has already received the output of the first transaction. This results in a double-spending.

As a long-term security problem for the Internet, the *distributed denial-of-service* (DDoS) *attack* is still a threat to blockchain networks. DDoS attacks create a huge amount of traffic on blockchain networks so that valid transactions cannot be processed, giving opportunities for invalid transactions to become successful.

On the other hand, blockchain networks can be used by criminals to *commit crimes*. One such example is ransomware. A typical ransomware is sent out as an email attachment. If the email receiver clicks the attachment, the ransomware starts running as a background process on the receiver's computer system. What it does is that it encrypts the files in the receiver's system so that the victim loses access to the contents of the files. The ransomware demands the receiver to pay funds to a blockchain account of the attacker within a given time frame. Otherwise, there will be no way to restore the encrypted files forever.

Blockchains can also be used by criminals to run underground markets. Bitcoins are used as the currency and hidden services for such markets. Criminals use underground markets to sell drugs, weapons, and other controlled items. Due to blockchain's anonymous nature, it is difficult to track down the sellers and deals.

8.11 Summary

This chapter discusses distributed ledger technologies (DLTs) which include blockchain and directed acyclic graphs. The chapter discusses the benefits of DLTs when they are adopted by current information systems. Detailed descriptions are given on how blockchain and DAG works and what the differences between blockchain and DAG are. The chapter also discusses the Internet of Things (IoT), the weaknesses of current IoT system implementations, why blockchain can help overcome the weaknesses of IoT, and how to integrate blockchain and IoT. Applications of DLTs and practical considerations of DLTs in enterprise environments are also discussed. Vulnerabilities of blockchain are described as well.

References

1. S. Nakamoto, Bitcoin: A peer-to-peer electronic cash system (2008), [Online]. Available: https://bitcoin.org/bitcoin.pdf. Accessed 23 Mar 2019
2. S. Haber, W.S. Stornetta, How to time-stamp a digital document. J. Cryptol. 3(2), 99–111 (1991)
3. D. Chaum, Blind signatures for untraceable payments, in *Advances in Cryptology Proceedings of Crypto 82*, ed. by D. Chaum, R. L. Rivest, A. T. Sherman, (Plenum (Springer-Verlag), New York, 1983), pp. 199–203

4. T.A. Mahler, Oncologist + gold = revolution? (2018), [Online]. Available: https://medium.com/blockwhat/96-oncologist-gold-revolution-c08a8dc26880. Accessed 23 Mar 2019

5. W. Dai., b-money (1998), [Online]. Available: http://www.weidai.com/bmoney.txt. Accessed 20 May 2019

6. A. Back, Hashcash – A Denial of service counter-measure (2002), [Online]. Available: http://www.hashcash.org/hashcash.pdf. Accessed 23 Mar 2019

7. V. Vishnumurthy, S. Chandrakumar, E.G. Sirer, KARMA : A secure economic framework for peer-to-peer resource sharing (2003), [Online]. Available: https://www.cs.cornell.edu/people/egs/papers/karma.pdf. Accessed 23 Mar 2019

8. Bexam: The next generation blockchain/DAG hybrid platform, (2019), [Online]. Available: https://bexam.io/. Accessed 6 May 2019

9. Ethreum, The blockchain app platform (2019), [Online]. Availablehttps://www.ethereum.org/. Accessed 23 Mar 2019

10. cointelegraph.com, Proof-of-work explained (2019), [Online]. Available: https://cointelegraph.com/explained/proof-of-work-explained. Accessed 23 Mar 2019

11. S. King, S. Nadal, PPCoin: Peer-to-peer crypto-currency with proof-of-stake (2019), [Online]. Available: https://decred.org/research/king2012.pdf. Accessed 23 Mar 2019

12. B. Asolo, Delegated proof-of-stake (DPoS) Explained (2019), https://www.mycryptopedia.com/delegated-proof-stake-dpos-explained/. Accessed 23 Mar 2019

13. M. Castro, B. Liskov, Practical Byzantine Fault Tolerance, Proceedings of the Third Symposium on Operating Systems Design and Implementation, New Orleans, USA, February 1999 (2019), http://pmg.csail.mit.edu/papers/osdi99.pdf, Accessed 23 Mar 2019

14. IOTA Foundation, IOTA basics overview (2019), https://docs.iota.org/docs/iota-basics/0.1/introduction/overview. Accessed 30 Mar, 2019

15. M. Green, Hash-based signatures: An illustrated primer (2019), https://blog.cryptographyengineering.com/2018/04/07/hash-based-signatures-an-illustrated-primer/. Accessed 10 May 2019

16. E. Hop, Exploring the IOTA signing process (2019), https://medium.com/iota-demystified/exploring-the-iota-signing-process-eb142c839d7f, Accessed 10 May 2019

17. R. Minerva, A. Biru, D. Rotondi, Towards a definition of the internet of things (IoT), 27 May 2015 (2015), https://iot.ieee.org/images/files/pdf/IEEE_IoT_Towards_Definition_Internet_of_Things_Revision1_27MAY15.pdf. Accessed 30 Mar 2019

18. A. Reyna, C. Martín, J. Chen, E. Soler, M. Díaz, On Blockchain and its integration with IoT. Chall. Oppor. Futur. Gener. Comput. Syst. **88**, 173–190 (2018)

19. ConsenSys, Blockchain and the energy industry, (May 25, 2018). https://media.consensys.net/the-state-of-energy-blockchain-37268e053bbd. Accessed 30 Mar 2019

20. Visa Inc, Visa acceptance for retailers (2019), https://usa.visa.com/run-your-business/small-business-tools/retail.html. Accessed 12 May 2019

21. K. Li, The blockchain scalability problem & the race for visa-like transaction speed. (Jan 30, 2019). https://hackernoon.com/the-blockchain-scalability-problem-the-race-for-visa-like-transaction-speed-5cce48f9d44. Accessed 12 May 2019

22. Tradeblock.com (2019), https://tradeblock.com/bitcoin/historical/1h-f-tsize_per_avg-01101. Accessed 12 May 2019

23. M. Mamun, How does hyperledger fabric work? (2019), https://medium.com/coinmonks/how-does-hyperledger-fabric-works-cdb68e6066f5. April 16, 2018. Accessed 12 May 2019

24. C. Ferris, Answering your questions on hyperledger fabric performance and scale, (January 29, 2019), https://www.ibm.com/blogs/blockchain/2019/01/answering-your-questions-on-hyperledger-fabric-performance-and-scale/

25. Ledger Insights Ltd, Samsung Tech to Speed up Hyperledger Fabric (2019), https://www.ledgerinsights.com/samsung-hyperledger-fabric-speed-blockchain/. Feb. 2019. Accessed on 15 May 2019

26. J. Heal, C. Rivet, March 5, 2019, Ethereum block generation time falls following Constantinople upgrade (2019), https://finance.yahoo.com/news/ethereum-block-generation-time-falls-080011370.html. Accessed on 15 May 2019
27. How Will Ethereum Scale?, (2019), https://www.coindesk.com/information/will-ethereum-scale. Accessed on 15 May 2019
28. Ethereum Block Size historical chart, (2019), https://bitinfocharts.com/comparison/ethereum-size.html. Accessed on 15 May 2019
29. The Anatomy of a Transaction, (2019), https://domschiener.gitbooks.io/iota-guide/content/chapter1/transactions-and-bundles.html. Accessed on 15 May 2019
30. IOTA Converters, (2019), https://laurencetennant.com/iota-tools/. Accessed on 15 May 2019
31. Transaction Speed – Bitcoin, Visa, Iota, Paypal, (2019), https://steemit.com/cryptocurrency/@steemhoops99/transaction-speed-bitcoin-visa-iota-paypal. Accessed on 15 May 2019
32. W. Mougayar, Tokenomics—A business guide to token usage, utility and value (Jun 10, 2017), https://medium.com/@wmougayar/tokenomics-a-business-guide-to-token-usage-utility-and-value-b19242053416. Accessed on 15 May 2019
33. S. Sayeed, H. Marco-Gisbert, On the Effectiveness of Blockchain against Cryptocurrency Attacks, UBICOMM 2018 : The Twelfth International Conference on Mobile Ubiquitous Computing, Systems, Services and Technologies, pp.9–14 (2018)

Chapter 9
Emerging Hardware Technologies for IoT Data Processing

Mahdi Nazm Bojnordi and Payman Behnam

No man has a good enough memory to be a successful liar.

Abraham Lincoln

Contents

M. N. Bojnordi (✉) · P. Behnam
University of Utah, Salt Lake City, UT, USA
e-mail: bojnordi@cs.utah.edu

© Springer Nature Switzerland AG 2020
F. Firouzi et al. (eds.), *Intelligent Internet of Things*,
https://doi.org/10.1007/978-3-030-30367-9_9

9.1   Challenges for Data Processing in the Era of IoT

Recent years have witnessed an ever-increasing need for big data processing in nearly all forms of computing systems from server computers and data centers to mobile and Internet of things (IoT) devices. The huge demand for big data processing has been mainly due to an unprecedented increase in the public use of social networks (e.g., Twitter, Instagram, and Facebook), digital video hosts (e.g., YouTube that performs an average of 72 hours' video upload per minute), smart phone applications, and IoT systems [1]. In particular, IoT plays a significant role in big data explosion [2] and is likely to have a profound impact on how computer systems will be designed and used in the coming decades. For example, one important sector of IoT-based big data processing is healthcare that builds upon the biological and social data analytics spanning the latest achievements in data mining, machine learning, computational intelligence, and statistical methodologies. Similar to all other sectors of IoT, today's healthcare applications encounter significant challenges for storing and moving big data within their computing platforms. These challenges have been one of the main motivations towards forming a paradigm shift in the design of memory systems for efficient data processing.

9.1.1   IoT System Architecture

IoT systems heavily rely on hardware-software interfaces that enable various forms of data communication among interconnected components. A typical IoT system comprises various hardware and software components used for identification, sensing, communication, computation, service, and semantic [3]. The IoT nodes need to be identified by name and address in the system. The electronic product code (EPC) and ubiquitous code (uCode) methods may be used for device identification, while IPv4 and IPv6 are normally used for addressing within the IoT network [4, 5]. IoT nodes interact with the user and the environment through actuators and sensors. Modern IoT systems employ smart sensors and wearable sensing devices to collect data in various forms such as temperature, audio, image, and video [6]. The system allows heterogeneous devices to be connected within a communication infrastructure that includes various technologies such as Wi-Fi, Bluetooth, RFID, and Near Field Communication (NFC). The key component of the IoT system is computation that is performed at the processing elements of the nodes and servers. User interfacing applications, real-time operating system (RTOS), hardware drivers and firmware, and the cloud data processing programs are executed on the IoT computational platforms that may include one or many microprocessors, microcontrollers, field-programmable gate array (FPGA) units, graphics processing unit (GPU) boards, and ASIC accelerators. Unlike the IoT servers that are designed

Fig. 9.1 Data communication among the main IoT components

for high-performance data processing, the IoT nodes are optimized for computation at low power consumption in the presence of communication noise. To serve the user requests, semantic and service components are necessary to operate in tandem. The semantic component receives the requests and understands the details of requested services by the user (Fig. 9.1). The service component then receives the details and serves the requests accordingly while maintaining the service quality high [7].

9.1.2 Energy Efficiency as a Paramount Concern

During the past two decades, power has become the central design problem that limits the performance of computer systems. Data movement is identified as one of the most significant contributors to energy dissipation in all classes of microprocessors [8–10]. In particular, the energy cost of moving data across the memory hierarchy in next-generation microprocessors is expected to be orders of magnitude higher than the energy cost of performing a floating-point operation [11, 12]. For example, Fig. 9.2 illustrates the relative energy expended for reading data from DRAM and performing a double precision addition on a graphics processing unit (GPU) implemented at the 22 nm technology node. The energy required to fetch the two operands from DRAM is 50× greater than the energy required to move the operands from the edge of the GPU chip to its center, which itself is another 10× higher than the cost of the actual addition.

Novel hardware and software techniques are required to bridge this significant energy gap between data movement and computation in modern computer systems. This requirement will pose a more critical challenge in designing future computer systems, because the gap between the energy cost of data movement and computation is expected to widen in next technology generations [8, 9]. Thus, minimizing data movement is a first-order design constraint for future computer systems. Notice that near-data processing with the help of recent innovations in 3D die-stacking alleviates the amount of chip-to-chip communication significantly [13]. These solutions, however, become less effective for extremely large workloads that span across multiple chips and cannot improve the energy efficiency of data movement in the processors and memory packages. Nevertheless, it is unclear how current three-dimensional (3D) die-stacking solutions can amortize the implementation costs and

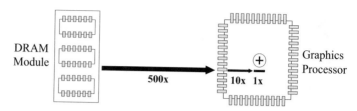

Fig. 9.2 The relative energy cost of computation and data movement between DRAM and GPU

fabrication complexities for the commodity computing systems while still relying on the conventional memory architectures.

9.1.3 Bandwidth Limitation for Big Data Processing

Historically, the microprocessor core counts and the DRAM capacity have doubled almost in every 2–3 years [14], which translates to a significantly higher rate than the one experienced for bandwidth improvements. As a result, most computing systems started to experience the *bandwidth wall* problem, where the off-chip bandwidth has become a bottleneck for both performance and throughput. The demand for memory bandwidth has been growing as the utilization of on-chip computational units increases through out-of-order execution, multithreading, task pipelining, and hardware specialization for data-intensive applications [15]. Moreover, to hide the memory access latency in microprocessors, various forms of data prefetching [16, 17] have been used that necessitate consuming more memory bandwidth. Conventional interfacing technologies and the limited number of input/output pins in the integrated circuit packages constrain the off-chip bandwidth significantly. Recent achievements in high-speed serial links [18] and through-silicon vias (TSVs) [19] can alleviate the issues by increasing the aggregate memory bandwidth at the cost of introducing a new set of reliability issues and fabrication complexity.

9.2 Recent Innovations for Bandwidth and Energy

This section provides a brief overview of a few state-of-the-art solutions for the energy-efficiency and bandwidth problems in the contemporary computing systems.

9.2.1 Heterogeneous Computing

Hardware accelerators have proven successful in achieving significant energy-efficiency and speedup over the general-purpose processors mainly due to better utilizing the computational and memory resources [20]. Along these lines, heterogeneous computers have been proposed that integrate more than one kind of processing core, each of which optimized for accelerating certain tasks. The main objectives in heterogeneous computing are (1) to enhance the application development through a seamless and flexible programing interface and (2) to improve the performance and energy-efficiency of the user applications by executing parts of the code on dedicated hardware. For example, consider a coprocessing architecture that comprises a central processing unit (CPU) to realize complex serial tasks, such as the sine function, and a graphics processing unit (GPU) that is specifically designed for accelerating massively data-parallel operations, such as pixel and vector processing. Other types of processing units, such as digital processing unit (DSP), field-programmable gate array (FPGA), and deep neural network (DNN) accelerators, are examples of popular technologies used for heterogeneous computing. A key challenge in designing heterogeneous computers is to strike a balance between the expected performance potentials and the cost of integrating disparate technologies. Typically, an ideal balance between cost and versatility may only be achieved if the heterogeneous cores require a minimal complexity and overhead for communication.

9.2.2   In-Package Die Stacking

One of the key solutions to the bandwidth and energy-efficiency problems is to reduce the high cost of data movement in computer systems. Minimizing the high cost of data movement in computing systems has been the main motivation for the recent innovations in 3D die stacking of silicon dice with disparate technologies within the same package. The 3D stacking technology has enabled energy-efficient solutions for near-data processing by integrating multiple dice of high-density memory layers and processor cores within the same package to amortize the high cost of off-chip data movement. For example, Micron's hybrid memory cube (HMC) stacks multiple DRAM layers on a flexible logic layer that communicates through energy-efficient and fast TSVs (Fig. 9.3) [21]. Intel integrates up to 16 GB of memory in a multichannel DRAM (MCDRAM) with four times higher bandwidth than DDR4 in Knights Landing processors [22]. As compared with off-chip memory systems, in-package integration provides up to ten times more bandwidth with a significantly lower power and smaller footprint, which make it an attractive solution for accelerating a variety of data-intensive applications from scientific and engineering domains [23–26, 116].

Fig. 9.3 An illustrative example of the 3D stack of logic and memory layers

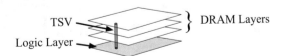

9.2.3 Emerging Memory Technologies

Power dissipation and the lack of technology scalability have become serious threats to the future of the conventional charge-based memory systems, such as SRAM and DRAM. Recently, resistive memory technologies have emerged as a promising alternative to the conventional memories. The emerging memories are nonvolatile, free of leakage power, and largely immune to radiation-induced transient faults. The resistive switching effect has been observed in a wide range of materials such as perovskite oxide (e.g., $SrZrO_3$, $LiNbO_3$, and $SrTiO_3$), binary metal oxide (e.g., NiO, CuO_2, TiO_2, and HfO_2), solid electrolytes (e.g., $AgGeS$ and $CuSiO$), and certain organic materials [27, 115]. Resistive RAM (RRAM) is one of the most promising resistive memory devices under commercial development that exhibits excellent scalability for less than 10 nm [28, 29], high-speed switching in the order of nanoseconds [30, 31], low power consumption in the order of pico-Joules [32], high endurance of performing trillions of writes [33], and high dynamic resistance range [34, 35].

The resistance of an RRAM element may be programmed to high or low using a sufficiently high voltage [33] or current [36] at runtime. A smaller voltage and current are used to read the current resistance state of the element. Numerous array topologies have been proposed in the literature to optimize the RRAM read and write operations within memory arrays [37, 38]. Figure 9.4 shows three example topologies for resistive memory cells using a single resistive element. The double bitline cell (a) is based on the one-transistor, one-resistor (1T-1R) topology that employs an access transistor controlled by a *wordline*. Once activated, the transistor establishes a current path through the resistive element between *bitline* and $\overline{bitline}$. The cell's content may be read or written by applying an appropriate voltage between *bitline* and $\overline{bitline}$. Similarly, double wordline cell (b) implements a 1T-1R topology, where the cell's content is read or written through *bitline* and $\overline{wordline}$. Unlike the 1T-1R memory cells, the crosspoint cell (c) does not require an access transistor. The cell's content is read and written through *bitline* and *wordline* [39]. The crosspoint structure achieves a better density than the 1T-1R memory cells; however, the absence of access transistor per cell creates a set of significant challenges in designing large memory arrays, such as half selected cells per access [40] and sneak current [41].

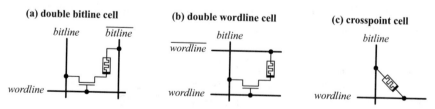

(a) double bitline cell   **(b) double wordline cell**   **(c) crosspoint cell**

Fig. 9.4 Illustrative examples of the double bitline (**a**), double wordline (**b**), and crosspoint (**c**) memory cell topologies

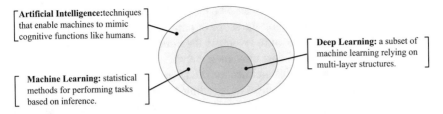

Fig. 9.5   Deep machine learning as a significant fraction of artificial intelligence

9.2.4   Machine Learning Accelerators in the IoT Era

The core application of most IoT devices is to detect different human behaviors, sense the ambient contexts, and produce the appropriate reaction. Machine learning has emerged as a key technique to enable these applications through extracting sensor data, identifying meaningful context, and performing intelligent tasks for face detection [42], image classification [43], and speech recognition [44]. As shown in Fig. 9.5, machine learning is a significant fraction of artificial intelligence (AI). Deep machine learning techniques, such as convolutional neural network (CNN), have emerged as the most successful class of machine learning that rely on multilayer neural networks for computation.

Due to the increasing demand of computation, memory, and energy consumption of machine learning applications, engineers and researchers have considered designing efficient ways to accelerate machine learning workloads. Software libraries have been proposed to accelerate deep learning tasks such as speech recognition (speech-to-text and speech-to-command) and computer vision (face detection and image classification) on low-power mobile GPUs [45, 46]. Recent research work on wear bench application shows that out-of-order processor cores may be adequate to achieve a high performance for deep learning workloads on wearable IoT devices [47]. To maximize the forward progress of IoT applications in unstable and intermittent power supply environment, nonvolatile processor architectures have recently been considered in the literature [48].

As an important class of deep learning, CNN has proven successful in image classification and face recognition [43, 49]. A typical deep convolutional neural network may require millions of parameters to be learned and stored during the training phase and to be retrieved and used for inference tasks. In addition to the stringent memory and storage requirements of the IoT nodes, accessing these parameters by various layers of the neural network necessitates consuming significant amounts of energy and time. Interestingly, high-precision parameters are not important to gain high accuracy in the outcome of a neural network; as a result, numerous techniques have been proposed in the literature that focus on trading the computation precision for achieving better energy efficiency and performance [50–54]. In addition, high-performance and energy-efficient hardware accelerators have been considered for computation and memory-intensive neural network workloads. For example, Dian-Nao [55] introduces a parallel multiply-and-accumulate (MAC) unit to exploit the scope of parallelism in CNN and deep neural network (DNN). This architecture leverages the concept of tiling and prefetching to reduce the long latency of data movement between main memory and the MAC units. In a newer version of this architecture, DaDian-Nao [15] extends the original design by alleviating the challenges of needing huge-memory bandwidth in the CNN and DNN workloads. Both architectures suffer from a limit performance for large-scale workloads with excessive memory bandwidth. Eyeriss [56] introduces a spatial architecture that maximizes data reuse through feeding inputs and weights to multiple processing elements with local storage and compute units. The design relies on a hierarchical memory organization that reduces the cost of data movement from main memory to the processing element. Similarly, ShiDian-Nao [57] employs a systolic array to maximize the reuse of input data and intermediate results in computing convolution.

Another important class of energy-efficient accelerators focuses on mapping the fundamental operations of the machine learning tasks onto analog functional units inside memory. As a result, these accelerators are able to gain significant speed and energy-efficiency over fully digital architectures. These accelerators that are often called analog neuromorphic accelerators rely on leveraging a connectionist model inspired by the human brain [58] to design the physical structure of memory and processing elements. Sheri et al. propose a spiking neural network based on memristive synapses to implement a single-step contrastive divergence algorithm for machine learning tasks [59]. Each synapse of the proposed design comprises two memristive elements representing limited-precision positive weights. Prezioso et al. report the fabrication of a memristive single-level perceptron system that takes ten inputs to produce three outputs. The circuit is used to classify a 3×3 black-and-white image [60]. The memristive Boltzmann Machine [61] and In-Situ Analog Arithmetic in Crossbars (ISAAC) [62] propose novel memory-centric accelerators that perform binary and multibit dot product operations within the emerging memory arrays. These accelerators propose to eliminate the need for data movement between memory and computational units. As a result, the memristive accelerators gain significant performance and energy efficiency. Similarly, PRIME [63] is proposed as a software/hardware platform for computing matrix-vector multiplications in resistive arrays for neural networks.

9.2.5　Approximate Computing

Ever since the power consumption was identified as one of the fundamental limitations for the microprocessors' performance, researchers have examined various techniques for making computer systems more energy-efficient. Important examples of such techniques have been considered in IoT systems that include low-power VLSI circuits for dynamic power and thermal management, dynamic voltage scaling, multiple-threshold voltage design, and energy harvesting. These techniques have become even more effective with the help of approximate computing that aims at balancing accuracy, area, delay, and power consumption based on the user's computational needs. Most IoT applications include multimedia processing that may largely tolerate computational errors without incurring a noticeable quality loss in output. Therefore, approximate computing is a natural fit for designing efficient IoT systems. Numerous circuits, architectural mechanisms, and design methodologies have been proposed in the literature that prove significant power and performance gains are attainable through applying approximate computing to different components of IoT systems [64–66]. Recent work on designing imprecise adders for IoT systems indicates that further performance gains and higher energy-efficiencies are attainable through incorporating design techniques that efficiently explore the design space of approximate units [67].

Designing the most efficient approximate circuit requires specializing both hardware and software for a given set of design objectives. For example, one can improve energy efficiency through hardware and software kernels that reduce the precision of computation [67–69] or reduce the power and energy consumption through lowering the supply voltage of the existing circuits [70]. The key to a successful approximate computation is to accurately identify which parts of the design are error-resilient. This may be done by software through kernel annotations and compiler techniques [71, 72] or a dedicated approximate data types [73]. Finally, a mechanism is often required to evaluate the quality of result and decide when to perform an approximation [74, 75].

Machine learning applications seemed largely amenable to approximate computing because of their massive stochastic computation load. Recent work [76] exploits the inherent redundancy of data and computation within deep CNN layers and applies a linear compression (singular value decomposition) technique on a pretrained model to speed up convolution operation during the inference tasks. Han et al. [77] exploit the sparsity of network parameters via pruning techniques to reduce the number of redundant weights. The parameters are represented in a compressed sparse row (CSR) format to increase the storage efficiency. The architecture is then extended to deep-compression [78] based on quantizing weights and applying the Huffman encoding to reduce the memory footprint significantly. Later, a hardware accelerator, called EIE [50], is designed for the compressed network that achieves substantial speedups and energy savings over prior work due to executing a set of sequential operations on the compressed data. It is proven that high precision weights are not important to achieving high accuracy in deep neural network. Gong et al. [79] propose to quantize the weights of the fully connected (FC) layer using vector quantization technique at the expense of 1% accuracy loss. Binarized neural network (BNN) [80] and XNOR-Net [81] extend the binarization further by binarizing both input and weights that gains significant reduction in memory footprint and execution time. Tang et al. [82] propose a resistive full-fledged BNN accelerator that employs binarized hardware for all the CNN layers. Moreover, Qiu et al. [83] propose an FPGA platform that accelerates the convolutional neural network for an embedded system.

Fig. 9.6 Relative system performance and energy consumption of various architectures

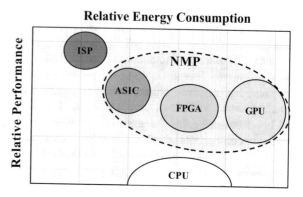

9.3 Near-Memory Processing

Most accelerators employ various techniques for in-memory processing that minimize data movement between memory and processor cores to provide significant energy savings and performance improvements. In-memory processing is an old concept that has been revisited recently by both industry and academia in the advent of big data computing and recent advances of technology—e.g., die stacking, emerging nonvolatile memories, and high-bandwidth memory interfaces. Based on the location of computation with respect to the memory cells, data-centric accelerators may be divided into near-memory processing (NMP) [84, 85] and in situ processing (ISP) [61, 62, 86]. Figure 9.6 illustrates the relative performance and energy consumption of various design approaches for hardware accelerators and general purpose processors based on the results from recent work on big data processing [84, 87], machine learning acceleration [43, 62, 88], and optimization problems [61, 89]. As shown in the figure, ISP and ASIC NMP provide significantly better performance and energy savings compared to other techniques. This superior energy-efficiency is achievable mainly because of exploiting the unprecedented parallelism at the level of memory arrays and cells while reducing data movement to a minimal amount through performing digital processing at the periphery of data arrays or analog computation within memory cells. Examples of these architectures are (1) computing bitwise Boolean functions, such as NOR, within DRAM arrays in DRISA [90], (2) utilizing the inherent dot-product capability of memory arrays to accelerate matrix-vector multiplication in ISAAC [62] and the memristive Boltzmann machine [61], and (3) performing associative search operations inside data arrays to realize TCAM-DIMM [91] and AC-DIMM [87]. In the rest of this section, we will explain the design of two ISP architectures based on memristive technology for big data processing in IoT systems. First, we introduce an energy-efficient memory system capable of accelerating binary neural network tasks in mobile IoT devices [92]. Then, we examine the architecture of an ISP system for large-scale data clustering for IoT servers and data centers [93].

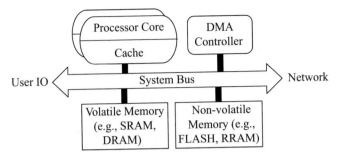

Fig. 9.7 Illustrative example of the system architecture of an IoT device

9.4 In Situ Processing for IoT Devices

IoT devices demand critical optimization for stringent ultra-low power requirements, which makes the realization of data-intensive applications, such as a full-fledged deep learning application, a significant challenge. Figure 9.7 shows an illustrative example of a generic system architecture for IoT devices and edge nodes. Depending on the application objectives and the design constraints, a single- or multicore processor is employed for executing the user programs. Typically, the memory system consists of both volatile and nonvolatile subsystems that are interfaced to the processor cores via a system bus. Single- or multiple-cache levels as well as a scratchpad memory may be used as fast and temporary memory. Nonvolatile memory is commonly used to store application programs and data permanently. Nowadays, FLASH is a widely used technology for building nonvolatile memory in wearable and mobile devices [94]. The emerging nonvolatile memory technologies such as FeRAM [95] and RRAM [96] are expected to replace the conventional FLASH memories in future [97, 98].

9.4.1 Deep Binary Neural Network

Given the strict power and performance requirements of IoT devices, binarized CNN seems a promising model of deep neural networks in edge computing. In a binary deep neural network, both the inputs and weights are binarized. For example, XNOR-Net is a binary neural network that consists of four main stages, namely *Batch Normalizations*, *Binary Activation*, *XNOR Convolution*, and *Pooling*. XNOR-Net was initially based on converting the parameters into either +1 or −1 using a sign function defined as the following:

$$x^b = \begin{cases} +1 & x \geq 0 \\ -1 & x < 0 \end{cases}$$

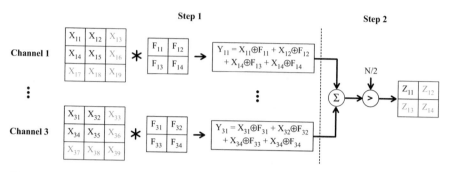

Fig. 9.8 Illustrative example of the two-step XNOR convolution

In the equation, x is a real-valued weight or activation data for which the binarized output (x^b) is computed. On difficulty of such bipolar quantization is the complexity of computation due to considering opposite signs. Instead, XNOR-Net maps all -1s to 0 to directly use logical operations for computation in the binary format. The binary values are then scaled to approximate the real-valued weights and to improve the accuracy of results. For example, a real-valued filter (**W**) is approximated by a**B**, where a is the scaling factor and **B** is an instance binary filter chosen from $\{+1, 0\}^{x \times w \times h}$. A consolidated scaling factor matrix (**K**) is then generated for all the input neurons in the binary activation layer. Notice that **K** is for approximating the convolution between the input (**I**) and weight (**W**) values.

$$\mathbf{I} * \mathbf{W} \approx (\text{sign}(\mathbf{I}) \circledast sing(\mathbf{W})) \odot a\mathbf{K}$$

In the above equation, $*$ indicates the real-valued convolution, $\odot$ represents the Hadamard product of two binary matrices, and $\circledast$ is the binary convolution based on the bitwise XNOR and addition (bit-count). Notice that the outcome of every bit-count operation can be a multibit value, which is passed through a threshold comparison function to ensure producing binary output for convolution. The resultant binary matrix is then multiplied by $a\mathbf{K}$ that computes a real-valued matrix to produce the output neurons of a binary convolution layer.

Figure 9.8 shows the two steps of XNOR convolution applied to an example input data ($X^{c \times h \times w} = 3 \times 3 \times 3$). X is first convolved with the filter $F^{3 \times 2 \times 2}$ using element-wise XNOR operations followed by summation to compute the intermediate output $Y^{1 \times 2 \times 2}$. Next, the intermediate values are summed, and the result is compared with $N/2$ to produce a single element of the output matrix. Notice that N is the total number of elements in the filter used for each layer.

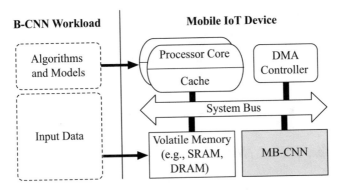

Fig. 9.9 Overview of an IoT system architecture using the MB-CNN memory-centric accelerator

9.4.2   The MB-CNN Architecture

Designing memory-centric accelerators for deep learning workloads in mobile IoT devices is challenging because of the increasing demand for more computational capabilities by the emerging applications. On the other hand, the hardware of IoT devices is significantly constrained by the stringent cost and power requirements. The memristive binary convolutional neural network (MB-CNN) is a memory-centric accelerator that addresses this challenge by enabling in situ binary convolution within the resistive crosspoint memory arrays. The key idea is to exploit the computational capabilities of the resistive crosspoints for performing the key operations of the XNOR-Net—i.e., binary XNOR and bit-count. MB-CNN may be used as a nonvolatile memory system that serves ordinary read and write requests. Furthermore, the structure of memristive arrays with an additional control logic allows the framework to perform XNOR convolution at low energy and performance costs. Figure 9.9 shows how MB-CNN may be employed in a mobile IoT system. The computational platform includes a microprocessor that executes the application programs on binary CNN models. A volatile memory module is employed to store the input data prior to execution. The system is complemented with an MB-CNN module that accelerates the XNOR convolution and stores the network parameters. The MB-CNN module is connected to the IoT system via an LPDDR3 standard memory bus [99]. All the network parameters, such as edge weights, are stored to the RRAM crosspoint arrays. These parameters are then read to complete multiple inference tasks. Prior to an XNOR convolution task, a direct memory access (DMA) controller is used to transfer data from DRAM to the MB-CNN module. For each layer of the neural network, a set of XNOR convolutions is computed and the intermediate results are reused for the next layer. At the end of this process, the software program is responsible to collect the final results from the MB-CNN module. All the communication between the processor cores and the MB-CNN module is carried out through LPDDR3 commands.

w	b	out
0	0	1
0	1	0
1	0	0
1	1	1

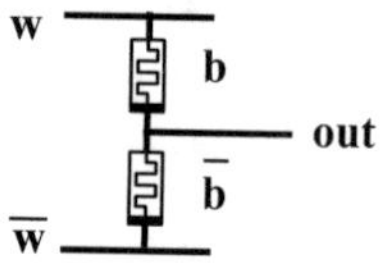

$$b = \begin{cases} 0 & \text{High Resistance} \\ 1 & \text{Low Resistance} \end{cases}$$

$$w = \begin{cases} 0 & \text{Gnd} \\ 1 & \text{Vdd} \end{cases}$$

Fig. 9.10 Performing XNOR operation using the 2R crosspoint cell

9.4.3 Memristive XNOR Convolution

A set of binary XNOR operations between the filter elements and the input channels followed by bit-counting and binary approximation are necessary to compute an XNOR convolution.

9.4.3.1 Computing XNOR Within RRAM Crosspoint

The MB-CNN accelerator exploits the computational capabilities of a novel two-resistor (2R) crosspoint that performs in situ XNOR. The basic operation relies on a differential bit representation of weights and inputs. Figure 9.10 shows how the differential form of a bit stored in the 2R cell helps performing an in situ XNOR operation. The true and complement values of each filter element are stored in a cell, denoted by b and $\bar{b}$. A logical 1 is represented by the low-resistance state (LRS) and 0 by the high-resistance state (HRS), the memristive element. Notice that the filter weights are computed during the training phase of the neural network; therefore, they remain constant for inference tasks. Similarly, an input element has to be represented in the differential format. Each input bit is applied to the 2R cell via two wires, denoted by w and $\bar{w}$. The 2R cell implements a simple resistive network that develops an output voltage (*out*), which is a binary value representing the logical XNOR between w and b.

In a crosspoint array, all of the memory cells within each column share a single bitline. Figure 9.11a shows multiple 2R memory cells connected to a shared bitline. Due to the differential representation of the values, each bitline is capable of performing a bit-count operation over all of the memory cells. Moreover, each 2R cell can compute the XNOR result of the binary values w and b. As the outputs of all memory cells are connected to the shared bitline, the final voltage of the bitline (*sum*) is an analog signal representing the sum of all partial results produced by the memory cells. To make sense of the bitline voltage, Fig. 9.11b illustrates an equivalent circuit to the bitline topology using four resistors. Notice that each memory cell has two memristive elements that are set to HRS or LRS in a complementary form. Similarly, the pairs of wordlines are Vdd and Gnd in the complementary format. The combination of the resistive states and the wordline voltages results in four possibilities that are considered in the equivalent circuit via

(a) Multiple cells sharing a bitline

(b) Equivalent circuit

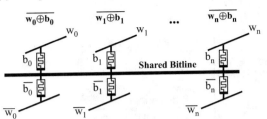

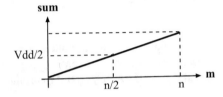

Fig. 9.11 Analog bit-count computation within the 2R crosspoint array. (**a**) Multiple cells sharing a bitline. (**b**) Equivalent circuit

Fig. 9.12 The bitline voltage versus the 1s in XNOR outputs

four resistors. For a bitline connected to n 2R cells, assume that m XNOR operations produce 1s in their outputs. (Notice that the binary outcome of an XNOR is high (1) only if Vdd is connected to the RRAM cell with low resistance; otherwise, a low voltage (0) appears at the output.) The resultant bitline voltage can then be computed through $sum = Vdd\frac{mH+(n-m)L}{n(H+L)}$, where H and L are the amount of high and low resistances of the RRAM cells. The bitline voltage is linearly proportional to the number of 1s produced by the XNOR operations—i.e., the bit-count.

By quantizing the bitline voltage, the final binary value for a single convolution can be computed. Figure 9.12 shows a linear relationship between the bitline voltage and the number of cells that produce a 1 at the output. MB-CNN employs a comparator that compares the sum against Vdd/2, thereby quantizing the final output to 1 if $sum \geq \frac{Vdd}{2}$ and 0 otherwise.

9.4.3.2 In Situ Bit-Counting

MB-CNN employs an in situ bit-counting technique that requires having all the operands connected to a single bitline during the XNOR convolution. Therefore, large crosspoint arrays may be required to compute the bit-count of modern deep learning workloads. Regrettably, such monolithic data arrays are largely impractical due to significant limitations in sensing mechanisms, excessive power dissipation, unacceptable latency of operations, and serious reliability problems. To address all these challenges, MB-CNN employs a novel hierarchical mechanism for computing partial bit-counts across multiple arrays. The partial counts are then aggregated into a single value and then quantized into a single bit. The hierarchical mechanism

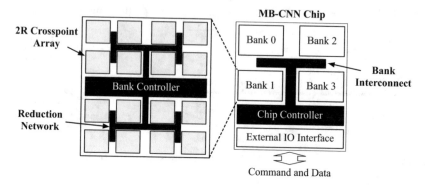

Fig. 9.13 Hierarchical organization of the MB-CNN architecture

needs to convert a bitline voltage (*sum*) into a multibit digital value. Unlike the conventional single-level sensing, the proposed mechanism employs an analog-to-digital converter (ADC) circuit to produce each partial bit-count.

9.4.4 The MB-CNN Architecture

The design of MB-CNN is based on the existing memory system architectures. Figure 9.13 shows the hierarchical organization of an MB-CNN chip that comprises an external IO interface and a memory core with a chip controller and multiple banks. The chip controller orchestrates all of the data movements between the IO interface and memory banks. MB-CNN banks operate independently and can serve a memory request or perform an XNOR convolution. For large problems that exceed the size of a single bank, multiple banks may be involved for an XNOR convolution. MB-CNN perform inference tasks only, while training is carried out once in the cloud to produce the network parameters for deployment in IoT devices.

9.4.4.1 MB-CNN Chip Control

Once the network parameters are available, an MB-CNN chip can be configured according to the number and size of the convolutional layers. A single bank may be used to store the parameters of one or multiple small layers, while a large layer may occupy more than one bank. The chip controller includes local nonvolatile RRAM arrays for tracking the banks that maintain the parameters of each layer. A typical B-CNN model includes multiple binary convolutional layers, each of which needs the software to make a call to the accelerator. First, the chip controller receives an initiation command to specify which layers are used next for computing the XNOR convolution. Then, the relevant banks will be configured accordingly such that the input data for a selected layer are streamed into the accelerator. The chip controller distributes the data stream among relevant banks internally. Local buffers are used to collect the convolutional results at the MB-CNN banks. At the end of every computation, the accelerator notified the software to read the results from the MB-CNN chip and to proceed with the subsequent layers operations.

9.4.4.2   Bank Organization

Each MB-CNN bank consists of a bank controller, a reduction network with an H-tree topology, and a set of data arrays. The bank controller is responsible for managing computed partial bit-counts within a local on-die buffer. Moreover, the controller is in charge of configuring the full adders and comparators of the reduction tree for computing the final sum. During each MB-CNN convolution, partial bit-counts are computed using the memory arrays. The counts are then merged into a single bit-count while being transferred through the reduction tree toward the bank controller. Figure 9.14 shows how four partial bit-counts (i.e., b_{0-3}) are merged in the reduction tree into a multibit digital value (b). Three nodes of the reduction tree are involved in serial addition. Each node employs a serial adder to add two single-bit operands and store the carry bit locally. The serial addition allows for low-cost and energy-efficient computation in the reduction tree. At the bank controller, a serial comparator is used to compute the difference between the final bit-count (b) and the quantization threshold ($n/2$). The values are represented in two's complement; therefore, a serial adder/subtractor may be used to compute the difference. The last bit to be computed by the serial comparator represents the sign of the result, which indicates whether the result is negative ($sum < n/2$) or positive ($sum \geq n/2$). The inverted version of this bit represents the binary result.

Each serial adder at tree nodes is reconfigurable using two flip-flops C_0 and C_1, each of which is used to mask a branch of the tree. Notice that the value of C_0 and C_1 determines whether the node performs a serial addition or only copies the value of one branch to upstream root. Valid combinations of the C_0 and C_1 flip-flops are (1, 1) for serial addition, (1, 0) for transferring the upper branch, and (0, 1) for transferring data from the lower branch. The carry bits of the serial comparator and

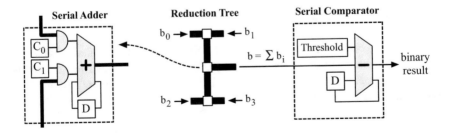

Fig. 9.14 Merging partial bit-counts in the MB-CNN reduction tree

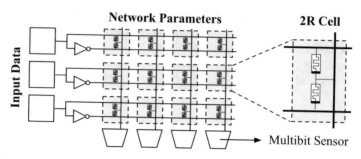

Fig. 9.15 In situ XNOR operations for the MB-CNN convolution

adders (D) are set to 0 after computing each XNOR convolution. All of the C_0 and C_1 flip-flops are configured by the chip controller based on the size of parameters for the convolutional layers.

9.4.4.3 Array Structure

Each MB-CNN memory array is implemented using an RRAM crosspoint comprising M rows and N columns. Figure 9.15 shows an example 3×4 MB-CNN data array. The RRAM cells are programmed to represent the binary network parameters (i.e., filter weights). To enable in situ XNOR convolution within the crosspoint arrays, a set of latches are provided at the periphery of the array to store the input data, which are applied to the array through horizontal wordlines. Each input bit requires two wires for its true and complement values. Along the lines of prior proposals on using multibit sensing mechanisms for analog computation [61, 62, 72, 100], a cost-efficient multibit sensor circuit is employed for quantizing the bitline voltage (*sum*). The proposed sensing circuit comprises a differential amplifier, a sample and hold unit [101], and a digital-to-analog converter [102]. Due to the exponential increase in the complexity of the sensor circuit with the number of output bits, its precision is limited to 5 bits only. As a result, each array can compute the sum of 32 partial XNOR convolutions.

9.4.4.4 Data Organization

One of the key challenges in performing an efficient MB-CNN convolution is how to lay out data within and across memory arrays. This section describes how the inputs and network parameters are mapped onto the MB-CNN accelerator. Figure 9.16 shows an illustrative example of the data organization MB-CNN. The size of feature map the ith convolution layer is $I^{c \times h \times w} = 32 \times 7 \times 7$, where c denotes the input depth and h and w are the height and width, respectively. The input is convolved with a kernel ($K_0^{c \times h \times w} = 32 \times 2 \times 2$). Assuming that there are 128 such kernels,

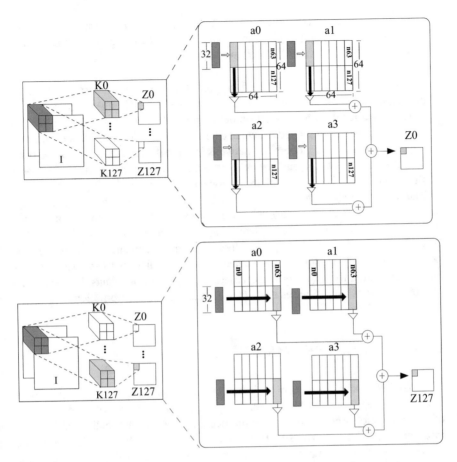

Fig. 9.16 Distribution of a convolution layer among four MB-CNN crosspoint arrays

the convolution layer will need to produce 128 different output feature maps of size $O^{h \times w} = 6 \times 6$. If we consider an array size of 64×64 for the memristive crosspoint arrays, then we need four such crosspoint arrays to map the entire convolution layer with a bank. The memristive arrays are denoted by a_0, a_1, a_2, and a_3 in the figure. Kernel K_0 is distributed among the first column of all the arrays with a maximum of 32 elements per array. Similarly, all the other kernels are mapped into the entire bank. Kernel K_n is distributed among the jth columns of all four arrays, where j is reminder of n by 64 and n is the position of the bitline in the memristive array. In this example, K_0 has a total of 128 ($32 \times 2 \times 2$) elements $\{w0_0, w0_1...w0_{127}\} \in K_0$, where $w0_0...w0_{31}$ are mapped to n_0 of a_0. Similarly, $w0_{32}...w0_{63}$ are mapped to n_0 of a_1 and so on. K_{64} is mapped to the lower half of n_0 in all arrays where $\{w64_0, w64_1...w64_{127}\} \in K_{64}$. In a similar fashion, $K_1... K_{127}$ are mapped to $n_1...n_{63}$ of $a_0...a_3$. Notice that the chip controller feeds 128 elements of the input I to the bank to be convolved with the kernels. The chip controller initiates streaming data to the crosspoint per XNOR convolution. The inputs are then distributed among the four arrays. Due to using a 5-bit sensor, only 32 rows (cell segment) of the arrays are driven by the input data. All the other rows remain inactive and do not contribute to the in situ computation.

As mentioned in Sect. 4.4.2, the 5-bit partial results from the arrays are given to the serial adders of the reduction tree to compute the final outputs $\{z_0, z_1 ... z_{35}\}$, where zn is the result of a convolution between input I and kernel Kn. An internal control mechanism is employed to switch a new set of 32 cells into each sensor every five cycles. The switching operation happens in a pipelined fashion inside the bank to produce one output element (zn) every seven cycles. Ultimately the first element of all output feature becomes available at local output buffer after 7×128 cycles.

The chip controller feeds the next set of inputs to the memristive arrays for convolution once the current subset is reused by all convolution filters. Therefore, producing the complete output of a convolution layer takes $7 \times 128 \times 6 \times 6 = 32{,}256$ cycles. Notice that more concurrent computation is possible through increasing the number of sense amplifiers per memristive arrays and by replicating the same kernel parameters across multiple banks. However, that improved performance requires more chip area and power consumption.

9.4.5   Potentials of the MB-CNN Accelerator

The MB-CNN accelerator can be integrated in mobile systems with single- or multicore processors. This section examines the energy and performance potentials of the accelerator used by single- and multicore processors that realize the MIPS64 instruction set architecture (ISA). For better evaluations, a GPU-based solution and an ASIC accelerator for processing-in-memory (PIM) are considered as the baseline systems for comparisons. The GPU solution is based on the Nvidia Tegra X1 low-power system with 256 processing cores [53]. The low power GPU is mainly used to implement the floating-point convolutions in the first and last layers of an end-to-end inference task. The PIM ASIC solution integrates additional gates near the RRAM arrays to implement the XNOR trees and bit-counters that fetch data from the arrays and compute the XNOR convolution. Like the MB-CNN hardware, the outcome of each XNOR convolution is transferred to software for completing the layer computation. Moreover, the PIM baseline is optimized so that it occupies the same area as that of MB-CNN; however, it does not support in situ XNOR computation.

Figure 9.17 shows the relative execution time and system energy of the XNOR-Net inference across various system configurations, namely CPU, GPU, PIM, and MB-CNN. For each configuration, three design points with respect to the number of processor cores—i.e., single (S), dual (D), and quad (Q)—are considered. MB-CNN outperforms all of the baseline systems. As compared with CPU, MB-CNN achieves $4.17\times$, $4.25\times$, and $3.71\times$ performance improvements for the single-, dual-, and quad-core systems, respectively. Although the PIM accelerator achieve

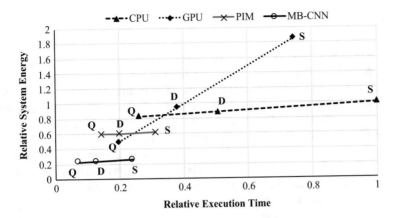

Fig. 9.17 Relative execution time and system energy of MB-CNN compared to the baseline architectures with single (S)-, dual (D)-, and quad (Q)-core processors

a better performance than the GPU-based systems, MB-CNN outperforms the PIM-like ASICs by 1.29×, 1.58×, and 2× in the single-, dual-, and quad-core systems, respectively. Notice that MB-CNN benefits from in situ computation within memory arrays that enables massive parallelism and eliminates unnecessary data movement.

Moreover, MB-CNN achieves average energy savings of 4×, 3.83×, and 3.64× over the CPU baselines with single-, dual-, and quad-core processors. The GPU-based systems consume more chip area and power to improve performance. However, they can only achieve a better energy saving that the PIM accelerator with a quad-core processor. This is mainly because of the reduced leakage energy in the GPU as the overall execution time on the processor decreased. MB-CNN achieves better energy savings over the PIM- and GPU-based accelerators by respective 2.46×, 2.61×, and 2.63× for the single-, dual-, and quad-core systems, respectively.

9.5 In Situ Data Clustering for IoT Servers

This section presents the memristive in-situ clustering (MISC) architecture as another example of in situ accelerators. The MISC architecture is designed to perform energy-efficient and fast data clustering within memristive arrays. The memory arrays are specifically re-structured to support the basic operations of MISC. Moreover, algorithmic techniques and special design strategies are considered to enable large-scale data clustering on MISC.

9.5.1 Data Clustering

Data clustering refers to partitioning a set of objects into meaningful groups (a.k.a. clusters) without using predefined labels [103]. Data clustering tasks are computationally difficult (NP-hard) problems. The entities of each cluster are more similar to each other than to those in other clusters. The class of k-means algorithms are the most prominent clustering techniques that have been successfully employed in numerous fields of science and engineering [104]. Algorithm 9.1 shows the basic steps of k-means clustering, where k centroids are defined to represent the clusters. A centroid is either a representative member of the cluster, such as the median of the cluster, or an additional data point computed based on the similarities among all of the cluster members (e.g., the arithmetic mean). The former has been proven to find better clusters than the latter due to its resistance against outlier members [103, 104]. Prior to partitioning the data, the k centroids are randomly selected for the clusters. The clustering task is carried out through two algorithmic steps (lines 3 and 4 in Algorithm 9.1) that are repeated after the initial step. Firstly, the clusters are formed by assigning data points to their closest centroids. Secondly, new centroids are computed for all of the clusters. These two steps are repeated for a constant number of iterations defined by the application or until convergence is reached and none of the members switches their clusters during the first step.

Algorithm 9.1 Basic k-Means Clustering

1: select k initial centroids randomly
2: **repeat**
3:    from k clusters: assign data points to their closest centroids
4:    recompute the centroid of each new cluster
5: **until** convergence is reached

9.5.2 Applications of Data Clustering

We can find numerous applications of k-means clustering in the literature. This section reviews only two representative examples for gene expression analysis (GEA) and text data mining.

9.5.2.1 Gene Expression Analysis

Recently, clustering has seen wide use in the field of medical research, such as cancer diagnosis and drug discovery. An accurate clustering algorithm can significantly improve the correctness of these applications. Lu and Han [105] have shown that data clustering techniques may be employed to classify cancerous cells

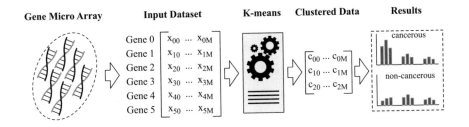

Fig. 9.18 Clustering gene samples to detect cancerous cells

based on the abundance of gene expression data. Interestingly, certain clustering techniques may achieve a higher accuracy rather than the traditional morphological- and clinical-based methods. A gene is defined as part of a deoxyribonucleic acid (DNA) that represents the basic unit of heredity transferred from parent to an offspring. Gene expression is the process of transcribing a gene's DNA sequence into ribonucleic acid. This process changes during biological phenomena, such as cell development. For example, in the case of diseases such as cancer, the genes of normal body cells undergo multiple mutations to evolve cancerous cells. Figure 9.18 shows how this anomaly is now possible to be detected through gene expression analysis (GEA). Genes are first sampled to create an input dataset. Thousands of gene samples are often required to achieve an acceptable output accuracy. The dataset is then processed by a clustering engine (e.g., k-means) to generate the clustered data. Finally, the clustered data are examined to produce the final results.

9.5.2.2 Document Clustering

An important branch of text mining is based on clustering text documents to organize paragraphs, sentences, and terms into meaningful clusters. This process improves information retrieval, document browsing, and data analytics [106]. Figure 9.19 shows an example flow of clustering text documents. First, the text corpus is converted into numerical vectors through a data-preprocessing mechanism. The vectors represent the features of the corpus and are used to group similar terms into the same clusters. Normalized term frequency (TF) is a commonly used feature vector for text clustering. The TF vector represents the number of word occurrences in every document divided by the total number of words. For each word, an inverse document frequency (IDF) is defined as the logarithmic ratio of the total number of documents to those documents that contain the word. The two metrics are then multiplied to compute a TF-IDF score matrix. Finally, the documents are partitioned into multiple groups with similar members using the clustering algorithm (i.e., k-means).

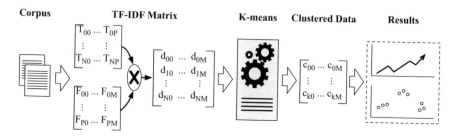

Fig. 9.19 TF-IDF text mining with k-means algorithm

9.5.3 Data Clustering with Rank-Order Filters

MISC proposes to leverage rank-order filtering to enable energy-efficient data clustering within memory arrays. Rank-order filters are nonlinear digital components widely used in signal and image processing. They are mainly used to filter out noise from input signals. A general rank-order filter may be characterized by the number of input signals (N) and an index (i) that determines which input signal must appear at the output. The filter needs to send the ith largest (or smallest) input signal to the output. A *median filter* may be viewed as a particular case of the rank-order filter where i is set to $N/2$. The median filter may be used to compute the cluster centroids for the k-medians algorithm. However, rank-order median filters are memory and compute intensive operations. Numerous hardware and software optimizations have been proposed for median filters in the literature, which pursue two different approaches. The first approach is word-based search that sequentially examines all of the objects to find the median. The second approach is based on a bit-serial process to computes the majority of selected bits from all of the objects in parallel. When applied to large-scale datasets, both approaches suffer from excessive memory traffic and high data movement costs.

9.5.3.1 Bit-Serial Median Filter

MISC relies on the bit-serial median filter for clustering. In principle, one can find the median of a list by sorting the data points. This technique, however, is complex and inefficient. In 1981, Danielsson [117] proposed the first bit-serial algorithm for median filtering that eliminates the need for sorting. Thereafter, numerous hardware and software implementations of the bit-serial median filter have been examined that rely on the *majority function*. Notice that the majority function defines a mapping from N binary data to a single binary output. The output is 0 if $N/2$ or more inputs are 0; otherwise, it is set to 1.

Figure 9.20 shows four major steps of the bit-serial algorithm to find the median of five numbers. Initially, all numbers are represented in their binary forms (1). Starting from the most significant bit (MSB) to the least significant bit (LSB), the

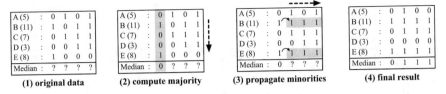

Fig. 9.20 Computing the median of five numbers using the bit-serial algorithm

algorithm computes majority (2) and propagate minorities (3). The majority vote computation is a vertical process that results in a single bit computed for the selected column. The minority propagation is a horizontal process that depends on the result of the majority function from the previous step. During the horizontal process, the minority bits of the selected column are identified and are used to replace all of the bits on their right-hand side. Repeating steps (2) and (3) for all bit positions results in computing the median of all five numbers (4).

9.5.4 Memristive k-Median Clustering

The key idea of MISC is to exploit the computational capabilities of the memristive arrays to perform the necessary computation for bit-serial median filtering in the memory cells. Therefore, MISC can reduce the latency, bandwidth, and energy overheads associated with streaming data out of the memory arrays during the clustering process. By eliminating the need for transferring data to/from memory arrays, MISC unlocks the unexploited massive parallelism in bit-serial median algorithm for data clustering.

9.5.4.1 The MISC Architecture

The MISC accelerator is designed as a memory module that consists of multiple chips. Figure 9.21 shows the hierarchical organization of MISC with respect to the CPU and main memory. Every MISC chip comprises a hierarchy of data arrays interconnected with a reconfigurable reduction tree. The memory cells are capable of storing data bits and computing the basic operations required for bit-serial median filters. The on-chip interconnection network allows for retrieving or merging partial results from the data arrays. The MISC module is connected to the processor via a standard double-data rate memory interface [107]. This modular organization of the proposed accelerator allows the user to selectively integrate MISC in those computer systems that execute data clustering workloads. The MISC memory architecture supports two operational modes: the storage mode to serve ordinary read and write requests and the compute mode that is for in situ data clustering. For a given

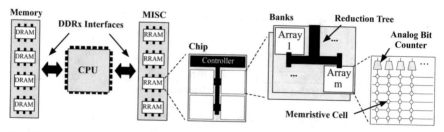

Fig. 9.21 Illustrative example of a multicore processor interfaced with the proposed memristive data clustering accelerator

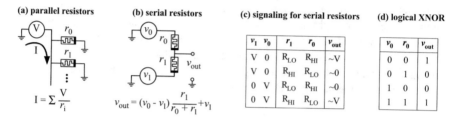

Fig. 9.22 Illustrative example of computing the median of five input numbers. (**a**) Parallel resistors. (**b**) Serial resistors. (**c**) Signaling for serial resistors. (**d**) Logical XNOR

computation task, three steps are followed by hardware and software. First, the MISC module is configured by software for solving the clustering problem. Next, the in situ computation will be initiated after transferring the input data from the main memory to the accelerator chips. Finally, the MISC controller notifies the CPU to collect the results.

9.5.4.2 The Design Principles for MISC

MISC requires three major operations to fully implement the bit-serial median filter within memory arrays. The operations are (1) computing the majority of bits within a selected column, (2) determining which rows hold the minority bit, and (3) replacing the LSBs of those rows with the minority bit. MISC realizes these operations using two basic topologies for memristor elements. As shown in Fig. 9.22, the serial and parallel topologies of the resistive elements are used to perform binary XNOR and to compute the majority vote of multiple bits.

Computing the Majority Vote The majority function is computed through parallel memristive cells connected to a single bitline. Assuming that each memory cell employs its high- and low-resistance states to represent **1** and **0**, respectively, the number of **1**s determines the amount of current (I) flowing through the bitline. One can determine the number of 1s by measuring this current and comparing it with a threshold. If the number of **1**s is greater than the half of bits, the output is set to 1;

otherwise, the output is 0. This functionality matches the definition of the majority vote, which is leveraged to implement the vertical computation step in the bit-serial median filtering algorithm.

Performing In Situ Bit Comparison Figure 9.22(b) shows an illustrative example circuit for performing in situ bit comparison using two serially connected memristive elements. Depending upon the input voltage (V) and the states of memristive elements (r_0 and r_1), the output voltage (v_{out}) varies between V and 0. According to parts (c) and (d) of Fig. 9.22, v_{out} represents the results of a binary XOR/XNOR on v and r. This novel functionality is used in the bit-serial median filter for finding the minority bits, as well as in the k-medians algorithm for searching and selecting all the members of a dynamically formed cluster prior to finding new centroids.

9.5.5 MISC Building Blocks

The MISC accelerator is designed based on three fundamental building blocks: a memory cell, a majority unit, and a network reduction unit. The building blocks are designed and optimized to achieve high memory density, low-energy consumption, and capacity for massive parallel computation at the memory cells.

9.5.5.1 Memory Cell

An example physical layout for the MISC memory cell is shown in Fig. 9.23. The cell is capable of (1) serving ordinary reads and writes, (2) performing in situ XNOR between the cell contents and an external input, and (3) propagating the minority bit to its adjacent cell on right. The cell comprises four transistors and four memristive elements that can be viewed as a combination of four conventional 1T-1R RRAM cells. Three wordlines and four bitlines are employed to perform read, write, and compute operations on the cell. Each memristive element of the MISC cell can be read or written through a set of three bitlines and wordlines. Also, it is possible to use the cells as four individual 1T-1R memory cells to store data. For example, R in the data bit is accessed using $\{I, C, \overline{C}\}$; R in the XNOR part is accessed through

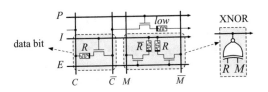

	Description
R	data bit value (high or low)
P	pass bit value to the next cell
E	enable XNOR computation
I	include bit value in computation
M	majority bitten
C	compute bitline

Fig. 9.23 Illustrative example of the proposed memory cell

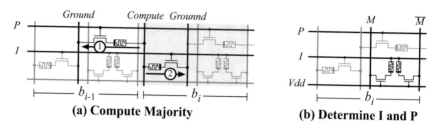

Fig. 9.24 The main operations of adjacent MISC cells. (**a**) Compute majority. (**b**) Determine I and P

$\{E, \overline{M}, I\}$; $\{E, M, I\}$ can be employed to access $\overline{R}$ in XNOR; and bitlines from adjacent cells are used to access low through $\{I, C, \overline{C}\}$.

Computing the Median Inside MISC Array Computing bit-serial median requires multiple steps, each of which involves selecting cells through bitlines and wordlines. Figure 9.24 shows the main operations of adjacent MISC cells in a row. On every iteration, only one cell of each memory row will be processed. First, P and I are initialized with nonzero binary values to determine if the cell should be included in computation. This is necessary to ensure that irrelevant data points are not included in median computation. To compute the majority vote of a bit position (b_i), bitlines C from columns b_i and b_{i-1} are connected to ground and bitline C of the bit position b_i is connected to Vdd. As shown in Fig. 9.24a, two current paths are possible between Vdd and ground. One path is established through column b_i based on the content of data bit (1). Another possible path includes the memristive element (low) of the neighboring column (b_{i-1}) that determines the amount of current pulled from the computed bitline in case of bit propagation (2). These two paths are controlled by P and I such that only one (or none) of the columns per every row contributes to the bitline current. Therefore, only three combinations of the P and I values (i.e., 00, 01, and 10) are used for computation. However, the P and I values may change during the computation of a row. First, none of the cells are selected if $P I = 00$; second, column b_i is selected to be included in the majority computation if $P I = 01$; and third, column b_{i-1} contributes to the majority vote computation if $P I = 10$. At every step, the majority is computed by measuring the total current driven through the compute bitline (C) and comparing it with a threshold.

One difficulty for the in situ bit-serial median computation is propagating the minority bit within each row. This operation may result in forming long chains of MISC cells per row, which impacts the area, delay, and power dissipation significantly. The MISC architecture, however, avoids forming long chains by allowing only 1s to be propagated from b_{i-1} to b_i. Firstly, 1 and 0 are represented with the low- and high-resistance states, respectively. Secondly, because of the significant difference between the high- and low-resistance states in RRAM [108], the currents contributed to the bitline by those memristive cells in high-resistance states can be omitted. This optimization is considered by dedicating a low memristive element in b_{i-1}, which is included in the current summation only if $P = 1$. After completing the majority vote at the current bit position (b_i), P and I are recomputed for the next bit position (b_{i+1}). Figure 9.24b shows how P and I are recomputed using b_i. The true and complement values of the newly computed majority vote are applied to M and $\overline{M}$. Then, E is connected to Vdd to enable the XNOR part of the cell. The result of XNORing M and R is produced on the wordline I. The wordline is connected to a control circuit that detects a 1-to-0 transition and locks it to 0 till the end of computation. Moreover, the control circuit sets P to **1** only if M is equal to **1**.

Updating the Cell Recall that the MISC cell stores the true and complement values of the data bits; therefore, updating the contents of every cell requires additional writes. MISC employs a two-phase update mechanism that writes all **1**s in the first phase and then all of the **0**s. This process does not incur significant overhead in a data clustering problem because the dataset is written in the memory once and is read by the algorithm multiple times. Moreover, the performance and energy benefits of in situ computing surpass this overhead significantly.

9.5.5.2 Analog Bit Counter and Reduction Network

Theoretically, solving a large-scale data clustering problem with MISC needs computing the majority vote of a large number of data points stored in a single memory array. Building large MISC arrays is impractical due to significant sensing and reliability issues. Instead, MISC stores data points in multiple limited sized arrays, and only a fraction of the cells within each column is processed using the analog bit counters. The multibit sensors are similar to those used in the MB-CNN arrays. Multiple MISC array computations are performed in parallel to gain significant performance. Again, a hierarchical merging mechanism is proposed to compute the majority vote of many data points stored in multiple MISC arrays. An interconnection network comprising reduction units merges the partial bit-counts computed per arrays into a single majority bit. The main purpose of the reduction tree is to merge the partial counts computed by the analog bit counters.

A reconfigurable reduction tree is used inside each bank to interconnect the data arrays and the chip controller. The tree is capable of selectively merging the partial counts from the data arrays into a single count value. Figure 9.25 shows the MISC reduction unit with nine possible ways of reading data from the children arrays A, B, C, and D. Each MISC reduction unit is configured using a 2-bit mode register (m). By sharing the modes values among the reduction units of each layer, nine useful configurations are possible for reducing the partial results in MISC. Similar to MB-CNN, the nodes are programmed to appropriate operational modes prior to a computation task. MISC makes it now possible to read the individual arrays that are used for serving ordinary read requests or to read the sum of values provided by every two or four adjacent arrays. Such flexibility has been essential to achieve significant energy efficiency in solving problems that partially occupy the MISC

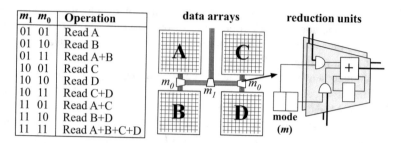

m_1 m_0	Operation
01 01	Read A
01 10	Read B
01 11	Read A+B
10 01	Read C
10 10	Read D
10 11	Read C+D
11 01	Read A+C
11 10	Read B+D
11 11	Read A+B+C+D

Fig. 9.25 Proposed configurable reduction tree for banks

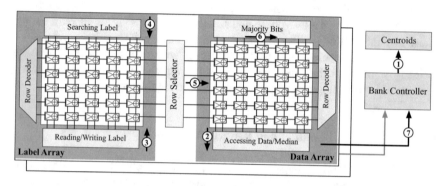

Fig. 9.26 Illustrative example of the MISC array comprising the cells and peripheral circuits

banks. Indeed, software is responsible for computing the mode bits for any given problem size. The mode bits are then streamed into the accelerator during the chip initialization phase.

9.5.5.3 MISC Array Organization

Each MISC bank includes a controller consisting of buffers for maintaining local data (e.g., centroids), serial adders, comparators, and logic for controlling iterative tasks, such as partitioning and recomputing the centroids. The bank controller makes it possible to read, write, and compute a set of selected data arrays efficiently. Figure 9.26 shows the structure of an MISC array with two subarrays used to maintain the data points and cluster labels.

Seven major steps are followed by the bank controller to initialize centroids, for new clusters, and compute medians.

- *Initializing Centroids*: Every k-medians clustering task begins with randomly initializing the centroids, which are maintained by the bank controller in local buffers (1). The index of each centroid in this table is used as a label for the corresponding cluster.

- *Forming New Clusters*: The bank controller forms new partitions by reading the data points from the data subarrays and comparing them with the centroids (2). The index of the closest centroid to every data point is used as the new cluster label for these data and will be written to the label array (3). This is accomplished through a set of serial comparators at the bank controller. As the data points are read out, the serial comparator determines the index of the closest centroid to the data.
- *Computing Medians*: The centroid of each cluster must be recomputed by applying the bit-serial median algorithm to all the elements of every cluster. This requires the bank controller to keep track of the cluster members at all time. The label array uses the same structure as the data array to carry out the required book keeping for all of the data points. At the beginning of every median computation, the label arrays are searched for matching entries using the cluster labels one after another (4). The outcome of every search operation is the matching lines in the label subarray connected to a row selector unit to determine the I and P values for the data array (5). Next, the median bits are computed by iteratively performing the vertical majority vote computation followed by the horizontal minority propagation (6). The median bits are streamed to the bank controller for updating the centroids as they are serially computed by the MISC arrays (7). This process ends after a certain number of iterations defined by the software. One other possibility for ending the program is to stop the process if all of the newly computed centroids are the same as the old ones. In other words, the computation is repeated until convergence is reached.

9.5.5.4 MISC Data Representation

MISC needs to represent the data points in a fixed-point positive format due to the limits of the bit-serial median algorithm on negative or real numbers. The software performs all the necessary data conversion and preprocessing for clustering real numbers and negative values prior to loading the data points into the MISC chips.

Clustering Real Numbers MISC converts the real valued numbers to fixed-point data prior to clustering. A 64-bit fixed-point format achieves virtually the same results obtained with a double-precision IEEE floating point format for a wide range of applications and datasets. However, for more sensitive applications, MISC is flexible enough to compute the medians of wider bit representations by increasing the number of vertical majority vote computation and applying minimal changes to the control logic. Figure 9.27 shows an example clustering tasks for five real valued numbers. A preprocessing step is considered to convert floating point to the fixed point. The input floating point data are scaled by a factor of 2^3. Then, the bit serial median algorithm is used to compute the median. Finally, the median value is identified.

Handling Negative Numbers The median computation by the bit-serial median algorithm assumes that the input data are positive integers. This may not be

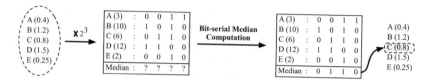

Fig. 9.27 Illustrative example of handling real valued numbers in the MISC framework

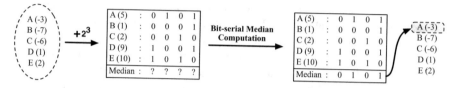

Fig. 9.28 Illustrative example of handling negative numbers in the MISC framework

necessarily true for real-world applications. The dataset may include negative numbers. MISC addresses this issue by representing data in a biased notation. A bias value of 2^n is added to all of the data points, regardless of being negative and positive. Then the clustering algorithm is performed for the positive values. Finally, the median value is identified in the original dataset. Figure 9.28 shows an example of computing the median of five integer values. The bias is 2^3, which is added to all the data points.

9.5.5.5 Handling Even Number of Data Points

Another limitation of the bit-serial median computation algorithm is to compute the median of an even number of data points. MISC addresses this problem by including a virtual data point in the median computation process. The members of a given cluster may be spread across different arrays and banks of the accelerator. Moreover, the number of cluster members may change per each iteration due to cluster reformation. Therefore, MISC has to find the number of cluster member first through sending a cluster label to the label array and counting the matches. The same analog bit counter is used to find the number of matches in all label arrays. If the cluster contains an even number of elements, the median of that cluster is computed in two steps. Figure 9.29 shows the two steps for an example problem. First, a virtual data point with value **0** is included in the cluster so that the number of data points becomes odd. Then, the first median (M1) is computed. In the second step, a virtual data point with all **1**s is included to compute the second median (M2). MISC used the average of M1 and M2 as true median of this cluster.

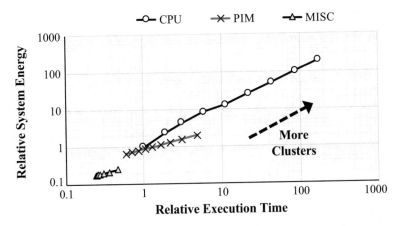

Original Data				
A (1)	:	0 0 0 1		
B (4)	:	0 1 0 0		
C (6)	:	0 1 1 0		
D (9)	:	1 0 0 1		
Median :	? ? ? ?			

Including 0		
X(0)	:	0 0 0 0
A (1)	:	0 0 0 1
B (4)	:	0 1 0 0
C (6)	:	0 1 1 0
D (9)	:	1 0 0 1
Median :	0 1 0 0	

Including 15		
X(15)	:	1 1 1 1
A (1)	:	0 0 0 1
B (4)	:	0 1 0 0
C (6)	:	0 1 1 0
D (9)	:	1 0 0 1
Median :	0 1 1 0	

Fig. 9.29 Illustrative example of computing the median of an even number of data points

Fig. 9.30 The system energy and execution times of MISC, PIM, and CPU with respect to the problem size

9.5.6 Potentials of the MISC Accelerator

MISC is another software-hardware approach to large-scale data clustering with significant energy savings and performance potentials. The simulation results on a clustering library with real datasets [109–111] and two applications pertaining to k-means clustering prove the significant energy-efficiency of MISC. This section provided the highlights of these potentials when compared to a baseline CPU and an ASIC processor-in-memory (PIM) accelerator. Figure 9.30 illustrates the impact of an increase in the number of clusters on the overall system energy and execution time of the CPU, PIM, and MISC. Each design point represents the relative execution time and system energy averaged on three runs of the library for 8 MB data from breast cancer, indoor localization, and US census datasets. The results indicate that the energy and execution time of data clustering increase as the number of clusters grows; however, such increase is much more significant for the PIM and CPU baselines. Overall, the MISC accelerator achieves 22–290 k× and 8–81× better energy-delay products compared to the CPU and PIM baselines, respectively.

References

1. G. Bello-Orgaz, J.J. Jung, D. Camacho, Social big data: recent achievements and new challenges. Inf. Fusion **28**, 45–59 (2016)
2. E. Ahmed, I. Yaqoob, I.A.T. Hashem, I. Khan, A.I.A. Ahmed, M. Imran, A.V. Vasilakos, The role of big data analytics in Internet of Things. Comput. Netw. **129**, 459–471 (2017)
3. A. Al-Fuqaha, M. Guizani, M. Mohammadi, M. Aledhari, M. Ayyash, Internet of things: a survey on enabling technologies, protocols, and applications, in *IEEE Communications Surveys & Tutorials*, 2015
4. N. Koshizuka, K. Sakamura, Ubiquitous ID: Standards for Ubiquitous computing and the Internet of Things, in *IEEE Pervasive Comput.*, 2010
5. N. Kushalnagar, G. Montenegro, C. Schumacher, Pv6 over Low-Power Wireless Personal Area Networks (6LoWPANs): overview, assumptions, problem statement, and goals, in *Internet Eng. Task Force (IETF)*, 2007
6. M. Kheirkhahan, S. Nair, A. Davoudi, P. Rashidi, A. Wanigatunga, D. Corbett, T. Mendoza, T. Manini, S. Ranka, A smartwatch-based framework for real-time and online assessment and mobility monitoring. J. Biomed. Informatics **89**, 29–40 (2019)
7. P. Barnaghi, W. Wang, C. Henson, K. Taylor, Early progress and back to the future, in *International Journal on Semantic Web and Information Systems (IJSWIS)*, 2012
8. G. Kestor, R. Gioiosa, D.J. Kerbyson, A. Hoisie, Quantifying the energy cost of data movement in scientific applications, in *IEEE International Symposium on Workload Characterization (IISWC)*, 2013
9. D. Pandiyan, C.-J. Wu, Quantifying the energy cost of data movement for emerging smart phone workloads on mobile platforms, in *IEEE International Symposium on Workload Characterization (IISWC)*, 2014
10. N. Chatterjee, M. O'Connor, D. Lee, D.R. Johnson, S.W. Keckler, M. Rhu, W.J. Dally, Architecting an energy-efficient DRAM system for GPUs, in *IEEE International Symposium on High Performance Computer Architecture (HPCA)*, 2017.
11. "The top ten exascale research challenges," Report of the Advanced Scientific Computing Advisory Committee Subcommittee, 2014.
12. I. Akturk, U.R. Karpuzcu, Amnesiac: Amnesic automatic computer, in *Proceedings of the Twenty-Second International Conference on Architectural Support for Programming Languages and Operating Systems*, 2017
13. R. Balasubramanian, J. Chang, T. Manning, J.H. Moreno, R. Murphy, R. Nair, S. Swanson, Near-data processing: insights from a micro-46 workshop. IEEE Micro **2**, 36–42 (2014)
14. K. Lim, J. Chang, T. Mudge, P. Ranganathan, S.K. Reinhardt, T.F. Wenisch, Disaggregated memory for expansion and sharing in blade servers, in *International Symposium on Computer Architecture*, 2009
15. Y. Chen, T. Luo, S. Liu, S. Zhang, L. He, J. Wang, L. Li, T. Chen, Z. Xu, N. Sun, Dadiannao: a machine-learning supercomputer, in *Proceedings of the 47th Annual IEEE/ACM International Symposium on Microarchitecture*, 2014
16. J.C. Beyler, et al., ESODYP: an entirely software and dynamic data prefetcher based on a Markov model, in *12th Workshop on Compilers for Parallel Computers*, 2006
17. X. Yu, C.J. Hughes, N. Satish, S. Devadas, IMP: indirect memory prefetcher, in *Proceedings of the 48th International Symposium on Microarchitecture*, 2015
18. J. Jeddeloh, B. Keeth, Hybrid memory cube new DRAM architecture increases density and performance, in *Symposium on VLSI Technology (VLSIT)*, 2012.
19. J. Kim, J.S. Pak, J. Cho, E. Song, J. Cho, H. Kim, T. Song, J. Lee, H. Lee, K. Park, et al., High-frequency scalable electrical model and analysis of a through silicon via (TSV). IEEE Trans. Compon. Packag. Manuf. Technol. **1**(2), 181–195 (2011)
20. R. Hameed, W. Qadeer, M. Wachs, O. Azizi, A. Solomatnikov, B.C. Lee, S. Richardson, C. Kozyrakis, M. Horowitz, Understanding sources of inefficiency in general-purpose chips. ACM SIGARCH Comput. Archit. News **38**(3), 37–47 (2010)

21. D.H. Woo, N.H. Seong, D.L. Lewis, H.-H.S. Lee, An optimized 3D-stacked memory architecture by exploiting excessive, high-density TSV bandwidth, in *16th International Symposium on High Performance Computer Architecture (HPCA)*, 2010
22. P. Mike, An Intro to MCDRAM (High Bandwidth Memory) on Knights Landing, Intel HPC Developer Conference, 2016
23. J. Ahn, S. Yoo, O. Mutlu, K. Choi, PIM-enabled instructions: a low-overhead, locality-aware processing-in-memory architecture, in *ACM/IEEE 42nd Annual International Symposium on Computer Architecture (ISCA)*, 2015
24. D. Kim, J. Kung, S. Chai, S. Yalamanchili, S. Mukhopadhyay, Neurocube: a programmable digital neuromorphic architecture with high-density 3D memory, in *ACM/IEEE 43rd Annual International Symposium on Computer Architecture (ISCA)*, 2016
25. K. Hsieh, E. Ebrahimi, G. Kim, N. Chatterjee, M. O'Connor, N. Vijaykumar, O. Mutlu, S.W. Keckler, Transparent offloading and mapping (TOM): enabling programmer-transparent near-data processing in GPU systems. ACM SIGARCH Comput. Archit. News **44**(3) (2016)
26. M.N. Bojnordi, F. Nasrullah, ReTagger: an efficient controller for DRAM cache architectures, in *Design Automation Conference (DAC)*, Las Vegas, NV, 2019
27. F. Pan, S. Gao, C. Chen, C. Song, F. Zeng, Recent progress in resistive random access memories: materials, switching mechanisms, and performance. Mater. Sci. Eng.: R: Rep. **83**, 1–59 (2014)
28. C. Ho, C.-L. Hsu, C.-C. Chen, J.-T. Liu, C.-S. Wu, C.-C. Huang, C. Hu, F.-L. Yang, 9nm half-pitch functional resistive memory cell with <1μa programming current using thermally oxidized sub-stoichiometric wo x film, in *in Electron Devices Meeting (IEDM)*, 2010
29. B. Govoreanu, G. Kar, Y. Chen, V. Paraschiv, S. Kubicek, A. Fantini, I. Radu, L. Goux, S. Clima, R. Degraeve et al., 10 × 10 nm 2 hf/hfo x crossbar resistive RAM with excellent performance, in *Electron Devices Meeting (IEDM)*, 2011
30. A.C. Torrezan, J.P. Strachan, G. Medeiros-Ribeiro, R.S. Williams, Sub-nanosecond switching of a tantalum oxide memristor. Nanotechnology **22**(48), 485203 (2011)
31. B.J. Choi, A.C. Torrezan, K.J. Norris, F. Miao, J.P. Strachan, M.-X. Zhang, D.A. Ohlberg, N.P. Kobayashi, J.J. Yang, R.S. Williams, Electrical performance and scalability of pt dispersed SiO_2 nanometallic resistance switch. Nano Lett. **13**(7), 3213–3217 (2013)
32. S. Lai. Current status of the phase change memory and its future, in *In Electron Devices Meeting*, 2003
33. C. Cheng, C. Tsai, A. Chin, F. Yeh, High performance ultra-low energy RRAM with good retention and endurance, in *In Electron Devices Meeting (IEDM)*, 2010
34. C. Cheng, A. Chin, F. Yeh, Novel ultra-low power RRAM with good endurance and retention, in *In VLSI Technology (VLSIT)*, 2010
35. H. Akinaga, H. Shima, Resistive random access memory (RERAM) based on metal oxides, in *Proceedings of the IEEE*, 2010
36. S. Sheu, M. Chang, K. Lin, C. Wu, Y. Chen, P. Chiu, C. Kuo, Y. Yang, P. Chiang, W. Lin, C. Lin, A 4Mb embedded SLC resistive-RAM macro with 7.2 ns read-write random-access time and 160ns MLC-access capability, in *IEEE International Solid-State Circuits Conference Digest of Technical Papers (ISSCC)*, 2011
37. M. Zangeneh, A. Joshi, Design and optimization of nonvolatile multibit 1T1R resistive RAM, in *IEEE Transactions on Very Large Scale Integration (VLSI) Systems,* 2014
38. J.J. Yang, M.D. Pickett, X. Li, D.A.A. Ohlberg, D.R. Stewart, R.S. Williams, Memristive switching mechanism for metal/oxide/metal nanodevices. Nat. Nanotechnol. **3**, 429–433 (2008)
39. D. Niu, C. Xu, N. Muralimanohar, N. Jouppi, Y. Xie, Design trade-offs for high density cross-point resistive memory, in *ACM/IEEE International Symposium on Low Power Electronics and Design*, 2012
40. C. Xu, X. Dong, N. Jouppi, Y. Xie, Design implications of memristor-based RRAM cross-point structures, in *In Proceedings of the Design, Automation & Test in Europe Conference & Exhibition (DATE)*, 2011

41. M. Zidan, H. Fahmy, M. Hussain, K. Salama, Memristor-based memory: the sneak paths problem and solutions. Microelectron. J. **44**(2), 176–183 (2013)
42. Y. Taigman, M. Yang, M. Ranzato, L. Wolf, DeepFace: closing the gap to human-level performance in face verification, in *In Proceedings of the IEEE Conference on Computer Vision and Pattern Recognition (CVPR)*, 2014
43. A. Krizhevsky, I. Sutskever, G.E. Hinton, ImageNet classification with deep convolutional neural networks, in *Advances In Neural Information Processing Systems*, 2012
44. X. Lei, A.W. Senior, A. Gruenstein, J. Sorensen, Accurate and compact large vocabulary speech recognition on mobile devices, in *Interspeech*, 2013
45. M. Motamedi, D. Fong, S. Ghiasi, Fast and energy-efficient CNN inference on IoT devices, *arXiv preprint arXiv:1611.07151,* 2016
46. L. Oskouei, S.G.H. Salar, M. Hashemi, S. Ghiasi, Cnndroid: Gpu-accelerated execution of trained deep convolutional neural networks on android, in *Proceedings of the ACM on Multimedia Conference*, 2016
47. S. Mehta, J. Torrellas, WearCore: a core for wearable workloads?, in *International Conference on Parallel Architecture and Compilation Techniques (PACT)*, 2016
48. K. Ma, X. Li, K. Swaminathan, Y. Zheng, S. Li, Y. Liu, Y. Xie, J.J. Sampson, V. Narayanan, Nonvolatile processor architectures: Efficient, reliable progress with unstable power. IEEE Micro **36**(3), 72–83 (2016)
49. Y. LeCun, B. Boser, J. Denker, D. Henderson, R. Howard, W. Hubbard, L. Jackel, Backprop-agation applied to handwritten zip code recognition. Neural Comput. **1**, 541–551 (1989)
50. S. Han, X. Liu, H. Mao, J. Pu, A. Pedram, M. Horowitz, W. Dally, EIE: efficient inference engine on compressed deep neural network, in *Proceedings of the 43rd International Symposium on Computer Architecture (ISCA)*, 2016
51. E. Denton, W. Zaremba, J. Bruna, Y. LeCun, R. Fergus, Exploiting linear structure within convolutional networks for efficient evaluation, in *In Advances in Neural Information Processing Systems; Curran Associates, Inc.*, 2014
52. C. Zhang, P. Li, G. Sun, Y. Guan, B. Xiao, J. Cong, Optimizing FPGA-based accelerator design for deep convolutional neural networks, in *In Proceedings of the ACM/SIGDA International Symposium on Field-Programmable Gate Arrays*, 2015
53. F. Iandola, S. Han, M. Moskewicz, K. Ashraf, W. Dally, K. Keutzer, SqueezeNet: AlexNet-level accuracy with 50× fewer parameters and <0.5 MB model size, in *arXiv:1602.07360*, 2016
54. W. Chen, J. Wilson, S. Tyree, K. Weinberger, Y. Chen, Compressing neural networks with the Hashing Trick, in *In Proceedings of the ICML*, 2015
55. T. Chen, Z. Du, N. Sun, J. Wang, C. Wu, Y. Chen, O.D. Temam, Diannao: a small-footprint high-throughput accelerator for ubiquitous machine-learning, in *In ACM Sigplan Notices*, 2014
56. Y. Chen, J. Emer, V. Sze, Eyeriss: a spatial architecture for energy-efficient dataflow for convolutional neural networks, in *In Proceedings of 43rd the ACM/IEEE Annual International Symposium on Computer Architecture (ISCA)*, 2016
57. Z. Du, R. Fasthuber, T. Chen, P. Ienne, L. Li, T. Luo, X. Feng, Y. Chen, O. S. Temam, ShiDianNao: shifting vision processing closer to the sensor, in *In ACM SIGARCH Computer Architecture News*, 2015
58. R. Kozma, R.E. Pino, G.E. Pazienza, *Advances in Neuromorphic Memristor Science and Applications* (Springer Publishing Company, 2012)
59. A.M. Sheri, A. Rafique, W. Pedrycz, M. Jeon, Contrastive divergence for memristor-based restricted Boltzmann machine. Eng. Appl. Artif. Intell. **37**, 336–342 (2015)
60. M. Prezioso, F. Merrikh-Bayat, B. Hoskins, G. Adam, K.K. Likharev, D.B. Strukov, Training and operation of an integrated 12 neuromorphic network based on metal-oxide memristors. Nature **521**(7550), 61–64 (2015)
61. M.N. Bojnordi, E. Ipek, Memristive Boltzmann machine: a hardware accelerator for combina-torial optimization and deep learning, in *IEEE International Symposium on High Performance Computer Architecture (HPCA)*, (IEEE, Barcelona, 2016)

62. A. Shafiee, A. Nag, N. Muralimanohar, R. Balasubramonian, J.P. Strachan, M. Hu, R.S. Williams, V. Srikumar, ISAAC: a convolutional neural network accelerator with in-situ analog arithmetic in crossbars. ACM SIGARCH Comput. Archit. News **44**(3), 14–26 (2016)
63. P. Chi, S. Li, C. Xu, T. Zhang, J. Zhao, Y. Liu, Y. Wang, Y. Xie, Prime: a novel processing-in-memory architecture for neural network computation in RERAM-based main memory, in *In Proceedings of the 43rd International Symposium on Computer Architecture*, 2016
64. M. Gao, Q. Wang, M.T. Arafin, Y. Lyu, G. Qu, Approximate computing for low power and security in the internet of things. Computer **50**(6), 27–34 (2017)
65. Z. Wen, P. Bhatotia, R. Chen, M. Lee and others, ApproxIoT: approximate analytics for edge computing, in *38th International Conference on Distributed Computing Systems (ICDCS)*, 2018
66. D. Liu, C. Yang, S. Li, X. Chen, J. Ren, R. Liu, M. Duan, Y. Tan, L. Liang, FitCNN: a cloud-assisted and low-cost framework for updating CNNs on IoT devices. Futur. Gener. Comput. Syst. **91**, 277–289 (2019)
67. S. Tajasob, M. Rezaalipour, M. Dehyadegari, M.N. Bojnordi, Designing efficient imprecise adders using multi-bit approximate building blocks, in *Proceedings of the International Symposium on Low Power Electronics and Design*, 2018
68. R. Venkatesan, A. Agarwal, K. Roy, A. Raghunathan, MACACO: modeling and analysis of circuits for approximate computing, in *Proceedings of the International Conference on Computer-Aided Design*, 2011
69. A. Ranjan, S. Venkataramani, X. Fong, K. Roy, A. Raghunathan, Approximate storage for energy efficient spintronic memories, in *In Proc. DAC*, 2015
70. D. Mohapatra, V. Chippa, A. Raghunathan, K. Roy, Design of voltage-scalable meta-functions for approximate computing, in *In Proc. DATE*, 2011
71. A. Sampson, W. Dietl, E. Fortuna, D. Gnanapragasam, L. Ceze, D. Grossman, EnerJ: approximate data types for safe and general low-power computation, in *in Proc. Int. Conf. Programm. Lang. Design Implement*, 2011
72. H. Esmaeilzadeh, A. Sampson, L. Ceze, D. Burger, Neural acceleration for general-purpose approximate programs, in *In Proceedings of the 45th Annual IEEE/ACM International Symposium on Microarchitecture*, 2012
73. J. Bornholt, T. Mytkowicz, K. McKinley, Uncertain<T>: a first-order type for uncertain data, in *ACM SIGARCH Computer Architecture News*, 2014
74. M. Samadi, J. Lee, D. Jamshidi, A. Hormati, S. Mahlke, Sage: self-tuning approximation for graphics engines, in *Annual IEEE/ACM International Symposium on Microarchitecture*, 2013
75. W. Baek, T.M. Chilimbi, Green: a framework for supporting energy-conscious programming using controlled approximation, in *in Proc. ACM SIGPLAN Conf. Programm. Lang. Design Implement*, 2010
76. E. Denton, W. Zaremba, J. Bruna, Y. LeCun, R. Fergus, Exploiting linear structure within convolutional networks for efficient evaluation, in *In Advances in Neural Information Processing Systems*, 2014
77. S. Han, J. Pool, J. Tran, W. Dally, Learning both weights and connections for efficient neural network, in *In Advances in Neural Information Processing Systems; Curran Associates, Inc.*, 2015
78. S. Han, H. Mao, W. Dally, Deep compression: compressing deep neural networks with pruning, trained quantization and Huffman coding, in *arXiv:1510.00149*, 2015
79. Y. Gong, L. Liu, M. Yang, L. Bourdev, Compressing deep convolutional networks using vector quantization, in *arXiv:1412.6115*, 2014
80. I. Hubara, M. Courbariaux, D. Soudry, R. El-Yaniv, Y. Bengio, Binarized neural networks, in *In Advances in Neural Information Processing Systems*, 2016
81. M. Rastegari, V. Ordonez, J. Redmon, A. Farhadi, XNOR-Net: ImageNet classification using binary convolutional neural networks, in *arXiv:1603.05279*, 2016
82. T. Tang, L. Xia, B. Li, Y. Wang, H. Yang, Binary convolutional neural network on RRAM, in *In Proceedings of the 22nd Asia and South Pacific Design Automation Conference (ASP-DAC)*, 2017

83. J. Qiu, J. Wang, S. Yao, K. Guo, B. Li, E. Zhou, J. Yu, T. Tang, N. Xu, S. Song, et al, Going deeper with embedded FPGA platform for convolutional neural network, in *In Proceedings of the 2016 ACM/SIGDA International Symposium on Field-Programmable Gate Arrays*, 2016
84. R. Balasubramonian, J. Chang, T. Manning, J.H. Manning, R. Murphy, R. Nair, S. Swanson, Near-data processing: insights from a MICRO-46 workshop, in *IEEE Micro*, 2014
85. M. Alian, S. Min, H. Asgharimoghaddam, A. Dhar, D. Wang, T. Roewer, A. McPadden, O. O'Halloran, D. Chen, J. Xiong, D. Kim, Application-transparent near-memory processing architecture with memory channel network," in *Annual IEEE/ACM International Symposium on Microarchitecture (MICRO)*, 2018
86. A. Nag, R. Balasubramonian, V. Srikumar, R. Walker, A. Shafiee, J. Strachan, N. Murali-manohar, Newton: gravitating towards the physical limits of crossbar acceleration," in *IEEE Micro*, 2018
87. Q. Guo, X. Guo, R.I.E. Patel, E.G. Friedman, Ac-DIMM: associative computing with STT-MRAM, in *ACM SIGARCH Computer Architecture News*, 2013
88. J. Su, J. Liu, D.B. Thomas, P.Y. Cheung, Neural network based reinforcement learning acceleration on FPGA platforms, in *ACM SIGARCH Computer Architecture News*, 2017
89. Y. Zhu, Y.P.Y. Zhang, Large-scale restricted Boltzmann machines on single GPU, in *IEEE International Conference on Big Data*, 2013
90. S. Li, D. Niu, K. Malladi, H. Zheng, B. Brennan, Y. Xie, Drisa: a dram-based reconfigurable in-situ accelerator, in *Proceedings of the 50th Annual IEEE/ACM International Symposium on Microarchitecture*, 2017
91. Q. Guo, X. Guo, Y. Bai, E. Ipek, A resistive TCAM accelerator for data-intensive computing, in *In Proceedings of the 44th Annual IEEE/ACM International Symposium on Microarchitecture*, 2011
92. A. Pal Chowdhury, P. Kulkarni, M. Nazm Bojnordi, MB-CNN: memristive binary convolutional neural networks for embedded mobile devices. J. Low Power Electron. Appl. **8**(38), 1–27 (2018)
93. Y. Karthik Rupesh, P. Behnam, G. Reddy Pandla, M. Miryala, M. Bojnordi, Accelerating *k*-medians clustering using a novel 4t-4r RRAM cell, in *IEEE Trans-actions on Very Large Scale Integration (VLSI) Systems*, 2018
94. Y. Li, S. Lee, K. Oowada, H. Nguyen, Q. Nguyen, N. Mokhlesi, C. Hsu, J. Li, V. Ramachandra, T. Kamei, et al, 128Gb 3b/Cell NAND flash memory in 19nm technology with 18MB/s write rate and 400Mb/s toggle mode, in *In Proceedings of the 2012 IEEE International on Solid-State Circuits Conference Digest of Technical Papers (ISSCC)*, 2012
95. D. Takashima, Y. Nagadomi, T. Ozaki, A 100 MHz ladder FeRAM design with capacitance-coupled-bitline (CCB) cell, in *IEEE J. Solid-State Circuits*, 2011
96. R. Simpson, M. Krbal, P. Fons, A. Kolobov, J. Tominaga, T. Uruga, H. Tanida, Toward the ultimate limit of phase change in Ge2Sb2Te5. Nano Lett. **10**(2), 414–419 (2010)
97. A. Benoist, S. Blonkowski, S. Jeannot, S. Denorme, J. Damiens, J. Berger, P. Candelier, E. Vianello, H. Grampeix, J. Nodin, et al, 28 nm advanced CMOS resistive RAM solution as embedded non-volatile memory, in *In Proceedings of the 2014 IEEE International Reliability Physics Symposium*, 2014
98. M. Ueki, K. Akeuchi, T. Yamamoto, A. Tanabe, N. Ikarashi, M. Saitoh, T. Nagumo, H. Sunamura, M. Narihiro, K. Uejima, et al, Low-power embedded ReRAM technology for IoT applications, in *In Proceedings of the 2015 Symposium on VLSI Circuits (VLSI Circuits)*, 2015
99. M. Oskin, F.T. Chong, T. Sherwood, Active pages: a computation model for intelligent memory, in *in Proc. 25th Annu. Int. Symp. Comput. Archit.*, 1998
100. M. Qureshi, M. Franceschini, L. Lastras-Montaño, J. Karidis, Morphable memory system: a robust architecture for exploiting multi-level phase change memories, in *In ACM SIGARCH Computer Architecture News*, 2010
101. B. Razavi, *Principles of Data Conversion System Design*. (Wiley-IEEE Press, 1995)
102. W. Kester, I. Analog Devices, *Data Conversion Handbook* (Analog Devices, Norwood, MA, 2005)

103. A. Vattani, The hardness of k-means clustering in the plane, in *[Online]*. Available: http://cseweb.ucsd.edu/avattani/papers/kmeans_hardness.pdf, (2009)
104. A.K. Jain, R.C. Dubes, Algorithms for clustering data, (Prentice-Hall, Upper Saddle River, 1988)
105. Y. Lu, J. Han, Cancer classification using gene expression data. Inf. Syst. **28**(4), 243–268 (2003)
106. P.G. Anick, S. Vaithyanathan, Exploiting clustering and phrases for context-based information retrieval, in *ACM SIGIR Forum*, 1997
107. I. Micron Technology, 8Gb DDR3 SDRAM, in *[Online]*. *Available:*http://www.micron.com//get-document/documentId=416, 2009
108. C.H. Cheng, A. Chin, F.S. Yeh, Novel ultra-low power RRAM with good endurance and retention, in *in Proceedings Symp. VLSI Technol.*, 2010
109. P. Mertins et al., Proteogenomics connects somatic mutations to signalling in breast cancer. Nature **534**(7605), 55–62 (2016)
110. J. Torres-Sospedra, et al., UjiindoorLoc: a new multi-building and multi-floor database for Wlan fingerprint-based indoor localization problems, in *Int. Conf. Indoor Positioning Indoor Navigat (IPIN)*, 2014
111. M. Lichman, UCI Machine Learning Repository, in *[Online]*. *Available:*http://archive.ics.uci.edu/ml, 2013
112. ITRS, International Technology Roadmap for Semiconductors, in *2013 Edition. Accessed: Nov. 22, 2017. [Online]. Available:*http://www.itrs.net/Links/2013ITRS/Home2013.html, 2013
113. P.J. Crawley, G.W. Roberts, High-swing MOS current mirror with arbitrarily high output resistance, in *Electron. Lett.*, 1992
114. R.L. Geiger, P.E. Allen, N.R. Strader, *VLSI design techniques for analog and digital circuits* vol 90 (McGraw-Hill, New York, 1990)
115. H.-S. Philip Wong, H.-Y. Lee, S. Yu, Y.-S. Chen, Y. Wu, P.-S. Chen, B. Lee, F.T. Chen, M.-J. Tsai, Metal–oxide RRAM, in *Proceedings of the IEEE*, 2012
116. P. Behnam, A. Pal Chowdhury, M. Nazm Bojnordi, R-cache: a highly set-associative in-package cache using memristive arrays, 2018
117. P.E. Danielsson, Getting the median faster. Comput. Graphics Image Process. **17**(1), 71–78 (1981)

Chapter 10
IoT Cyber Security

Brian Russell

*For happiness one needs security, but joy can spring like a
flower even from the cliffs of despair.*

Anne Morrow Lindbergh

Contents

B. Russell (✉)
TrustThink, LLC, San Diego, CA, USA
e-mail: russell_brian@trustthink.net

© Springer Nature Switzerland AG 2020
F. Firouzi et al. (eds.), *Intelligent Internet of Things*,
https://doi.org/10.1007/978-3-030-30367-9_10

10.1 Introduction

In 2013, Linux.Aidra was released on the Internet. The malware automatically identified and exploited vulnerable routers. Since then, dozens of new or derived botnets have been discovered that prey on basic weaknesses in Internet of Things (IoT) products. BASHLITE, Remaiten, Mirai, and others scan the Internet for open ports and then run exploits against known vulnerabilities in connected IoT products. This automated malware is programmed with well-known username and password combinations that are very often left unchanged as IoT products are deployed into networks.

Automated botnets have gained significant media attention because of the damage that can be achieved when using them to take down Internet sites and services. But, botnets almost never include zero-day exploits and rarely represent the true capabilities of cyber criminals. More advanced exploits and capabilities exist that are often kept secret until the time is right to use them. Bad actors are even commercializing their exploits. Malware-as-a-Service, for example, now allows less-skilled adversaries to purchase or rent exploits and tools developed by savvy criminals. As we continue to see a risk in the use of malware-as-a-service, we will likely see a continued rise in attempts to compromise IoT products and systems.

Effective cyber security must become a fundamental goal within all IoT products and systems. This chapter details methods that can be used to understand and mitigate the myriad cyber security threats faced by IoT developers and operators. We discuss the complex threat environment that IoT systems operate within and provide guidance for implementing cyber security measures across the core architectural elements of an IoT system: the edge, the network, the cloud, and the user.

10.2   A Complex Threat Environment

We borrow from the domain of Systems Engineering and the Systems Engineering Body of Knowledge (SEBoK) [1] to define an IoT system as a "system-of-systems." According to the SEBoK, a system-of-systems has unique engineering and design considerations when compared to a standard "system." These include dynamic and reconfigurable boundaries and interfaces, complex performance measurements given the need to operate across different stakeholder systems, and metrics that are often difficult to define and quantify given independent management of component systems. For purposes of simplicity, we will use "system-of-system" and system interchangeably in this chapter.

Complex IoT system-of-systems incorporate sensors and actuators, gateways, attached storage devices, mobile applications, enterprise applications, cloud analytics, workflow management, notification systems, and various cloud services. IoT systems also incorporate Application Programming Interfaces (APIs) and Software Development Kits (SDKs) that allow users to mix-and-match capabilities within a network. The use of these APIs and SDKs makes it easy for an enterprise to create new features based on machine-to-machine interactions that were not previously intended by the IoT product manufacturer.

IoT systems also rely upon multiple points of integration. Dozens of communication protocols exist to bind together devices and applications over both wireless and wired networks. Messaging protocols such as the Constrained Application Protocol (CoAP) or Message Queue Telemetry Transport (MQTT) protocol provide IoT systems with the ability to pass data in a standardized format. Even legacy bus protocols such as ModBus or a vehicle's CAN Bus can be seen as components within an IoT system.

Mobile applications play a vital role in most IoT systems. Mobile apps can be used to interface directly with IoT products or the gateways and cloud services that support those products. These might connect via protocols such as Bluetooth or Wi-Fi. Cloud services also interact with IoT products. APIs collect data and communicate management commands to and from devices at the edge.

As IoT systems begin to provide new autonomous capabilities across different industries, their reliance on sensor inputs becomes critical. Autonomous vehicles (AV), for example, rely upon a suite of cameras, sensors, radar, lidar, Global Positioning System (GPS), and vehicle-to-vehicle communications in order to support critical functions such as mission planning, behavioral planning, and motion planning. These inputs can be spoofed and denied resulting in unexpected and even dangerous behavior from the vehicle.

This web of components and interfaces makes the job of cyber security professionals difficult at best. A malicious actor needs to only identify and exploit a weakness in a single component of this system-of-systems to gain a foothold into the IoT. A weakness left open in a single insecure product, network device, cloud service, mobile application, or even protocol can leave the door wide open to an attacker gaining access. Once inside, bad actors can begin moving toward higher value targets, for example, databases that store sensitive information. This "lateral movement" through the system is difficult to protect against and one of the reasons that traditional perimeter security measures such as firewalls are no longer sufficient to protect an organization's data. An incident in 2014 against target shows the exposure to unexpected attack vectors now constantly faced by IoT system security engineers.

In 2014, attackers gained access to target networks via credentials stolen from a Heating, Ventilation and Air Conditioning (HVAC) vendor. Fazio Mechanical Service – the HVAC vendor was the victim of a phishing campaign which led to the compromise of their network login credentials, giving the attackers access to Fazio systems. Fazio also had an established connection with Target in order to support electronic billing and contract management. This allowed attackers to leverage the login credentials stolen to access Target networks and identify a method to implant the Trojan.POSRAM malware onto target's payment card systems [2].

As can be seen, the target payment card system was breached only after bad actors identified a weakness exposed by a third-party vendor. The original weakness was not even a technology weakness. Instead the attackers targeted the humans associated with the system in an attempt to trick them into giving up their credentials. Indeed, technological implementations are not the only aspects of a system that must be secured. Users of systems must also understand their unique weaknesses and how their actions can impact the security of the enterprise. Although the target breach occurred 5 years ago, we have seen additional security research that demonstrates that users still represent a weakness to IoT system security. Smart cameras, for example, are still being compromised through unauthorized access to their owner's usernames and passwords, allowing strangers to view them in their homes [3].

Another consideration is the need to protect the integrity of the data generated within IoT systems. Even video is susceptible to manipulation when sufficient cyber security safeguards are not applied. Systems that collect and act upon video footage must protect the video from manipulation. This includes the deletion, modification, or even replacement of collected footage.

Security expert Josh Mitchell demonstrated in 2016 the ability to remotely compromise police video cameras in order to download, edit, and then re-upload video footage [4]. Consider this threat in terms of analytics systems that process data from distributed sensors and video recorders. If the source data is untrusted, then actions based on that data are of limited worth.

Unfortunately, video camera products are one of the worst offenders in terms of IoT cyber security weaknesses. Well-known default passwords, insecure configurations, known-insecure network services, and software vulnerabilities have exposed many IoT products to attack by the various threat actors in the world whose goal is to compromise systems.

Fig. 10.1 NTP amplification attack

10.2.1   Threat Actors and Risk Likelihood

Not every system is targeted for attack by the same bad actors. Different people and organizations have different motivations. Some attackers aim to simply make a name for themselves, others chase opportunities for financial wealth, while others seek to disrupt the operations of organizations that operate in a manner for which they disagree. Each of these threat actors has a finite amount of resources, limited capabilities, and some level of risk tolerance. At the far end of the threat actor spectrum lives a threat actor that is different from the rest. The nation-state threat is often characterized as an adversary with virtually unlimited resources and a high degree of subject matter expertise. These are often government organizations that employ teams of engineers capable of reverse engineering software and developing zero-day exploits based on vulnerabilities that they identify for themselves or introduce through a compromised supply chain.

When determining the level of risk associated with any particular threat, one of the primary considerations is the threat actor(s) involved. The likelihood of a risk being realized is based heavily on the capabilities and motivations of the attackers that will target a system. The perpetrators of the target breach were highly motivated and knowledgeable, for example, and the financial prize provided significant motivation. Most organizations will never be targeted by nation-state actors; however, all organizations must be cognizant that there are individuals or groups of individuals with resources and training that are capable of executing complex campaigns to capture sensitive data, disrupt operations, or cause physical or monetary damage. Figure 10.1 describes common threat actors (Table 10.1).

It helps when designing cyber security architectures to understand the types of attackers that might be targeting your systems. To gain a better understanding of the attacks that might be levied against systems, an understanding of the motivations and abilities of attackers is required.

10.2.2   Threat Types

We previously discussed in this chapter the threat of data modification and the need to implement integrity protections to ensure data is legitimate. That threat is simply

Table 10.1 Threat profiles

Threat actor	Motivations and capabilities
Nation-State	A well-funded government with an established offensive cyber security program. Motivations include large-scale financial gain or disruption, access to intellectual property, political influence, or physical damage to critical infrastructure. Nation-state actors are often involved in the creation of zero-day exploits.
Cyber Terrorist	May be well-funded. Attacks may be focused on inflicting damage or disruption, or aimed at inducing fear within a population.
Hacktivist	May use existing tools to exploit known weaknesses in a system. May write automated attack scripts and target specific organizations or industries. Mostly politically motivated.
Organized Crime	Most likely financially driven. Will often use existing tools to exploit known weaknesses in systems.
Insider/System User	A trusted insider to the system or organization. Often has at least minimal privileges within the system or at times elevated privileges. May feel unappreciated or wronged.

one in many that an IoT system must be designed to guard against. Additional types of threats faced by IoT systems include:

Identity Spoofing Consider the ability for an attacker to misrepresent himself or herself as trusted entities within a system. If an attacker is able to compromise the identity of a system user, then he or she will be able to perform actions that include modification of configurations, theft of data, or insertion of malicious content. Since IoT systems include human users and operators, those people that use or manage the system are at risk of having their identities spoofed. Additionally, the IoT introduces the ability for adversaries to insert rogue devices into the network. Rogue devices can be used by bad actors to inject false data into the data stream, or even to monitor the activities of the system or its users.

Tampering with Data Tampering can occur at any point in the lifecycle of data. We have discussed tampering at the source – that is by gaining access to an IoT device at the edge and manipulating the generated data. Data is often distributed across the network and ingested by analytic processing systems. Consider an attackers' opportunity to access and modify data within a network attached storage (NAS) device, or as it is streamed to the cloud. Cloud storage and processing systems must also be considered as attack vectors for data tampering.

Repudiation Repudiation is an important concept in terms of IoT systems. As these systems begin to use analytics to support coordination of autonomous actions, being able to ensure that one organization or product cannot repudiate inputs to an action or the actions themselves is critical to being able to assign liability for malfunctions or accidents induced by the system. If a system performs an action that results in harm (physical, financial), then a trail of evidence must be available to review and determine root cause. Without proper non-repudiation controls within a system, that trail will not be trusted.

Information Disclosure One of the most well-known cyber security controls is confidentiality. Confidentiality protections such as the application of encryption can prevent information disclosure in traditional Information Technology (IT) systems. The same is true with IoT systems. Proper placement of cryptographic protections can ensure that sensitive information is not disclosed to unauthorized parties. Without proper confidentiality controls in place, there are many methods an attacker might use to eavesdrop or steal information from a system.

IoT systems introduce a new concept related to information disclosure, however. It is not simply the risk of application-layer information that must be protected against. Instead, metadata that could be used to track or surveil the location or usage patterns of system users must also be protected. Disclosure of this metadata can have dire consequences. This is why complex IoT systems such as Advanced Metering Infrastructure (AMI) and Connected Vehicles (CV) engineer robust privacy and anonymity controls into their applications. Within an AMI, for example, system designers must guard against leakage of usage patterns from the smart meter. Within Connected Vehicle implementations, system designers must guard against the ability of someone to track the location of a vehicle or to be able to determine a vehicle owner based on the vehicle's identity.

Denial or Degradation of Service IoT systems are beginning to provide safety-critical and health-critical capabilities. These systems must be designed to ensure that they can withstand the threat of denial or degradation of service. A traditional distributed denial of service attack coordinates the flooding of networks by thousands of computers. There are ways for attackers to increase the magnitude of the attack as well. For example, amplification attacks take minimal input to generate maximum output against a victim machine. Amplification attacks take advantage of inherent features of well-known protocols. A Network Time Protocol (NTP)-based amplification attack is a good example.

NTP amplification attacks were enabled by the use of the NTP monlist command. Monlist returns a list of the last 600 Internet Protocol (IP) addresses that access the NTP server. This enabled a small input to generate a maximum-sized output directed at a victim server. Now consider these attacks being levied against the cloud services that support your IoT systems.

We must also include concern for any degradation of service when discussing life-critical and safety-critical IoT systems. Many IoT systems require near real-time processing capability and any reduction in that ability can have severe consequences. For example, researchers [5] have shown that it is possible to blind cameras on AVs for an infinite amount of time using lasers. This reduces an AV's ability to use those cameras for the functions of lane detection, object detection, and even traffic sign identification.

Bypassing Physical Security A relatively unique threat to IoT systems is the ability of attackers to gain physical access to IoT devices in order to evaluate their security posture and perform reverse engineering. In some cases, devices can simply be bought online and analyzed in the comfort of an attackers' lab. In other cases, firmware that runs a device can be found online and analyzed for vulnerabilities using firmware analysis tools. Devices may also be installed in exposed locations, leaving them open to theft. When an attacker has physical access to an IoT machine, he or she has many methods to compromise that machine. Physical security controls must be considered whenever an IoT system includes machines that are physically exposed to untrusted humans.

10.3 Cyber Security Controls for IoT Systems

The selection of security controls that can mitigate the threats described in this chapter must be made across the various layers of an IoT system. Controls are not simply technological in nature. Process-based controls and security controls applied to the human users of IoT systems must also be incorporated. In the following sections, we detail cyber security controls that can be applied at discrete layers within the IoT system (Fig. 10.2).

We begin with a framework for the secure development of a new IoT system and then discuss specific controls that can be applied to the devices at the edge of an IoT system, within the network layer that supports IoT systems and within the cloud services that enable IoT features. We finish by examining security controls that can be used to help secure the human element of any IoT system.

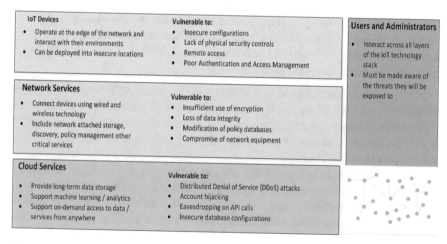

Fig. 10.2 Cyber security controls

10.3.1 Establishing a Secure IoT System Development Methodology

Much of today's software is exposed to multiple threats when deployed. Pressure to meet Sprint and Release commitments often overrides the proper handling of security requirements within an application or device. It is easy to push off security functionality in favor of satisfying a functional requirement or staying on track with cost and schedule. The approach defined in this section provides a framework for ensuring that software and firmware are developed on a secure foundation and that cyber security requirements are not dropped from a sprint backlog. The recommendations here are based on adaptation of the Microsoft Agile Development model [6] as well as secure agile approaches from the Open Web Application Security Program (OWASP). Each development project begins with a threat model, which is created during project initiation and updated continuously as the project matures. The threat model requires an examination of the functionality of the application or device in order to understand the threats and vulnerabilities associated with the system.

10.3.1.1 Threat Modeling an IoT System

Organizations typically have a limited amount of resources that can be applied to the cyber security of their systems. Leaders must make informed decisions as to the impact of a particular threat as well as the likelihood of that threat occurring to be able to determine where to invest funds for mitigations. Threat modeling is the process of documenting a system, identifying the threats to that system and then rating the threats based on their impact and likelihood. Without threat modeling, security engineers operate somewhat in the dark as to what risks require the most attention, and may possibly invest in security tools that don't even apply to the most serious concerns. Microsoft's threat modeling approach – a component of the Microsoft Agile SDL – begins with documenting the assets within a system, creating and analyzing a system architecture, decomposing the system to identify data flows, and identifying high value assets within the system. These are the assets that a malicious actor might attempt to compromise – for example, a database or gateway. Once complete, the process includes steps to identify, document, and rate the threats to the system (Fig. 10.3).

Simply because a threat to your system exists does not mean that action must be taken. There may be esoteric threats that would likely never occur. In these instances, it is often better to invest funds in mitigating threats that would be much more likely to occur. Risk analysis allows security engineers to assign a risk score to each risk in the system. The risk score (Risk = Likelihood of Occurrence * Impact of Occurrence) provides a quantifiable view of all the risks of a system and allows them to be objectively prioritized. The Microsoft Agile SDL includes a process termed DREAD that security engineers can use to assign risk scores.

Fig. 10.3 Threat modeling flow

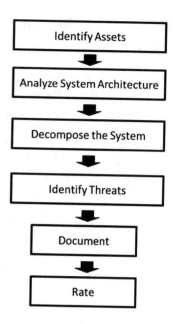

- *D*amage – what amount of damage (physical, monetary, reputation) would occur?
- *R*eproducibility – can the attack be reproduced easily?
- *E*xploitability – how difficult is it to execute the attack?
- *A*ffected users – how many customers or stakeholders will be impacted?
- *D*iscoverability – is the threat well known? Can anyone discover it?

Once risks are quantified and prioritized, leaders can choose to mitigate the risk, defer the risk, or even to accept the risk.

10.3.1.2 Documenting Cyber Security Requirements

Identifying the specific security user stories to include in your backlog requires an analysis of the specific threats faced by your system. Once understood, engineers can begin to document the cyber security requirements that must be satisfied to secure the system. Determining when to spend time on specific requirements can be difficult though in an agile development process, since agile projects focus on showcasing potentially shippable products at the end of each sprint. Some might consider cyber security requirements as a "tax" on system development, with a handful of these requirements being added to each sprint backlog. A better approach is to review and categorize all cyber security requirements during the creation and/or update of the product backlog and release roadmap. Microsoft's Agile SDL provides guidance on handling security requirements within the cadence of an agile development. Three types of requirements are defined and aligned to the sprint planning process.

- *Every-Sprint Requirements*: These requirements are included within each sprint. They reduce your team's velocity. An example of an "every-sprint" user story is: As a user I can expect that my software is free from known vulnerabilities identified in the OWASP Top 10. This user story ensures code is analyzed each sprint and cannot be deemed complete until all OWASP Top 10 vulnerabilities are remediated. Alternatively, a development team might consider adding this type of requirement to their "definition of done." Either way, the team's velocity is decreased as these important security requirements must be handled each sprint.
- *Bucket Requirements*: These requirements are allocated to sprint backlogs as velocity permits over the course of product development. The "bucket" that contains each of these requirements must be emptied by the end of the project. Proper management of these bucket requirements is essential to ensure that there is a manageable amount of work left as the project completion nears. Considerations regarding the release schedule of the project must also be taken into account. If an IoT system continuously releases new features, then these requirements should be incorporated during each release.
- *One-Time Requirements*: These requirements are included in a sprint one-time only. Activities such as setting up a secure configuration management repository or designing the security architecture can be included as one-time requirements. Although these requirements are only included in a single sprint, they are usually critically important to the cyber security posture of the product or system. An example of a one-time user story that drives architectural decision making might be: *As a device administrator, I can be assured that my device's private key material is stored in protected hardware at all times, so that potential compromise of device cryptographic keys is mitigated.*

10.3.1.3 Establishing a Cyber Security Culture

The cyber security of an IoT product or system is everyone's responsibility. Members of the DevOps team must understand this, and all staff must be on the lookout for poor security hygiene and bad operational security (OPSEC) practices. There are a number of ways to instill a culture of cyber security within your organization. This begins with appointing an executive to be accountable for the security of the system. This executive must be given the authority to ensure that sufficient investments are made to provide the team with the resources they need to properly execute their cyber security responsibilities. There should also be security evangelists identified within an organization. In some cases, these are funded security professionals. In constrained environments, developers with an interest in cyber security can be identified and trained in secure software development. The role of the evangelist is to motivate the rest of the team to take security seriously and to work with executive leadership to identify gaps in the secure development process.

Best practices traditionally advocated within Extreme Programming (XP) are useful for establishing cyber security processes across the development team. Pair programming, for example, is a valuable cyber security tool. Teams can rotate security-aware developers with the rest of the development team to instill that awareness across the team over time. The process of pair programming also has additional benefits. Higher quality code can be developed given a second pair of eyes. Also, skills can be shared and learned across different members of the team to eliminate single points of failure.

Each development team should also codify a set of secure coding guidelines. OWASP provides a useful Secure Coding Cheat Sheet that can be used as a template to customize your own secure coding guidelines [7]. In addition to establishing secure coding guidelines, make sure to incorporate peer reviews into your process. This can be handled organically when pair programming is used, but at a minimum ensure that peer reviews occur each sprint.

Secure configuration management (CM) practices must also be put in place for IoT system development. Ensure that your CM tool is locked down to allow only authorized access. Do not embed API keys directly in code – instead use configuration files that can be updated separately from the code. Make sure that critical Secure Shell (SSH) keys that provide administrative access to cloud servers are accounted for and only provided to authorized users. Keep an eye out for any sensitive data that may be posted to public repositories.

Do not forget the need to provide training to your development teams. Organizations such as SAFECode and SANS provide secure development training courses that can provide required skills to the development team. SANS also offers certifications such as the GIAC Secure Software Programmer (GSSP)-Java and GSSP-.NET certifications. Additionally, have your development teams review best practices from industry organizations to gain a better understanding of the latest recommendations for securing connected products. Table 10.2 provides a listing of some well-written guidance documents that can provide valuable information for security teams.

Table 10.2 References for industry best practices

Industry guidance	Applicability	Available at
IoT Security Foundation (IoTSF) IoT Security Framework	Enterprise IoT systems	https://iotsecurityfoundation.org/wp-content/uploads/2016/12/IoT-Security-Compliance-Framework.pdf
European Network and Information Security Agency (ENISA) Baseline Security Recommendations for IoT	Enterprise IoT systems	https://www.enisa.europa.eu/publications/baseline-security-recommendations-for-iot
Cloud Security Alliance (CSA) Future Proofing the Connected World	IoT Machine Development	https://downloads.cloudsecurityalliance.org/assets/research/internet-of-things/future-proofing-the-connected-world.pdf
OWASP/CSA Secure Medical Devices IoT Deployment	Deployment of Medical IoT devices	https://downloads.cloudsecurityalliance.org/assets/research/owasp/OWASP_Secure_Medical_Devices_Deployment_Standard_7.18.18.pdf

10.3.1.4   Conducting Code Audits and Automating Processes

Code auditing today should occur continuously. Static and dynamic analysis tools provide feedback that can identify weaknesses in code. Dynamic analysis is used to identify weaknesses during the operation of the software. Static analysis tools review the source code for known weaknesses. Depending on the sensitivity of the code being developed, manual source code reviews can also be required. Systems today often contain millions of lines of code (LoC), so any manual review must be focused on high-value areas of the code. For example, a manual review might be required based on the results of a static or dynamic analysis tool output. Architectural reviews should also be conducted regularly. As time progresses, the features and functions of an application or system may change substantially. Revisiting the security architecture can ensure that open weaknesses are not unintentionally exposed.

Binary analysis should also be performed. There are many binary analysis tools available. Binwalk is an open source tool that supports analysis of the files contained within firmware. The tool is available for use within Kali Linux distributions [8].

Premium binary analysis products also exist. VDOO, for example, provides a firmware analysis tool that maps results to cyber security standards and best practices available from organizations such as ENISA, Cloud Security Alliance, IoT Security Foundation (IoTSF), and the National Institute of Standards and Technology (NIST) [9].

Additional testing such as fuzzing should also be performed. Fuzz testing is a black box testing method that introduces unexpected inputs into software. By injecting "bad" data such as malformed data automatically to the software, anomalies can be identified in the way that the software handles unexpected inputs. Tools such as Peach [10] allow security engineers to quickly begin the process of fuzzing their software. Fuzzing should include the initial connection setup, state changes such as power on or power off, and http or protocol fields such as headers or length/value fields.

Automation has been a critical enabler of quality code development for years now. Automated test tools evaluate developer adherence to coding standards and quickly pinpoint bugs in code. The same is true for evaluating the cyber security posture of software used within your IoT system. Automated analysis tools can be integrated into your Continuous Integration (CI) environment to check for adherence to secure coding guidelines. Static analysis tools, dynamic analysis tools, firmware analysis tools, and even fuzz testers can now be incorporated within the CI environment. These tools can be integrated with requirements management platforms such as JIRA to automatically open ISSUES that must be addressed by the development team.

10.3.1.5 Gaining Visibility into Your Supply Chain

It is difficult to understand the cyber security posture of your IoT system without having full visibility into the libraries and third-party components that together make up your software and firmware. As an IoT system developer, make sure to establish processes to document all components used within your system – a bill of materials (BOM).

Communicating the BOM to your customers is also important. This allows the customers that implement your products to understand their risk exposure by providing visibility into the dozens or hundreds of libraries used across their enterprise systems. This also provides customers with a valuable tool to determine what patches must be applied to their IoT systems to ensure that they do not remain vulnerable. Today, there are two competing standards that can be used to create a bill of materials. Software Package Data Exchange (SPDX) Tools is available from The Linux Foundation [11]. Cisco offers a propriety tool known as SWID.

The US National Telecommunications and Information Administration (NTIA) has spearheaded a project to create a standardized bill of materials (BOM) for software. The premise is that third-party software introduces risk into the supply chain and must be properly tracked to be able to close vulnerabilities quickly. A good example of this problem was seen in 2018 when a bug was found by researchers GraphicsFuzz in the Qualcomm Adreno 630 graphics driver [12]. This bug in the Adreno driver was able to be exploited to cause a reboot of the Samsung Galaxy S9.

> Adreno 630 is a graphics card used in the Qualcomm Snapdragon 845 System-on-Chip (SoC) used in a variety of smartphones. A bug in the 630 driver could have impacts on not only the Samsung Galaxy S9 but also on other mobile phones that make use of that driver. This means that when a bug is found in the driver associated with one mobile phone, other mobile phone manufacturers should take action to patch their systems as soon as possible. Without good visibility into the libraries used within your systems though, that would be difficult at best.

10.3.1.6 Working with the Security Research Community

In 2018, the IoTSF conducted a study to analyze the vulnerability disclosure programs across consumer IoT companies. The study found that 90% of consumer IoT companies did not have a formal vulnerability disclosure process in place [13]. This means that security researchers have not pre-defined method of communicating bugs and vulnerabilities found in those company's products. A vulnerability disclosure process establishes formal procedures for accepting inputs from external sources. Make sure to provide a Pretty Good Privacy (PGP) public key on your organization's website to support encrypted submissions.

The security research community has led the creation of a grassroots vulnerability research function. Researchers across the globe are constantly analyzing products to identify new vulnerabilities. As a developer of IoT products and systems, make it easy for those researchers to share their information by establishing a formalized vulnerability disclosure program. The IoTSF also provides guidance on establishing a vulnerability disclosure program [14].

Bug bounty programs are also an option for proactively engaging with the security community. Organizations like BugCrowd (https://www.bugcrowd.com/product/bug-bounty/) can support the creation of a bug bounty program. The organization HackerOne (https://www.hackerone.com/) successfully established a bug bounty for the Department of Defense (DoD) in 2016. The first Hack the Pentagon event identified 138 unique vulnerabilities [15].

10.3.2   Integrating Safety and Security Engineering

IoT systems integrate the cyber and physical domains. As such, both safety and security engineering disciplines must be incorporated in order to mitigate both safety and security risks.

Safety and security engineering have some similarities. IoT systems must be engineered to adhere to security and safety goals simultaneously. A safety engineers' primary modeling tool is the fault tree. Fault tree analysis (FTA) identifies common mode failures.

There are similarities between fault trees and a common cyber security modeling tool known as attack trees. Fault trees, however, do not take into account the creative approaches used by attackers to cause system disruption. IoT system designers must be able to evaluate both attack trees and fault trees. This helps to analyze how a motivated attacker might target the safety controls within a system, for example.

Establish a safety engineering function within your development team to analyze the safety impacts of your system. Given the intended use of the system, is there a potential for physical harm if the device stopped working, or malfunctioned. Even if the machine itself is not safety-critical, consider whether there are any other safety-critical devices or services that depend upon outputs from the machine. Consider how potential harm from machine failure could be minimized or avoided.

A good example of the integration of safety and cyber security is the definition of a fail-safe state that is entered upon detection of a cyber security event. For example, with an AV the vehicle may be programmed to enter into this fail-safe mode that brings it to a safe stop at the side of the road upon detection of a serious event. Alternatively, the system may simply report the event to the cabin occupant and instead fail-operationally, allowing the driver to take control and decide on a course of action.

10.3.3 Safeguarding Stakeholder Privacy

Privacy in IoT systems is about more than simply safeguarding Privacy Protected Information (PPI). IoT systems can expose information about users that allow tracking of user habits or even the locations of individuals. IoT system designers have to consider privacy at the beginning and throughout an IoT implementation. The best method of integrating privacy protections into a system is through the concept of Privacy by Design (PbD). PbD incorporates strategies and activities that protect user privacy and give users control over how their data is used. To learn more about implementing the foundational principles of PbD, review the paper by Ann Cavoukian, PhD: Privacy by Design – The Seven Foundational Principles [16].

There are a number of activities that must be conducted based on these PhD principles. Organizations should prepare a Privacy Impact Assessment (PI) for each system and routinely review privacy impacts based on updates to the system. All data collected by a device and/or processed within a system should be documented and categorized to identify PPI. After documenting this data collection, make an active effort to limit the amount and types of data collected to only what is actually needed for system operation. Establish policies and procedures that are fair to customers as well. Notify customers regarding data collection and the expected use of their data. Provide a mechanism for customers to opt-in or opt-out of data collection. Also, setup a breach notification program that alerts customers to a data breach within acceptable timelines per local and national regulations.

10.4 Securing the IoT Edge

When we visualize the "IoT," we usually envision devices installed at the edge of a network. These devices interact with their physical world by collecting data or performing actions. Devices may be anywhere in the world and may be installed in remote locations. This gives attackers the advantage of being able to physically access their targets. The ability to introduce unexpected behavior in physical environments makes IoT devices a prime target for attackers. Product developers must keep in mind the need to:

- Protect from adversaries obtaining firmware and reverse engineering devices
- Protect from adversaries gaining access to a device either locally or remotely
- Design devices to be resilient – to operate even in the face of attacks
- Protect from adversaries installing malicious software or hardware into devices (e.g., Skimmers or backdoors)

Many of the automated attacks against IoT devices have been successful because of a small number of common weaknesses at the edge. In 2018, a team from the Cloud Security Alliance composed of the author, Michael Roza, Aaron Guzman and Hananel Levine identified the common weaknesses exposed at the edge [17]. A summary of the common weaknesses identified include:

- *Insecure default credentials.* This includes well known username/password combinations. These are often published in data dictionaries used by password cracking tools and automated malware to quickly compromise a device. Also included are hard-coded passwords that cannot be changed and password shared across many devices in a family.
- *Insufficient encryption.* This includes not encrypting communications between IoT devices and peer devices, gateways, or servers. This also includes a lack of encryption on storage mediums allowing attackers to compromise sensitive information. This also includes failure to protect the keys used by cryptographic algorithms to secure information.
- *Weaknesses in authentication.* This includes a failure to authenticate remote access to the device or cloud service. This also includes failure to require multi-factor authentication to access cloud services that are directly integrated with the device.
- *Use of vulnerable network services.* Telnet, FTP, UPnP, and other well-known insecure network services can leave open insecure ports that are targeted by automated malware and other exploits.
- *Software vulnerabilities.* Just as Information Technology (IT) systems require secure software, so do IoT devices and services. Standard vulnerabilities such as cross site scripting, command injection and buffer overflows can be exposed by not sufficiently testing the security of a product.
- *Weaknesses in firmware or software update processes.* This includes allowing unsigned firmware to be loaded to a device, or not validating the signature of a firmware file. These inactions allow bad actors to upload modified firmware files with new and potentially malicious capabilities that provide an easy path for gaining root privileges on an IoT device.
- *Failure to enforce least privilege.* This includes failure to restrict access limit privileges assigned to non-admin/root accounts, allowing bad actors to take advantage of the excess privilege to perform unintended functions.

The threats associated with IoT devices is real. Once at attacker gains access to the firmware of an IoT device, there are high quality tools available that can reverse engineer and identify or insert vulnerabilities. For example:

- Firmwalker: This tool supports firmware file system searches [18].
- IDA-Pro: Supports disassembly and debugging of software [19].
- Binwalk: Signature-based file scanning and extraction [20].

If attackers can gain physical access to a machine, there are many opportunities to capture the firmware. For example, JTAG or UART interfaces may provide unauthenticated access to extract firmware. Attackers may also use tools such as the ones mentioned above to modify firmware and change the intended behavior of a device.

Defending against firmware modification is possible using techniques such as tamper resistance and firmware integrity protections. To sufficiently guard against the myriad attacks possible against IoT devices though, a layered security approach must be implemented. In the following sections we describe techniques that can be used to secure IoT devices at the edge. We categorize these recommendations based on the need for:

- A Hardware Security Element that Supports Trusted Operations
- Configure a Secure Real-Time Operating System
- Implement Physical Security Controls
- Deploy Confidentiality Protections
- Implement Strong Authentication and Access Controls
- Harden Network Services
- Implement Logging
- Integrate Framework Security

10.4.1 Use a Hardware Security Element to Support Trusted Operations

The cyber security posture of an edge machine begins with the security features provided at the platform layer. This includes a hardware security module, a secure real-time operating system, and physical protections against tamper. Many hardware security features were previously only found in hardware security modules (HSMs); however, there are many products on the market today that meet strict size, weight, and power requirements while still offering secure hardware features. MCUs on the market today come from vendors such as FreeScale, Atmel, NXP, microchip, and ST Microelectronics. Security features provided by a hardware security module should include:

Trusted Boot A cryptographic boot loader can be implemented to validate the integrity of the operating system prior to boot. The trusted boot process initializes the trusted execution environment. The process involves validation of signed hashes applied to operating system components including the kernel and system partition. A hardware root of trust applies the signature of the hash which is then verified prior to loading.

Trusted Execution Environment A hardware root of trust can be used to segregate untrusted and trusted portions of an IoT machine. Trusted applications can be installed that are validated prior to allowing to run on the machine. The Trusted Computing Group (TCG) and others have worked on adapting TEE specifications to the IoT environment. The TEE incorporates hardware capabilities of the machine including storage, peripheral access, and the secure element. The TEE also incorporates trusted OS drivers and exposes a client API that can be used by untrusted applications to interact with trusted components [21].

Secure Firmware Processes Firmware should be encrypted before transmitting to an IoT machine. Encryption provides confidentiality to protect Intellectual Property (IP) within the firmware. Machines must authenticate firmware images prior to updating the image. Firmware is signed using a privacy key and validated with the associated public key. This requires a secure key management process to ensure that the private keys are not compromised. Compromise of the private firmware signature key would allow anyone to sign firmware which would then allow that firmware to be uploaded and run on the IoT machines.

Cryptographic Co-processing Hardware-based cryptographic co-processors support higher-speed processing of cryptographic operations. This is useful when encryption processes are used in near real-time messaging constructs. There are also specialized security co-processors that can be integrated into devices. An example is the microchip CEC 1302 model crypto processor. The CEC 1302 comes with a random number generator (RNG) which is essential for the creation of cryptographic key material. There is also a 32KB secure BOOT ROM and dedicated crypto processors for SHA and AES. In addition, there is a public key accelerator. Developers can use these crypto processors to offload cryptographic operations from the core processor within the device and to enable secure storage of cryptographic primitives.

Tamper-Based Zeroization Zeroization is the process of deleting critical data files upon detection of a tamper event. Zeroization can erase key material within the device that protects sensitive data (e.g., a Key Encryption Key – KEK). Selective zeroization allows designers to choose the specific files to delete upon tamper.

Entropy Source A high value entropy source is required for random number generation supporting cryptographic operations. This is required when generating cryptographic keys to be used for protocols such as Transport Layer Security (TLS).

Secure Key Storage Cryptographic keys are the foundation for many of the other security services used by IoT machines. Hardware security elements store keys in hardware protections which limit the ability of attackers to extract those keys and use them to impersonate machine identities or eavesdrop on communications.

10.4.2 Configure a Secure Real-Time Operating System

IoT machines often run a scaled down version of an operating system, known as a real-time operating system (or RTOS). The RTOS is responsible for various low-level services such as task scheduling and hardware abstraction. The RTOS for an IoT edge device can provide many of the same features as a secure MCU/SoC. Some RTOS have minimal security features however and should be avoided, even for low-risk commercial devices.

Developers can select a device RTOS that is certified for use in specific industries (airborne, industrial control systems, medical devices, transportation systems).

Examples of RTOS include those from VXWorxs, FreeRTOS, and Windows 10 IoT Core, among many others. Security features to look for in the chosen RTOS include:

Microkernel The likelihood of security vulnerabilities expands when code becomes bloated with unneeded features. An important component of a secure RTOS is a microkernal that has been optimized for use. Optimization should include removal of all unneeded features from the kernel. A microkernel may provide interprocess communication (IPC), memory management, and central processing unit (CPU) management. Features outside of these are then pushed up to run within the user space, including device drivers.

Kernel Separation IoT Machines that operate in high threat environments can make use of a separation kernel. A separation kernel implements secure partitions within the machine and controls the flow of information across those partitions. Information flows across partitions must be explicitly allowed in order to execute [22].

Secure Memory RTOS that make use of a separation kernel can implement secure memory protections. For example, the INTEGRITY RTOS available from Green Hills Software uses the separation kernel to prevent processes from accessing data from outside a specified partition.

Trusted Applications Applications that run within a Trusted Execution Environment can be specified as trusted applications and the operating system can validate the integrity of the application prior to executing.

Process Isolation Process isolation techniques prevent one process from interacting with another process. Each process is assigned a separate memory space [23].

Application Sandboxing Sandboxing allows the RTOS to manage the privileges of an application. For example, a sandboxed application may not be able to access the sandbox of another application. This effectively limits the spread of malware that might have entered through one application.

Hardware Abstraction One of the primary functions of an operating system is management of hardware resources. An RTOS may implement a Hardware Abstraction Layer (HAL). A Secure RTOS can restrict which processes and applications have access to the hardware on the machine.

10.4.3 Implement Physical Security Controls

Since IoT machines operate so often in unprotected spaces, the physical security of those machines becomes important. Physical Security Controls designed into IoT machines at the edge should include:

Tamper Protections For machines that warrant additional protections, implement tamper protections that restrict access to the computing hardware. These can range from simple seals or locked covers to piezo-electric circuits depending on the threat environment.

Authenticated or Disabled Debug Ports The ability to disable debug/test ports or at minimum apply password restrictions (JTAG, UART, etc.)

10.4.4   Deploy Confidentiality Protections

Artifacts from your threat model provide detailed information on the data flows within your IoT system. These artifacts will guide security engineers in determining whether sensitive information is generated, processed or stored at points within the system. Once sensitive data flows are identified, confidentiality controls can be put in place to secure data. One of the key methods of securing data is through cryptography.

Cryptography works at various layers of the communication stack. At Layer 2 of the OSI model, link layer encryption protects network packets completely. Layer 3 network encryption protects the Internet Payload (IP) packet payload – for example, using the IPsec protocol. Layer 4 Session Layer encryption encrypts UDP and TCP communications. At Layer 7 Application layer encryption can be applied to enable end-to-end encrypted and authenticated communications.

IoT protocols make extensive use of cryptography. Machines may use cryptography to support authentication of devices to peers or to network services, for data integrity, for non-repudiation, or for confidentiality of data at rest or in transit. Encryption is the process of transforming plaintext data to ciphertext data. Ciphertext is unintelligible to those that are not authorized to have access to the data. A decryption key is required to transform the ciphertext back to plaintext form. Encryption enables data confidentiality (Fig. 10.4).

One of the most used protocols for data encryption is Transport Layer Security (TLS). TLS uses certificates to enable a secure channel between two endpoints. TLS operates with Transport Control Protocol (TCP)-based communications. For example, a web server hosted on an IoT machine to enable http-based remote management would implement TLS. This provides confidentiality of the data between the remote management station and the machine. TLS also provides authentication services. Typically, on the Internet, a web server scenario is provisioned with an X.509 certificate that authenticates the server to the remote management client. To support authenticated remote management for IoT machines, a certificate is provisioned to the remote user. That certificate is used to authenticate the remote

Fig. 10.4 Encryption flow

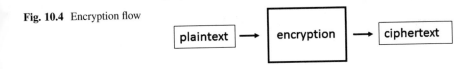

user to the IoT machine to ensure that only authorized users are able to access the web server and make configuration changes.

Short range Radio Frequency (RF) protocols also make use of cryptography. Bluetooth-LE for instance uses cryptography for pairing devices, bonding (where pairing keys are stored for the future), device authentication, and encryption.

ZigBee makes use of the IEEE 802.15.4 protocol for security services. 802.15.4 defines security protections at the APS, Network and MAC layers of the wireless frames. At the APS and Network layers, the AES algorithm in Counter mode is used with CBC MAC for message integrity. The MAC layer also uses AES in Counter mode for confidentiality and AES in CCM when integrity is required.

Key Management is an important aspect of cryptography. Take, for example, the ZigBee protocol. There are three primary types of keys that can be employed within a ZigBee network. Master keys are often pre-installed by the vendor and protect the exchange between two ZigBee nodes as they generate link keys. Link keys support node-to-node communications, and network keys support broadcast communications. All of these keys need to be sufficiently secured across their entire lifecycle.

Remember also that there are tools available that can harvest data and attempt to crack encryption keys. When encryption keys are weak this becomes trivial to accomplish. AirCrack, for example, is an example of an 802.11 cracking tool. AirCrack can be used against WEP, WPA, and WPA2- Pre-Shared Key (PSK) security and features several modes of operation for cracking keys including the use of dictionary attacks and brute force.

Popular IoT messaging protocols also require cryptographic protections. Message Queue Telemetry Transport (MQTT), for example, requires higher-layer cryptographic services as there are not inherent encryption features built into the MQTT protocol. The use of the Transport Layer Security (TLS) protocol is recommended when using MQTT. MQTT by itself simply passes the username and password of the node or broker in cleartext.

Another popular IoT messaging protocol – the Constrained Application Protocol (CoAP) – is UDP-based; it employs Datagram Transport Layer Security (DTLS) for cryptographic services. DTLS can use various types of keys, including preshared keys or symmetric keys which can also be used to establish group communications. DTLS can use public keys which are not bound to certificates and can also use X.509 certificates.

A Note on confidentiality of API Keys: establish policies that API keys and other credentials not be stored in public-facing source control systems (e.g., gitlab/github) and establish API key management procedures. Do not hardcode API keys into firmware, mobile applications or any client-based application. Monitor at least quarterly to verify that API keys and other credentials are not stored in public-facing source control systems.

10.4.5   *Implement Strong Authentication and Access Controls*

Cryptography also supports authentication. Require authentication for access to all services and ports. Establish good password processes. Do not hard-code passwords into the device and do not provision duplicate identities or passwords across multiple devices. Require password changes on a quarterly basis.

When possible, use certificates as an authentication method instead of passwords. When managed properly, certificates provide a higher degree of assurance that the authentication token has not been compromised. A private key is used to authenticate and the associated public key stored in a Certificate Authority (CA) provisioned certificate "vouches" for the signer. Keys and certificates are used for a variety of processes across an IoT ecosystem as detailed in Table 10.3.

10.4.5.1   Authorization and Access Control

Access within an IoT system can be further secured by implementing authorization and access controls. Define roles to limit access to IoT devices. For example, assign a role to devices that categorizes them as either privileged or standard. Additionally, assign anyone that accesses a device with a role: either as a user or administrator.

Implement the concept of Least Privilege Adopt the concept of least privilege within the system and on the device. Limit the applications that run as root. For example, do not run network services such as web servers as root. Further, disable the root password. Users that require administration access to the device should instead use sudo. Using sudo allows an audit trail to be created and attributed directly to the user. With root login, it is difficult to attribute actions to a specific user.

Table 10.3 Authentication and access controls for devices/software

Device/software	Key/certificate options	Discussion
IoT Devices/Sensors	None Symmetric Keys Raw Public/Private Keys Certificates and key pair Code Signing Certificates	Many existing sensors do not have the capability to support encrypted communication and therefore would not be provisioned with key material. Certificates can also be used by machines and users to authenticate to web servers, applications and network services.
Web Servers	TLS Certificates	TLS Certificates are used by clients to verify that the web server is authentic and to support encryption of data traffic between clients and server
Mobile Applications	Code Signing Certificates	Code signing certificates support secure software updates
Application Software	Code Signing Certificates	Code signing certificates support secure software updates

Use Geofencing Systems can be configured to track IoT device locations and take actions when a device leaves an authorized location. For example, by applying Global Positioning Satellite (GPS) tracking to an IoT device, alerts can be triggered when the device leaves a configured boundary. Devices can also be configured to automatically deactivate services upon leaving a geofenced boundary.

Establish Time-of-Use Time-of-use restrictions can be applied to IoT machines to restrict access to the machine outside of a configured time window, or for the machine to interact with other machines or system services.

10.4.6 Harden Network Services

Harden one of the primary entry points into IoT devices and systems. Disable all services on IoT devices that are not explicitly required by your applications. Do not use network services that are known-vulnerable. Instead design devices to use more secure network services that require authentication and can support encrypted communications. Common services that should be avoided when possible include:

- Telnet which operates on port 23. Telnet is a well-known port targeted by many automated malware variants include Mirai.
- File Transfer Protocol (FTP) which operates on port 21. FTP allows for insecure data transfers.
- Universal Plug and Play (UPnP) supports network discovery of devices.
- Server Message Block (SMB) operates on ports 139 and 445. It supports file sharing.

Other protocols and services can be implemented within your device; however, each service should be configured securely.

Secure Shell (SSH) Access Secure access to the SSH service. Require certificate-based authentication. Do not allow log-in over SSH using the root account.

X-Windows Many IoT machines do not require a Graphical User Interface (GUI). Although most RTOS operate as headless implementations, some IoT machines do require the configuration of a GUI. Depending on your use case, ensure that X-windows is disabled if you do not require user interaction with a GUI.

Network Time Protocol (NTP) Without a common time reference, correlating activities occurring across multiple points within a system is difficult. Configure IoT machines to use the network time protocol (NTP) to synchronize on a standard time reference.

10.4.7   Implement Logging and Behavioral Analytics

Logging allows system administrators to have visibility into the actions occurring within IoT machines. Enable logging on IoT machines and configure log files to be automatically transmitted to an aggregation point. This could be a local gateway or the cloud. Not all devices support complex auditing and logging, however. When this is the case, use behavioral analytics to identify anomalies in operation. For example, look for use of unexpected wireless or messaging protocols being used by the device or unexpected message sequences.

10.4.8   Implement Framework Security

IoT machines can be developed using common frameworks and platforms. This allows machines to inherit security controls from that framework and enables interoperability across device types. Take advantage of the various security services offered by these frameworks.

Onboarding By provisioning an initial set of credentials to IoT machines, administrators can securely bootstrap those machines into the operational environment. Onboarding provides the initial set of device credentials which can be transitioned to operational device credentials, for example, by wrapping an operational key pair in the original device credential (e.g., a bootstrap key).

Configuration Management Frameworks/platforms provide simple UIs and APIs that allow a system administrator to configure the operational details of a device or set of devices. This includes configuration items such as passwords, device names or even the ability to quickly reset devices.

Asset Management Frameworks/Platforms may provide the ability for a user/administrator to track IoT assets within a system. This is enabled through APIs that support services such as geolocation as well as maintenance of a device hardware and software profile (e.g., versions, third-party libraries).

Discovery Frameworks/platforms often support secure discovery services. This enables IoT devices to find peers or locate appropriate services in the cloud.

Secure Connections This may include out-of-the-box support of cryptographic libraries that enable protocols and algorithms for secure data at rest and data in transit, for example, DTLS and TLS.

Cloud Gateways Frameworks/platforms often provide cloud gateways that IoT devices can communicate with securely, providing a link between a local network and the cloud to support global interactions.

10.5   A Secure Network

The concept of edge networking is being transformed. Cloud Service Providers (CSPs) are pushing analytics capabilities directly into the physical device and Fog computing now provides analytics, storage, protocol abstraction, service discovery and policy management nearer to the edge. This allows new operating models that enable automation across product lines and makes the network an integral component of an IoT system. In this section, we discuss techniques to enhance the cyber security posture of IoT networks. We discuss recommendations for implementing IoT networks that:

- Limit a bad actors' ability to gain unauthorized access to network services.
- Limit the ability of bad actors from being able to move laterally through a network.
- Keep network services available at all times
- Protect the confidentiality and integrity of data processed and stored within the network.

Network designers should consider the following techniques for implementing a secure and available IoT network.

10.5.1   Secure Wireless Sensor Network (WSN) Configuration

Wireless Sensor Networks (WSNs) may be used extensively in AI-driven IoT systems. These networks use RF protocols such as ZigBee or ZWave to communicate between peers and gateways. Configure WSN gateways in a cluster formation to support handling of heavy loads and automated fail-over. Configure monitoring to alert on low batteries within edge machines. In certain instances, it may be prudent to tune the power levels of these edge devices to avoid signal leakage outside of a defined geographic boundary. Implement physical security controls to restrict access to the machines.

10.5.2   Segment the Network

A method for limiting lateral movement across a system is to segment IoT machines from sensitive enterprise applications on the network. To segment the network, identify categories of machine types and configure Virtual Local Area Networks (VLANs) for each category.

10.5.3   Implement Zero-Trust/Software-Defined Perimeter

Zero-trust configurations allow system owners to restrict access at the network layer. Machine, users and even applications authenticate themselves to the network prior to gaining access. Once access is granted, these users, applications, and machines are only provided the privileges needed to accomplish their authorized activities. This least-concept model of network access is useful to restrict access to the enterprise should a machine be compromised.

10.5.4   Protect the Perimeter

Even with the concept of zero-trust, security controls that protect the perimeter should still be implemented within an IoT system. Firewalls have been around a long time. Firewalls Access Control Lists (ACLs) can be configured to restrict access across a perimeter and offer capabilities such as threat management and virtual private networking (VPN). Forward firewall log data to a Security Information Event Management (SIEM) system for correlation and security analysis.

Intrusion Detection and Prevention systems (IDS/IPS) also play a vital role in network security. IDS/IPS monitor the network and report on potential security issues. An IPS can also be configured to block unauthorized connections. Place IDS sensors at strategic points within the network architecture. An example of an open source IDS is snort [24].

10.5.5   Secure Discovery Services

A quick look at a site such as Shodan [25] shows how easy it can be to identify IoT machines configured with various protocols. Many of these protocols provide discovery services can be useful for dynamic configuration of networks, but also enable easy reconnaissance efforts by potential attackers. Disable discovery services or at a minimum require authenticated access to discovery services within networks.

10.5.6   Implement Asset Management

One of the most valuable cyber security measures to apply to a network is to implement an asset tracking program. Configuration management tools should be used to track all hardware and software used within a network. Strict change control procedures should also be implemented to ensure that unknown devices and services are not installed on the network. This basic security measure will provide the data necessary to better understand the threat profile of an IoT system. This includes tracking of third-party libraries installed within each IoT product to allow tracking of patches associated with not only the devices but also the subcomponents used within each of the devices.

10.5.7 Implement Vulnerability Tracking

Your networks are populated with many devices and software. Each exposes their own potential vulnerabilities. These vulnerabilities are found over time, by the research community. As vulnerabilities are identified, vendors implement patches to resolve the vulnerabilities. If system administrators do not patch software, then the software remains exposed to those vulnerabilities. Time is of the essence here; once a vulnerability is known, exploits begin being crafted in the wild. Keep track of vulnerabilities. Monitor sources such as MITREs Common Vulnerabilities and Exposures (CVE) which publishes vulnerabilities through the National Vulnerability Database (NVD) [26]. Also, use the Common Vulnerability Scoring System (CVSS) which assigns ratings for IT vulnerabilities.

Many industries have Information Sharing and Analysis Centers (ISACs) that share information on attacks.

10.5.8 Audit and Monitoring

Automated malware can wreak havoc on an IoT system. Task network security tools to search for new botnet activity and immediately remove infected IoT devices upon detection. Many well-known botnets use telnet to spread so monitor for activity on ports 23 and 2323. Botnets such as Reaper, Mirai, IoTroop, and Satori can spread on many other ports, including 21 (FTP), 80 (http), and ephemeral ports such as 37215. Also monitor for Command and Control (C2C) communications that enable coordinated use of botnets. For example, JenX uses port 127 for C2C communications.

System designers should also plan for monitoring of their network of IoT devices and services. There are, however, many challenges associated with this. For example, some IoT devices may not generate any security audit logs. Other IoT devices typically do not support formats such as syslog and may require custom connectors. In addition, gaining timely access to audit logs from remotely deployed and RF-based IoT devices may provide difficult. For example, devices may not be reachable over a low-bandwidth RF link or may only wake during times of operation. Finally, confidence in the integrity of audit logs collected from IoT devices may be limited given potential lack of cryptographic controls.

An auditing and monitoring framework should collect security-relevant events. These might include *(source: Russell, VanDuren, PACKT Publishing, Practical IoT Security)*:

- *Device not reachable*: A device that is disconnected from the network may have been compromised and brought offline or suffering from a denial of service attack.
- *Time-based anomalies*: Determine standard time-of-use characteristics for the system and monitor for use outside of these time restrictions.
- *Spikes in velocity*: Increased velocity can mean that a device has been hijacked by automated malware and is being used in botnet activity.
- *New protocols on the network*: If seeing abnormal protocols/ports are being used, this may indicate repurposing of an IoT machine for nefarious purposes.
- *Authentication anomalies*: Repeated authentication failures may mean that an IoT machine or service has been targeted by a malicious actor.
- *Attempted elevation of privilege*: Attempts to increase privilege by accessing the root account or attempting sudo access may indicate that the machine has been compromised.
- *Drops in velocity*: A drop in velocity may indicate network congestion or attempted denial of service against the IoT machine or service.
- *Rapid change in device physical state*. A rapidly increased Central Processing Unit (CPU) usage may mean that an attempt to drain the device battery is being conducted by executing multiple new applications or that the device is operating outside of standard parameters.
- *Communications with unexpected destinations*: Most IoT machines are configured to communicate with a limited number of devices and gateways. When communication from an edge machine to a non-normal service or device is detected, it may indicate attempted lateral movement through the system. Even if these attempts are low in volume they should be investigated.
- *Unexpected audit results*: Logging should be enabled on all computers and machines in the system. These logs must be audited on a regular basis. Abnormal audit findings may indicate attempted or successful compromise.
- *Purged audit trails*: A deleted audit log may be indicative of a trusted insider attempting to cover his or her tracks. Modification or deletion of audit logs should be restricted to an "audit group" but even members of this trusted group must be monitored for malicious activity.
- *Sweeping for topics*: Topic sweeping using publish/subscribe protocols may be indicative of a user attempting to gain access to data he or she is not authorized to view.

10.5.9   Vulnerability Scanning

Vulnerability scanning should be performed at least annually and preferably continuously. If hiring an outside consultant, the processing begins by reviewing agreements and assumptions, providing a detailed listing of IP addresses to be scanned, planning the approach, setting up user accounts, and assembling tools and network access. The scanning process is executed using the selected tools and an extensive review process is conducted to verify the discovered vulnerabilities, prioritize findings, and identify remediation. Finally, remediation is the process of fixing the problems, for example, by patching or making configuration changes to devices and services.

There are many tools available to conduct vulnerability scanning. OWASP maintains a listing of vulnerability scanners and this list is updated regularly [27]. One of the most well-known tools used for this activity is Nessus.

10.5.10   Penetration Testing

Vulnerability scans and penetration testing are two different activities. People often mistake one for the other. Vulnerability scans look for known vulnerabilities in a network and the scanning process is well suited for automation. Penetration tests are designed to exploit weaknesses in a network or system. Penetration tests include not only electronic methods but also non-electronic methods. For example, "dumpster diving" to collect information that may be useful in reconstructing passwords or aiding in social engineering attacks. There is a high degree of creativity that goes into a penetration test.

There are three basic types of penetration test.

- A white box test involves a tester that has intimate knowledge of the system under test. The tester is provided with information such as IP addresses, ports, and services in use, network diagrams and may even be provided with user accounts.
- A black box test assumes the tester has no inside knowledge of the system under test. This type of test requires the most creativity from the tester as he or she must be able to amass information used to attack the system.
- A gray box test is a hybrid of the black and white box tests. The tester may be provided some limited amount of information, for example, IP addresses and a basic description of the applications.

Once a penetration test has been set for a system, the tester will begin by enumerating the system to collect and later analyze data. Information on the devices and computers that are on the network, and their operating systems, versions and more can be found using network scanners. This can also include the open ports that are being listened on and the associated services being used.

A tester will examine the system for basic weaknesses. She will look for default passwords used across IoT machines and gateways, or even the networking equipment within the system. The tester will examine the system for untrusted services such as telnet, ftp, and UPnP and look for out-of-date firmware, software, and third-party libraries with known vulnerabilities that can be exploited quickly.

A Penetration Tester will also deploy tools to monitor traffic within the system. He will use tools such as Wireshark, SecBee, AirCrack, KillerBee, BlueMaho, etc. depending on the target protocols used within the system. For testers with advanced RF knowledge, a tool such as HackRF can be configured to monitor the RF spectrum from 1 MHz to 6 GHz. This allows the tester to look for unencrypted or unexpected communications within the system. Password crackers can also be used to attempt to find weak passwords and keys.

A Penetration Tester will use creativity to examine the hardware of the IoT machines at the edge. This includes evaluation of the device operating location to determine if the device can be taken without an alert being generated or someone noticing. If the device can be surreptitiously swapped out, then perhaps the firmware can be extracted and analyzed. Tamper protections will also be evaluated to determine any controls that prevent opening the device casing. Testing and debug ports will be evaluated to determine if any allow unauthenticated access to device internals or the command line. Tools such as Shika [28] can be used to interface with device hardware and attempt to dump memory or download firmware.

A good penetration tester will also examine the system for process weaknesses. This includes analyzing the process of updating firmware, if no authentication or integrity check is performed by a machine prior to loading new firmware, then he or she will attempt to reconfigure and upload a malicious firmware version. Additional processes associated with the handling or administration of devices will also be examined. When applicable social engineering skills will be put to use to attempt to trick trusted insiders into giving out knowledge needed to access the system.

As can be seen, a thorough penetration test will do much more than a vulnerability scan. A creative tester can probe for system weaknesses at multiple points in the architecture, including the human elements. This very closely mimics the length and effort that a malicious actor might go to try and gain unauthorized access to the system.

10.6 A Secure Cloud

The cloud has become a key component of any IoT system. Cloud services provide the web servers that allow IoT services to collect and process information and implement unique capabilities. The cloud also provides data storage, analysis, and work-flow management features. Capabilities such as machine-learning are vital to injecting intelligence into IoT systems and these are often configured to run in the cloud.

The Cloud also serves as an integration point for edge machines and gateways. Multi-protocol interfaces support connectivity from the edge. These interfaces must be authenticated and encrypted. Standard interfaces from the edge to cloud include HTTP, MQTT, and AMQP, although custom interfaces can be implemented as well.

- *HTTP*: Encrypt all HTTP connections using Transport Layer Security (TLS) 1.2 or later. Use certificates to authenticate TLS connections.
- *MQTT*: Encrypt and authenticate MQTT communications using TLS 1.2 or later as well. MQTT communications must ensure that the native username/password field is encrypted and never sent in cleartext.
- *CoAP*: Secure UDP-based communications such as CoAP using DTLS.

There are many threats that must be guarded against within the cloud. For example, most CSPs provide an identity registry that stores the identifies and configurations of devices connected to the cloud. Amazon Web Services (AWS) terms these "Thing Shadows." Proper access controls must be applied to ensure that unauthorized access to these configuration repositories is not allowed. Similarly, access to keys must be restricted within the cloud. The use of Hardware Security Modules (HSMs) in the cloud is often warranted. Unauthorized database access must also be guarded against.

There are a number of recommendations listed here to lock down an IoT cloud. Some of these make use of capabilities offered directly by a CSP and others require the use of third party processes and tools.

10.6.1   Evaluate the Security of the CSP

Each CSP implements its own security architecture and it can be difficult to objectively evaluate the security state of each provider. To aid in this evaluation, download and review the Cloud Security Alliance (CSA) Cloud Controls Matrix (CCM) which provides a set of evaluation criteria to methodically review the security state of a CSP [29]. This registry contains self-assessments and third-party assessment results covering dozens of CSPs. When choosing a CSP to host IoT system applications, review these assessment results and make comparisons within the evaluation process to ensure selection of an optimal CSP partner. Once selected, work with the CSP to establish Service Level Agreements (SLAs) that detail minimum service levels and include penalties when a CSP does not meet minimum availability or performance metrics.

10.6.2 Design the Cloud Service to be Resilient and Available

IoT services should be designed to operate in the face of adversity and to be highly available. A robust cloud architecture will support these goals. For example, implement load balancers in front of virtual servers to evenly distribute load across applications and effectively handle peak traffic. You can also contract with Content Delivery Networks (CDNs) to help improve the performance of your services. Make use of CSP availability zones and regions and ensure that system infrastructure spans multiple zones/regions. This will support disaster recovery. If a particular availability zone or region goes dark, then failover will switch operations to the other zones/regions. You can also implement rate-limiting on IoT APIs to defend against malicious attacks coming from trusted devices that may have been compromised.

10.6.3 Securely Configure the Cloud Network

Each CSP offers the ability to configure firewall policies that restrict access from external sources and from sources internal to the cloud. Make use of these policies to restrict communications between subnets and nest sensitive resources such as databases within subnets. For example, do not allow devices to interact directly with cloud databases, instead require them to go through applications which are then configured via policy as trusted by database subnets.

Additionally, ensure that all management access to servers in the cloud is secured. This includes Secure Shell (SSH) access to manage virtual machines. Implement strict policies and procedures for managing the SSK keys that allow access to each VM. This includes restricting ownership and access to SSH keys within their VM directories.

10.6.4 Apply Encryption to Cloud Communications

All cloud interfaces should be encrypted. Use Transport Layer Security (TLS) whenever possible. When using User Datagram Protocol (UDP)-based protocols, implement the Datagram TLS (DTLS) security system. Limit TLS cipher suites used within your cloud configurations to those recommended in the latest version of NIST SP 800-52. This will reduce the risk of a successful cipher suite downgrade attack being run against your IoT systems.

10.6.5 Manage Cloud Identities

Management of IoT devices across a global organization will quickly become challenging if a standard naming approach is not agreed upon and implemented. Guard against future scalability issues by implementing a central directory. Use globally unique names for all devices.

10.6.6 Require Multi-Factor Cloud Authentication

A cyber security best practice today is to use multi-factor authentication for all access. Although this is often difficult to achieve at the device level, this requirement should always be implemented for access to the cloud services that support edge devices. Various MFA approaches can be implemented to include requiring SMS-based texts or email validation of codes prior to logging in. Look to protocols such as those from Fast Identity Online (FIDO) (e.g., FIDO 2) to implement cost-effective MFA methods.

10.6.7 Audit Cloud Services

Many CSPs will allow your team to conduct limited-scope penetration testing of the services running in the cloud. Ensure that this is allowable and if so schedule regular penetration tests against the cloud installation. Also monitor for adherence to all SLA metrics and promptly notify a CSP if any metric fails to meet requirements.

10.6.8 Monitor the Cloud

Make use of native CSP monitoring services and integrate into a wider enterprise monitoring strategy. Make use of third-party managed security service providers (MSSP) when necessary and feasible to monitor 24/7 for suspected malicious activity. CSP monitoring services can often provide near real-time inputs on actions occurring on the API and within the cloud itself.

10.6.9 Implement Cloud Identity Management

Establish an identity and access management (IAM) process for all credentials used to authenticate to the cloud, including those provisioned to edge machines.

Prefer the use of certificate and token-based approaches instead of passwords whenever feasible. Do not generate private keys for edge machines in the cloud unless absolutely necessary. Instead, generate private keys on machine and request certificates from your Public Key Infrastructure (PKI) Certificate Authority (CA). Do not share credentials (passwords or certificates) across devices.

Authentication of cloud interfaces consists of both authentication at the gateway or cloud service (e.g., "the server") as well as authentication of the edge machine. Authentication considerations vary somewhat based on the protocol employed. Use certificates for intra-cloud authentication between cloud servers and gateways.

Establish a global naming convention for IoT machines and services used in the system. IoT device identities should be unique across an enterprise. Establish device attribute metadata for each machine to support ease-of-lookup when needed. Establish group management capabilities to group IoT machines based on their profiles or usage characteristics.

Some CSPs provide "Identity Registries" which maintain the unique identifiers of each device and support lifecycle management of devices identities to include the provisioning, update and deletion of edge IoT device identities. Require authentication for write access to device identity registries.

10.6.10   Use Zero-Touch Provisioning

Standard such as Device Identifier Composition Engine (DICE) from the Trusted Computing Group (TCG) specify methods for provisioning cloud trust directly at manufacture time. Zero-touch provisioning allows implementers acquiring these cloud-secure MCUs to be provided with a list of Electronic Serial Numbers (ESNs) and associated public keys to register devices in their specific environment and securely bootstrap products into an operational environment. Contract with your MCU vendor enable zero-touch provisioning.

10.6.11   Role-Based Access Controls

Establish roles for applications, gateways, users and machines that interact with or operate within the cloud. Assign privileged roles to applications and users that are authorized to manage or configure edge machines. Assign standard roles to single purpose devices that require limited-scope read/write permissions or to gateways that provided limited functionality such as simply storing data. Example roles include:

- Gateway
- Privileged Gateway
- Management Application

- Device
- Privileged Device
- System Application
- Privileged Application
- Auditor
- Third Party

10.6.12 Secure Data in the Cloud

Categorize data elements generated by edge machines and transmitted to the cloud for storage and processing. Implement authenticated access to cloud data storage and apply encryption to database fields associated with sensitive information.

Configure cloud gateways to accept communications from only trusted devices and require encrypted communications. Document authorized ports and protocols on cloud gateways and disable non-authorized ports/protocols. Blacklist any unauthorized device that attempts to communicate and log and report the action. Keep all cloud infrastructure updated and patched.

10.6.13 Secure Web Services

A 2018 report by Barracuda Networks [30] found that vulnerabilities in mobile applications and web services allowed attackers to steal the credentials of IoT devices. IoT machines operate in an interconnected ecosystem, and weaknesses in any part of that system can result in compromise of individual system components.

Implement encrypted communications between the web application and the IoT machine. Use token-based authentication for access. Implement secure pairing methods for short-range RF transactions. Keep mobile devices and applications that access edge machines updated and configure storage of all key material in hardware. Develop web services in accordance with OWASP security guidance.

- https://www.owasp.org/index.php/REST_Security_Cheat_Sheet
- https://www.owasp.org/index.php/Web_Service_Security_Cheat_Sheet V18 of ASVS 4.0
- https://github.com/OWASP/ASVS/blob/master/4.0/en/0x23-V18-API.md

Also secure the Application Programming Interface (API) keys used by your applications by not exposing the API keys within public configuration management tools/repositories.

10.7   Secure System Users and Administrators

Any system is only as secure as the weakest link in that system. The weakest link is usually the user of the system that is untrained in the basics of cyber security. Users can be turned from a weakness to an advantage by educating them on the threats faced by the systems they use and the ways they can help to keep systems and data secure.

10.7.1   User Training

Establish training for system users. Training for users should include the following topics at minimum:

- The risks associated with use of their IoT systems. This includes a discussion of the threat actors that target their systems and the motivations for attempting to gain access to systems. Users should have a clear understanding why their systems are at risk of being compromised and the impact that a system/data breach would have on their organization and the customers of the organization.
- The types of IoT devices that are authorized to be used within the system. This includes any restrictions on devices that may be brought from home and added to the system, or any policies that limit use of certain types of devices within the system.
- The presence of any safety-related risks associated with the IoT system. This is especially important in human/robot collaborative systems where interaction with automated machinery may occur.
- General guidance on avoiding phishing attempts and other attempts to trick users into divulging sensitive account information.

10.7.2   Administrator Training

System administrators should also be trained on the proper function and security of their IoT systems. System administrator training should be extensive and should include:

- A detailed technological overview of the IoT system that includes a discussion of all sensitive data collected by the system, key architectural components of the systems, high value components of the system.
- Policies for allowable use of the IoT system. This includes a detailed description of the policies for integrating new device types into the network. For example, any requirement that a product be certified as secure by a third party. This can include procedures for bringing new devices online, procedures for monitoring

the security state of IoT machines, and the rest of the services. Also include proper procedures and timing of firmware and software updates within the system and proper procedures for patch management. Proper procedures for patch management of the system.

- General guidance on maintaining security awareness, to include training on the identification of malicious software and attempts to trick administrators into divulging passwords or other sensitive information (e.g., phishing attacks).
- Procedures for investigating and responding to incidents involving the IoT system. This includes requirements for escalating issues to senior technical and managerial leadership within the organizations and procedures for interacting with third-party organizations to perform investigations and remediations – for example, cloud service providers. This also includes procedures for conducting a root cause analysis (RCA) as well as requirements for maintaining a secure chain of custody during an investigation.
- Policies for performing penetration testing and vulnerability scanning of the system. This includes a description of any tools that should be used in the conduct of vulnerability analysis.
- Procedures for proper disposal of IoT system equipment. This includes any procedures associated with sanitization of data stored on IoT products or attached storage systems.

Proper training of users and administrators is required in order to ensure an IoT system is operated securely. Without proper training, users and administrators may fall victim to countless phishing attacks or may improperly dispose of IoT equipment. By educating users on why the IoT systems they use are targeted by bad actors and how users can play a role in defending their systems, they can become powerful allies that can aid in helping to secure large complex IoT systems.

10.7.3   Incident Response Planning

Another aspect of the human element of cyber security is associated with incident response planning. Planning for the response to incidents involving IoT systems is similar in nature to traditional I.T. incident response planning. The four phases of an incident response include (1) planning; (2) detection and analysis; (3) containment, eradication, and recovery; and (4) post-incident activity. The *planning* process involves:

- Identification and assignment of the Incident Response Team
- Establishment of a communication plan
- Development of a training plan that allows the team to execute and respond to simulated incidents
- Creation of a coordination plan. The coordination plan is essential for IoT systems as the myriad business-area owners responsible for each IoT system

or subsystem must be identified. These business area owners will be contacted immediately during an incident response to begin coordination.

Detection and analysis activities are related to the ongoing monitoring of system operations for anomalies. This also includes threat intelligence sharing to ensure that the Security Operations Center (SOC) team is looking for the latest attempts to exploit devices and services. Additionally, ongoing monitoring includes monitoring of activities occurring in the cloud and within the Radio Frequency (RF) environment on-premise.

Containment, eradication, and recovery include the actions related to quarantine of infected devices or computers. This is necessary to contain the threat and limit the spread of an infection or an adversaries' attempts to move laterally through the system. This also includes the process of eradicating the incident and returning to normal operations while continuously monitoring for a repeat of the incident.

Post-incident activity includes performing an RCA to determine the exact cause of the incident as well as how the implemented security defenses failed to achieve their goals. This also includes information sharing across not only the other security professionals within the organization but also trusted third-party organizations and importantly an update to the incident response plan to take into account the realized incident.

10.8 Conclusion

This chapter described approaches to enhance the cyber security of an IoT system. The chapter explored the many components which make up an IoT system and the vulnerabilities associated with the components. We discussed the unique threats to IoT systems and then provided recommendations for developing secure IoT products and systems. We then described specific controls that can be applied to IoT devices at the edge, to the network, and to the cloud as well as to the human users and administrators of IoT systems. Overall, we gave an in-depth analysis of vulnerabilities which exist in IoT systems and ways to mitigate such risks to ensure the integrity of the IoT system.

References

1. https://sebokwiki.org/wiki/Systems_of_Systems_(SoS)#Difference_between_System_of_Systems_Engineering_and_Systems_Engineering
2. https://www.zdnet.com/article/anatomy-of-the-target-data-breach-missed-opportunities-and-lessons-learned/
3. https://www.wftv.com/news/9-investigates/-no-one-wants-to-see-that-man-says-someone-was-watching-him-through-his-nest-camera/920933458
4. https://www.wired.com/story/police-body-camera-vulnerabilities/

5. https://pdfs.semanticscholar.org/e06f/ef73f5bad0489bb033f490d41a046f61878a.pdf
6. https://msdn.microsoft.com/en-us/library/windows/desktop/ee790621.aspx
7. https://www.owasp.org/index.php/Secure_Coding_Cheat_Sheet
8. https://tools.kali.org/forensics/binwalk
9. www.vdoo.com
10. http://www.peachfuzzer.com/resources/peachcommunity/
11. https://spdx.org/tools
12. https://www.xda-developers.com/the-snapdragon-samsung-galaxy-s9-has-a-gpu-stability-bug-that-can-be-exploited-to-trigger-remote-reboots/
13. https://www.iotsecurityfoundation.org/tag/iot-vulnerability-disclosure/
14. https://iotsecurityfoundation.org/wp-content/uploads/2017/01/Vulnerability-Disclosure.pdf
15. https://defensesystems.com/articles/2018/07/13/government-bug-bounties.aspx
16. https://iapp.org/media/pdf/resource_center/Privacy%20by%20Design%20-%207%20Foundational%20Principles.pdf
17. https://gitlab.com/brianr/CloudSA_IoT_WG/wikis/top_iot_vulnerabilities
18. https://github.com/craigz28/firmwalker
19. https://www.hex-rays.com/products/ida/index.shtml
20. http://binwalk.org
21. https://www.securetechalliance.org/wp-content/uploads/TEE-101-White-Paper-FINAL2-April-2018.pdf
22. https://en.wikipedia.org/wiki/Separation_kernel
23. https://docs.pivotal.io/nist/sc/sc-39.html
24. www.snort.org
25. https://www.shodan.io/
26. https://nvd.nist.gov/download.cfm
27. https://www.owasp.org/index.php/Category:Vulnerability_Scanning_Tools
28. http://int3.cc/products/the-shikra
29. https://cloudsecurityalliance.org/star/registry
30. https://www.computerweekly.com/news/252456406/IoT-application-vulnerabilities-leave-devices-open-to-attack

Part II
IoT Technologies for Smart Healthcare

Chapter 11
Healthcare IoT

Bahar Farahani, Farshad Firouzi, and Krishnendu Chakrabarty

> *We have a moral obligation to get healthcare to people who*
> *need it.*
>
> Ron Williams

Contents

B. Farahani (✉)
Shahid Beheshti University, Tehran, Iran

F. Firouzi · K. Chakrabarty
Department of ECE, Duke University, Durham, NC, USA

© Springer Nature Switzerland AG 2020
F. Firouzi et al. (eds.), *Intelligent Internet of Things*,
https://doi.org/10.1007/978-3-030-30367-9_11

11.1   Modern Healthcare Challenges

Many parts of the world are facing significant healthcare challenges to manage the rapidly increasing aging population, people with chronic issues, disease epidemics, child mortality, poor living conditions with lack of sanitation, lack of clean water, and rising pollution. While the need for medical care has increased over the last several years, the traditional hospital-centric model of care requiring patients to visit a physician when sick is still the norm. Managing chronic conditions requires that patients visit a clinic or hospital so that a physician may monitor disease progression and adjust treatment based on clinical observations. Generally, hospitals are based on a reactive, physician-and-disease-centered model that does not include patients as active participants in the medical process. The main obstacles and challenges faced by hospital-centered healthcare practices are listed below [1, 2].

- *Time and perspective limitations* – Because the number of people dealing with illness or disabling conditions continues to grow, physicians are unable to spend significant quality time with patients. Shorter exam times leave physicians with less perspective into the patient's daily life including diet, sleep, physical activity, social interactions, etc. Understanding all of these aspects is important in appropriately diagnosing and treating patients effectively.
- *Adherence monitoring* – Physicians are often ill-equipped to monitor patient compliance with prescribed treatments such as rehabilitative exercise, medication, or dietary restrictions. Non-compliance with treatment can increase the risk of future hospitalization which increases a patient's healthcare cost and the economic burden [3].
- *Increasing geriatric population* – The number of older adults (60+ years) worldwide is anticipated to grow by more than 200%, from 841 million people in 2013 to over 2 billion people by 2050; therefore, additional medical facilities and healthcare resources will be required to treat a larger geriatric population [1].
- *Urbanization* – The world health organization (WHO) forecasted that by 2015, 70% of the global population would live in urban areas, implying that large cities would require greater healthcare infrastructure to care for rising populations. Additionally, large urban areas are a more likely center of disease epidemics, allowing contagious diseases to quickly infect densely populated areas.
- *Healthcare workforce shortages* – As the need for healthcare services increases, the demand for physicians, surgical staff, nurses, caregivers, and medical labora-

tory staff able to support the healthcare system in both urban and rural areas also increases. One possible solution to this challenge is to increase the prevalence of telemedicine.

- *Rising medical costs* – One of the largest challenges in the medical field is the rapidly rising cost of healthcare. For example, the cost of diabetes care in the United States jumped to approximately $245 billion in 2016, an increase of 21% since 2007 [4].

11.2 What Is IoT-Driven Healthcare: Transitioning from Hospital-Centric to Patient-Centric

Patient-centered care (PCC) is an emerging healthcare model focused on the individual medical needs of patients, which originally was coined by the Picker/Commonwealth Program developed by the Picker Institute in 1988. The Institute of Medicine defined PCC in a 2001 landmark report as [5] "Healthcare that establishes a partnership among practitioners, patients, and their families (when appropriate) to ensure that decisions respect patients want, needs, and preferences and that patients have the education and support they need to make decisions and participate in their own care."

While many initiatives have provided evidence of PCC's success on smaller scales, the potential of PCC has yet to be realized on a larger scale due to its conflicts with the existing hospital-centric model. PCC is not meant to eliminate the need for hospitals or clinics; on the contrary, PCC leverages these institutions in a shared model of patient care through the utilization of IoT. As discussed in previous chapters, IoT is a convergence of telecommunication, sensors, actuators, cloud computing, and Big Data through the Internet to provide specific services. IoT can be customized to meet the challenges of modern healthcare. Healthcare system can be categorized into three main areas: i) large healthcare institutions (i.e., hospitals), ii) small clinics and pharmacies, and iii) non-clinical environments (i.e., patient homes, communities, rural areas lacking healthcare). To understand IoT's role in the healthcare field, one must first understand the operations of each area. Because IoT has the potential to make a specific operation less costly and more reliable or timely, it would eventually impact operations in other areas, resulting in a more stable, self-sustaining healthcare ecosystem. Next, we will explore the areas where IoT technology could take on a key role [1].

- *Hospitals:* Hospitals are growing to admit and treat an increasing number of patients. In order to support operations, hospitals constantly rely on advanced technology. The addition of smart ambulances, integrated with IoT, would allow in-ambulance diagnosis, enabling medical staff to arrange for appropriate treatment before the patient even arrives at the hospital. Smart ambulances would require dependable diagnostic medical sensors, the secure communication link with the hospital, and intelligent workflow management tools designed to help hospitals prepare operating rooms or other needed tools in advance. In addition, surgery could become more efficient and proactive if IoT were utilized to exchange information and orchestrate activities among medical staff, physicians, and medical devices in the operating room. In a similar vein, IoT could play a role in other hospital areas such as the intensive care unit (ICU), primary care unit (PCU), or other specialized areas.

- *Clinics*: Primary care physicians, the main source of healthcare for many patients, could also benefit from IoT in many ways. Doctors could more efficiently access virtual laboratory reports prior to examining a patient. IoT would also enable physician offices to verify insurance coverage in real time or provide secure, bidirectional communication for making appointments without the need for telephone calls. IoT would also serve as a cost-efficient solution for connected pre-visit screenings, enabling physicians to invest more quality time with individual patients. Doctors could also prescribe low-cost IoT smart devices designed for self-testing at home.

- *Anywhere (non-clinical settings)*: There is currently much excitement around IoT's role in telehealth or remote care, which benefits patients in exciting ways:

  - *Mobile clinics:* Interest in mobile clinics is growing worldwide. In African countries mobile clinics have proven to provide high-quality, low-cost care for at-risk populations without access to basic healthcare in rural areas. Because mobile clinics have limited medical facilities, IoT could greatly improve their infrastructure and capability for care by enabling them to collaborate with hospitals remotely to make diagnoses or other treatment determinations [6].

  - *Telemedicine*: Telemedicine is the primary stakeholder in IoT today. It is estimated that the telemedicine market is worth approximately $23, 244 million worldwide. The telemedicine market is expected to expand by 18.8% to reach a value of $66,606 million by 2021 [7]. Because most of the world's population has access to smartphones able to connect with other sensors, IoT has much to offer in this area. Smartphones enable patients to self-test health and generate data that can predict future health or disease issues. In addition, doctors and medical staff can see patients virtually through teleconferencing. In the future, IoT could further improve telemedicine with the addition of networking sensors for tele-screening. An increase in virtual care services would enable hospitals to function with fewer healthcare staff.

  - *Wellness*: The well-known proverb that "an ounce of prevention is work a pound of cure" is especially true in the area of healthcare. Many people are beginning to understand that preventative care and maintaining wellness is preferable to suffering or requiring preventable medical care later in life. Around the world, patients are taking advantage of IoT connectivity via sports technology, wearable fitness trackers, and self-testing medical devices that provide continuous health and wellness monitoring. IoT could be useful in creating an economic framework that reduces health insurance costs through wellness rewards, based on health monitoring. Establishing a more efficient, patient-centered system could be one more way to encourage patients to become active partners in healthcare.

– *Smart homes*: Telemedicine services are useful for reducing the number of hospital or clinic visits for elderly, disabled, or chronically ill patients. However, a technical infrastructure in their homes would provide efficient, inconspicuous support, enabling patients to live more independently while enjoying a greater quality of life and well-being. Ambient assisted living (a smart home's framework) utilized in Europe serves as an ambient intelligence that improves a person's capability through sensitive, responsive digital environments. IoT has already found its way into the home automation market as homeowners control systems and appliances using smartphones. IoT could also be applied to the healthcare needs of geriatric patients. For example, smart toilets could handle regular urine testing without requiring a physician visit. Smart homes could also adapt lighting or noise levels to meet the sensory needs of individuals with autism [8].

– *Smart cities/connected communities:* People are migrating to urban areas all over the world, increasing the demand on city infrastructures. Efforts are being made globally to create smart cities and connected communities by connecting and integrating infrastructures in order to make urban areas more affordable, efficient, and sustainable. IoT architecture is being used to provide information at the right places and times so that citizens can make decisions more quickly. Because healthcare is an integral part of urban areas, it has been included in the infrastructure of smart cities in many ways. For example, widespread environmental sensors can stream data to centralized locations accessible by the healthcare system. Environmental information about pollutants, temperature, humidity, water contamination, or allergens could be distributed to citizens in a timely manner, helping prevent health issues. For example, asthma patients could be notified to avoid areas with high pollen or dust counts. IoT would also be benefit the healthcare system by integrating smart traffic lights to assist with ambulance routing, saving lives through cooperation with transportation infrastructure. IoT's future is bright when it comes to healthcare and wellness; however, it requires a smart ecosystem enabling interconnection among stakeholders including physicians, patients, hospitals, cities, and communities.

11.3 Benefits of Adopting IoT Healthcare

Healthcare research brings together a wide range of disciplines and fields, in such a way that scientist in medicine, microbiology, biomedical engineering, computer science, and Big Data analytics frequently find themselves working and collaborating on related projects. Considering all these areas of expertise together, it becomes clear that many gaps remain between them; these gaps present major technological challenges in the way of the development of a unified and highly adaptive framework for healthcare. In our view, the most direct way to develop this framework is to construct an IoT-based solution, in order to facilitate breakthroughs in all areas mentioned above.

Advances in the IoT help create significant advances in healthcare. For instance, technologies such as microfluidic biochips and wearable biosensors can improve clinical diagnostics in a variety of applications, from the laboratory to the hospital. In the foreseeable future, IoT-enabled devices will allow health providers to routinely assess patients who suffer from breast, lung, and colorectal cancers and perform point-of-care molecular testing as an aspect of standard care. This will help provide physicians with the information they need to create truly data-driven treatment plans, significantly improving the chance of a successful recovery.

When these tests are time-stamped, location-tagged, and also tagged with data on the testing environment and other situational information, as well as personal information such as age, weight, height, and gender, a data fabric will begin to take shape, spotlighting not only the patient's condition but also overall patterns in the population as a whole (e.g,, helping predict an outbreak of an epidemic). In short, IoT-enabled healthcare (e-Health) can move disease research forward, enable more accurate diagnoses at the point of care, and speed up the development of beneficial pharmaceuticals. It should be noted that IoT e-Health is not just the simple stack of different worlds. Instead, these pieces are networked together in order to assess, predict, and adapt in close to real time.

As the above vision is realized, the benefits of adopting IoT e-Health can be summed up as follows (see Fig. 11.1) [1, 9]:

- *All-encompassing* – IoT e-Health has a holistic solution for everyone needs and it is useful in diverse areas including beauty, health monitoring, exercise, and patient safety.
- *Resiliency* – The e-Health framework can also be self-learning and resilient to inaccuracies.
- *Seamless fusion and integration* – IoT e-Health overcomes technological barriers to enable diverse or complex technologies to collaborate.
- *Big Data processing and analytics* – IoT e-Health can effectively process, analyze, and manipulate the tsunami of multi-modal, multiscale, distributed, and heterogeneous data created by network sensors within a reasonable timeframe, resulting in the timely receipt of actionable information.
- *Personalized service or content* – Big Data analytics and IoT can widen the possibilities to meet the needs of personalized healthcare or treatments, playing a key role in personal well-being. For example, Big Data analytics and machine learning can be utilized to predict health conditions such as heart attacks, cancer development, or infections before they actually occur, enabling patients and physicians to act quickly.
- *Lifetime monitoring* – Patients can benefit from the receipt of comprehensive, past, present, and future health data.

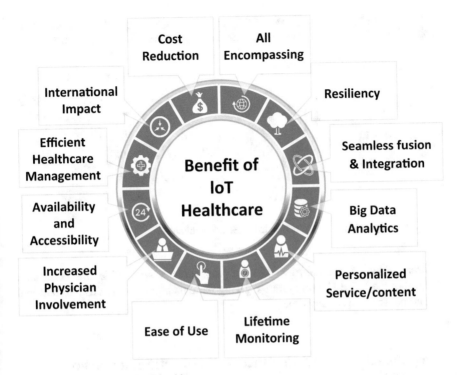

Fig. 11.1 Main benefits of IoT healthcare

- *Ease of use* – IoT e-Health can be easily adopted by users since they only require clicks on wearable devices and/or some simple operations on smartphone applications.
- *Cost reduction* – Because IoT e-Health is able to integrate diverse technologies, there is no need to pay for separate technologies and patients are better equipped to monitor their own health, enabling them to only pay for physician consultation when health status drops out of a recommended range.
- *Increased physician involvement* – Because physicians can obtain patient health data in real time, fewer intensive exams are needed. Additionally, doctors are able to monitor a higher number of patients through healthcare IT systems if the organizational structure can evolve to optimize the use of real-time data through telemedicine.
- *Accessibility and availability* – Patients, caregivers, and healthcare professionals can access e-Health data or services at any time without any geographical barriers.
- *Online assistance* – Allows 24/7, real-time connection with healthcare professionals such as physicians, nurses, etc.
- *Efficient healthcare resource management* – IoT guarantees that patients can learn about their health status and equips physicians to easily monitor patient status, driving more efficient use of healthcare resources.

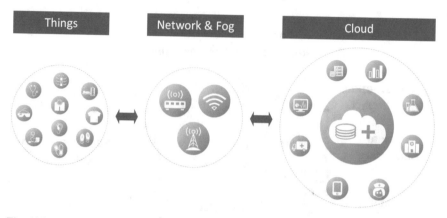

Fig. 11.2 A layered view of fog-driven IoT e-Health architecture

- *International partnerships* – Health professionals worldwide are connected through the IoT e-Health systems, giving patients greater access to international medical facilities at any time.

11.4 Fog-Driven IoT Healthcare Architecture: A Layered View

In this section, we explain the general architectural elements required for IoT e-Health systems. As shown in Fig. 11.2, this system consists of three main layers: i) *things layer*, ii) *network layer*, and iii) *cloud layer* [1, 9].

11.4.1 Things Layer

With a rich set of connected IoT medical devices, patients can monitor their health data in real time on any computer or mobile device, and their information is securely synchronized with a cloud-based e-Health platform. The main key requirements of connected medical devices are [10]:

1. *Unobtrusiveness*: An essential requirement in the design of wireless medical sensors relates to their lightweight and miniature size. These characteristics allow both noninvasive and unobtrusive continuous monitoring of health. In this contact, energy harvesting as well as flexible and printed batteries hold promise for wearable devices.
2. *Security and privacy*: The essential design fundamentals of a connected IoT medical device and eventually the entire ecosystem are security and privacy. For

example, data integrity must be ensured where the sensors must fulfill the privacy requirements provided by the law.
3. *Interoperability*: The main objective is to reform the chaotic and at times dysfunctional nature of information exchange among healthcare stakeholders such as hospitals, doctors, and patients.
4. *Low-power communication*: It is very important to consider the trade-off between communication, computation, and energy consumption to be able to design an optimal system. Many constraints and policies such as sampling rate can impact the selection of the communication technology. For example, instead of sending raw electrocardiogram (ECG) data from sensors, we can perform feature extraction on the wearable devices and transfer only information about the particular event.

The state-of-the-art IoT e-Health devices are typically classified into two categories:

1. *Physical sensors*: In general, any wired/wireless medical device can be used in an e-Health ecosystem to track the physical wellness of patients and digitally monitor their health.

    (a) *Electrocardiogram (ECG)*: Electrical activity of the heart. ECG is a continued waveform showing the contraction and relaxation phases of the cardiac cycles.
    (b) *Heart rate*: The speed at which the heart beats.
    (c) *Heart sounds*: A record of heart sounds, produced via stethoscope (chest microphone).
    (d) *Electroencephalogram (EEG)*: Electrical activity generated by the brain. These signals enable doctors to better monitor and tackle brain diseases such as seizure.
    (e) *Electromyogram (EMG)*: Electrical activity produced by skeletal muscles.
    (f) *Blood pressure*: Arterial pressure measured in the vessels and chambers of the heart which shows the performance of the heart and the resistance of the vessels. Blood pressure is represented by two numbers: i) systolic, highest level your blood pressure reaches when the heart beats, and ii) diastolic, lowest level blood pressure reaches when the heart relaxes between beats. According to American Heart Association, there are five blood pressure ranges as follow:

        i. *Normal*: Blood pressure numbers of less than 120/80 mmHg.
        ii. *Elevated*: When the readings consistently range from 120 to 129 systolic and less than 80 mmHg diastolic.
        iii. *Hypertension stage 1*: When pressure consistently ranges from 130 to 139 systolic or 80 to 89 mmHg diastolic.
        iv. *Hypertension stage 2*: When blood pressure ranges at 140/90 mmHg or higher.

v. *Hypertensive crisis*: When blood pressure exceeds 180/120 mmHg. In this case, patients experience signs of possible organ damage, e.g., chest/back pain, shortness of breath, change in vision, or difficulty speaking.

(g) *Blood glucose (sugar)*: Concentration of glucose present in the blood. For non-diabetics, it should be between 70 and 130 milligrams per deciliter.

(h) *Hemoglobin*: It is a protein in red blood cells. Oxygen entering the lungs adheres to this protein, allowing blood cells to transport oxygen.

(i) *Respiration rate*: It is the number of breaths you take per minute. The normal respiration rate for an adult at rest is 12 to 20 breaths per minute.

(j) *Blood oxygen saturation level*: Oxygen saturation is concentration of oxygen that is dissolved or carried in a given medium. In healthcare, it represents the fraction of [oxygen]-saturated hemoglobin relative to total hemoglobin (unsaturated + saturated) in the blood. Note that a healthy individual at sea level should exhibit oxygen saturation between 96% and 99%.

(k) *Skin conductance (SC)*: It is also known as electrodermal response, electro-dermal activity (EDA), or as galvanic skin response (GSR). Skin resistance mainly changes with the state of sweat glands in the skin. As perspiration increases, more sweat glands begin to conduct electricity, which in turn increases skin conductance.

(l) *UV radiation*: It is intended to measures your exposure to ultraviolet radiation that can adversely impact body skin and eyes which can ultimately lead to cancer.

(m) *Context-aware body movement*: Body motion can reveal so many information about an individual. For example, it can show whether a person is in light sleep or deep sleep. Using these sensors in diseases such as Parkinson, caregivers can determine how effective is the medicinal dosage based on the body movement. These wearable sensors can also be used in several applications such as fall detection, activity monitoring, or measuring calories burned.

(n) *Hydration level*: This wearable enables individuals to manage hydration status and address several important questions including when to drink, what to drink, and how much to drink. There are several approaches to implement this functionality in wearables; however, the most common approach is to use infrared light to measure water in the blood.

(o) *Ultrasound scanning*: Nowadays, there is also an opportunity to perform 3D ultrasound scanning via wearable devices which can be exploited in several health applications. For example, it allows pregnant mothers to track the movement of the baby. Three-dimensional wearable scanners can also be used to perform breast scans and localize breast lesions.

(p) *Pregnancy wearable*: There are several sensors for pregnancy. For example, wearable sensors can measure the electrical activity of the uterine muscle of mothers, which in turn can be used to compute baby contractions. There are also several wearables which allow to boost and hear the sound of your baby's heartbeat.

2. *Virtual sensors*: Virtual sensors use software and mobile applications to gain patient's health and contextual data from the environment. A virtual sensor includes many categories such as remote monitoring, remote consultation, diagnostic, patient health record, nutrition, and medical reference applications.

11.4.2   Network Layer

There is a need to enable multi-protocol data communication technology between devices at the edge as well as gateways, fog nodes, and the cloud servers. As discussed in the third chapter of this book, common groups of IoT network technology are as follows:

- *Body area network (BAN) and personal area network (PAN)*: The scale of this network is a few meters. This network is used to connect the edge nodes (e.g., sensors, actuators, devices, control systems, and assets) to gateways and fog nodes. BAN/PANs are usually wireless and more constrained by antenna distance (and sometimes battery life) than LANs. A notable example of BAN/PAN is the Bluetooth technology, mostly used for connecting wearable devices to portable devices like smartphones, laptops, tablets, etc.
- *Home area network (HAN)*: The scale of this network is a few tens of meters. At this scale, common wireless technologies for IoT include ZigBee and Bluetooth Low Energy (BLE).
- *Field area network (FAN)*: The scale of this "open space" network is several tens of meters to several hundred meters. FAN typically refers to an outdoor area larger than a single group of house units. 6LoWPAN is one of the most well-known technologies for FAN.
- *Local area networks (LAN)*: The scale of this network is up to 100 m. This term is very common in networking, and it is therefore also commonly used in the IoT space when standard networking technologies (such as Ethernet or Wi-Fi) are used.
- *Metropolitan area network (MAN) and wide-area networks (WAN)*: The scale of MAN is up to a few kilometers and the scale of WAN is more than a few kilometers. These networks provide connectivity for data and control flow between the IoT gateway (or fog nodes) and the cloud. WANs can be wired (e.g., using fiber-optic cable) or wireless such as 3G/4G mobile networks or even satellite networks. It should be noted that with the introduction of low-power wide-area network (LPWAN) such as NB-IoT and LoRa, there is a possibility to directly connect endpoint IoT devices (without any gateway) to cloud servers without the need to use 3G or Wi-Fi.

In this context, there is a vast variety of protocols. Figures 11.3 and 11.4 show the IoT e-Health protocol stack. Note that the selection of the best connectivity and the communication protocol depends on the application and the specific use-case

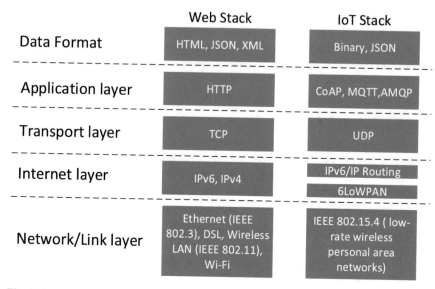

Fig. 11.3 IoT e-Health protocol stack compared to Web stack

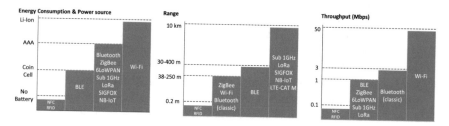

Fig. 11.4 Power source, range, and throughput of various network technologies

[11]. For example, a Wi-Fi connection is ideal when transferring many documents. However, BLE works well for short-range, low-power communications.

Today's e-Health cloud architectures are not designed to adequately handle the volume, variety, and velocity of data generated by e-Health devices. To tackle this issue, there is a need to revisit the network architecture, pushing certain data, processing, and services away from the massive centralized infrastructure of the cloud to the edge of the network where the data originates. An edge node (sometimes nicknamed fog node) is defined as a device with integrated computing, storage, and networking [12]. The edge node is inserted between the cloud and all IoT e-Health devices add two important features to the system [1, 9]:

- *Real-time analytics and decision making*: Some important applications of IoT e-Health such as myocardial infarction (MI) detection cannot tolerate latency. In these time-sensitive applications, it is a necessity to process and act on health data in seconds. In such applications, it is not practical to transfer patients' sensitive medical data, vital signs, and bio-signals across a wide geographical area in the presence of various environmental conditions and store and process them in different data centers or the cloud. Instead, moving intelligence to the edge is a promising approach to eliminate latency and evolve IoT e-Health solutions. In this approach, an edge node with localized processing capability enables us to respond more quickly than the cloud by making time-sensitive decisions closer to the source of data. Thereby, this solution results in a more efficient solution that can better handle low-latency demands of e-Health applications.

- *Traffic reduction on overburdened networks*: Considering the limited network bandwidth, it is not practical and in certain use cases even not necessary to transfer enormous volume of big raw data from millions of e-Health devices to the cloud. Edge computing reduces the data transport costs, which can be significant for data-intensive applications, such as genomic-association analysis, generating several GBs of raw time-series data within a day. In this regard, edge nodes can process, filter, and compact the medical data before delivering it to the cloud to dramatically minimize bandwidth requirements.

Other important tasks of a fog node are explained below:

- *Two-way connectivity*: Fog nodes establish a secure reliable bidirectional data flowing between e-Health devices and the cloud platform. An edge node gathers feeds in real time from health devices using an appropriate protocol and, after processing, sends the corresponding summary periodically to the cloud to facilitate the long-term data sets aggregation, exploration, analysis, and globally intelligent decision making. On the other hand, it might also need to receive commands, configuration data, etc., from the cloud. Note that the edge node is also dealing with the compliance challenges associated with connectivity such as protocol translation, security, switching, routing, and networking analytics. For example, nodes might not be assigned with a public IP address. Therefore, to enable reachability from the cloud, an edge node can rely on different mechanisms such as WebSocket, MQTT (message queue telemetry transport), and IP tunneling.
- *Time-series data capture*: Edge nodes can either use an interrupt or a polling mechanism for data acquisition. Depending on the application, time precision might also be required to be able to extract the trend over time. In such a case, accuracy is increased if the time stamp is generated in close proximity to the e-Health device generating the data. In this regard, edge nodes time-stamp the incoming data and store it in a historian database.
- *Transient data storage*: Fog nodes are required to provide short-time historical storage for e-Health device data. For example, filtering outliers of data (i.e., in the case of deviation from normal) depend on previous samples of the data.
- *Device management*: This includes device discovery, device registration, and device control.
- *Edge processing*: A rich set of applications can be executed on the edge node. For example, edge nodes are capable of on-demand data cleaning, data normalization, filtering, data reducing, compressing, integrity check, and formatting, data

sharing, data purging, and data buffering. An edge node may also include, as an example, signal processing, concurrent streaming, event handling, embedded web server, embedded WebSocket server, etc.

- *Streaming edge analytics*: In some e-Health applications such as anomaly detection, it is a major necessity to learn actionable insight and actionable information in real time close to the local context. Keeping this in mind, edge nodes should be able to analyze the stream of device and sensor data with millisecond response time. To do so, edge nodes can incorporate lightweight feature extraction, data mining, time-series pattern recognition, machine learning, rule-based event processing, and automated reasoning.

- *Data delivery*: Edge nodes can rely on either of the following message-exchange techniques to deliver the IoT data: (i) message-based, (ii) request-based, and (iii) publish-subscribe.

- *Security and data protection*: To protect patient data, the fog node offers multi-layer security measures for authentication, encryption, and access control to fully meet the requirements of FDA standards.

- *Flexible integration*: Considering the availability of different device vendors and OEMs, edge nodes should implement a wide-range of interface standards to maintain interoperability. To address the integration concern, edge nodes should be compatible with a large variety of communication protocols and peripherals (e.g., UART, SPI, and USB), PAN and WSN protocols (e.g., RFID, BLE, Zigbee, Wi-Fi, 3G/4G, and Ethernet), and wired protocols (e.g., Ethernet).

- *Protocol translation*: Another challenge arises from the fact that there is a large number of communication protocols at different levels of abstraction as follow:

  - *Network layer*: An IoT e-Health network is scattered among various networking protocols (e.g., BLE, ZigBee, Wi-Fi). To bridge the gap among these protocols, the edge node needs to convert and translate the incoming stream to an appropriate format and propagate it to the destination network.
  - *Message layer*: A large number of application-level protocols (e.g., MQTT, CoAP, and XMPP) or processing messages exists. Thereby, it is very crucial that edge nodes despite the underlying differences of standards be able to transfer messages among different protocols.
  - *Data annotation layer*: Different organizations proposed distinct standards for integration, exchange, and retrieval of e-Health information (such as HL7). Whenever it is required, edge nodes should be capable of understanding, processing, and translating the data.

11.4.3 Cloud Layer

The cloud platform can benefit from a multi-layer architecture that consists of the following layers (see Fig. 11.5) [1, 9].

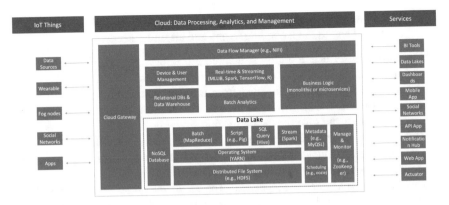

Fig. 11.5 The multi-layer architecture of the e-Health cloud

1. *Cloud gateway*: This layer includes many built-in features needed to create a connection between IoT things such as e-Health devices, sensors, actuators, fog/edge nodes, BI tools, dashboards, social networks, external databases, applications, and the cloud. This layer delivers ultimate flexibility to select an appropriate communication method based on different protocols (e.g., MQTT, WebSocket, Representational State Transfer APIs, ODBC, JDBC, etc.) that suits the requirements of the given health application.

2. *Data lake*: James Dixon, the CTO of Pentaho and the creator of the term data lake, defines this term as "If you think of a traditional database as a store of bottled water cleaned and packaged and structured for easy consumption, the data lake is a large body of water in a more natural state. The content of the data lake stream in from a source to fill the lake, and various users of the lake can come to examine, dive in, or take samples." The main advantages of a data lake are as follow:

   (a) It is capable of deriving values from many different data sources.
   (b) It can store and converge both structured and instructed data from sensor data, to e-Health documents, to social media data.
   (c) It can efficiently handle a growing amount of data by leveraging a distributed file system such as the hadoop distributed file system (HDFS).
   (d) It can process a large and diverse set of data.
   (e) It is very flexible in a way that it can be extended by several distributed applications to enable different access and process patterns of the stored data: batch (MapReduce), SQL Query (Hive, Impala, Spark SQL), Script (Pig), Stream (Spark), and many other processing engines.
   (f) It changes the old Early-binding ETL (Extract: retrieving raw data: Transform: structuring the raw data and storing it in a data repository; Load: loading the structured data for analysis) paradigm of the traditional databases and data warehouses to process the data. Indeed, a data lake follows a

Late-binding ELT approach, leading to more flexibility and faster access to all data at any time responding to any and all future needs.

3. *Data warehouse*: It is a highly structured repository used mainly for reporting and representing an abstracted picture of the e-Health system. The data stored in this repository can be uploaded from the data lake or from the operational systems (such as sales). However, note that before storing any data, we need to process, model, and give the data a specific structure.

4. *Data flow manage and orchestration*: This layer is responsible for managing, automating, and orchestrating the cloud sources (e.g., data ingestion, data storage, data processing, and visualization).

5. *User, device, and data management*: The cloud integrates data from multiple sources. It captures data from many fog nodes and stores the data in a safe and secure manner. In this way, the data is always there to be accessed by those engaged in patient care. This platform seamlessly integrates with non-sensor sources such as EHRs, e-prescriptions, web sources, and more. As a result, patients, physicians, or any other member of a patient's care team can access vital health data when needed. This significantly increases collaboration across all disciplines, increasing the efficiency of the healthcare plan. Moreover, cloud-based platforms offer a unified schema to capture and query transactions. In doing so, the versatility to create new applications is increased. This module is also used to manage users, groups, devices, and fog nodes and access permissions and roles.

6. *Big Data analytics*: This is a key component for analyzing medical data. The use of analytics allows the platform to use event- and rule-based processing, data mining, machine learning, and automated reasoning-based algorithms on stored historical records. This way, the platform can make meaningful insights about patient health. Having these early health insights could be a game-changer for a patient who can begin to take preventative action against an otherwise fatal ailment. The configurations of the connected e-Health devices can also be adjusted using the extracted insights. For instance, users may alter the frequency and type of information collected, as well as the multimedia (images and videos) resolution. Note that, there are two different engines in the analytical module. The first one handles all requests that are subject to (near) real-time constraints. The second one deals with batch data and extracting historical intelligence.

7. *Output integration*: This is typically based on an Enterprise Service and Integration Bus such as Apache Camel with a rich set of protocols (e.g., REST API, Message Broker, Websocket, etc.) that enables connection with any system, application, or portal. It should also be noted that this module can exploit in-memory databases (such as Redis and HBase) to answer fast incoming queries by merging and caching the results from warehouse, data lake, analytics, etc. However, this database only stores very recent data and results.

11.5 Key Services and Applications of IoT Healthcare

In this section, we will briefly review a few of the applications made possible within the IoT e-Health ecosystem [1, 9].

11.5.1 Mobile Health (m-Health)

Cloud technology makes it possible for patients to access health information no matter where they are. m-Health can be accessed via smartphone applications or a web-based cloud dashboard. Medical professionals or caregivers can also make use of the IoT platform and m-Health smartphone applications through P2P video and audio, which enable patient/caregiver interaction anywhere and at any time. This technology allows patients to more conveniently receive diagnoses and treatment such as prescription medication refills without a hospital or clinic visit. Because healthcare professionals can access a patient's holistic health database through the cloud platform at any time, they are able to receive the most effective and appropriate treatment possible. Therefore, a holistic IoT e-Health system would grant the patient's better access to the most appropriate care available.

11.5.2 IoT in Ambient Assisted Living

An IoT e-Health ecosystem would also enable geriatric or disabled individuals to live longer and more independently. The geriatric population is growing significantly and current estimates suggest that 20% of the world's population will be age 60 or older by 2050. Increased age brings a greater risk of chronic illnesses such as cancer, type II diabetes, stroke, etc. The integration of IoT technology through ambient assisted living (AAL) allows real-time, location-specific monitoring of patient living parameters (i.e., heart rate) as well as environmental conditions [8].

11.5.3 IoT Medication

IoT can also be useful in determining compliance with medication regiments and in the prevention of fatal adverse drug reactions (ADR) [13]. A combination of smart pill bottle technology, wearable audio sensors, and classification capability is able to accurately determine medication compliance. Research shows that the ADR rate in hospitals is approximately 6.5% worldwide [14]. IoT medication, in partnership with NFC-equipped smart pill bottles, a cloud-based electronic medical records

(EMR), and a knowledge-based system could help to prevent the administration of incorrect drugs and their subsequent adverse reactions.

11.5.4 IoT to Assist Individuals with Disabilities or Special Needs

In 2011, the world health organization (WHO) completed its initial survey concerning disabilities and discovered that more than 1 billion people, almost 15% of the world's population, lives with some kind of disability. IoT e-Health could significantly improve quality of life to this population through automated, dependable, resistive technologies. For example, smart gloves with cost-efficient inertia sensors have been developed to enable those living with hearing loss to communicate with others who are not fluent in american sign language (ASL). Smartwatches can be utilized to assist patients with speech disorders to learn speech functions in rural areas. IoT systems are also able to improve wheelchair access in smart cities for those with mobility issues, collect information about individual special needs remotely, or enable schools to make special needs education more obtainable and productive for disabled children. One example of an IoT device available today is the Wireless Nano Retina Eyeglasses that enable communication between retinal implants to support real-time vision fine-tuning for blind patients.

11.5.5 Smart Medical Implants

Beyond wearable medical devices, IoT e-Health is also introducing very sophisticated, reliable implantable medical devices designed to improve or restore human bodily functions. Examples of some of these electronic implants include deep brain stimulation (DBS) systems that use controlled electrical pulses to stimulate brain areas to lessen involuntary movements caused by neurological disorders like Parkinson's disease or essential tremors, pacemakers needed for regulating heart rhythm, and cochlear implants that use electrodes inside the inner ear to restore hearing. These electronic implants utilize tiny circuits that house analog front ends, micro-controllers, and battery power management. The IoT model continues to be explored to help make medical implants more efficient, secure, and context-specific. For example, research continues to find ways to improve DBS using inertia sensors worn on the arms or legs. IoT also provides a fundamental structure for the programming and tele-management of cochlear implants so that patients do not have to make unnecessary trips to implant centers to receive device programming services.

11.5.6   IoT for Early Warning Score (EWS)

Because a large variety of bio-sensors create a huge volume of medical data from thousands of patients, it is not practical to monitor every patient directly. An IoT-based early warning score (EWS) is able to assist medical professionals by accurately predicting or quickly detecting the deterioration of a patient's health [15]. Basically, EWS processes and analyzes six basic vital signs including respiratory rate, body temperature, pulse rate, systolic blood pressure, oxygen saturation level, and degree of consciousness. These signs are then translated to a composite patient deterioration risk score. Each processed vital sign is scored so that the size of the score corresponds to the level of divergence from its established norm. When all the scores are combined, a composite score is created, reflecting the total level of deterioration risk for a particular patient. The EWS approach is in widespread use in hospitals around the world, and many studies have indicated that EWS is able to forecast patient complications approximately 24 hours before the actual complication arises. In this context, IoT e-Health systems allow medical professionals to constantly monitor vital signs remotely and then calculate the patient's deterioration risk level remotely as well. This is a revolutionary model able to spot deterioration early enough to save lives and reduce mortality rates.

11.5.7   IoT-Based Anomaly Detection

There are two apparent drawbacks related to EWS. First, EWS does not evaluate all biosignals such as hydration and sweat levels. In addition, EWS is based on supervised machine learning and may not be able to accurately capture deviations that are outside the supervisor's real knowledge. Overcoming these drawbacks could be addressed by utilizing an anomaly detection system that learns and evolves continuously over time [16]. For example, hierarchical temporal memory (HTM) is an unsupervised machine learning technique, which can be used to detect anomalies (see Fig. 11.6). This model enables medical professionals to quickly recognize anomalies pointing to a potentially severe problem such as a stroke or heart attack. The image below depicts the main steps such a system uses to perceive anomalies.

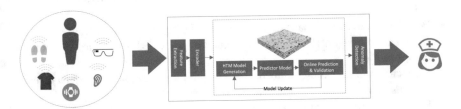

Fig. 11.6 IoT-based anomaly detection system

In short, timed bio-signals are monitored by sensors in the IoT e-Health device and those signals are transmitted to fog nodes. The fog nodes then sort, process, extract, and compress the data before the processed, timestamped data is sent to the cloud through a secure connection. Next, the data is converted to a sparse distributed representation (SDR). The SDRs are then put through the HTM model, which attempts to mimic neocortex brain activities by learning temporal SDR patterns constantly. Over time, the HTM creates a complex, adaptive model that is able to predict the next incoming SDR sequence. If the next actual sequence does not match the predicted sequence, an anomaly alert is triggered. Reducing data variance induced false-positive results requires that HTM calculate a time-varying average of the error that is compared with the distribution of all errors. This comparison enables HTM to forecast the accuracy level of the predicted anomaly or alert. The outputs generated by the risk analysis and warnings can be viewed via a dashboard and sent to patients and medical professionals tasked with providing the best possible treatment.

11.5.8 Population Health Management

IBM has estimated that available medical data will double every 73 days by 2020 [17]. Big Data analytics can be used to help comprehend the data and extract deep insights in order to improve medical outcomes and reduce costs through more customized care plans and early intervention. IBM has also reported that the Medical Center of Columbia University used Big Data analytics to review medical data from patients suffering from strokes in order to predict serious complications up to 48 hours before traditional models. The Rizzoli Orthopedic Institute also used Big Data analytics to better understand the clinical differences in families with genetic bone diseases, resulting in a 30% decrease in yearly hospitalizations. Similarly, the Hospital for Sick Children reviewed diverse Big Data analysis methods to monitor and process patient vital signs and predict infections up to 24 hours faster than traditional methods. Additionally, machine learning methods have been considered for automatically monitoring psycho-physiological stress from bio-signals including skin conductance and an accelerometer. Generally, Big Data analytics is useful in answering the important questions listed below:

- How can alarm-based screenings be used to predict which patients are at the most risk?
- Which treatments and plans of care generate the best patient outcomes?
- How can bio-signals and medical data be processed to accurately predict complications?
- How can an automatic diagnostic system use classification techniques to help patients even if a medical professional is not present?
- How can a recommendation system using collaborative filtering systems be created to suggest the most beneficial treatment or HCC gene-based treatment based on similar measures between patients or treatments?

11.6   Major Challenges of IoT Healthcare

IoT e-Health seamlessly connects patients, clinics, and hospitals across a vast variety of locations to coordinate and orchestrate healthcare. There are, however, many research issues that must be carefully addressed before it can become viable for mainstream deployment [1, 9].

11.6.1   Interoperability, Standardization, and Regulation

IoT has generated concerns when it comes to standardization. Manufacturers, end users, and service providers all require operating standards within and among IoT focus areas. The issue of standardization is complex because IoT seeks to capture a diverse range of disciplines that are monitored and regulated by different regulatory agencies. The issue is further complicated by the stringent guidelines mandated by guiding medical authorities. For example, in the United States standardizing wireless medical devices requires a multi-agency regulatory collaboration including the food and drug administration (FDA), centers for medicare and medicaid services (CMS), and the federal communications commission (FCC). This means that companies must carefully evaluate the policies and procedures required by all three agencies. IoT e-Health will also need to maneuver through a complicated structure comprised of multiple agencies before IoT e-Health products will become readily available in the market. These issues are not unique to the United States; e-Health will come up against similar issues in other areas of the world [1, 9].

11.6.2   Heterogeneity

IoT healthcare applications require a broad range of contextual data obtained using both different and heterogeneous health sources. Heterogeneity is generally perceived in two forms [18]:

- *Data heterogeneity* – Multimodal sensors that are different in structure, format, and semantics result in significant heterogeneity in data. These data sets can be inconvenient to share or reuse because they lack formal descriptions.
- *Sensor heterogeneity* – Integrating several sensors that function at differing frequencies and require differing network protocols creates issues around interoperability. In addition, combining devices and sensors can create increased levels of interference. When interference and the ranges as well as the working frequencies of coexisting wireless networks overlap, the healthcare network system can be substantially affected, hindering the availability of important data.

11.6.3 Interfaces and Human Factor Engineering

It is important to consider additional factors such as human acceptability of a particular system and the level to which that system supports human interactions. The interface of sensors, front-end technology, tablets, smartphones, computers, and other interactions is one of the main factors in IoT e-Health. In order for individuals with little experience using technology to be able to use IoT e-Health devices, it is important that end users be able to train themselves how to use high-tech tools intuitively. Generally, end users have little knowledge about sensor syncing, wireless networks, or other technical operations. Therefore, when devices are utilized in remote environments, it is important that establishing the e-Health system be as forthright and autonomous as possible. For example, geriatric populations are one of the largest stakeholder groups in IoT e-Health, and device interfaces must be user-friendly, requiring minimal assistance from experts. Successful human factor engineering includes participatory design that encourages end users to participate with the design team by providing continuous feedback regarding ease of use, dislikes, likes, and comfort levels.

11.6.4 Scalability

IoT on a smaller scale requires that portable devices contain data collection sensors and that centralized servers process user requests to ensures all users are able to access medical services using compact devices like smartphones. This functionality can be scaled up to serve an entire hospital so that patients can utilize medical services, receive status updates, and benefit from continuous monitoring. In addition, this model could be scaled up to serve a whole city, assuming antenna and sensors in the city are able to gather data. Smart Big Data algorithms and APIs could be used to process data and evaluate user requests while smart interfaces can provide users with the status of requests in real time. A smart, e-Health-based city could collect and process all data via smartphones and mobile applications, sending feedback to patients so that each is aware of personal health status and has access to results of medical testing or examinations seamlessly. Utilizing IoT medical services saves patients' time waiting for appointments or results and provides clear access to basic medical services and resources. Scaling up from a small network to the level of a smart city has the potential to improve efficiency, support relationship building, and promote greater trust between medical professionals and patients [1, 9].

11.6.5 Power Consumption

The consumption of energy is another important factor in IoT healthcare. The finite battery life of sensors can negatively impact an application's life cycle. Charging or replacing batteries in many real-life IoT systems is complicated and inefficient, especially if multiple sensors are utilized. Generally, an IoT device's battery life span is dependent upon the network (i.e., transmission range, duty cycle, and utilization of communication channels) and the level of complexity of signal or data processing.

11.6.6 Intrusiveness

There are some IoT e-Health applications that necessitate that a subject continually carries or wears the sensor, which can become an inconvenient task. Therefore, additional effort should be made to keep from violating the subject's quality of life.

11.6.7 Design Automation Challenges

The design challenges for IoT e-Health systems arise from a combination of the following characteristics [9]:

- *Cross-domain*: IoT e-Health is about the intersection of many fields that spans bioengineering, embedded system, to network design, and to data analytics. Therefore, modeling, design, verification, and monitoring of such a heterogeneous system require multi-disciplinary knowledge.
- *Heterogeneous*: IoT e-Health spans the cyber and physical worlds. Therefore, it involves many components such as hardware and software, network, etc. As a result, it is very important to pay detailed attention to interfacing and interoperability of such a holistic system.
- *Dynamic environments*: IoT e-Health incorporates a significant dynamic environment. Therefore, the system should be able to evolve continually.
- *Distributed systems*: IoT e-Health is built on top of many layers and physically and/or temporally separated components that are tightly networked.
- *Large-scale*: IoT e-Health is a swarm of connected devices, network components, and computation systems that must deal with data volume, variety, velocity, and veracity.
- *Human aspects of the design*: Since IoT e-Health is used in close collaboration with humans, it is very important that the design of such a system considers the role of humans as well as human interfacing.

- *Learning-based*: IoT e-Health should be designed based on suitable data-driven learning techniques to handle the varying dynamics of cyber and physical components.
- *Time-aware*: Spatial and temporal variations in the dynamics of the cyber and physical components of an IoT e-Health must also be addressed.

11.6.8 Data Management

In the healthcare sector, IoT e-Health faces many of the same data management challenges as in other fields. One distinguishing factor, however, is the fact that e-Health data originates from medical sensors worn by human subjects, and the human body is a constantly changing system. Thus, from an IoT e-Health perspective, an ongoing flux of data will continually flow inward from edge sensors via fog computing nodes. On a positive note, sensors and computing are both declining in cost, making Big Data more cost-effective to be collected in a brief timeframe. IoT e-Health has evolved to deal with the complicated nature of these data, even as their variety, volume, and velocity have continued to increase. At the same time, IoT faces a challenge almost unknown 10 years ago: that of data variety. Dozens of healthcare applications targeted at end users use their own data format; for example, ECG data is often encoded in XML, while camera-based IoT devices typically record data in a variety of image formats. Meanwhile, various edge computer manufacturers use their own data formats, which can also vary by customer. Data models on the cloud also vary widely, creating a desperate need for standardization. Difficulties related to data volume and velocity, on the other hand, are more related to the ability of the fog node hardware to acquire, analyze, store, and transmit data from medical devices (which could be located at hospitals or clinics or carried with the patient) at high fidelity and resolution. This creates a clear demand for fog administrators capable of supervising the data flux between computing in the cloud and the fog [1, 9].

11.6.9 Context Awareness

A complete, detailed picture of a particular patient's context enables a system to determine its actual need for assistance more appropriately. In the area of IoT healthcare, events and data must be accurately interpreted in order to obtain a reasonable, holistic understanding of contextual data. For example, based on a patient's condition and context, not every detected fall indicates a critical case or requires a reaction by a caregiver or health provider. In general, the creation of context-aware healthcare applications is challenging due to the issues of data acquisition (e.g., determining which data is important enough to capture) and data analysis methods, including presentation of context-based services and information.

When designing context-aware monitoring systems, it is important to give proper consideration to reasoning, interpretation, and observation of the patient's condition from multiple perspectives including environmental, behavioral, and physiological areas. A system also has to consider all pertinent context dimensions including human activities, objects, location, time, frequency, and posture. Available historical data such as health documentation of disease, diagnoses, treatment plans, daily behavior, noted health changes, and environmental conditions (i.e., humidity and temperature) are also influential to system intelligence [18].

11.6.10 Availability and Reliability

Enabling accessibility to suitable health data in a timely fashion is another important issue for consideration in IoT healthcare. Dependable data delivery affects the availability of health information, and lack of data can negatively affect a patient at a critical point. The reliability of data accessible via wireless networks in healthcare is dependent upon several factors such as device range, network coverage, availability of power, routing protocols, and device or network failure. Issues of reliability can be categorized into data acquisition, communication, data analysis, data measurement, and data governance.

11.6.11 Data Transmission

Choosing a method for transmitting data collected from sensors to back-end cloud servers for processing and analysis is a frequent challenge in the design and implementation of IoT healthcare applications. The transmission of data in a network is classified into four different schemes: anycast, broadcast, multicast, or unicast. Broadcast or multicast schemes can improve reliability because both concurrently supply packets to multiple receivers. However, this policy can increase network traffic and can cause transmission delays. Unicast schemes include the least amount of network traffic and deliver packets to a solitary receiver; however, a procedure for locating an alternate receiver is required in case a transmission should fail. The anycast scheme includes lower overhead traffic in comparison to multicast or broadcast schemes and is a newer routing method that sends data packets to the closest receiver. It is considered more dependable than the unicast scheme when it comes to locating new receivers, but anycast requires more complex network routing and devices. More comprehensive considerations born out of the challenge of data transmission include the amount of transmitted data, transmission technology and frequency, and normal packet size. Each of these considerations can have a significant impact on a system's availability, dependability, effectiveness, energy use, and network traffic [18].

11.6.12 Security and Privacy

IoT e-Health devices, like all networked devices, will present some level of the potential risk to the security and privacy of end users, through the use of unauthorized authentications. This is an especially significant concern in the area of healthcare, where personal safety could be put at risk. In fact, the entire life cycle of IoT e-Health is built around privacy and security, from specification generation and all the way to implementation and deployment. Even so, a holistic multi-layered set of strategies will be necessary in order to overcome the complex security challenges of engineering an IoT healthcare ecosystem. This approach can be described as follows [1, 9]:

1. *Device layer*: Connected devices such as sensors, medical devices, gateways, fog nodes, and mobile devices, when are involved in capturing, aggregating, processing and transferring medical data to the cloud. Widespread forms of attacks in the device layer include *tag cloning, spoofing, RF jamming, cloud polling*, and *direct connection*. In a cloud polling attack, network traffic is redirected in order to inject commands directly to a device, through the use of man-in-the-middle (MITM) attacks as well as changes to domain name system (DNS) configuration. The most effective defense against this attack is an ongoing policy of evaluation and verification of certifications, at the device level, in order to ensure that every certificate actually belongs to the e-Health cloud. A direct connection attack, meanwhile, involves the use of *Service Discovery Protocol* like universal plug and play (SSDP/UPnP), or the on-board properties of BLE, to locate and target IoT devices. This type of attack is best prevented by a policy of ignoring and blocking unauthenticated requests at the device level, through the use of robust cryptographic algorithms, along with a key management system. Other device-layer security measures include identity, authentication, and authorization management, secure booting (i.e., prevent unauthorized applications to be executed), application sandboxing, whitelisting, fine-grained access control capability of resources, protection of data during capture, storage, and transit, traffic filtering feature, fault tolerance, password enforcement policies, secure pairing protocols, and secure transmission mechanisms. It is also important to take into account the extremely limited memory, processing capabilities, power resource, network range, embedded operating systems, and thin embedded network protocol stacks of many devices while implementing security algorithms in an IoT Health system [1, 9].

2. *Network layer*: In this layer, a multitude of diverse network protocols, including Wi-Fi, BLE, and ZigBee can be leveraged to establish appropriate connections among sensors. Eavesdropping, Sybil attacks, sinkhole attacks, sleep deprivation attacks, and Man-in-the-Middle attacks are all common at this level. Thus, the use of trusted routing mechanisms is crucial, as is the use of message integrity verification techniques (using hashing mechanisms like MD5 and SHA) and point-to-point encryption techniques based on cryptographic algorithms. These algorithms fall broadly into two groups: symmetric algorithms such as AES,

DES, Blowfish, and Skipjack and asymmetric public-key algorithms such as the Rabin's Scheme, NtruEncrypt, and Elliptic Curve Cryptography. As a rule, symmetric algorithms are less computationally intensive, making them for low-power 8-bit/16-bit IoT devices. At the same time, problematic key exchange mechanisms and confidentiality issues often create difficulties [1, 9].

3. *Cloud layer*: A large body of literature exists on the security issues involved in the deployment of cloud applications. Any provider of e-Health products and services will need to establish an efficient, effective set of tactics for proactively combating the negative impacts of attacks. Widespread vulnerabilities in the cloud include Denial-of-service (DoS) attacks, SQL injections, malicious code injections, Spear-Phishing attacks, sniffing attacks, path traversals, unrestricted file uploading (remote code execution), cross-site scripting (XSS), Trojan horses, viruses, and brute-force attacks using weak password recovery methods [1, 9].

4. *Human layer*: The fundamental principle of IoT e-Health security is that individuals should receive training on how and when to avoid disclosing private healthcare information. If a knowledgeable group of attackers gain physical access to an end user's IoT e-Health device, those attackers could directly pull data from the device's internal memory and firmware and modify its settings to obtain partial or complete control over it. In addition, it will be crucial to train users to avoid common security pitfalls such as sharing physical or electronic keys, choosing weak passwords, or purchasing used medical equipment [1, 9].

11.7 Case Study: Collaborative Machine Learning-Driven Healthcare Internet of Things

To justify the proposed multi-layer architecture for healthcare, as a case study, we discuss how an online arrhythmia detection can be mapped to the architecture. In this case study, we build a collaborate solution that distributes the intelligence between endpoint IoT health device, fog node, and the cloud. This strategy enables us to have the best of different worlds and to have a tradeoff between accuracy, communication latency (transmission time), processing time, and energy consumption. Note that in the cloud we can execute very complex algorithms, but it has two drawbacks. First, cloud-based arrhythmia detection is not real time. Second, the device needs to transfer a large amount of raw data (ECG signals) to the cloud resulting in high power consumption which is not suitable for wearable devices. On the other hand, endpoint IoT devices and fog nodes can provide real-time decision making; however, unfortunately we cannot execute holistic machine learning techniques on those devices due to their limited processing capabilities. Thereby, this can result in inaccuracy. To tackle this issue, we use a collaborative technique [19]. In this collaborative solution, we distribute the intelligence across device, edge, and the cloud. We utilize a shallow neural network to fast ECG-based anomaly detection in the endpoint IoT devices. Although the accuracy of this type of machine learning is not high, it can increase the lifetime of the battery. In the fog node, we utilize an advance convolutional neural network (CNN) to identity anomalies with the maximum accuracy. Fog nodes (e.g., smartphones) usually have enough power and processing resource to be able to run CNNs. Finally, we perform all the machine learning trainings in the cloud. The reason is that cloud has unlimited resources to ingest and process a large amount of data collected from several patients and the results (machine learning models) are dispatched to the edge of the system. With this approach we can make a trade-off between response time, power consumption, and accuracy.

- *Machine learning on chip (IoT device)*: As mentioned, endpoint IoT devices suffer from enough processing power. Therefore, we design and execute a light machine learning algorithm to meet this limitation. To do so, we train and implement a three-layer neural network to detect any heart anomalies based on ECG signals. In particular, we implement the following functions in device:

  - *Signal pre-processing*: In this stage we remove any unwanted noise from ECG recordings. In particular we perform:

    - DC noise removal
    - High-frequency noise removal
    - Low-frequency noise removal
    - Power line interference removal

  - *Feature engineering*: A typical normal ECG signal is depicted in Fig. 11.7. Since our goal is to implement the ECG signal processing on an IoT wearable device, we need to extract the lowest possible number of working features. This includes four features:

    - Backward time difference of two consequent heartbeats (pre-RR)
    - Forward time difference of two consequent heartbeats (post-RR)

Fig. 11.7 A typical normal ECG signal

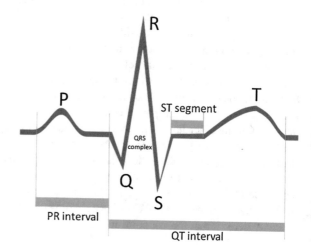

- The average of 10 consequent pre-RR values (local-RR)
- Similar to local-RR but for the last 20 minutes instead of the last 10 values (Global-RR)

  - *Neural network*: As mentioned above, we utilize a shallow neural network with a few layers and neuron as our machine learning on chip (IoT devices). Our rich experiments show that the accuracy of this network for arrhythmia detection is above 90%.

- *Edge/fog intelligence*: Utilizing a fog layer allows delay-sensitive applications (e.g., arrhythmia detection) to make online real-time decisions. The fog enables us to develop more complex machine learning algorithms offering better accuracy compared to shallow neural networks which we used in endpoint IoT devices. In this case study, we design and implement a convolutional neural network that have been proven to be very effective in several areas such as image classification. To do so, the fog node converts the ECG signal to an equivalent image and feeds it into a CNN. In this case, we use an AlexNet CNN. Our experimental results show that CNN is able to significantly improve the performance of arrhythmia detection. Indeed, AlexNet can detect the anomalies in ECG signals with the accuracy of 97%. As a result, the patient can switch between machine learning on chip (endpoint IoT device) and fog node depending on the working conditions and requirements. Note that the wearable ECG monitoring device (endpoint IoT device) can be connected to the fog node (e.g., smartphone) via Bluetooth. However, transferring raw ECG signals can drain the battery.
- *Big Data analytics in the cloud*: Although we moved the decision making task to endpoint IoT devices and fog nodes, we need to train the machine learning algorithms (shallow neural network and CNN) in the cloud. The reason is that training is a huge computational burden for such devices. In addition, in the cloud we have access to more data (ECG signals) from several patients resulting in better machine learning modes. The cloud enables us to continuously train and improve the models, and periodically, we can dispatch the model parameters (e.g., coefficient, weights, etc.) to edge.

11.8 Summary

As the Internet of Things (IoT) paradigm becomes more widespread, a host of novel opportunities have arisen. Technologies such as miniature wearable biosensors, along with advances in Big Data, especially with respect to efficient handling of large, multiscale, multimodal, distributed, and heterogeneous data sets, have opened the floodgates for e-Health and m-Health services that are more personalized and precise than ever before. However, IoT hints at an even greater change in healthcare paradigms; it promises greater accessibility and availability, personalization and tailored content, and improved returns on investments in delivery. Even so, as IoT e-Health broadens the horizons of fulfillment in terms of existing healthcare needs, quite a few major hurdles remain before consistent, suitable, safe, flexible, and power-efficient solutions can be deployed to address many medical demands. In this chapter, we presented a holistic hierarchical multi-layer IoT e-Health ecosystem, where various applications such as early warning systems can be mapped to those layers. We then finally discussed and addressed the main benefits as well as major challenges of IoT e-Health such as data management, scalability, regulations, interoperability, device-network-human interfaces, security, and privacy.

References

1. B. Farahani et al., Towards fog-driven IoT e-Health: Promises and challenges of IoT in medicine and healthcare. Futur. Gener. Comput. Syst. **78**, 659–676 (2018)
2. F. Firouzi et al., *Internet-of-Things and Big Data for smarter Healthcare: From Device to Architecture, Applications and Analytics* (Elsevier, Chicago, IL, 2018)
3. M.C. Sokol et al., Impact of medication adherence on hospitalization risk and healthcare cost. Med. Care **43**, 521–530 (2005)
4. Association, A.D., Economic costs of diabetes in the US in 2012. Diabetes Care **36**, *1033–1046* (*2013*). Diabetes Care, 2013. **36**(6): p. 1797.
5. J.M. Corrigan, E.K. Swift, M.P. Hurtado, *Envisioning the National Health Care Quality Report* (National Academies Press, Washington, D.C, 2001)
6. C.F. Hill et al., Mobile health clinics in the era of reform. Am. J. Manag. Care **20**(3), 261–264 (2014)
7. *Global Telemedicine MarketGrowth,* Trends and Forecasts. Available from: http://www.mordorintelligence.com/industryreports/global-telemedicine-marketindustry
8. P. Rashidi, A. Mihailidis, A survey on ambient-assisted living tools for older adults. IEEE J. Biomed. Health Inform. **17**(3), 579–590 (2012)
9. F. Firouzi et al., Keynote paper: From EDA to IoT e-Health: Promises, challenges, and solutions. IEEE Trans. Comput. Aided Des. Integr. Circuits Sys. **37**(12), 2965–2978 (2018)
10. H. Elayan, R.M. Shubair, A. Kiourti, *Wireless sensors for medical applications: Current status and future challenges,* in *2017 11th European Conference on Antennas and Propagation (EUCAP),* (IEEE, Piscataway, NJ, 2017)
11. *Texas Instruments Wireless Connectivity.* Available from: www.ti.com/wirelessconnectivity
12. F. Bonomi et al., *Fog computing and its role in the internet of things,* in *Proceedings of the first edition of the MCC workshop on Mobile cloud computing,* (ACM, New York, NY, 2012)
13. H. Kalantarian et al., A wearable sensor system for medication adherence prediction. Artif. Intell. Med. **69**, 43–52 (2016)
14. A.J. Jara et al., *A Pharmaceutical Intelligent Information System to detect allergies and Adverse Drugs Reactions based on internet of things,* in *2010 8th IEEE International Conference on Pervasive Computing and Communications Workshops (PERCOM Workshops),* (IEEE, Piscataway, NJ, 2010)
15. A. Anzanpour, et al., *Internet of things enabled in-home health monitoring system using early warning score.* In *Proceedings of the 5th EAI International Conference on Wireless Mobile Communication and Healthcare.* (ICST (Institute for Computer Sciences, Social-Informatics and . . ., 2015).
16. M. Sarrafzadeh, F. Dabiri, H. Noshadi, *Fast Behavior and Abnormality Detection.* 2016, Google Patents

17. *IBM Watson Health.* Available from: http://www.ibm.com/smarterplanet/us/en/ibmwatson/health/
18. H. Mshali et al., A survey on health monitoring systems for health smart homes. Int. J. Ind. Ergon. **66**, 26–56 (2018)
19. B. Farahani, M. Barzegari, F.S. Aliee, *Towards Collaborative Machine Learning Driven Healthcare Internet of Things.* In *Proceedings of the International Conference on Omni-Layer Intelligent Systems.* (ACM, 2019)

Chapter 12
Biomedical Engineering Fundamentals

Ram Bilas Pachori and Vipin Gupta

If it weren't for electricity, we'd all be watching television by candlelight.

George Gobel

Contents

R. B. Pachori · V. Gupta (✉)
Discipline of Electrical Engineering, Indian Institute of Technology Indore, Indore, India
e-mail: vipingupta@iiti.ac.in

© Springer Nature Switzerland AG 2020
F. Firouzi et al. (eds.), *Intelligent Internet of Things*,
https://doi.org/10.1007/978-3-030-30367-9_12

12.1 Introduction of Bioelectricity and Biomechanics

In the current scenario, the innovation in technology is increasing as per the requirements of our lives. This fact is also true for the area of health-care services and medicine. The recent advancement in health-care system leads to effective diagnosis and better treatment of diseases with the help of biomedical engineering. Biomedical engineering includes two major fields, medicine and engineering. The engineering field has assisted health-care technology by providing tools and techniques such as biosensors, signal processing, image processing, and artificial intelligence. These tools and techniques help health-care technology in the research, diagnosis, and treatment of various diseases [1]. The field of biomedical engineering also includes many new areas of research such as bioelectricity and biomechanics.

Bioelectricity is also known as electrophysiology [2]. Bioelectricity has the same principle which the electricity has in the atmosphere and solid-state materials. One of the major differences in bioelectricity and electricity is that the living systems derive their electrical energy from the difference of ionic concentration which is present across cell membranes as compared to man-made electrical systems. Therefore, the energy sources in living systems are distributed in space along the membrane, and this energy can be utilized by involving a flow of current across the membrane. In other words, the systems designed by humans have a localized energy source, for example, a battery, which conducts the currents through a conductor, whereas living systems have distributed sources of energy. The bioelectricity is quantified with the help of potentials and currents which values are functions of position. The animals and people have huge volumes with conducting solution through which ionic currents can flow. Hence, the study of bioelectricity is important for understanding the electrical phenomena in different parts of a living system [3].

On the other hand, the biomechanics is a study of human movement which is defined as the interdisciplinary that describes, analyzes, and assesses human movements [4]. Biomechanics includes the fields of engineering mechanics, biology, and physiology. The knowledge of biomechanics helps us to understand the normal and pathologic gait, mechanics of neuromuscular control, and mechanics of growth and form. This understanding plays a significant role during the development of medical diagnostic and treatment procedures. The human athletic performance has also been enhanced with the help of biomechanics [5].

There are broad varieties of physical movements involved in biomechanics such as the lifting of a load by a factory worker and the performance of a superior athlete. These cases have used the same physical and biological principles, but the specific movement tasks and level of detail change from case to case. Thus, the biomechanics is all about the highest level of assessment of human movements [4].

12.2 Biosensors

The biomedical field basically depends on the monitoring of physical parameters and chemical properties for effective outcomes. The analysis of these physical parameters is performed in centralized laboratories, which require both capital and skilled labor. However, these methods of analysis of physical parameters seem to be accurate, but they have certain disadvantages such as time consuming and inability to monitor concentrations at any instant in real-time situations. Therefore, the development of biosensors has played an important role in instant monitoring of biochemical under real-time situations which involve invasive and noninvasive methods that offer an economic, fast, and easy analytical tool. The applications of biosensors in the biomedical field have revolutionized the biomedical field with the concept of self-monitoring. Biosensor can be defined as a device which monitors the products of an enzymatic reaction in order to obtain the potentiometric response [6].

A biosensor generally has two main components: a molecular recognition or bioreceptor component and a transducer component [7, 8]. Figure 12.1 shows a typical biosensor in which an analyte is used to provide information to bioreceptor. The bioreceptor component can be an enzyme, antibody, nucleic acid, microorganism, and whole cell or tissue. The transducer component can be optical, electrochemical, and mass-based. The types of biosensors based on these two main components include temperature, light, spectrophotometry, fluorescence, and immunosensors. The description of these biosensors is given below [9].

12.2.1 Temperature Sensors

Temperature sensors are most widely used in biological systems. The temperature sensors which are especially used for the biomedical application must exhibit high sensitivity and fast response. The semiconductor-based temperature sensors fulfill

Fig. 12.1 Block diagram of a
typical biosensor [9]

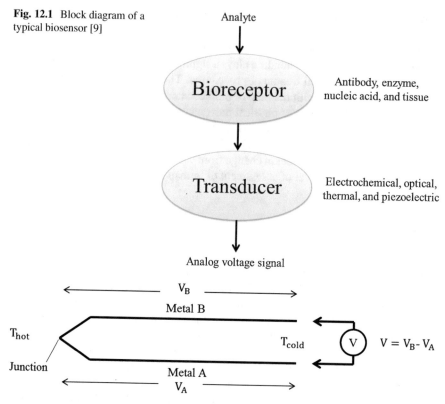

Fig. 12.2 Working principle of thermocouple [9]

the criteria of higher sensitivity compared to the others. Their response is also very fast because they are typically operated in direct contact with the medium, usually water. Hence, the semiconductor-based temperature sensors, namely, thermocouple, thermistor, diode, and transistor temperature sensors, are generally used for biomedical application and whose descriptions are as follows [9].

12.2.1.1 Thermocouple

A thermocouple consists of two dissimilar metals joined together as depicted in Fig. 12.2 [9]. The T_{hot} represents the hot junction temperature where two metals joined together while the temperature at the open junction is the cold junction temperature which is represented by T_{cold}. The temperature difference between T_{hot} and T_{cold} causes flow of heat and this heat flow creates a flow of electric current which is known as the Seebeck effect [9]. The metals used for thermocouple have some degree of resistance which will generate voltage drop V_A and V_B across metals. The difference of these voltages provides the output voltage V [9].

12.2.1.2 Thermistor

The conventional resistors may also be used as temperature sensors because the voltage drop across a resistance is inversely proportional to the temperature. A special type of resistance which is very sensitive to temperature is known as thermistor. The relationship between temperature and resistance for thermistor can be approximated through the use of the following curve-fitting equation as follows [9]:

$$\frac{1}{T} = A + B \ln(R) + C[\ln(R)]^3 \tag{12.1}$$

where T = degrees Kelvin (K), R = resistance of thermistor (Ω), and A, B, and C are curve-fitting constants. The abovementioned expression in (12.1) is called the Steinhart–Hart equation.

12.2.1.3 Diode Temperature Sensor

In the category of diodes, the Zener diode is specifically used for temperature sensing. Figure 12.3 represents the current–voltage (I-V) curve of a typical Zener diode [9]. The Zener diode is a unique type of diode which has a reverse bias configuration. The reverse bias operation of a Zener diode using negative quadrants is shown in Fig. 12.4 for understanding the working principle of Zener diode as a temperature sensor [9]. It can be observed from Fig. 12.4 that the Zener voltage is constant for a certain range of Zener currents (0.5–5 mA). This Zener voltage changes with the environmental temperature and it is linearly proportional to the temperature. On the basis of this phenomenon, we can use a Zener diode to sense temperature within a certain range of current. It is also clear from Fig. 12.4 that the high Zener current produces self-heating effect [9]. The typical working temperature range of Zener diode sensors is −400 °C to +1200 °C, which are approximately

Fig. 12.3 I-V characteristic of a Zener diode [9]

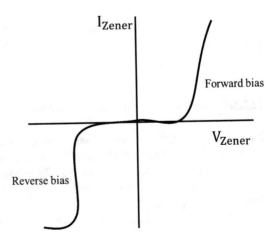

Fig. 12.4 Reverse bias V-I characteristic of a Zener diode [9]

Fig. 12.5 V-I characteristics of bipolar transistor [9]

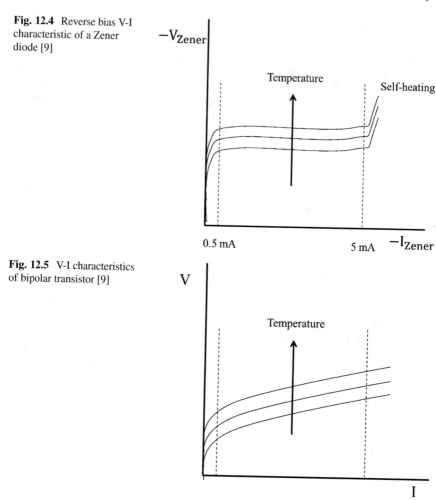

similar to the range of a thermistor. The sensitivity of Zener diode is also similar to that of a thermistor. The only benefit of a Zener diode temperature sensor is its linear operation.

12.2.1.4 Transistor Temperature Sensor

The collector current and base–emitter voltage characteristic of a transistor is very similar to reverse bias V-I characteristic of Zener diode and it can be seen in Fig. 12.5. The base–emitter part (P-N) of a bipolar transistor is actually a diode (P-N), and if we join the base and collector terminals, the bipolar transistor behaves very similar to a diode. The operating temperature range is same as that of a Zener diode and it also gives linearity over a range of temperatures [9].

12.2.2 Light Sensors

Light sensors are very important in many biosensor applications and are commonly used in conjunction with fluorescent dyes. Light is basically a part of electromagnetic radiation which is visible to the eyes of humans, and it is known as visible light. The word light is also used for some other electromagnetic radiations which are not visible to the eyes of humans such as ultraviolet (UV) or infrared (IR). The existence of light is basically in tiny energy packets which are known as photons. The properties of waves and particles are exhibited by photons. In light, the waves are sinusoidal and its peak-to-peak distance is called wavelength (λ). The wavelength of light determines its color in visible light range. A light contains a single wavelength (monochromatic) or multiple wavelengths (polychromatic). The speed of light in vacuum is always constant and its value is 3×10^8 m/s [9].

The light sensors which are made out of semiconductors are photoresistor, photodiode, and phototransistor, and the descriptions of these light sensors are as follows [9].

12.2.2.1 Photoresistor

A photoresistor is a photoconductive cell which conducts only when it is exposed to light. The semiconductor materials used for making photoresistor are cadmium sulfide (CdS), lead sulfide (PbS), and cadmium selenide (CdSe). Figure 12.6 shows a conventional photoresistor which conducts with the exposure of light. The photoresistor is usually S-shaped in order to increase the area of light exposure. In the photoresistor, the holes and electrons are bound together, and when the light (i.e., photons) is exposed to these photoresistor materials, this process creates extra electrons. Therefore, these extra electrons provide extra energy which can make the material more conductive and lowers its resistance. The mechanism in photoresistor is somewhat similar to that of a thermistor [9].

Fig. 12.6 Photoresistor [9]

Light sensitive material

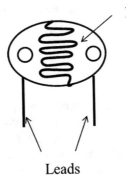

Leads

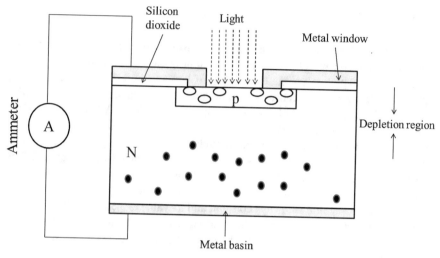

Fig. 12.7 Photodiode in photovoltaic mode [9]

12.2.2.2 Photodiode

A diode which is sensitive to photons is known as photodiode. A photodiode can be used without or with the applied voltage in order to sense light. A photodiode is constructed with a very thin P-type semiconductor which is diffused into the N-type semiconductor. The P-side of photodiode is exposed to light during operation. The mechanism of photodiode without applied voltage is shown in Fig. 12.7. This mode of operation of photodiode is also known as photovoltaic. In Fig. 12.7, the N-type semiconductor contains free electrons and P-type semiconductor contains holes. The electrons and holes repel each other. Thus, a small depletion region is formed between them. This depletion region resulted in a "less conductive" region. The sufficient supply of photons filled the depletion region with extra holes and electrons. Therefore, the depletion region will start conducting and a noticeable electric current will flow between these two semiconductors which can be observed with an ammeter [9].

12.2.2.3 Phototransistor

A phototransistor is basically a transistor which produces high current as compared to photodiode when exposed to light. The phototransistor can be constructed in two different manners which are NPN and PNP phototransistors. In an NPN phototransistor, the base current is replaced with light which provides significant energy to jump the electrons and holes from emitter to collector and vice versa. Figure 12.8 shows an NPN phototransistor which is fabricated by diffusing P-type semiconductor (base) into the N-type semiconductor (collector), followed by

Fig. 12.8 An NPN
phototransistor [9]

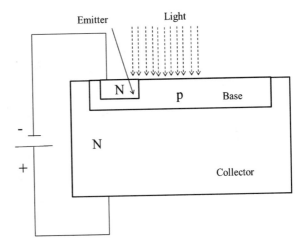

diffusing the N-type (emitter) into the P-type. A phototransistor has also a built-in
amplifying ability [9].

12.2.3 Spectrophotometry

A spectrophotometer measures the light intensity which is transmitted or absorbed
through a material. This material may be a liquid solution or gas in a container. This
measure can be performed for a specific color (wavelength) or a range of colors
(wavelengths). If the measurement is observed for a specific color, then it is known
as photometry. On the other hand, if the measurement is observed for a range of
colors, then it is known as spectrometry. This measure provides us a light intensity–
wavelength curve which is called a spectrum (or spectra) [9].

The principle of absorption is most commonly used in spectrophotometer
because it has certain applications in biomedical field. Figure 12.9 shows a
schematic of a simple spectrophotometer. In Fig. 12.9, the source of light is a lamp
which generates light of all colors in approximately equal proportions resulting in a
white light source. This white light is transmitted through a monochromator which
consists of a prism and a slit. The slit passes a particular color of light and this
selected beam of light finally passes through a rectangular container that holds a
liquid solution or gas mixture. The liquid solution or gas mixture attenuates the light
intensity and the attenuated light intensity hits on the surface of photodiode which
provides a current corresponding to attenuated light intensity. The absorbance A can
be calculated by comparing this light intensity (I) with that from the light source
(I^0) [9]:

$$A = \log\left(\frac{I^0}{I}\right) \tag{12.2}$$

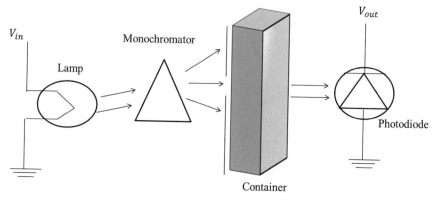

Fig. 12.9 A simple spectrophotometer [9]

where l = attenuated light intensity and l^0 = light intensity from the light source.

12.2.4 Fluorescence

The fluorescence principle is based on the absorption spectrophotometry in which we change the solute in a solution with fluorescent dyes. The color of emitted light from the container is shifted to longer wavelength. The term fluorescence was derived from the mineral fluorite, which is largely calcium fluoride. In fluorescence, the light irradiation excited the molecules and placed them in unstable excited states. The excited molecules lose their excessive energy due to their unstable nature, and this process requires emitting of the photons at the identical wavelength as that of initial light irradiation [9].

The commonly used example of fluorescence is a fluorescent lamp. In a fluorescent lamp, the charged tube of mercury vapor is used to produce ultraviolet (UV) light upon applying electrical voltage. The fluorescent coating is applied to the inner surface of the tube in order to absorb UV light and emit visible light [9].

12.2.5 Immunosensors

Biosensors which use antibodies or antigens as bioreceptors are called immunosensors. Immunosensors are widely developed for medical and veterinary diagnostics, food safety, and environmental monitoring because antibodies are very specific to proteins, viruses, bacteria, cells, etc. In comparison to other biosensors, the immunosensors are provided good sensitivity and specificity. Immunosensors have become very popular recently [9], although the use of antibodies in biological assays has been a very common analysis in laboratory.

12.3   Basics of Signals and Systems

The signal definition plays a very important role in understanding the behavior of signal processing algorithm and its interpretation. The signal can be represented as a function of independent variables and these independent variables can vary from one to many. In other words, signal can be considered as a physical quantity which varies with respect to these independent variables and this physical quantity also contains some kind of information and behaves as a function of one or many independent variables.

12.3.1   Types of Signals

The major classification of signals is as follows [10].

12.3.1.1   Continuous, Discrete Time, and Digital Signals

The signals which have continuous amplitude and continuous time are known as continuous signals. These signals are also known as analog signals and such signals are defined at any point of time. These signals are generally denoted by $f(t)$ where f is a function which depends on the continuous variable t which is continuous in nature. Figure 12.10 shows an example of continuous-time signal.

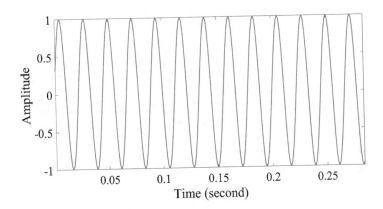

Fig. 12.10 Continuous-time sinusoidal signal

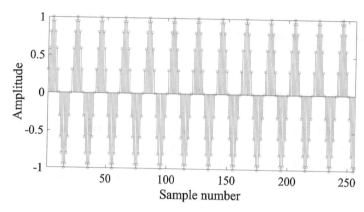

Fig. 12.11 Discrete-time sinusoidal signal

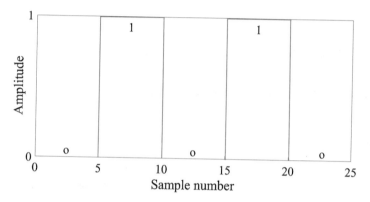

Fig. 12.12 Digital signal

Discrete-time signals have discrete time and continuous amplitude. In these signals, discretization of time is performed using sampling theorem on analog signals. Many signals are discrete signals based on the nature of their measurement. For example, if we measure the weight of a person every day for 1 month, then the plot of weight with respect to whole 30 days can be considered as a discrete-time signal. Such signals are represented by $x[n]$. The small n indicates time index or discrete time which is corresponding to the actual time $t = nT$, where T is the sampling interval. Here, n is also known as normalized time. Discrete-time signals can be also represented in the form of sequences. Figure 12.11 shows an example of discrete-time signal.

The digital signals are those signals which have discrete time and discrete amplitude. These signals have a finite number of values. For example, binary digital signal will have only two values either zero or one. The analog to digital converter (ADC) process can be used to obtain digital signal from the analog signal. Figure 12.12 shows an example of digital signal.

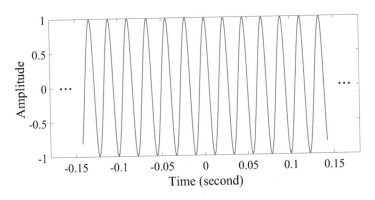

Fig. 12.13   Periodic cosine signal

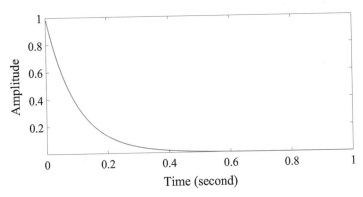

Fig. 12.14   Aperiodic exponential signal

12.3.1.2   Periodic and Aperiodic Signals

The signal which follows repetition after a time interval is known as periodic signal. For a given signal $x(t)$, it can be mathematically expressed as follows:

$$x(t) = x(t + T) \tag{12.3}$$

Here, T is known as the period of the signal.

Sine and cosine waves are examples of periodic signals and Fig. 12.13 shows an example of periodic cosine signal.

On the other hand, the aperiodic signal does not satisfy the abovementioned condition in Eq. (12.3). An example of aperiodic signal is shown in Fig. 12.14.

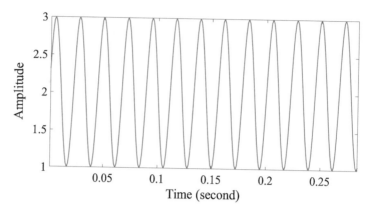

Fig. 12.15 Deterministic cosine signal

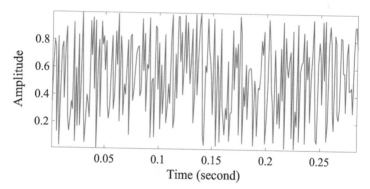

Fig. 12.16 Random noise signal

12.3.1.3 Deterministic and Random Signals

Deterministic signals are those signals which can be represented by mathematical expression and such kind of signals are well determined at any point of time. Sine, cosine, and exponential signals are examples of deterministic signals. Figure 12.15 shows an example of a deterministic signal.

On the other hand, random signals are nondeterministic signals which include uncertainty in the signal values at some point of time. For representation of such kind of signals instead of mathematical representation, they require probabilistic models. Random noise is an example of random signal. Figure 12.16 shows a random noise generated in Matlab.

12.3.1.4 Even and Odd Signals

Even signals $x(t)$ satisfy the following condition:

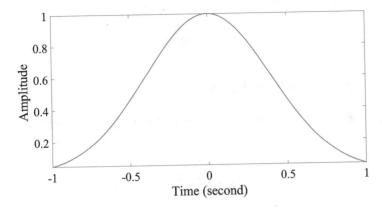

Fig. 12.17 Even signal (Gaussian window)

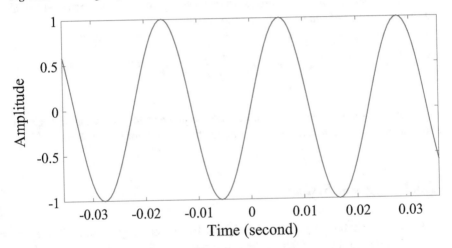

Fig. 12.18 Odd sinusoidal signal

$$x(t) = x(-t) \tag{12.4}$$

On the other hand, odd signals satisfy the following condition:

$$x(t) = -x(-t) \tag{12.5}$$

Figures 12.17 and 12.18 represent the even and odd signals, respectively.

It should be noted that any signal $x(t)$ can be represented in terms of even signal and odd signal.

$$x_{even}(t) = \frac{x(t) + x(-t)}{2} \tag{12.6}$$

$$x_{\text{odd}}(t) = \frac{x(t) - x(-t)}{2} \tag{12.7}$$

12.3.1.5 Energy and Power Signals

Energy signals have finite energy and the energy of the signal $x(t)$ can be defined as follows:

$$E = \int_{-\infty}^{\infty} x^2(t)dt \tag{12.8}$$

The power signals are those signals which have finite power and the mathematical expression for power can be given as follows:

$$P = \lim_{T \to \infty} \frac{1}{2T} \int_{-T}^{T} x^2(t)dt \tag{12.9}$$

It should be noted that any signal cannot be power and energy signal together and it is also possible that a signal may be neither energy nor power signal.

12.3.2 Types of Systems

Systems are required to process the signals for various applications. There are various types of systems which can be categorized as follows [10]:

12.3.2.1 Linear and Nonlinear Systems

A system which follows homogeneity and additivity principles is known as linear system. On the other hand, a nonlinear system does not follow these principles.

For two input signals $x_1(t)$ and $x_2(t)$, the homogeneity and additivity principles are as follows:

$$L\{a_1 x_1(t) + a_2 x_2(t)\} = a_1 L\{x_1(t)\} + a_2 L\{x_2(t)\} = a_1 y_1(t) + a_2 y_2(t) \tag{12.10}$$

Here, $L\{a_1 x_1(t) + a_2 x_2(t)\}$ is the overall response of the system and $a_1 L\{x_1(t)\} + a_2 L\{x_2(t)\}$ represents the individual response of systems of signals $x_1(t)$ and $x_2(t)$, respectively. The overall response of system is equal to the response of individual systems for a linear system where the sum of these individual responses is not equal to the overall response in a nonlinear system.

Example of a linear system is as follows:

$$y(t) = 7x(t) \tag{12.11}$$

Example of a nonlinear system is as follows:

$$y(t) = x(t) + 7 \tag{12.12}$$

12.3.2.2 Time-Invariant and Time-Variant Systems

A system can be considered as time-invariant if input–output characteristics of the system do not vary with time. On the other hand, a time-variant system does not follow such characteristics.

A system with input signal $x(t)$ and output signal $y(t)$ is time-invariant when

$$L\{x(t-\tau)\} = y(t-\tau) \tag{12.13}$$

where τ is shifting a parameter.

Example of time-invariant system is

$$y(t) = \cos\{x(t)\} \tag{12.14}$$

Example of time-variant system is

$$y(t) = x(3t) \tag{12.15}$$

12.3.2.3 Linear Time-Invariant and Linear Time-Variant Systems

If a system satisfies linear and time-invariant properties, then it is known as linear time-invariant system and a system which satisfies linear and time-variant properties is called a linear time-variant system.

12.3.2.4 Static and Dynamic Systems

A system which does not require memory is known as static system and a system which requires memory is called as dynamic system.

Example for memory-less static system is as follows:

$$y(t) = 3x(t) \tag{12.16}$$

Example of a dynamic system is as follows:

$$y(t) = 3x(t) + x(t-3) \tag{12.17}$$

12.3.2.5 Causal and Noncausal Systems

For causal system, the output depends on the present and past values of the input signal. On the other hand, the noncausal system output also depends on the future values of the input signal.

Example of causal system is as follows:

$$y(t) = x(t) + x(t-2) \tag{12.18}$$

The following is the example of a noncausal system:

$$y(t) = x(t+2) \tag{12.19}$$

12.3.2.6 Invertible and Non-invertible Systems

A system can be considered as an invertible system if the input signal can be obtained on the output signal of the system. When input signal cannot be obtained on the output of the system, then a system is known as non-invertible system.

Example of an invertible system is as follows:

$$y(t) = 3x(t) \tag{12.20}$$

Example for a non-invertible system is as follows:

$$y(t) = 0 \tag{12.21}$$

12.3.2.7 Stable and Unstable Systems

In stable system, bounded input signal provides bounded output signal, whereas in unstable system, we will get unbounded output signal for the bounded input signal.

Example of stable system is given as follows:

$$y(t) = x(t) \tag{12.22}$$

Example of unstable system is as follows:

$$y(t) = \int x(t)dt \tag{12.23}$$

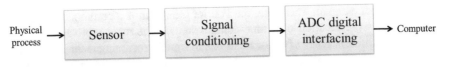

Fig. 12.19 Schematic block diagram of signal acquisition [11]

12.3.3 Signal Acquisition

Signal acquisition is a process in which we study how the physical signals collected from the sensors get into the computers or digitized for the processing of signal in computers and machines. The main blocks of signal acquisition process contain signal conditioning which is mainly possible with a sample and hold circuit and ADC by which a physical analog signal can be converted into a digital signal. The block diagram of signal acquisition process along with sensor and computer interface units is shown in Fig. 12.19. In this Fig. 12.19, sensor senses the physical signal and then signal conditioning is applied with the help of sample and hold circuit, and in order to get this signal in digital domain, an ADC is used. Thus, the converted signal has a number of bits which represent the analog signal at a particular instant of time and which can be stored in a computer with this interfacing mechanism [11].

The typical signal acquisition process has some additional processing units because the sensors have multiple channels. Therefore, a multiplexing unit with sample and hold circuit which quickly scans all the channels and provides data to sample and hold circuit in the short interval of time is required. On the other hand, each ADC has a certain dynamic range of working. The violation of dynamic range of ADC leads to approximation errors during analog to digital conversion process. Thus, it is necessary to amplify the signal to increase the resolution of the ADC. The isolation is also a part of signal acquisition process because the electric and magnetic fields may affect the signal properties. Therefore, a good signal acquisition process should be properly isolated in order to get less interference of external factors. In addition to these abovementioned units, an anti-aliasing filter just after the multiplexer unit is also required because the outputs from the multiplexer are very closely placed in time and the sample and hold circuit with ADC will also take some time to complete the analog to digital conversation process [11].

12.3.4 Time- and Frequency-Domain Representations

The digital signals stored in computer have significant information which enable us to extract the desired information present in the signal. These signals physically exist in time domain and we can analyze the behavior for most of the signals by visual inspection. However, the frequency-domain characterization is equally important for the analysis of a signal. Therefore, the Fourier transform is a commonly used tool for spectral representation of a time-domain signal. The main motivation behind the uses of different types of transformations in signal processing techniques is due to the fact that transforms can highlight certain characteristics present in signal in different domains.

According to Fourier, a continuous periodic signal $x(t)$ can be formed by combining a number of scaled and phase-shifted sinusoidal components. The frequencies of these components are in multiple of the fundamental frequency (ω_0) for the signal $x(t)$. Hence, the synthesis equation for a general periodic signal $x(t)$ can be written as [12]:

$$x(t) = \sum_{k=0}^{\infty} g_k \cos\left(2\pi k f_0 t + \Psi_k\right) \tag{12.24}$$

where g_k and Ψ_k are sets of constants and $f_0 = \frac{\omega_0}{2\pi}$. Suppose $p_k = g_k \cos(\Psi_k)$ and $q_k = -g_k \sin(\Psi_k)$, then Eq. (12.24) with the help of trigonometric expansion can be written as [12]:

$$x(t) = \sum_{k=0}^{\infty} [p_k \cos(2\pi f_k t) + q_k \sin(2\pi f_k t)] \tag{12.25}$$

Equation (12.25) can also be written as [12]:

$$x(t) = \sum_{k=-\infty}^{\infty} A_k [\cos(\omega_k t) + j.\sin(\omega_k t)] \tag{12.26}$$

where $j = \sqrt{-1}$, $A_k = \frac{p_k \pm j q_k}{2}$, and it is a complex number for $k > 0$ and $k < 0$, respectively. Equation (12.26) with the help of Euler's relation can be written as [12]

$$x(t) = \sum_{k=-\infty}^{\infty} A_k e^{+jk\omega_0 t} \tag{12.27}$$

Here, the magnitude of coefficient, $|A_k| = g_k = \sqrt{\left(p_k^2 + q_k^2\right)}$ and phase $\angle A_k = \Psi_k = \tan^{-1}(q_k/p_k)$. Equation (12.27) is known as Fourier synthesis equation for a periodic continuous signal $x(t)$.

Conversely, the Fourier analysis equation for a periodic continuous signal $x(t)$ with time period T_0 can be written as [12]:

$$A_k = \frac{1}{T_0} \int x(t) e^{-jk\omega_0 t} dt \tag{12.28}$$

It should be noted that k has only integer values and A_k is a discrete function in Eq. (12.28).

Fourier transform is a linear transform which plays a very important role in digital signal processing applications, and fast Fourier transform (FFT) algorithm is commonly used in analyzing the spectral content of any deterministic signal due to less computational complexity.

The discrete Fourier transform (DFT) allows the decomposition of discrete-time signals into sinusoidal components whose frequencies are multiples of a fundamental frequency. The amplitudes and phases of the sinusoidal components can be determined using the DFT and is represented mathematically as [12]

$$X(k) = \frac{1}{N} \sum_{n=0}^{N-1} x(n) e^{-j\left(\frac{2\pi kn}{N}\right)} \tag{12.29}$$

For a given signal $x(n)$ whose sampling period is T with N number of total samples (NT is therefore the total duration of the signal segment). The spectrum $X(k)$ is determined at multiples of fs/N, where fs is the sampling frequency.

On the other hand, the spectrum can also be obtained using Fourier–Bessel series expansion (FBSE) [13–15]. In FBSE, the Bessel functions are used as basis sets for signal representation, and these basis functions are aperiodic and decay over time. These features make FBSE-based representation suitable for analysis of nonstationary signals, while DFT has certain limitations for these kinds of signals. FBSE has been successfully applied for nonstationary and biomedical signals [16–25].

The FBSE of $u(n)$ using zero-order Bessel functions can be expressed as follows [25]:

$$u(n) = \sum_{k=1}^{L} M_k J_0 \left(\frac{\beta_k n}{L}\right), n = 0, 1, \ldots, L - 1 \tag{12.30}$$

where M_k are FBSE coefficients of $u(n)$ which can expressed as follows [25]:

$$M_k = \frac{2}{L^2 (J_1 (\beta_k))^2} \sum_{n=0}^{L-1} nu(n) J_0 \left(\frac{\beta_k n}{L}\right) \tag{12.31}$$

where $J_0(.)$ and $J_1(.)$ represent zero- and first-order Bessel functions, respectively. The ascending order positive roots of zero-order Bessel function ($J_0(\beta) = 0$) are represented by β_k with $k = 1, 2, \ldots, L$. The order k of the FBSE coefficients is corresponding to continuous-time frequency f_k (Hz) and it can be computed by the expression given as follows [25]:

$$\beta_k \approx \frac{2\pi f_k L}{f_s} \tag{12.32}$$

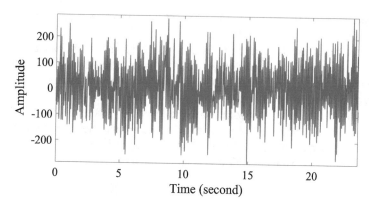

Fig. 12.20 EEG signal of a normal person during eyes-closed condition

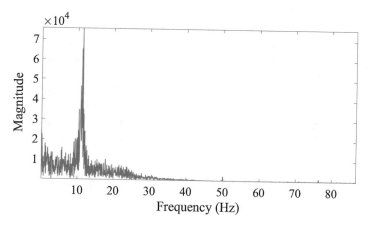

Fig. 12.21 Magnitude spectrum of Fourier transform

where $\beta_k \approx \beta_{k-1} + \pi \approx k\pi$ and f_s is sampling frequency. From Eq. (12.32), the order k can be expressed as follows [25]:

$$k \approx \frac{2 f_k L}{f_s} \qquad (12.33)$$

It can be observed from Eq. (12.33) that order k should be varied from 1 to L in order to cover the entire bandwidth of signal $u(n)$. Hence, the magnitude spectrum of FBSE is the plot of magnitude of FBSE coefficients ($|M_k|$) versus frequencies (f_k).

The time- and frequency-domain representations of an eyes-closed normal EEG signal obtained from Bonn University EEG database are shown in Figs. 12.20, 12.21, and 12.22, respectively. The sampling frequency of this EEG signal is 173.61 Hz [26].

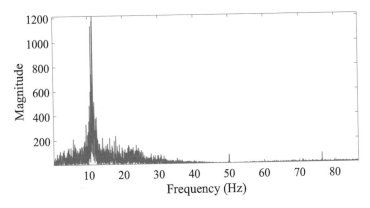

Fig. 12.22 Magnitude spectrum of FBSE

12.3.5   Finite Impulse Response (FIR) and Infinite Impulse Response (IIR) Filters

The response of a FIR filter depends on current and past inputs. Thus, the filter will not produce outputs if it has not received any inputs. The impulse response of this kind of filter is unequal to zero for a finite range. On the other hand, the response of an IIR filter is based on current inputs, past inputs, and past outputs. The dependency of this filter on past outputs generates outputs even after the filter has stopped receiving inputs. The impulse response of an IIR filter is unequal to zero for infinite range. The mathematical forms of FIR and IIR filters for the input signal $x(n)$ and output signal $y(n)$ are as follows [27, 28]:

$$\text{FIR filter}: \quad y(n) = \sum_{i=0}^{M} G_i x\,(n-i) \tag{12.34}$$

$$\text{IIR filter}: \quad y(n) = \sum_{i=0}^{M} G_i x\,(n-i) - \sum_{j=1}^{p} H_j y\,(n-i) \tag{12.35}$$

where G and H are the filter coefficients. The physical structure which will realize Eqs. (12.34) and (12.35) are shown in Figs. 12.23 and 12.24, respectively. In Figs. 12.23 and 12.24, Z^{-1} represents the unit delay element.

The main reason for the description of FIR and IIR filters in this chapter is because the biomedical signals have small amplitude. These signals are contaminated by various artifacts and interferences which change the properties of the signals. One of the commonly present interference in biomedical signals is power line frequency of 50 or 60 Hz. The FIR and IIR filters are used in order to reduce the noise due to power line frequency of 50 or 60 Hz [29].

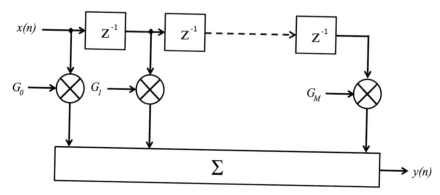

Fig. 12.23 Physical structure of FIR filter [28]

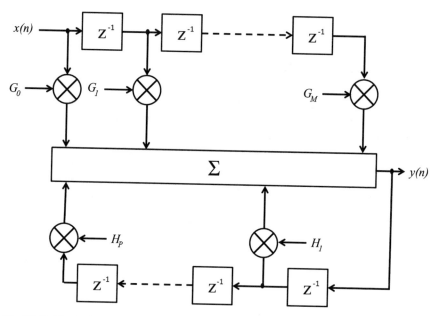

Fig. 12.24 Physical structure of IIR filter [28]

12.4 Types of Biomedical Signals

The electric activities present in the cell which create a potential difference across the cell membrane are used for a number of biomedical signal measurements. These biomedical signals are categorized based on the functioning of different parts of biological system and the descriptions of these biomedical signals are as follows [30].

12.4.1   Electroencephalogram (EEG)

In the biological system, the monitoring and control over the different parts are processed through the brain. The action potentials are used to generate neural activity in the brain and this brain activity can be recorded with the help of electrodes. The signals obtained with these electrodes are known as EEG signals.

The history behind the use of EEG signal is based on an experiment performed in 1929 in which a German psychiatrist named Hans Berger was performing an experiment on his daughter's head to verify the hypothesis that the brain exhibits electrical activity. He observed that the electrical activity increased when she was trying to solve some difficult multiplications. Thus, he deduced from this experiment that the wave patterns observed in the brain recordings reflected the depth of the brain activity [30].

It has been observed that the approximate range of nerve cells in the brain is in the order of 1011. The potential of a nerve cell in steady state is typically around −70 mV, and it is generally negative. On the other hand, the action potential peak is +30 mV and it approximately falls for 1 ms. Thus, the nerve impulse has a peak-to-peak amplitude of approximately 100 mV. In the gray matter, each neuron releases the action potentials during the process of sensing inputs transmitted from other neurons or external stimuli. The spatially weighted sum of all these action potentials at the surface of the skull can be measured by EEG signal. The instrument which is used to record EEG signals is less expensive and accurately measures the brain's electrical activity from the skull. These EEG recordings can be possible in unipolar or bipolar manner. The depolarization signals from the nerve cells may attenuate while passing through the skull because it has complex impedances. Thus, the collections of these signals are possible with quality contact of electrodes with the skull in order to overcome the impedance mismatch created by the hair and dead skin on the skull. The collected EEG signals from the surface of the skull are amplified to represent these signals on electric potential versus time graph [30].

The electric activity in a brain is simultaneously present at many different locations of the head. The most common recording technique of EEG signals utilized 21 electrodes to record these simultaneously occurring electrical activities. The number of these electrodes varies from 64 to 256 for other measuring techniques. The frequency range of amplifiers used to record EEG signals should cover the range from 0.1 to 100 Hz for the analysis of all activities [30].

The EEG signals are broadly used for the diagnosis of various diseases or disorders such as epilepsy, sleep disorders, neurodegenerative diseases, and brain death. The EEG signals of a normal person with eyes-open condition and an epileptic patient during seizure are depicted in Figs. 12.25 and 12.26, respectively [26]. These signals are also obtained from Bonn University database.

The EEG signals are also used in research of brain functional activity. The analysis of evoked potentials (EPs) and event-related potentials (ERPs) of the brain using EEG signal is most common of them. In such applications, the responses of EEG signals are recorded providing specific stimuli such as auditory and visual

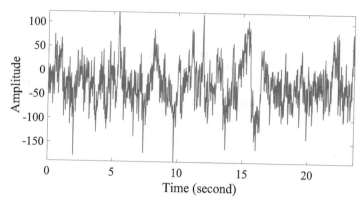

Fig. 12.25 Normal EEG signal with eyes open

Fig. 12.26 Seizure EEG
signal from patient

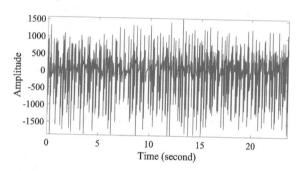

inputs. The EPs and ERPs are particularly used to investigate the response of a brain corresponding to specific stimulation. The level of attention and stress can also be monitored with the help of EPs during experiment.

The major limitation present in EEG signals is that the EEG signals cannot reveal the information about the structure which is responsible for originating these signals. The limitation is due to the fact that the EEG signals are the spatial sum of all action potentials transmitted from billions of neurons at different depths below the cerebral cortex. Therefore, the functional magnetic resonance imaging (fMRI) is used where the functional information from the structures deeply situated in the brain is required [30].

12.4.2 Electrocardiogram (ECG)

The electrical activity recorded from the heart is known as ECG. The ECG signal is used for the clinical diagnosis of heart diseases. The cellular electrical excitation due to cardiac muscle contraction can be recorded by ECG signals. The functioning of these cells can be indicated by its electrical activation, while the depolarization

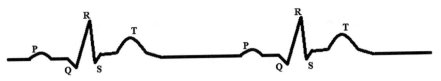

Fig. 12.27 A sample ECG signal with P, Q, R, S, and T wave representation

indicates the shortening of muscle cells. The repolarization and depolarization generate electric potential differences on the muscle cells, which can be recorded using electronic recording instrument. Thus, the ECG signal is due to controlled repetitive electric depolarization and repolarization patterns of the heart muscle cells.

In early history, the ECG signal recording was possible by the efforts of a Dutch scientist Willem Einthoven in 1903. He designed a galvanometer to record the action potentials. The galvanometer was directly coupled to an ink pen. This pen was moved directly on paper as a voltage leading to a deflection of galvanometer was given. Nowadays, the electrodes are directly coupled to amplifiers and filters in order to record ECG signals.

The characterization of ECG signal is usually possible by five waves. These waves are denoted by letters P, Q, R, S, and T. These P, Q, R, S, and T waves can be seen in Fig. 12.27. The ECG signal is also characterized sometimes by a sixth wave with letter U. The P wave in ECG signal is due to depolarization of the atrium, while the Q, R, S, and T waves are caused by the ventricle. The time duration for P wave in ECG signal is approximately for 90 ms and the amplitude for this wave does not usually exceed 2.5×10^{-4} V. During P wave, the atrium contracts to fill the ventricle due to the depolarization. The QRS complex in ECG signal is occurring for time duration of 80 ms with amplitude of about 1 mV. The QRS complex represents the depolarization of the septum and Purkinje fiber conduction. The septum is a wall which separates the left and right ventricle. In simple language, the QRS complex shows the depolarization of ventricular wall from bottom to top and from inside to outside. It should be noted that the quiet time between the P wave and the QRS complex is generally used as a reference line. The repolarization effects of ventricular wall from outside to inside which is also opposite to depolarization represent with a pulse called T wave. During the repolarization process, the atrium is relaxed and filled back. The repolarization process can be distinguished from the depolarization process with the fact that the repolarization process takes longer time as compared to depolarization process. The action potential gradient of the repolarization process is also straightforward wherein it incorporates a smaller gradient in the time derivative of the cell membrane potential. The U wave also shows sometimes a portion of the ventricular repolarization [30].

The ECG signals are used for the diagnosis of various cardiovascular diseases such as myocardial infarction and coronary artery disease (CAD). The ECG signals of a normal person and a patient suffering from CAD are shown in Figs. 12.28 and 12.29. These ECG signals of a normal person and a patient are obtained

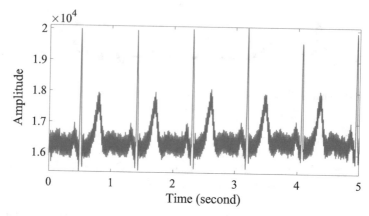

Fig. 12.28 Normal ECG signal

Fig. 12.29 CAD ECG signal

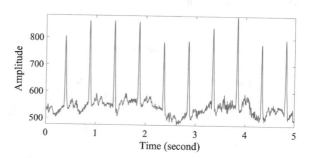

from Fantasia open access database and St. Petersburg Institute of Cardiological Technics 12-lead Arrhythmia Database, respectively [31]. The sampling frequencies of normal and CAD ECG signals are 250 and 257 samples per second, respectively.

12.4.3 Electromyogram (EMG)

The recording of muscle's electrical activities is known as EMG signal. Moreover, the EMG signal is a signal which records the electrical activities produced by the depolarization of muscle cells during muscle contraction. This recording also contains the nerve impulses that initiate the depolarization of the muscle.

In 1907, the first time recording of action potentials produced by human muscle contraction was reported by Hans Piper. The EMG signal has emerged as vital signal in the biomedical field because a number of neuromuscular disorders can be diagnosed using EMG signals.

The recording of the electrical activities of muscle tissue is possible with two methods. In first method, the electrodes are applied on the skin and the signals are recorded from surface of the skin. The second method actually uses the insertion of

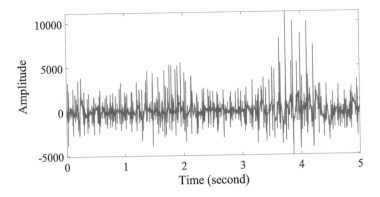

Fig. 12.30 Normal EMG signal

needles with electrodes into the muscle. The EMG signals are the spatially weighted sum of the electrical activities collected from the surface of the skin due to a number of motor units. However, the information present in EMG signal is the combined information of the entire muscle groups. In general, the EMG signal is used to identify the muscle groups which are involved in a particular motion or action.

The EMG signals of specific motor unit can be measured with subcutaneous concentric needle electrodes after implanting it on the muscle. The depolarization of the muscle cells which are present surrounding the needle electrodes can be recorded with these electrodes. Moreover, the electrical activity of a single motor unit can be directly measured with these types of electrodes, and if the needle has more than one electrode, then the bipolar measurement is also possible. There is a short burst activity happening during needle electrode insertion for recording of EMG signals. These burst activities may be repeated several times when an axon of a nerve is touched. The EMG signals also have muscle potential spikes which may be present during muscle contraction. These spikes are not true action potentials of muscle cells because the muscle excitation is usually due to the presence of calcium, potassium, and chlorine ions. Thus, the electrical potential measured from the surface or inside the skin is a triphasic potential phenomenon [30].

The presence of amplitude in excitation potential is sometimes due to the distance between the muscle fibril and the electrode. This amplitude will reduce with the square of distance to the source. The typical range of muscle potential is between 2 and 6 mV with range of time duration of 5–8 ms. The processing of raw EMG signal is performed in a different way as compared to other biomedical signals because these signals often have many noise. The raw EMG signals for a normal person and myopathy patient are shown in Figs. 12.30 and 12.31. These signals are obtained from PhysioBank ATM [31]. The sampling frequency of these signals is 4000 Hz.

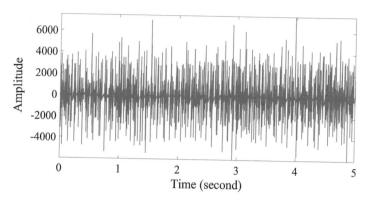

Fig. 12.31 Myopathy EMG signal

12.4.4 Electrooculogram (EOG)

A signal that measures the skin around the eyes is known as EOG signal. The EOG signal is used to determine the gaze and the dynamics of the eye motion. The electrodes used for recording of EOG signals are implanted on the sides of the eyes in order to measure horizontal motions of the eyes. The vertical motions of the eyes are measured with the placement of electrodes above and below the eyes. The motions of the eyes are measured with potential difference between each pair of these electrodes for both cases with the help of differential amplifiers. The presence of this potential difference is due to eyeball movement and it is generated by the cornea and retina. The range of this potential is often between 0.4 and 1.0 mV. The sampling frequency of EOG signal is often in the range of 0–100 Hz and it can be identified by the mechanical limitations of the eye motion.

There are various disorders which can be detected by the EOG signals such as laziness of the eyes in tracking moving objects. In laziness detection, the subject tracks the moving object on a monitor with their eyes and the EOG signals are captured during this event. The diagnosis is based on the lag between the cursor movement and the captured EOG signals.

In another application, the EOG signals help the severely paralyzed patients. In the United States, it is observed that the number of patients who are paralyzed due to spinal injuries is about 150,000. The EOG signals from patients help them to communicate with their caretakers and computers. This communication process requires a large board with an array of commands and placed opposite to the patient.

The gaze angle obtained with EOG signals identifies the command the patient is trying to execute. Similar kinds of systems find importance for navigation of aircrafts and boats.

The EOG signal is very closely related to a signal known as electroretinogram (ERG) signal. This signal is the potential difference among the retina and the eyeball surface. The EOG signal is frequently used to represent ERG signal [30].

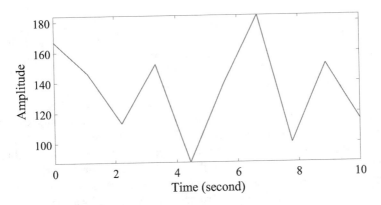

Fig. 12.32 EOG signal recorded with the left eye

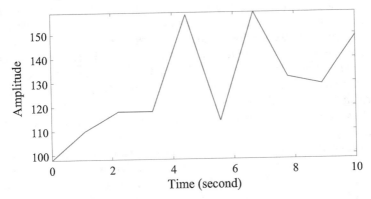

Fig. 12.33 EOG signal recorded with the right eye

The EOG signals of the left and right eyes obtained from PhysioBank ATM are shown in Figs. 12.32 and 12.33, respectively. The sampling frequency of these signals is 1 Hz [31].

12.4.5 Magnetoencephalogram (MEG)

The magnetic field activities of brain neurons are captured by MEG signals. The fact behind the involvement of MEG signal to capture brain activities is based on the electromagnetic theory. The change in electric field causes a magnetic field proportional to electric field. Thus, the change in electric charges of the neurons produces a proportional magnetic field which can be used to measure brain activities. The MEG signal can measure the extracranial magnetic fields created by intraneuronal ionic current flow inside the appropriately oriented cortical pyramidal cells.

The main reason behind the use of MEG signal over EEG signal is that the EEG signals have significant noise because of the muscles' electrical activities being very close to the electrodes, whereas the MEG signal can record from DC to very high frequency (>600) without skin contact. The MEG signal is also capable of detecting neuronal electrical activities from deep inside the brain, while the neuronal electrical activities close to surface of the brain are often captured by EEG signals. Moreover, the MEG signal has less distorted signals which provide much better spatial and temporal representation of the brain. A major advantage of MEG signal is that it can provide an exact location and timing of cortical generators for event-related responses and spontaneous brain oscillations. The MEG signal provides a spatial accuracy of a few millimeters along with submillisecond accurate temporal resolution under optimal conditions. These accurate configurations provide the much effective spatiotemporal tracking of distributed resolution in case of cognitive tasks or epileptic discharges. The weak magnetic field in MEG signal recording machine is sensed by large superconducting quantum interference devices (SQUIDs). The SQUID sensors are able to deliver both natural and evoked physiological responses in MEG signal due to weak strength of magnetic field which is about picotesla (pT). The interference present in MEG signal is mainly due to earth's magnetic field and this interference can be filtered by the MEG signal recording machine. The analysis of MEG signals is possible in a similar way as the EEG signals due to resemblance. Thus, the same processing techniques which are used for EEG signals can be utilized for MEG signals [30].

The MEG signals of left, right, forward, and backward movements from subject S01 are shown in Fig. 12.34, respectively. These signals are obtained from BCI competition IV dataset 3 [32]. The dataset contains ten channels of MEG signals, namely, LC21, LC22, LC23, LC31, LC32, LC41, LC42, RC41, ZC01, and ZC02. These signals are recorded with two subjects S01 and S02 with a sampling frequency of 400 Hz. A total of 400 samples are present in a signal resulting in a 1-s time duration.

12.4.6 Other Biomedical Signals

Biomedical signals are not limited to the abovementioned category of signals. There are many other biomedical signals which are used for clinical and research purpose. The signals which are used for the diagnosis of heart sounds are known as phonocardiogram (PCG) signals. In PCG signals, the heart sounds are observed during the inside and outside flow of the blood in the heart compartments. These signals are often recoded with the help of mechanical stethoscopes which amplify the heart sounds. However, the mechanical stethoscopes have an uneven frequency response and this frequency response distorted the heart sound signals. Thus, an electronic stethoscope can overcome this problem and provide a less distorted heart sound signal.

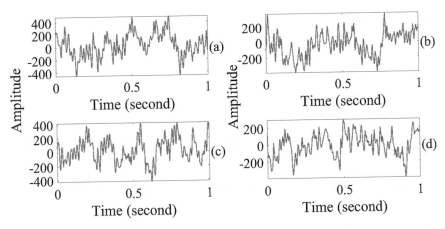

Fig. 12.34 MEG signal of (**a**) left, (**b**) right, (**c**) forward, and (**d**) backward movements recoded with LC21 electrode from S01 subject

A typical application of PCG signals is to detect heart murmurs. In murmurs, the heart sounds are usually due to imperfections in the heart valves or the heart walls. These murmurs are also present in infants due to flow of blood from one side to the other side of the heart through a hole. This hole in infants is usually filled in a few weeks after birth which will stop the heart murmur.

Another signal which records the electrical activity of the stomach is known as electrogastrogram (EGG) signal. The midcorpus of the stomach generates this electrical activity with intervals of approximately equal to 20 s in humans. This signal consists the rhythmic waves of depolarization and repolarization of stomach muscle cells. These waves are related to the spatial and temporal organization of gastric contractions. The external (cutaneous) electrodes can record the EGG signal [33].

12.5   Physiological Phenomena and Biomedical Signals

The biomedical signals can also represent the physiological phenomenon. Hence, the physiological parameters can be reflected by biomedical signal parameters. These biomedical signal parameters can be obtained with an adequate knowledge of their physiological causes for diagnosis purpose. Figure 12.35 shows a block diagram approach to extract physiological parameters from recorded biomedical signals. These parameters are extracted using signal processing techniques and the radically different biomedical signals may have information of the same physiological parameter (heart rate, respiratory rate, etc.) [34].

Fig. 12.35 Block diagram
for physiological parameter
extraction process [34]

12.5.1 Vital Phenomena and Their Parameters

There are various physiological phenomena as well as biomedical signals. Therefore, we focus on some of the vital phenomena which are frequently used in
clinical practice such as heartbeat, blood circulation, blood oxygenation, and body
temperature. A brief description of these phenomena is given as follows [34].

12.5.1.1 Heartbeat

The heart is used for pumping blood into the circulatory system using its rhythmic
contractions which create pulsating waves of blood pressure and blood flow. The
cardiac cycle obtained with this heart rate is very important for diagnosis purpose
[34].

There are three widely used methods to register this cardiac activity. In the first
method, the ECG signal is used to show the rhythmic waves and peaks which are
corresponding to heart muscle excitation with heart rate. An optical biomedical
signal named as optoplethysmogram (OPG) is used in the second method to
represent a smoother waveform reflecting pulsating blood absorption of incident
light. In the third method, the PCG signal represents two consecutive temporary
signal deflections corresponding to heart sounds which are induced by consecutive
closures of heart valves. The cardiac activity recorded with the ECG signal has
nearly instant response at the corresponding sensor location, while the time delayed
response is observed in the recording of OPG and PCG signals due to the pulse wave
propagation velocity and sound velocity, respectively [34].

Although the spontaneous cardiac activity is inherently present in many pacemaker tissues of the heart, the heart rate level and its change are mostly controlled
by the autonomic nervous system. This control is possible with the sinoatrial
node, which is the main pacemaker in the heart. The activities of sympathetic and
parasympathetic nervous systems directed to the sinoatrial node are characterized by
discharges synchronous with each cardiac cycle. These activities can be modulated
with central oscillators present in the central nervous system and peripheral oscillators which depend on respiratory movements and arterial pressure fluctuations.
The balance between these activities determines the instantaneous heart rate. The
central and peripheral oscillators create noisy fluctuations in the corresponding
instantaneous heart rate. However, these types of fluctuations can also be observed
at different timescales [34].

The estimation of energy expenditure in the body is the most efficient measure
which can be calculated with the help of the heart rate level because heart rate

increases with increase in oxygen consumption at an instant. The heart period which is a reciprocal of heartbeat is generally referred to as RR interval (the time interval between two consecutive R peaks in ECG signal). The only criterion for considering the RR interval is that the sampling frequency should be very high (>500 Hz) to assess the sinoatrial node activities.

The heart rate variability (HRV) is another standard term which describes heart period oscillation as well as the oscillation between consecutive instantaneous values of heart rate. The HRV is very closely related to the mechanism of the autonomic nervous system which gives immediate response to any physiological states such as respiration phase, sleep stages, and emotional activities. The HRV is also good in representing functional integrity of a physiological process (thermal, hormonal, neural, etc.) Therefore, the assessment of HRV gives early signs of pathological developments such as cardiovascular diseases [34].

12.5.1.2 Respiration

In the respiration process, the lung plays a major role which delivers oxygen to the bloodstream and releases carbon dioxide from the blood through a rhythmic expansion and contraction process. The assessment of respiratory cycle performs an important role in the diagnosis of various diseases. There are numerous methods to register the respiration on which the three well-established methods are discussed here [34].

In the first method, the mechanorespirogram signal is a mechanical biomedical signal used to record the circumference changes of the abdomen and chest during breathing. A periodic waveform showing respiratory rate is observed through this process during normal breathing. On the other hand, this waveform disappears during holding of breath. The amplitude deflection in this signal increases during snoring in order to overcome an increased respiratory resistance. The recorded mechanorespirogram signal from the abdomen and chest may differ in amplitude and phase due to different strengths of abdominal and chest breathing. The waveform recorded from abdominal breathing is delayed with respect to the waveform recorded with chest during breathing [34].

The lung sounds are also present in PCG signal during normal breathing due to air turbulences in the lung branching airways. These sounds have much lower amplitude; due to this reason, it cannot be easily distinguished over time. In addition, an overlapping signal component is also recognizable during the inspiration phase of snoring sounds. This signal component is present due to elastic oscillations of the pharyngeal walls which may lead to a temporal closure of the airways [34].

In the third method, a mechanical biomedical signal is used which records the airflow through the mouth considering nasal airflow is stopped using a tube with a woven screen inside which acts as a flow resistance. This method is commonly used in clinical practice. In this method, the airflow is considered positive and negative corresponding to inspiration and expiration during normal breathing, respectively. The flow is zero during holding of breath. The high-frequency oscillations of the flow can be obtained during inspiratory phase of snoring. The amplitude of flow is also increased during the phase of both inspiration and expiration of snoring. These oscillations and increased amplitude of flow are due to intermittent closures of the airways and the aforementioned intensified respiratory efforts [34].

In addition to these methods, the thermal biomedical signal is known as thermorespirogram signal in which variations of the air temperature are observed in front of the nostrils during breathing. The temperature increases during expiration and decreases during inspiration phases of breathing. The registration of the respiratory activity is also possible with an electric signal known as electroplethysmogram signal. In this method, the inflated and deflated lung changes the thoracic electrical impedance which can be observed by electroplethysmogram signal. The optical biomedical signal known as optoplethysmogram signal can also be used to register the respiration activities. This signal reflects the peripheral blood volume changes over the respiratory cycle [34].

During respiration, the breath-holding condition deserves some extended description. This condition is generally known as a Greek word apnea (breathlessness) in which a complete or partial cessation of effective respiration occurs. This breath-holding condition can also be possible during sleep at night and it is known as sleep apnea. The sleep apnea is usually detected by polysomnography [34].

12.5.1.3   Blood Circulation

The blood circulation mainly depends on systemic and pulmonary circulation in which the first one comprises the rhythmic transport of the oxygenated blood to the body and the deoxygenated blood back to the heart, whereas the second one is used for the transportation of the deoxygenated blood to the lungs and oxygenated blood back to the heart. In addition to assessment of cardiac cycle with heart rate, a simultaneous registration of blood circulation is also required for the highly relevant diagnosis purpose with the help of circulatory parameters, namely, blood pressure, blood flow, and arterial radius. The brief description for the registration of these circulatory parameters is as follows [34].

Blood Pressure

The unobtrusive and long-term monitoring is difficult in blood pressure registration. The characteristics such as systolic value, diastolic value, and the pressure pulse waveform are used to assess the blood pressure. There are basically some invasive and direct methods as well as noninvasive and indirect methods to register the artifacts of free blood pressure values [34].

In invasive and direct methods, the blood pressure is directly recorded in the vessel by inserting a catheter with a mounted internal pressure sensor or a fluid-filled and rigid catheter for transmitting the blood pressure characteristics to the external pressure sensor. Although these methods are precise and direct, they are not popular due to their invasiveness and related complications for routine use. On the other hand, the noninvasive methods are popular for the determination of blood pressure characteristics. These methods include the auscultatory method, oscillometric method, volume clamp method, and tonometric method [34].

In auscultatory method, the Korotkoff sounds are detected by a stethoscope to determine systolic and diastolic values. In this method, an inflatable cuff encircles an extremity (upper arm) and the cuff pressure is increased until a complete cessation of downstream blood circulation is observed. The first release of the cuff pressure after cessation resulted in the Korotkoff sound which indicates the time instant when the upper (systolic) part of the blood pressure pulse wave passes under the cuff and the cuff pressure is equal to systolic value. On the other hand, the transition from muffling to silence indicates the time instant when the lower (diastolic) part of the pulse wave passes and at this time instant the cuff pressure is equal to diastolic value [34].

The second method is the successor of the ancient mercury sphygmomanometer. It is based on the principle that the pulsatile blood flow generates radial oscillations of the arterial vessel wall. These radial oscillations are transmitted to the cuff encircling an extremity and then to a pressure sensor kept inside it. During the deflation, the intra-arterial blood pressure exceeds the cuff pressure and the oscillations of the vessel walls are strengthened due to turbulent flow of blood and progressing arterial decompression. The cuff pressure during the initial increase in oscillation amplitude is proportional to the systolic value and the diastolic value is proportional to cuff pressure value at the time of subsequent rapid decrease in the oscillations. In this method, the maximal amplitude of the oscillations for the vessel walls and cuff pressure is obtained when the cuff pressure passes the mean arterial pressure [34].

The volume clamp method is another noninvasive method in which a miniaturized cuff fixes on a finger. This cuff is equipped with as optical transmission sensor. This method is based on the principle that the radius (volume) of the finger artery tends to increase at the time of the blood pressure (volume) pulse and this increased radius is detected by transmitted light intensity. Afterward, the cuff pressure (volume) is increased just enough to keep the radius and transmural pressure constant. The resulting cuff pressure is proportional to blood pressure waveform because the cuff pressure follows the intra-arterial pressure up to a constant factor at constant transmural pressure. A pneumatic feedback system is also used in this method for cuff pressure (volume) control so that a maximum pulsatile change of the vessel radius is achieved when transmural pressure approaches zero. The cuff pressure pulsations roughly equal to intra-arterial pressure at the time of zero transmural pressure. The main advantage of this method is that it does not require previous calibration with patients to attain absolute blood pressure values [34].

The last one method is the tonometric method which is a successor of the ancient sphygmograph. In this method, a rounded probe over a superficial (radial or carotid) artery which has a backside support of bone is pressed, allowing the artery to be flattened in a reproducible way. The flattening removes the tangential forces in the arterial wall and the rounded probe is barely exposed to the artery pressure. During flattened artery, the applied force by the rounded probe is opposite and equal to the pulsatile force. This force exerts in such way that the pressure of blood exerts on the flattened arterial wall. This rounded probe is connected to a pressure sensor which reflects waveform of blood pressure. This method requires an initial calibration with the patient to compensate changes of arterial mechano-elastic function among patients in order to obtain absolute blood pressure values [34].

Blood Flow

The recording way of blood flow is analogous to blood pressure. The blood flow can also be recorded in invasive and noninvasive ways. The stroke volume, blood flow velocity, and pulsatile flow waveform are the parameters of interest in blood flow registration [34].

The invasive methods for blood flow monitoring have fewer acceptances due to invasiveness and related complications. Some of the invasive methods are indicator method, electromagnetic method, and transit-time ultrasonic method [34].

In indicator method, the oxygen is used as an indicator which is introduced into the stream of blood flow and the resulting arterial as well as venous concentrations from this indicator are measured based on Fick principle. Alternatively, a thermistor catheter is used to introduce a bolus of ice-cold saline into the right atrium. This catheter is also used to detect the resulting drop in temperature of the blood which is present in the pulmonary artery. The amount of indicator injected divided by the area under the blood temperature dilution curve represents the cardiac output in this method. In the second method, the blood vessel with flowing blood is exposed to electromagnetism. The blood vessel is placed in transverse magnetic field which induced a potential difference in the blood vessel with flowing blood. This potential difference is directly proportional to internal diameter of the vessel and the mean blood flow velocity which can be used to measure the blood flow. In the last transit-time ultrasonic method, an ultrasound beam (wave) passes through the blood vessel. There are two ultrasound receivers placed diagonally on either side of the vessel. The difference of time taken for the ultrasound to pass in one direction as opposed to the other is used to obtain waveform for the flow velocity of blood [34].

The noninvasive methods for the determination of blood flow parameters are frequently acceptable. There are many methods for the noninvasive registration of blood flow from which three of the most popular are echocardiographic method, impedance cardiography method, and pressure pulse contour method [34].

The echocardiographic method is based on ultrasonic Doppler effect. In this method, an ultrasound beam in the frequency range of a few MHz is backscattered from the moving blood cells. The blood velocity is related to frequency shift due to backscattered sound. In other words, the frequency increases when the blood moves toward the ultrasound probe. The volumetric blood flow can be computed from the velocity profile over the cross-sectional area of the vessel combined with the cross-sectional dimensions. In impedance cardiography method, an electric current is introduced and the resulting voltage is measured across the axial direction of the thorax while most of electric current follows the path of least resistance and seeks the path of blood-filled aorta. The thoracic impedance is represented by measured voltage. During the cardiac cycle, the volumetric changes of the aorta induced thoracic impedance changes allowing for the determination of the cardiac stroke volume in absolute units. This method is also known as electric field plethysmography. In another pressure pulse contour method for blood flow registration, a generalized transfer function is used to derive the aortic pressure from the radial pressure and then the aortic flow obtained with applying an aortic impedance model. This model indicates the ratio of aortic pressure and flow. During a cardiac cycle, the stroke volume is the integral of the flow waveform. This method requires a previous calibration to achieve absolute blood flow values [34].

Arterial Radius

The mean value of the arterial radius and its pulsatile waveform are interesting topic in physiological phenomena. There are various invasive and noninvasive methods for the monitoring of arterial radius like blood pressure and blood flow. The methods used to calculate arterial radius are somewhat similar which are used for blood pressure measurement. However, the measurement of arterial radius is more sensitive than that of blood pressure measurement due to the reason that the radius changes up to 10%, while blood pressure may change up to 50% [34].

The invasive methods for the measurement of arterial radius are based on resistance/inductive strain gauges, photoelectric devices, and transit-time ultrasonic approach [34].

In the first method, the resistance/inductive strain gauges are fixed directly to the outer artery wall or even inserted into the artery in order to measure radius by the catheter. The second method is based on photoelectric devices in which a pulsating artery casts a shadow on a photocell. The transit-time ultrasonic approach is similarly used like those for blood flow registration. In this method, the two ultrasound transceivers are placed opposite to each other on the outer sides of the arterial wall. The time taken between the impulse emission and its reception on the opposite side is proportional to the arterial radius [34].

The most popular noninvasive methods for the registration of arterial radius are based on ultrasonic beams, optical plethysmography, and mechanical plethysmography [34].

In ultrasonic beam method, the reflections of the ultrasound waves are used and the time taken between the impulse emission from the ultrasound probe on the skin and reflected impulse reception from both arterial walls is calculated, which delivered the arterial radius. The method based on optical plethysmography is an indirect method to assess the local pulsatile volume of the transilluminated artery. In this method, the arterial radius increases with each blood pulse and the transilluminated region encloses with an increased ratio of blood which strongly absorbs the incident light as compared to the surrounding tissue. As a consequence, the intensity of transmitted light decreases for increased arterial radius or at the time of systole. This method not only assesses the pulsatile changes of the local blood volume but also assesses the basic level of blood absorption related to blood oxygenation. The recorded biomedical signal shows similarity with blood pressure from the carotid artery and mild similarity with blood pressure recorded with the ascending aorta but no similarity is obtained with the radial artery blood pressure. Another method is based on mechanical plethysmography which targets local skin curvature in order to assess a superficial artery such as the carotid artery on the neck. During arterial cardiac deflections, the curvature of the local skin changes and it can be assessed by a skin curvature sensor [34].

12.5.1.4 Blood Oxygenation

Blood circulation implies a rhythmic transportation of oxygenated and deoxygenated blood from the lung and back to the lung. Hence, the blood oxygenation level is a vital physiological parameter which is usually extracted with optical biomedical signals. In this method, the light absorption due to pulsatile arterial blood is measured at two wavelengths and interrelated by an algorithm. The blood oxygenation level is usually maintained at a fairly constant level. Thus, the monitoring of blood oxygenation level is very important in order to diagnose cardiac and vascular anomalies. This examination is more specific for anesthesiology to prevent an inadequate oxygen supply. The presence of oxygen in arterial blood is due to binding of oxygen molecules with hemoglobin and dissolving of oxygen with blood plasma in gaseous state. However, the quantity of oxygen in the blood is mainly due to hemoglobin oxygenation as blood plasma carries a very less amount of oxygen. Therefore, the oxygenated hemoglobin implies as a local oxygen buffer to maintain the partial pressure of oxygen in the plasma. On the other hand, the reduced hemoglobin reserves oxygen in the pulmonary capillaries by depleting partial pressure of oxygen in the plasma, resulting in oxygenated hemoglobin. Although blood plasma contains a negligible amount of oxygen, it plays an important role in delivering oxygen to the tissues and the storing of oxygen in the pulmonary capillaries by hemoglobin. It should be noted that the noninvasive assessment of blood oxygen level in the elderly is faced with progressing accuracy problems by optical method [34].

12.5.1.5 Body Temperature

The temperature of the human body is generally governed by the heat production and loss. During rest condition, the heat production is usually carried out by the inner organs such as kidneys, liver, heart, intestines, and brain under the scope of metabolic activity. The rest condition in metabolic activity generally consumes almost 50–70% of daily energy. During normal condition, the inner organs produce more than 50% of thermal energy and about 20% by the skin and muscles, whereas the contribution of the skin and muscles may reach to 90% during physical work [34].

On the other hand, the heat loss is represented by heat radiation, heat convection, and evaporation. The heat radiation and evaporation are proved more powerful at room temperature and warm environment. The heat loss cannot be efficiently realized by the proximal skin surface because its shape is too flat for efficient heat transfer to the environment. Hence, the heat loss mostly occurs from distal body parts such as fingers and toes which have high surface to volume ratio in order to conduct heat to the environment. In other words, the body consists of a heat-producing core which regulates the temperature at 37 °C homeostatically. The core body temperature reflects a circadian variation of about ±0.6 °C with a maximum in the early evening nearly around 6 p.m. and a minimum during 3 a.m. The regulating mechanisms involve a readjustment of the target temperature value of 37 °C during the whole day. This target value is instructed by the central nervous system of the brain (hypothalamus region), while the actual value registration of core body temperature by the thermal receptors is also carried out in hypothalamus region [34].

12.5.2 Parameter Behavior

The vital physiological phenomena of the heartbeat, respiration, blood circulation, blood oxygenation, and body temperature represent specific changes in their reflexive and tonic behavior. The typical behavior and interrelations of the physiological parameters are also a major concern in order to coordinate and integrate body functions. The physiological parameter behavior with their mutual coordination facilitates vital physiological functions, limited resources of body energy, limited space and time in organs and cells for life-supporting functions, environmental changes adaptation, adaptation to physical and mental stress, and regeneration task of the body with sleep [34].

The behavior and coordination of the physiological parameters can be explained with a feedback control loop represented in Fig. 12.36. The hypothesis behind the control loop is that the central nervous controls the physiological phenomenon or function through a quantitative feedback such as thermal, chemical, and pressure receptors. The desired performance can be obtained by minimizing the error calculated with the difference between the target and actual value of the physiological parameters. In this way, controlled body functions can be achieved with the help of the central nervous system. In Fig. 12.36, the controller comprises neurogenic, myogenic, and hormonal controls. The neurogenic control yields a fast response with the help of the autonomic nervous system while as myogenic control through muscle excitation. The slow response is obtained with hormonal control with the release of hormones [34].

The cardiovascular system is an example of this feedback control system in which when blood pressure drops below the normal value, the arterial stretch-

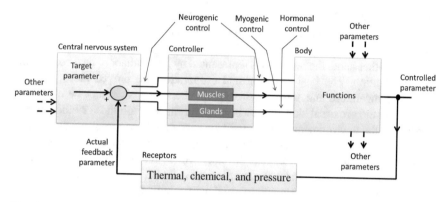

Fig. 12.36 Feedback control loop of the physiological parameters [34]

sensitive receptors (baroreceptors) give an imbalance signal of blood pressure to the brain. The difference of the actual and target values of the blood pressure starts neurogenic control inhibiting the vagus nerve parasympathetic activity which is connected to the sinoatrial node (pacemaker) of the heart. Afterward, the heart rate and contractility of the heart muscles increased. In parallel to this activity, myogenic control forces are applied to increase the total peripheral resistance through smooth muscle activation in the peripheral arteries and increase in regulatory action normalizing the blood pressure level [34].

Mutual interrelations of physiological functions and parameters are basically depending on control loops. In particular, physiological parameter interrelations are needed for the efficient use of energy in humans. The main interrelations of the physiological phenomena during inspiration are cardiorespiratory and cardiovascular interrelations [34].

In cardiorespiratory interrelations, an increase in the inspired air volume resulted in decrease in the left ventricular stroke volume as well as increase in heart rate to level off the cardiac output; due to this efficient blood supply is achieved [34].

In cardiovascular interrelations, a decrease in the systolic blood pressure overlapped with an increase in heart rate to level of the blood pressure [34].

The cardiorespiratory and cardiovascular interrelations are driven by a complex interaction of the circulatory and pulmonary systems with the hemodynamic and nervous systems [34].

In addition, a phenomenon known as biological rhythms which is a periodic and cyclic phenomena of living organs and organisms is described in order to explain the fact that organism needs to give a special performance as well as operating efficiency should be assured by regeneration. These rhythms are used to integrate and coordinate body functions. These rhythms can also be used to anticipate environmental rhythms around the body. This can help to reduce energy due to tuning and synchronization of rhythms, especially during rest or sleep. The exogenous and endogenous are the two types of biological rhythms. In exogenous rhythms, the rhythms are directly controlled by the environment around the body

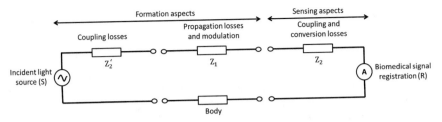

Fig. 12.37 Electrical model for the registration of induced optic biomedical signal [35]

such as the presence of light. On the other hand, the internal biological clocks drive the endogenous rhythms [34].

12.6   Sensing by Optic Biomedical Signals

The optic biomedical signals are induced biomedical signals in which an artificial light source is coupled to biological tissue. This source resulted in a transmitted light intensity which is strongly governed by the light absorption and scattering phenomenon in the biological tissue. The induced optic biomedical signal due to this phenomenon is proportional to the light absorption strength and is usually registered for diagnosis purpose such as blood oxygenation and blood volume. Consequently, the transmitted light intensity also shows multiple physiological parameters which are very useful for the assessment of the health state [35].

The optic biomedical signals are traditionally used to register blood oxygenation and heart rate. The recent advancement in medical technology has also the waveform analysis of optic biomedical signals which facilitate the derivation of respiratory rate. The state of vascular structures (arteries and veins) can also be indicated by the waveform of optic biomedical signals [35].

The model for the understanding of formation and sensing aspects of optic biomedical signals can be seen in Fig. 12.37. In Fig. 12.37, the incident artificial light source is represented by voltage source S which is applied on the skin and coupled to body tissue. The coupling losses are represented by electrical impedance Z_2'. The propagation of the coupled light throughout tissue modulated by diverse physiological phenomena and the electrical impedance for modulation is Z_1. Consequently, some light portion leaves the body and it is available for the detection purpose over the skin. This light is coupled with light sink at a certain distance from the light source in which coupling losses are represented by electrical impedance Z_2. Afterward, the transmitted light intensity is converted into an electric signal which resulted in registration of optic biomedical signals [35].

12.6.1 Formation Aspects

The formation aspects basically revealed the propagation light modulation in body tissues simultaneously with dynamic physiological phenomena. This modulation extracts the physiological information present in optic biomedical signals. The formation aspects of the induced optic biomedical signal include an artificial incident light source which entered the body through the skin, incident light coupling into body, and light propagation through body tissues to a distant light sink applied on the skin [35].

The emission of the artificial incident light depends on the accelerated charge in which energy is released during the transition of electrons from higher to lower energy levels resulting in light emission. The sources of light used for this purpose are of two types, namely, broadband and narrowband. The broadband light sources emit light in a relatively wide band of the electromagnetic spectrum such as incandescent lamps and noble gas arc lamps whereas a narrowband is covered by narrowband light sources such as lasers, fluorescent sources, and light-emitting diode (LED) [35].

In these light sources, the LED is the most popular and widely used light source in order to induce optic biomedical signals. The LED basically works on the principle of electroluminescence. A charge migration takes place to obtain the light photon [35].

After coupling of light source, the transmission of light through biological tissue is a major aspect. The optical light path is started with the light source and then it diffuses through tissue. The diffusion is subjected to changes in light intensity because of the light absorption, diffraction, reflection, scattering, and refraction. A large portion of light intensity has also dissipated and does not reach the skin where a light sink is placed due to this fact [35].

The interaction between light and tissue can be determined quantitatively such as quantitative strength and duration of the interaction and the spatial distribution of the tissue interaction. The interaction is limited to areas of tissue where coupled light is easily reached. This interaction depends on light and tissue characteristics in which light characteristics represent size of incident light, while light transmission is determined by tissue characteristics [35].

The propagation velocity (υ) of light in a biological medium which oscillates with frequency (f) and wavelength (λ) along its propagation path can be written as [35]:

$$\upsilon = \lambda \times f \tag{12.36}$$

where the electric and magnetic properties of propagation medium determine the value of υ and it can be computed as follows [35]:

$$\upsilon = \frac{c}{\sqrt{\mu_r \varepsilon_r}} \tag{12.37}$$

Here, c is the speed of light in vacuum ($c = 3 \times 10^8$ meter/second), μ_r is the relative magnetic permeability ($\mu_r \approx 1$ in biological media), and ε_r is the relative electric permittivity ($\varepsilon_r \gg 1$ in biological media).

The light energy (photon energy) can also exist and it can be given as follows [35]:

$$W = \frac{h \times \upsilon}{\lambda} \tag{12.38}$$

Here, W is the photon energy and h is Plank's constant ($h = 6.6 \times 10^{-34}$ Joule second).

The induced light is subjected to volume and inhomogeneity effects. In volume effects, the light absorption takes place which attenuates the propagation of light beam in homogenous medium. On the other hand, the heterogeneous medium causes scattering, diffraction, reflection, and refraction effects which attenuate and redirect the light beam in a particular direction in inhomogeneity effects. These effects are not fully independent to each other [35]. The basic inhomogeneity effects are shown in Fig. 12.38 [36]. In Fig. 12.38, the reflection of light occurs at biological tissue interface and the refraction of light occurs when the light enters in tissue that has different refractive index. The absorption and scattering of light also take place in between the tissue structure. The physical parameters such as refractive index, absorption coefficients, and scattering coefficients related to these inhomogeneity effects vary continuously at biological tissue boundaries. The different biological tissues have different strengths of absorption coefficients which determine penetration power and energy absorption into a specific tissue from a particular light source. The absorption degree is depending on the type of tissue and wavelength of light in many cases. The mainly light absorption takes place between the wavelength range of UV (<400 nm) and IR (>2 μm). Hence, the light cannot deeply penetrate in this spectral range and attenuation due to scattering is less in this range. The scattering causes broadening of light beams and the light beams decay as it travels through the tissue due to this scattering phenomenon. This scattering phenomenon dominates over absorption in the spectral range of 600–1600 nm and the forward and backward scattering of incident light within tissue are used in various optic biomedical applications such as Raman vibrational spectroscopy and surface-enhanced Raman scattering (SERS) [36].

The reflection and transmitted modes of light can also be used for various optic biomedical applications such as optical plethysmography. The different arrangements of light source and sink are used for reflection and transmitted modes of operation. Figures 12.39 and 12.40 show the different arrangements of light source and sink for reflection and transmitted modes of operation applied on a finger, respectively. In reflection mode, red and near-IR lights are generally used due to the fact that these lights can penetrate tissue to relatively large depths as compared to other lights. The arrangement of light source and sink for red and near-IR lights are shown in Fig. 12.41. In Fig. 12.41, the light source and sink for red and near-IR lights are arranged in reflection mode which yields different pathways by photons

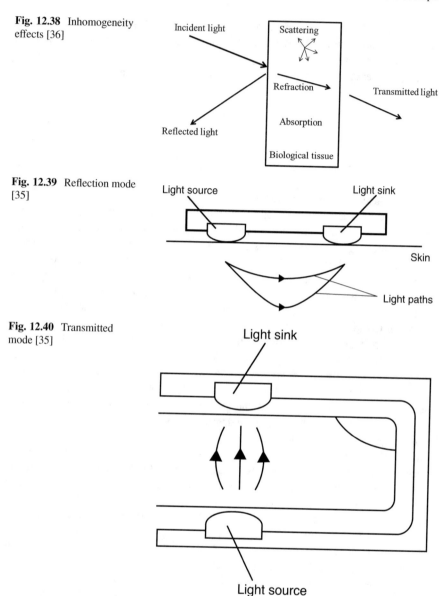

Fig. 12.38 Inhomogeneity effects [36]

Fig. 12.39 Reflection mode [35]

Fig. 12.40 Transmitted mode [35]

of both wavelengths and these paths vary with hemoglobin oxygen saturation which is denoted by S [35].

The light is also dynamically modulated by physiological phenomena in tissue due to the reason that a physiological phenomenon modulates optical properties of the tissue. There are various light absorbers present in tissue such as pulsatile arterial blood, nonpulsatile arterial blood, capillary blood, venous blood, bloodless

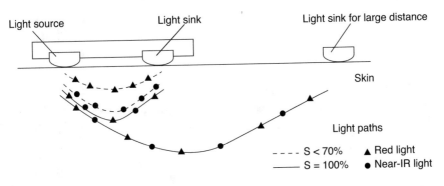

Light source   Light sink   Light sink for large distance

Skin

Light paths

- - - - S < 70%   ▲ Red light

——— S = 100%   ● Near-IR light

Fig. 12.41 Different arrangements of light source and sink for red and near-IR lights in reflection mode [35]

tissue, etc. The large volume of nonpulsatile arterial blood in the light path decreases the intensity of transmitted light. However, the transmitted light intensity passing through biological tissue experiences a relatively fast modulation from physiological point of view. The cardiac activity, respiratory activity, and blood oxygenation changes are basically responsible for the achievement of this fast modulation [35].

12.6.2 Sensing Aspects

The sensing aspects of the induced optic biomedical signal include transmitted light coupling with the light sink which is applied on the skin at a certain distance from the incident light source and its conversion into an electrical signal within light sink. In this way, the optic biomedical signals are registered with the help of transmitted incident light through tissues [35].

The fast fluctuations in local blood volume residing in the light propagation path modulate light absorption in tissue and slow fluctuations are present in the density of dominant chromophores in tissue. Due to this fact, there are three technologies which can mainly be used for the optic biomedical signals without considering the designed factor of optical sensors, namely, spectrometry, optical plethysmography, and optical oximetry [35].

In spectrometry, the light is absorbed by a chromophore in tissue which depends on the density of chromophore and the wavelength of applied light. Similarly, the light absorption spectrum gives a signature of the chromophore type as a function of wavelength. The amount of chromophore at the sensor is used for monitoring the local environment of the tissue [35].

The optical plethysmography detects the variations in the light absorption in tissue. The pulsating volume of arterial blood produces these variations in the illumination region due to transmitted light. The changing of optical path lengths

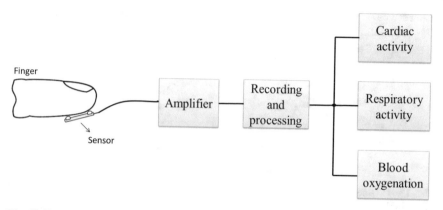

Fig. 12.42 Registration of optic biomedical signals from the finger in order to extract physiological parameters [35]

changes the total light absorption in the tissue. However, the total optical path length of the light beam is almost constant in tissue. The registered variations in the transmitted light intensity gives a signature of blood volume changes from the illumination region of transmitted light and these variations can be used for the monitoring of cardiac and respiratory activities [35].

The optical oximetry includes both existing technologies which are spectrometry and optical plethysmography for the assessment of blood oxygenation. In this technique, the level of the hemoglobin oxygen saturation in pulsatile arterial blood is calculated. The spectrometry technique in this method is used for evaluation of degree of hemoglobin oxygenation with the help of light absorption in the blood. On the other hand, the optical plethysmography is used for the separation of absorption by the pulsatile arterial blood from the nonpulsatile absorption with the help of the pulsatile nature of the transmitted light intensity. The registration of blood oxygenation during exploiting arterial pulsations is called pulse oximetry [35].

In spectrometry, the coefficient of absorption is the parameter of interest, whereas the path length is the parameter of interest for optical plethysmography [35].

The registration of optic biomedical signal for the extraction of physiological parameters, namely, cardiac activity, respiration activity, and blood oxygenation, is shown in Fig. 12.42. In Fig. 12.42, the optoplethysmogram signal is recorded from the finger in order to extract the physiological parameters. The amplifier is used to amplify the recorded signal, whereas the recording and processing block is used for the de-noising purpose. The recorded signal simultaneously offers the three different physiological parameters with multiparametric processing [35].

12.7   Analysis of Biomedical Signals

Biomedical signals are primarily used for the diagnosis and monitoring of specific pathological/physiological states. In some cases, the researchers have also used these signals for decoding and eventual modeling of specific biological systems. The recent advancement in technology allows the acquisition of multiple channels of biomedical signals. This process leads to additional signal processing challenges to identify meaningful interactions between these channels. The main aim of signal processing is generally noise removal, accurate signal modeling, extraction of components for analysis purpose, and feature extraction for deciding function or dysfunction of the heart and brain. The signal processing is used in biological applications due to several reasons. In most of the cases, the monitored biological signal contains an additive combination of signal and noise. The presence of noise can be due to instruments (sensors, amplifiers, filters, etc.) and electromagnetic interference (EMI). Therefore, the different conditions suggest different assumptions for noise characteristics, which will eventually lead to an appropriate choice of signal processing method [37].

12.7.1   Time-Domain Analysis

The time-domain analysis of biomedical signals is usually fast and easy to implement, because time-domain analysis does not need any transformation of biomedical signals. In time-domain analysis, several features based on different characteristics of signal are computed from biomedical signals. These time-domain features have been generally used in different areas of medical as well as engineering research. A major drawback of these features is due to the nonstationary nature of the biomedical signal, which changes the statistical properties over time. Therefore, the computed values of time-domain features may vary largely when the biomedical signal is recorded in interference and noisy environments. However, the time-domain features have been widely used for biomedical signal due to their lower computational complexity [38]. There are different time-domain characteristics which vary from one biomedical signal to the other.

In the time-domain analysis of EEG signals, the artifact which usually exceeded instantaneous amplitude as compared to normal instantaneous amplitude present in EEG signal is determined by amplitude thresholds. The muscle artifacts can be minimized with the use of slope or steepness threshold. The first-order derivative of EEG signal gives us the slope. In addition to first-order derivative, the second-order derivative is also used to measure the complexity present in EEG signal [30]. Apart from this measure, there are some other complexity measures such as fractal dimension and entropy which are also frequently used as features. The fractal dimension of the EEG signal decreases as the age increases in humans. Hence, it can be concluded with this fact that the fractal dimension is higher for a brain whose all parts are active, but it is lower for an old brain whose parts are considerably less active. The same behavior is also recorded for the other complexity measures from EEG signals of older people as compared to younger people. This phenomenon is also true for a person suffering from diseases like epilepsy and Alzheimer's. The complexity measures decrease due to the presence of such kind of diseases. Moreover, the reduction of 30% fractal dimension is used as diagnosis criteria for epilepsy from EEG signals [30].

The most common feature in time-domain analysis of ECG signals is the duration of the heart cycle. The heart cycle duration is basically a time span from one R wave to the next occurring R wave. The other generally used features for ECG signal are the duration of QRS complex and the time interval between T and P waves. The QRS complex is determined by the characteristic shape and relative stable time constant in the pattern [30].

The time-domain analysis of EMG signals is possible with several features. The mean absolute value (MAV), root mean square (RMS), zero crossing, v order, log detector, waveform length (WL), Willison amplitude (WAMP), and slope sign change (SSC) are commonly used for the time-domain analysis of EMG signals. The MAV feature provides the information about energy and fatigue present in EMG signals. On the other hand, the RMS feature represents the non-fatigue as well as fatigue contraction. The frequency information present in EMG signals is provided by time-domain features such as zero crossing, WAMP, and SSC. The v- order and log detector features estimate the muscle contraction force and fatigue [39].

12.7.2   Frequency-Domain Analysis

The frequency-domain analysis of biomedical signals is possible with Fourier transform which converts the time-domain representation of a signal into frequency domain. The frequency-domain analysis is mainly used to characterize the frequency contents present in a signal. The major limitation of this technique is that the Fourier transform works only for stationary signals because the time information is lapsed in frequency-domain analysis.

In frequency-domain analysis of EEG signals, the main feature is the computation of power of the particular frequencies from the power spectra of the EEG signal. The spectral analysis of EEG signal will quickly identify any irregular pattern of higher harmonics in the frequency spectrum. The spectrum of EEG signal is generally analyzed only over a consecutive short-time segment. This short-time segment of EEG signal is known as "epoch." The length of epochs decides the frequency resolution of EEG signal in frequency spectrum. However, the selection of longer time segments will result in lower time resolution which is a trade-off between time and frequency resolution.

The frequency components of EEG signal such as alpha, beta, delta, and theta waves are very informative and it can be easily extracted from the power spectra of the EEG signal. These frequency components are also used in the diagnostics of various diseases and disorders.

The analysis of EEG signals is also possible with frequency measure which is known as spectral edge frequency (SEF). This measure plays a significant role in the analysis of depth of anesthesia. It is found that a decrease in the value of SEF corresponds to a deeper level of anesthesia [30].

Similarly, median peak frequency (MPF) is another frequency measure in which frequency situated at 50% of the energy level is considered. The MPF gives the high-frequency contribution present in the frequency spectrum. This measure is also used in the analysis and classification of anesthesia depth [30].

In the frequency-domain analysis of ECG signal, the QRS complex is well localized in the high-frequency region. On the other hand, the low-frequency components are mainly due to P and T waves. The ST segment in the ECG signal mostly contains the low-frequency component. The frequency contents in a normal ECG signal and the deviating ECG signal have a significant difference because the normal heart rate is in the range of 60–100 beats per minute, whereas the fibrillation can exceed the range of 200 beats per minute. The depolarization and repolarization ramps in ECG signal are also changed under diseased conditions. This requires a much wider frequency bandwidth to identify different phenomena. The minor deviation of higher frequency in ECG signal creates a much larger number of harmonics which describe the frequency-domain features in the ECG signal. Therefore, a frequency span of 0–100 Hz usually represents normal ECG signal, whereas arrhythmias may require a high-frequency analysis up to 200 Hz. However, the high-frequency spectra will also be dominated by noise and it may not contribute any additional information [30].

A disease named as sinus tachycardia is also often detected in the frequency domain. A sinus tachycardia is detected when a sinus rhythm higher than 100 beats per minute appears. This similar condition may also occur during a physiological response to physical exercise or physical stress but it may lead to congestive heart failure in diseased cases. The case of sinus arrhythmia is also possible when the longest PP or RR intervals exceed the shortest interval by 0.16 s. This condition is frequent in teenage groups who have never suffered a heart disease [30].

The detection of fetal heart diseases using ECG signal during pregnancy is another area where the frequency-domain analysis plays a vital role. The ECG signals recorded from the leads placed on the abdomen of the mother are used to monitor the fetal heart diseases. The P and T waves obtained with the maternal ECG signal can easily be recognized in most cases. The maternal heart rate is usually lower than the fetal heart rate which is distinct from the mother's and the baby's ECG signals using filters designed in the frequency-domain.

The frequency-domain analysis of EMG signal leads to the fact that the frequency spectrum is mostly in the higher frequencies during fatigue, whereas the power spectrum is shifted toward lower frequencies after fatigue. This frequency shift indicates the muscle status such as rest and contraction states [30]. There are many features which are used for the frequency-domain analysis of EMG signals. The mean frequency (MF) measure is able to denote muscle fatigue during cyclic dynamics. The median (MD) frequency measure is a universal index for muscle force and fatigue. The peak frequency (PF) feature is used for identifying fatigue state. The mean and total power features are also used for the identification of fatigue. The first, second, and third spectral moments are alternative statistical measures for fatigue identification. The frequency ratio is also used as a feature to distinguish between rest and contraction states of muscles [39].

12.7.3 Time-Frequency Domain-Based Analysis

The time-frequency domain analysis is widely used for the analysis of nonstationary signals because it provides time and frequency information together for a given signal. There are several time-frequency domain analysis methods such as short-time Fourier transform (STFT), wavelet transform (WT), and Wigner-Ville distribution (WVD) [40].

The time-frequency analysis is used for detecting spike-like epileptic patterns in EEG signals because these patterns appear for a short time period or random in most suspected epileptic EEG signals. Due to random occurrence of these patterns, the frequency-domain analysis does not provide exact time information for these patterns. The choice of epoch length is an important issue in time-frequency domain analysis of EEG signal. The epoch lengths of 1–2 s duration are usually recommended for time-frequency domain analysis of EEG signal. The epochs of this time duration may provide stability in data features [30].

The time-frequency analysis of ECG signal identifies the typical pattern or wave. The time-frequency analysis provides the separation of the mother's and baby's ECG signals. It should be noted that the waveform of the fetal ECG signal is analogous to adult ECG signal [30].

12.7.4 Other Methods

In real situations, most of the signals are nonlinear and nonstationary in nature. The analysis of such type of signals is a tedious task. The predefined basis function may fail to provide solutions. This problem can be overcome by an adaptive or signal-dependent basis which is used for the representation of nonlinear and nonstationary signals. A method named Hilbert-Huang transform (HHT) is an adaptive and empirical method. HHT consists of two parts for signal analysis. One of them is empirical mode decomposition (EMD) and the second one is Hilbert spectral analysis (HSA). This method provides good results for time-frequency-energy representations of many signals [41].

Another method for analysis of real signals is higher-order spectra (HOS). A real signal most specifically a non-Gaussian signal can be decomposed into higher-order spectral functions in which each higher-order spectral function may contain different information about the signal [42]. These methods also provide significant contribution in biomedical signal analysis.

12.8 Modeling of Biomedical Signals

Biomedical signals can be modeled by mathematical functions. The procedure is started with finding a model which follows the laws of physics. The equations are solved for typical functions. The response of these equations is compared with developed physical model with the same typical functions. If these two responses are approximately equal, then we can use the developed model for analysis; otherwise, we have to improve our developed model [43].

12.8.1 Models for ECG Signal Representation

The ECG signals have pseudo-periodicity feature and features related to the constituent signals (P, QRS, and T). The modeling of ECG signals can be possible with parametric and nonparametric models. Most of them are parametric models such as impulse response of a pole-zero model and damped sinusoid model. The nonparametric models fail to exploit the nature of ECG signals. Hence, the parametric models overcome these problems.

The autoregressive (AR)/autoregressive moving-average (ARMA) model which is a parametric model is also used for the modeling of ECG signals. The amplitude-modulated (AM) sinusoidal signal model which is a special case of AR/ARMA model is also used due to its burst-like feature [44]. The model based on hidden Markov is also proposed to model every specific abnormal beat classification [45]. The dynamical model based on three coupled ordinary differential equations was used for generating synthetic ECG signals [46]. The Hilbert transform-based model is a recent approach for ECG signal modeling [47].

12.8.2 Models for EEG Signal Representation

The EEG signals have certain deviation or patterns as compared to the normal EEG signals during neurological disorders. These patterns occur for one or few seconds in the EEG signals. These patterns can be identified by modeling of EEG signals to detect various neurological diseases. The parametric modeling of EEG signal is the most common approach among them. The parametric model which is mostly used for EEG signal modeling is a rational transfer function with selected parameters. If the parameters lie in the denominator, then it is known as an all-pole or AR model, whereas if all the parameters lie in the numerator, then it is known as all-zero or moving-average (MA) model. A model with parameters lie in the numerator and a denominator is known as pole-zero or ARMA model [48]. The parametric modeling of ictal EEG signal using Prony's method is also possible. This method is based on the assumption that the original signal is a sum of damped complex exponential sinusoids, and it has good frequency resolution compared to AR model. This method suggests that the modeling of ictal EEG signal is based on the poles of EEG signal [49]. A method based on second-order linear time-varying AR (TVAR) with appropriate chosen length obtained using FBSE is used for parametric modeling of EEG signal [17].

12.8.3 Models for EMG Signal Representation

The main purpose for modeling EMG signal is to understand electrophysiological information for the detection of neuromuscular disorders. The EMG signal modeling can also be possible by AR model [50]. A modified method autoregressive integrated moving-average (ARIMA) model has been also proposed for EMG signals in the literature [51].

12.8.4 Models of Other Biomedical Signals

The parametric modeling of PCG signals for the detection of murmurs is possible with AR modeling. This model used dominant poles for pattern classification and spectral tracking [52, 53].

The modeling of respiratory sound signals can be possible with mechanical as well as electrical models. In these models, the vocal and respiratory tract of humans can be represented by tubes and pipes and their electrical equivalent circuits [54].

12.9 Applications

The biomedical signals have also been used in certain areas of applications based on signal processing techniques such as detection of heart-related diseases, neurological disorders, neuromuscular diseases, postural stability analysis, and other related disease. The description of these applications is illustrated below.

12.9.1   Detection of Heart-Related Disorders

The heart-related signals are very useful to detect cardiovascular diseases. These detections are generally based on heart sounds and ECG signals. The detection of congestive heart failure (CHF) is carried out using eigenvalue decomposition from HRV signals extracted with ECG signals [55, 56]. The heart valve disorders can be classified with the method based on tunable-Q wavelet transform (TQWT) [57]. The diagnosis of arrhythmia using flexible analytic wavelet transform (FAWT) from ECG signals has significant importance [58]. The FAWT method is used for the detection of myocardial infarction (MI), which is a condition that indicates injury of the heart cell [59]. The automated detection system of CAD is developed with FAWT method using ECG signals [60]. In another work [61], the detection of CAD is also possible with HRV signals involving the FAWT. A method for diabetic patients using RR interval signals obtained from ECG signal is developed for screening [23].

12.9.2   Detection of Brain-Related Diseases

The brain-related diseases and disorders such as epilepsy, Alzheimer's, Parkinson's, and sleep disorder can be detected by EEG and MEG signals. The detection technique based on EMD method using EEG signals for epileptic seizure has been proposed [62]. The phase space representation of intrinsic mode functions has been also utilized to classify epileptic seizure EEG signals [63]. The second-order difference plot of intrinsic mode functions has been also used for epileptic seizure classification [64]. The entropy of intrinsic mode functions has been used for the automated detection of focal EEG signals [65]. The detection of ictal EEG signals using fractional linear prediction has been also proposed [66]. The sleep stages have been also classified using time-frequency image of EEG signals [67]. The iterative filtering method has been also used to develop an automated system for sleep stage classification [68].

12.9.3   Detection of Neuromuscular Diseases

The diagnosis of neuromuscular diseases has been also possible using computer-aided method. The technique based on wavelet neural network applied on EMG signals has been proposed for neuromuscular disorder detection [69]. A technique based on discrete wavelet transform for EMG signal classification with comparison of decision tree algorithms has been proposed [70]. The detection of muscle fatigue using EMG signals with time-frequency methods has been presented [71]. The fatigue during dynamic contractions of the muscle has been also detected using

EMG signals [72]. The automated classification of hand movements using TQWT-based filter bank with EMG signals has been also proposed [73].

12.9.4 Postural Stability Analysis

The postural control is very useful for everyday movement and the central nervous system provides sensory information for postural control. This system is used to maintain a proper postural balance. Any postural imbalance may lead to instability, falls, and injury. The center of pressure signals are commonly used to examine the postural control [74]. These signals can be analyzed with various signal processing methods. The method based on FBSE applied on the center of pressure signals has been used for postural stability analysis [74]. The method for assessment of standing postural stability in children has been also proposed [75]. The EMD method with second-order difference plots has been used for postural time-series analysis [76].

12.9.5 Other Related Applications

The knee joint pathological conditions change the vibroathrographic (VAG) signals. These VAG signals provide the abnormalities associated with knee joints. The automated screening of knee joints using double density dual-tree complex WT has been proposed [77]. The detection of direction of eyes movement has been possible using EOG signals [78]. This detection provides help to various disabled persons.

References

1. J. Enderle, J. Brozino, *Introduction to Biomedical Engineering* (Academic Press, Burlington, 2012)
2. O.G. Martinsen, S. Grimnes, *Bioimpedance and Bioelectricity Basics* (Academic Press, London, 2011)
3. R. Plonsey, R.C. Barr, *Bioelectricity: a Quantitative Approach* (Springer Science & Business Media, New York, 2007)
4. A. Winter, *Biomechanics and Motor Control of Human Movement* (Wiley, Hoboken, 2009)
5. G.S. Firestein, R. Budd, S.E. Gabriel, I.B. McInnes, J.R. O'Dell, *Kelley's Textbook of Rheumatology E-Book* (Elsevier Health Sciences, London, 2012)
6. A.P. Turner, J.C. Pickup, Diabetes mellitus: biosensors for research and management. Biosensors **1**(1), 85–115 (1985)
7. L.J. Blum, P.R. Coulet, *Biosensor Principles and Applications* (M. Dekker, New York, 1991)
8. T. Vo-Dinh, B. Cullum, Biosensors and biochips: advances in biological and medical diagnostics. Fresenius J. Anal. Chem. **366**(6–7), 540–551 (2000)
9. J.-Y. Yoon, *Introduction to Biosensors: from Electric Circuits to Immunosensors* (Springer, Cham, 2016)

10. A.V. Oppenheim, A.S. Willsky, S.H. Nawab, *Signals and Systems*, 2nd edn. (Prentice Hall, Upper Saddle River, 1997)
11. S. Mukhopadhyay. NPTEL (2016). [Online]. http://textofvideo.nptel.ac.in/108105088/lec7.pdf
12. S.R. Devasahayam, *Signals and Systems in Biomedical Engineering* (Springer US, Boston, 2000)
13. J. Schroeder, Signal processing via Fourier-Bessel series expansion. Digit. Signal Process. **3**(2), 112 (1993)
14. S. Gupta, K.H. Krishna, R.B. Pachori, M. Tanveer, Fourier-Bessel series expansion based technique for automated classification of focal and non-focal EEG signals, in *International Joint Conference on Neural Networks (IJCNN)*, 2018, pp. 1–6
15. A.S. Hood, R.B. Pachori, V.K. Reddy, P. Sircar, Parametric representation of speech employing multi-component AFM signal model. Int. J. Speech Technol. **18**(3), 287–303 (2015)
16. R.B. Pachori, Discrimination between ictal and seizure-free EEG signals using empirical mode decomposition. Res. Lett. Signal Process. **14**, 2008 (2008)
17. R.B. Pachori, P. Sircar, EEG signal analysis using FB expansion and second-order linear TVAR process. Signal Process. **88**(2), 415–420 (2008)
18. R.B. Pachori, P. Sircar, A new technique to reduce cross terms in the Wigner distribution. Digit. Signal Process. **17**(2), 466–474 (2007)
19. R.B. Pachori, P. Sircar, Analysis of multicomponent AM-FM signals using FB-DESA method. Digit. Signal Process. **20**(1), 42–62 (2010)
20. R.B. Pachori, P. Avinash, K. Shashank, R. Sharma, U.R. Acharya, Application of empirical mode decomposition for analysis of normal and diabetic RR-interval signals. Expert Syst. Appl. **42**(9), 4567–4581 (2015)
21. S. Sood, M. Kumar, R.B. Pachori, U.R. Acharya, Application of empirical mode decomposition–based features for analysis of normal and CAD heart rate signals. J. Mech. Med. Biol. **16**(1), 1640002 (2016)
22. P. Jain, R.B. Pachori, Time-order representation based method for epoch detection from speech signals. J. Intell. Syst. **21**(1), 79–95 (2012)
23. R.B. Pachori, M. Kumar, P. Avinash, K. Shashank, U.R. Acharya, An improved online paradigm for screening of diabetic patients using RR-interval signals. J. Mech. Med. Biol. **16**(1), 1640003 (2016)
24. R.B. Pachori, P. Sircar, *Non-stationary Signal Analysis: Methods Based on Fourier-Bessel Representation* (LAP LAMBERT Academic Publishing, Germany, 2010)
25. A. Bhattacharyya, L. Singh, R.B. Pachori, Fourier–Bessel series expansion based empirical wavelet transform for analysis of non-stationary signals. Digit. Signal Process. **78**, 185–196 (2018)
26. R.G. Andrzejak et al., Indications of nonlinear deterministic and finite-dimensional structures in time series of brain electrical activity: Dependence on recording region and brain state. Phys. Rev. E 64 **6**, 061907 (2001)
27. M.V. Gils, Lecture 01: Introduction and recapitulation of essential techniques (2015). [Online]. https://mycourses.aalto.fi/course/view.php?id=9367&lang=fi
28. B. Mulgrew, P. Grant, J. Thompson, *Digital Signal Processing: Concepts and Applications* (Macmillan Education London, 1999)
29. V. Singh, K. Veer, R. Sharma, S. Kumar, Comparative study of FIR and IIR filters for the removal of 50 Hz noise from EEG signal. Int. J. Biomed. Eng. Technol. **22**(3), 250–257 (2016)
30. K. Najarian, R. Splinter, *Biomedical Signal and Image Processing* (CRC Press, Boca Raton, 2005)
31. A.L. Goldberger et al., PhysioBank, PhysioToolkit, and PhysioNet: components of a new research resource for complex physiologic signals. Circulation **101**(23), e215–e220 (2000)
32. M. Tangermann et al., Review of the BCI competition IV. Front. Neurosci. **6**, 55 (2012)
33. R.M. Rangayyan, *Biomedical Signal Analysis*, vol 33 (Wiley, Hoboken, 2015)
34. E. Kaniusas, *Biomedical Signals and Sensors I* (Springer, Berlin, Heidelberg, 2012)
35. E. Kaniusas, *Biomedical Signals and Sensors II* (Springer, Berlin, Heidelberg, 2015)

36. G. Keiser, *Biophotonics – Concepts to Applications* (Springer Singapore, Singapore, 2016)
37. M. Kutz, *Standard Handbook of Biomedical Engineering and Design* (McGraw-Hill, New York, 2003)
38. A. Phinyomark, P. Phukpattaranont, C. Limsakul, Feature reduction and selection for EMG signal classification. Expert Syst. Appl. **39**(8), 7420–7431 (2012)
39. T.N.S. Tengku Zawawi et al., A review of electromyography signal analysis techniques for musculoskeletal disorders. Indones. J. Electr. Eng. Comput. Sci. **11**(3), 1136 (2018)
40. B. Boashash, *Time-Frequency Signal Analysis and Processing: a Comprehensive Reference* (Academic Press, Amsterdam, 2015)
41. N.E. Huang, *Hilbert-Huang Transform and Its Applications*, vol 16 (World Scientific, Hackensack, 2014)
42. C.L. Nikias, J.M. Mendel, Signal processing with higher-order spectra. IEEE Signal Process. Mag. **10**(3), 10–37 (1993)
43. C.L. Phillips, J.M. Parr, E.A. Riskin, *Signals, Systems, and Transforms* (Prentice Hall, Upper Saddle River, 1995)
44. S. Mukhopadhyay, P. Sircar, Parametric modelling of ECG signal. Med. Biol. Eng. Comput. **34**(2), 171–174 (1996)
45. R.V. Andreao, B. Dorizzi, J. Boudy, ECG signal analysis through hidden Markov models. IEEE Trans. Biomed. Eng. **53**(8), 1541–1549 (2006)
46. P.E. McSharry, G.D. Clifford, L. Tarassenko, L.A. Smith, A dynamical model for generating synthetic electrocardiogram signals. IEEE Trans. Biomed. Eng. **50**(3), 289–294 (2003)
47. J.-C. Nunes, A. Nait-Ali, Hilbert transform-based ECG modeling. Biomed. Eng. **39**(3), 133–137 (2005)
48. J. Pardey, S. Roberts, L. Tarassenko, A review of parametric modelling techniques for EEG analysis. Med. Eng. Phys. **18**(1), 2–11 (1996)
49. O.A. Elsayed, A. Eldeib, F. Elhefnawi, Parametric modeling of ICTAL epilepsy EEG signal using Prony method. Int. J. Comput. Sci. Softw. Eng. (IJCSSE) **3**(1), 86–89 (2014)
50. D. Graupe, K.H. Kohn, A. Kralj, S. Basseas, Patient controlled electrical stimulation via EMG signature discrimination for providing certain paraplegics with primitive walking functions. J. Biomed. Eng. **5**(3), 220–226 (1983)
51. M.H. Sherif, R.J. Gregor, J. Lyman, Effects of load on myoelectric signals: the ARIMA representation. IEEE Trans. Biomed. Eng. **5**, 411–416 (1981)
52. A. Iwata, N. Suzumura, K. Ikegaya, Pattern classification of the phonocardiogram using linear prediction analysis. Med. Biol. Eng. Comput. **15**(4), 407–412 (1977)
53. A. Iwata, N. Ishii, N. Suzumura, K. Ikegaya, Algorithm for detecting the first and the second heart sounds by spectral tracking. Med. Biol. Eng. Comput. **18**(1), 19–26 (1980)
54. Z. Moussavi, Fundamentals of respiratory sounds and analysis. Synth. Lect. Biomed. Eng. **1**(1), 1–68 (2006)
55. R.R. Sharma, A. Kumar, R.B. Pachori, U.R. Acharya, Accurate automated detection of congestive heart failure using eigenvalue decomposition based features extracted from HRV signals. Biocybern. Biomed. Eng. **39**, 312 (2019)
56. M. Kumar, R.B. Pachori, U.R. Acharya, Use of accumulated entropies for automated detection of congestive heart failure in flexible analytic wavelet transform framework based on short-term HRV signals. Entropy **19**(3), 92 (2017)
57. S. Patidar, R.B. Pachori, Classification of heart disorders based on tunable-Q wavelet transform of cardiac sound signals, in *Chaos Modeling and Control Systems Design*, (Springer, Cham, 2015), pp. 239–264
58. M. Kumar, R.B. Pachori, U.R. Acharya, Automated diagnosis of atrial fibrillation ECG signals using entropy features extracted from flexible analytic wavelet transform. Biocybern. Biomed. Eng. **38**(3), 564–573 (2018)
59. M. Kumar, R.B. Pachori, U.R. Acharya, Automated diagnosis of myocardial infarction ECG signals using sample entropy in flexible analytic wavelet transform framework. Entropy **19**(9), 488 (2017)

60. M. Kumar, R.B. Pachori, U.R. Acharya, Characterization of coronary artery disease using flexible analytic wavelet transform applied on ECG signals. Biomed. Signal Process. Control **31**, 301–308 (2017)

61. M. Kumar, R.B. Pachori, U.R. Acharya, An efficient automated technique for CAD diagnosis using flexible analytic wavelet transform and entropy features extracted from HRV signals. Expert Syst. Appl. **63**, 165–172 (2016)

62. V. Bajaj, R.B. Pachori, Classification of seizure and nonseizure EEG signals using empirical mode decomposition. IEEE Trans. Inf. Technol. Biomed. **16**(6), 1135–1142 (2012)

63. R. Sharma, R.B. Pachori, Classification of epileptic seizures in EEG signals based on phase space representation of intrinsic mode functions. Expert Syst. Appl. **42**(3), 1106–1117 (2015)

64. R.B. Pachori, S. Patidar, Epileptic seizure classification in EEG signals using second-order difference plot of intrinsic mode functions. Comput. Methods Prog. Biomed. **113**(2), 494–502 (2014)

65. R. Sharma, R.B. Pachori, U.R. Acharya, Application of entropy measures on intrinsic mode functions for the automated identification of focal electroencephalogram signals. Entropy **17**(2), 669–691 (2015)

66. V. Joshi, R.B. Pachori, A. Vijesh, Classification of ictal and seizure-free EEG signals using fractional linear prediction. Biomed. Signal Process. Control **9**, 1–5 (2014)

67. V. Bajaj, R.B. Pachori, Automatic classification of sleep stages based on the time-frequency image of EEG signals. Comput. Methods Prog. Biomed. **112**(3), 320–328 (2013)

68. R. Sharma, R.B. Pachori, A. Upadhyay, Automatic sleep stages classification based on iterative filtering of electroencephalogram signals. Neural Comput. & Applic. **28**(10), 2959–2978 (2017)

69. A. Subasi, M. Yilmaz, H.R. Ozcalik, Classification of EMG signals using wavelet neural network. J. Neurosci. Methods **156**(1–2), 360–367 (2006)

70. E. Gokgoz, A. Subasi, Comparison of decision tree algorithms for EMG signal classification using DWT. Biomed. Signal Process. Control **18**, 138–144 (2015)

71. A. Subasi, M.K. Kiymik, Muscle fatigue detection in EMG using time–frequency methods, ICA and neural networks. J. Med. Syst. **34**(4), 777–785 (2010)

72. M. Gonzalez-Izal et al., EMG spectral indices and muscle power fatigue during dynamic contractions. J. Electromyogr. Kinesiol. **20**(2), 233–240 (2010)

73. A. Nishad, A. Upadhyay, R.B. Pachori, U.R. Acharya, Automated classification of hand movements using tunable-Q wavelet transform based filter-bank with surface electromyogram signals. Futur. Gener. Comput. Syst. **93**, 96–110 (2019)

74. R.B. Pachori, D. Hewson, Assessment of the effects of sensory perturbations using Fourier–Bessel expansion method for postural stability analysis. J. Intell. Syst. **20**(2), 167–186 (2011)

75. G.F. Harris, S.A. Riedel, D. Matesi, P. Smith, Standing postural stability assessment and signal stationarity in children with cerebral palsy. IEEE Trans. Rehabil. Eng. **1**(1), 35–42 (1993)

76. R.B. Pachori, D. Hewson, H. Snoussi, J. Duchêne, Postural time-series analysis using empirical mode decomposition and second-order difference plots, in *International Conference on Acoustics, Speech and Signal Processing (ICASSP 2009)*, 2009, pp. 537–540

77. M. Sharma, P. Sharma, R.B. Pachori, V.M. Gadre, Double density dual-tree complex wavelet transform-based features for automated screening of knee-joint vibroarthrographic signals, in *Machine Intelligence and Signal Analysis*, (Springer, Singapore, 2009), pp. 279–290

78. M. Merino, O. Rivera, I. Gómez, A. Molina, E. Dorronzoro, A method of EOG signal processing to detect the direction of eye movements, in *First International Conference on Sensor Device Technologies and Applications (SENSORDEVICES)*, 2010, pp. 100–105

Chapter 13
Smart Learning Using Big and Small Data for Mobile and IOT e-Health

Pei-Yun Sabrina Hsueh, Xinyu Hu, Ying Kuen Cheung, Dominik Wolff, Michael Marschollek, and Jeff Rogers

There are no secrets to success. It is the result of preparation, hard work, and learning from failure.

Colin Powell

Contents

P.-Y. S. Hsueh (✉)
Viome, Inc., Yorktown Heights, NY, USA

X. Hu · Y. K. Cheung
Department of Biostatistics, Mailman School of Public Health, Columbia University, New York, NY, USA

D. Wolff · M. Marschollek
Peter L. Reichertz Institute for Medical Informatics of TU Braunschweig and Hannover Medical School, Hanover, Germany

J. Rogers
IBM Watson Research Center, Yorktown Heights, NY, USA

© Springer Nature Switzerland AG 2020
F. Firouzi et al. (eds.), *Intelligent Internet of Things*,
https://doi.org/10.1007/978-3-030-30367-9_13

13.1 Introduction

In this chapter, we provide a snapshot of the state-of-the-art research in mobile and IOT e-health studies that leverage AI technologies for making sense of personal health measurement and assessment, as well as for delivering situational, actionable insights in care flows. In recent years, the proliferation of consumer and pervasive health technologies has enabled a whole new generation of sensor-based precision measurement technologies and mobile ecological momentary assessments that are able to capture patient-specific characteristics in context [1–3]. The captured physiomes (i.e., a collection of quantitative and integrated descriptions of the functional behavior of the physiological state of an individual [4]) can help detect physiological macro-phenotypes such as inflammatory response and fatigue [5], as well as critical conditions such as seizure and atrial fibrillation [6, 7]. The accumulated longitudinal records of such phonemes are also expected to capture patterns that can help distinguish individual physiological differences, e.g., being insulin-sensitive or insulin-resistant, which will make a difference in disease diagnosis and prognosis [5].

Moreover, the patient-specific characteristics include not only physiome patterns exhibited in physiological measures but also behavioral patterns and its associated contexts in daily-life settings, which explain more than 50% of the premature deaths in the United States for the past 50 years and become one of the major categories of health determinants [8]. Take cardiovascular health for an example: Public health experts have advocated on the importance of healthy behaviors for reducing cardiovascular risk, which estimatedly sums up to 41% of global cardiovascular disease burden [9, 10]. Past research has also demonstrated the benefit of healthy behavior (e.g., regular physical activity) on emotional well-being [11, 12].

Being able to capture behavioral patterns in context would have tremendous implications to healthcare applications. However, despite the success of capturing the varying patient-specific characteristics, early evidence of using them to improve outcomes yields only mixed results [3, 13–15].

On the one hand, simply providing technologies for self-monitoring is not sufficient to activate and sustain behavioral changes. Individuals in different stages of change would need different ways to communicate and different types of intervention. This, in turn, resulted in the failure of many implementation plans for integrating mobile and IOT e-health data to guide behavioral change that often require more intuitive reasoning and explanation capability [16].

On the other hand, a number of theoretical model-driven interventions have been shown to have positive impact on activating and sustaining health behavioral changes. For example, implementation intentions (i.e., representations of simple plans that can translate goal intentions into behavior under specific conditions) have a medium to large effect on goal attainment of physical activity behavior [17], especially when coupled with e-health capabilities [18–22].

As a middle ground, behavioral scientists and healthcare service researchers are now working together to employ e-health technologies as an interactive platform to conduct ecological momentary assessment and further provide tailored feedback from clinicians to patients, based on each person's ongoing performance [23–26]. For example, by using mobile and wearable devices to record a user's exercise patterns, caloric intake and health status, the user is then provided with personalized advices that help make sense of his own data and reflect on the impact of exercise or food on himself. As another example, mobile and IOT e-health data are used to personalize messages from clinicians [27, 28] and to support the tracking and communication of outcomes between clinical visits for conditions such as obesity [29], chronic obstructive pulmonary disease [30] and post-traumatic stress disorder [31].

Beyond self-monitoring and clinician-patient communication, mobile and IOT e-health technologies have also been studied for its benefits on delivering behavioral interventions to sustain healthy behavior change in a variety of self-care settings. The applications widely range from behavior change theory-driven lifestyle interventions for chronic disease management [32, 33] to self-regulation-based psychotherapeutic interventions for the management of mental health [34], stress [35] and substance use disorders [36].

13.1.1 Key Challenges in Smart Learning for Mobile and IOT e-Health

The increasing accumulation of mobile and IOT e-health data has enabled an immense potential of applying AI for smart learning. The recent emerging AI technologies (including but not limited to the neural modeling and reinforcement learning approaches) have offered promises to capture the patient-specific, dynamic and intricate relationships through an end-to-end learning framework. The framework is flexible enough to allow for the incorporation of additional exogenous patient data sources, ranging from patient-reported outcomes and contexts to sensor measurements such as heart rate variability (HRV) and electroencephalogram (EEG) signals.

Meanwhile, the detection of context-aware physiological and behavioral patterns over time enables healthcare professionals to understand how a treatment plan or care management program is working for a particular patient. More frequent checkpoints between clinical visits allow for more pertinent analysis of how a patient responds to a certain medication, treatment plan or engagement strategy. In the cases where constant monitoring is feasible, real-time analysis can be provided to care teams at any given moment. It will also allow for just-in-time intervention at the time of adverse events.

One great domain wherein smart learning for mobile and IOT e-health can make great impacts is to generate feedback and recommend behavioral interventions based on both the small (personal-level) and big (population-level) data collected. Two key challenges emerge. One is regarding how to generate adaptive strategies to guide a target user's choice of actions by learning from real-world experiences leading to desired health outcomes. The other regards how to make the learned strategies more interpretable by representing the decision rules in a simpler form that can be digested by users with ease as well as by providing the target user with actionable goal-setting guidance, e.g., the predictive sub-goal performance on outcome improvement. The multilayer challenges current smart learning methods, which are often biased and uninterpretable [37, 38], to deliver interpretable actionable insights to individuals.

To address the first challenge, researchers could leverage a long line of studies in adaptive clinical trial and adaptive treatment regime, which have established the benefits of incorporating sequential multiple adaptive randomized trial (SMART) methods to achieve best mean outcomes [39–42]. More recently, researchers have considered supporting just-in-time delivery of adaptive interventions [43, 44]. To address the second challenge, researchers have started developing interpretable machine learning models for healthcare applications [45]. However, most of these studies define interpretability as a measure of model characteristics (e.g., complexity) learned by the algorithms. The problem of such an operational definition is its disconnection with actual user interpretation [46, 47].

To address both challenges simultaneously, most recently policy learning methods such as reinforcement learning are being studied to lend support to the generation of easy-to-interpret adaptive strategies according to the observations of target users [48, 49].

Another great domain wherein smart learning for mobile and IOT e-health can make great impacts is to provide systematic support for operationalizing patient education and coaching programs in daily-life settings. The prevalence of low health literacy and self-efficacy in the population is one of the central problems in the healthcare system. Due to the service-oriented nature of strict time scheduling, physicians often cannot contribute significantly to patient education. A possible solution is the use of automated educational systems, which adapt the delivery of knowledge according to a learner's needs and level of knowledge. Although adaptive automated educational systems for other topics have been applied in practice, the adaptive automated patient education system practically does not exist.

Several challenges have emerged therein. The key challenge lies on its long development phase and non-standardized evaluation. Additionally, specific limitations in the domain of healthcare lie on the needed amount of personal information about the user and the number of related time-consuming assessments on the patient's side. Furthermore, there exist challenges on the bandwidth of healthcare experts needed for implementing such systems and the difficulties in explicating their knowledge.

13.1.2   Incorporating Domain Knowledge in Data-Driven Learning

To overcome these challenges, incorporating domain knowledge in the data-driven learning framework would help. First, structural knowledge bases, such as decision rules used in clinical practice guidelines (CPG), can help shorten the development cycle and reduce the amount of data needed for training. Second, it can help avoid the biased results because of its capability to better generalize and make the results more self-interpretable among domain experts.

However, how to leverage structural knowledge bases to guide data-driven learning remains to be a challenging issue. This is especially true in healthcare, which has amassed evidence-based, comprehensive knowledge bases over the years and is in need for interpretable decision rules to trust AI/ML-augmented recommendations [50–52].

Previous solutions could be roughly divided into two categories. One is the homogeneous method, which integrates knowledge-based representations (e.g., logic rules) with data-driven learning tasks into a unified inference framework [53–55]. The other is the heterogeneous method, which integrates the respective inference results from knowledge-based and data-driven modules [56]. While both categories of methods have its merits, each has its own gap to fill before knowledge bases and data-driven models can work together to generate real-world evidence in healthcare practice.

13.1.3   Structure of the Book Chapter

In the rest of this book chapter, we will use three case studies to illustrate the challenges and opportunities facing the development of a smart learning framework. Across the three case studies, we will especially focus on the lessons learned from applying smart learning for mobile and IOT e-health and embedding actionable insights in care flows for decision-making.

The aim of this chapter is twofold. First, we would like to provide an overview of the status quo of this emerging field of applying and developing a smart learning framework for mobile and IOT e-health technologies. Second, we would identify gaps and gather requirements to achieve the goal of generating real-world evidence in practice. After introducing the cases studies, we will discuss the best practices and barriers of implementing such a smart learning framework in real-life settings and further developing a hybrid infrastructure that will allow the generation of knowledge-augmented, data-derived insights for mobile and IOT e-health.

13.2 Predictive and Reinforcement Learning for Life Coaching

In the first theme of case studies, we introduce the potential of collecting heterogeneous personal health data via mobile/IOT e-health technologies and discuss how to handle the issues regarding the volume, variety, veracity and velocity of such data sources. In particular, we illustrate a number of smart learning approaches that help with not only association mining from population-level "big data" but also the identification of N-of-1 behavioral signature from personal-level "small data." We also emphasize on the importance of a smart learning framework and discuss how policy learning approaches, which are inspired by reinforcement learning tasks such as Q-learning, can be used to improve interpretability and actionable personalization insights. The case study will be presented in two examples of application: exercise behavior prediction and personalized stress management.

13.2.1 Background: Stress-Activity Data

Psychosocial stress contributes to heart diseases by affecting health-related behavior patterns [8]. Exercise helps reduce psychosocial stress, thus in turn alleviating heart disease risks [57]. There potentially exists a bidirectional relationship between stress and exercise behavior. Therefore, understanding how exercise behavior influences stress can provide insights of developing stress management tools; reversely, understanding how stress affects exercise behavior can help build healthy lifestyle interventions. To explore and test the bidirectional relationship, a single cohort, 12-month randomized controlled experiment has been conducted (refer to [58] for a complete description of the protocol and data curation process). The dataset used in this case study would be referred to as stress-activity data hereafter.

To achieve personalized stress management, how to collect real-time stress assessment and exercise statistics continuously is a crucial but challenging problem. There is a growing interest in using mobile and IOT technologies, such as smartphones and wearable devices, for collecting health-relevant data and delivering health guidelines [59]. The stress-activity data are collected from 79 participants who were followed for up to a 1-year period with physical activity objectively monitored by actigraphy and stress recorded via ecological momentary assessment on a mobile app. Physical activity can be continuously monitored via mobile devices (e.g., Fitbit) over a long time horizon. The Fitbit device tracks users' daily physical activity, including step counts, walked distance and burned calories. Fitbit wirelessly transmits individual activity data in real time, thus preventing loss of data for individuals who may otherwise not return the device. To collect stress data, an electronic diary in mobile devices was used to capture momentary and summary aspects of users' stress experience. Self-evaluated stress level was recorded at the end of each day via the electronic diary with specifically designed questions.

The range of self-evaluated stress level is from 0 ("not at all") to 10 ("extremely likely"). In addition, exogenous and environmental variables retrieved from public archives, such as day in a week, daylight time, temperature and precipitation, are also included. These assessment instruments have been used extensively in the literature with smartphone for EMA in recent studies [60].

The continuously collected personal health data is one type of the ecological momentary assessments (EMA), which are used to repeatedly sample individuals' experience and behavior in real time under their natural environments [61]. The advantage of collecting EMA data is that it reduces recall bias and allows for a closer investigation of the process dynamics that influence individuals' real-world state and behavior. EMA data collected from mobile and IOT e-health devices can be used to personalize messages to individuals [28] and to support their self-care management of certain concerns [62].

Powerful statistical methods have been proposed to model massive mobile health data and characterize the bidirectional relationship between stress and exercise behavior. With the advance of mobile and IOT technologies, data can be collected in real time over a long time horizon. For data collected from an individual, referred to as "small data," N-of-1 analytical methods are available to estimate individual effect and to provide personalized healthcare advices. To handle data collected from a large group of individuals, referred to as "big data," actionable learning methods (inspired by reinforcement learning) are able to learn population-level patterns and make recommendations to individuals. These methods offer a smart learning framework to model large-scale and complex mobile and IOT e-health data and shed light on the diverse and bidirectional nature of the relationship between stress and exercise behavior in general.

13.2.2 Case Study: N-of-1 Analytical Methods

Individual information can be collected via mobile and IOT e-health technologies for a long or even indefinite period of time. It is challenging to gain insights from a massive volume of data to understand how modifiable variables influence health outcomes of interest. The understanding of the relationship is traditionally obtained at a population level. However, the environment exposures, individual characteristics, and modifiable elements that affect the dynamics of the relationship are usually individual-specific. Therefore, individual-specific analytical approaches are desirable to design personalized guidance and tailor for individual needs. This motivates the use of N-of-1 methods instead of the traditional population-based approach. An N-of-1 trial is a clinical trial in which only a single individual is in the entire trial and in which random allocation can be used to decide the order of interventions given to the target individual. N-of-1 analytical methods utilize the data collected from N-of-1 trials and provide a personalized assessment that leads to better learning.

To gain insights of how stress influences exercise behavior, [63] has proposed to incorporate into the smart learning framework an N-of-1 method, which uses random forest and classification tree techniques to model the individual trajectories. The understanding of stress-behavior pathway is traditionally attained at the population level; thus, it ignores individual heterogeneity and may lead to biased estimation.

The proposed approach explores whether an N-of-1 model can better capture an individual's stress-behavior pathway, thus providing useful individual-specific pattern of stress to predict exercise behavior. This method consists of three steps: First, a classification decision tree is built for each participant. Second, a ranking of variable importance is measured by classification accuracy; only those variables that achieve significant decrease in classification accuracy would be kept. Third, another classification tree is built for exercise prediction using the selected variables. The experiment result shows that the proposed N-of-1 approach achieved an average lower classification error compared to a nomothetic approach. An example of an N-of-1 tree and a nomothetic tree is illustrated in Fig. 13.1. A nomothetic tree predicts exercise based on the data of all study participants, and an N-of-1 tree predicts for each of the three study participants.

In this case study, we follow sport medicine guidelines to define exercise as any consecutive 30-minute period within which 24 or more minutes were classified as moderate or vigorous intensity. The analyses on the stress-activity data show that

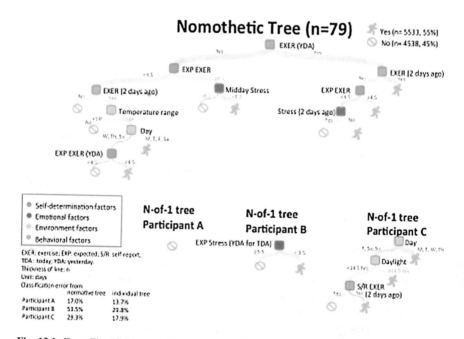

Fig. 13.1 From Fig. 13.2 in [63]. Top: A nomothetic model of exercise prediction using all study participants' data. Bottom: three "N-of-1" models of exercise prediction for three study participants

the average effect of a 30-minute bout of exercise on stress reported at the end of that day is negative and highly significant. This effect has a high variability at the individual level. In addition, there exists a significant negative average effect of anticipated stress for a given day on the probability of exercising that day. This effect also exhibits a high variability at the individual level.

Furthermore, a substantial proportion of individuals demonstrated significant bidirectional effect of stress and exercise. This indicates that the anticipated stress is associated with a lower chance for the individual to exercise, and exercising is associated with a lower reported stress on the same day. However, the significant effect in either direction is only demonstrated on a small number of individuals.

These findings highlight the limitation of the homogeneous assumption of individuals and the necessity of investigating stress-behavioral pathways at the individual level. This case study shows promise to research on personalized medicine and maintaining health practices at the individual level.

13.2.3 Case Study: Actionable Learning Methods

Mobile and IOT e-health technologies that are able to continuously collect data over an extended period of time could deliver interventions in an adaptive manner. In this way, individuals can be provided with personalized coaching plans based on their ongoing performance. Understanding the relationship between stress and exercise behavior will facilitate transforming behavior patterns into actionable interventions. In this case study, we leverage the stress-activity data to explore how to recommend a sequence of actions to individuals in order to maximize the overall stress reduction.

One solution to that is to model the data through a sequential decision-making process using reinforcement learning [64]. Particularly, Q-learning is a commonly used reinforcement learning technique, which uses a sequence of state-action-reward triples to determine a sequence of decision rules to maximize the cumulative reward. Such decision models have been developed and tailored in the healthcare area, with regression models to approximate the expected reward given action and state [39, 40, 42]. Another solution is to model the expected rewards through direct maximization when a particular form of decision rules is of interest [65–67].

However, as mentioned in the introduction of this chapter, there still exist multilayer challenges in making the adaptive strategies interpretable. The decision rules of assigning interventions, i.e., policies, generated by black-box learning algorithms can be difficult to explain, rendering them unappealing in practical settings in the healthcare domain.

To make recommendations for e-health users, it is hence imperative to develop actionable analytics methods that can provide highly relevant and person-specific insights. As healthcare decisions often involve unclear choices, we need methods that can account for human psychology to support the implementation of effective engagement technologies beyond simply addressing clinical efficacy. For example, [68, 69] have proposed a method to estimate an interpretable policy that guides the delivery of interventions. This method applies a regression model to approximate the expected reward with an additional threshold finding step, referred to as multistage threshold Q-learning (mTQL), in order to produce explainable policies. The estimated policy is a tree-structured policy that maps from historical observations to a next-step recommendation, with the split threshold as a sub-goal of intermediate reward change that helps differentiate whether a subject can obtain an optimal outcome by adopting a different strategy. By applying mTQL to study the relationship between stress and exercise behavior, the task of multi-stage stress management can be cast as a sequential decision-making task of recommending exercise based on micro-level feedback for stress reduction [68, 69].

In the stress-activity case study, the mTQL models stress reduction as the reward to maximize and exercise behavior as the action to intervene. The action defined as a binary variable to indicate whether the mean duration of daily moderate or vigorous physical activity (MVPA) bout over the time period was greater than 30 minutes or not. The stress reduction is defined as the stress level difference of current stage from a previous stage. The estimated policy is then used to understand how exercise patterns affect users' perceived stress levels and to perform coaching more effectively.

An example of the estimated policy is shown in Fig. 13.2. For individuals who have baseline stress level lower than a certain threshold, it would be important to observe whether being active in the first week followed by a decrease of mean stress level in the next week. If the mean stress level decreases less than a threshold, then it is better not to suggest the active action any further to accommodate the individual behavioral preferences and barriers. The analysis results show that dividing the study period into multiple stages and incorporating more sub-goals for micro-level feedback potentially help individuals achieving better stress reduction. Also, when an individual has started going astray from his usual path, a corresponding recommendation should be proposed for adaptation.

Moreover, another challenge is inherent in the large volume of mobile and IOT e-health data, which can often be collected continuously over time due to sensing technologies. In order to extract key features from these time series, [70] proposed a novel quantile coarsening analysis (QCA) and applied to analyze daily physical activity data. The key idea of QCA is to represent a time series by a small number of quantiles of time of activity. By construct of this representation, QCA is effective at capturing the time regions with high level of activities, in addition to the magnitude of activity. These are usual key features of behavioral patterns. In the analysis of about 20,000 time series of daily activity counts, [70] represented each time series of 1440 time points by 19 quantile features (1.3% of original data), used the quantiles as inputs in unsupervised learning, and was able to identify clusters that were distinctly associated with users' characteristics (e.g., employment status) and extraneous variables (e.g., weekday vs weekend). From a computational viewpoint, since quantile transformation requires only simple and scalable computations, QCA can be efficiently applied to large-scale data and facilitates "on-the-device" computations in mobile phones. While developed in the context of activity counts, QCA is a versatile analytical tool that can be applied to

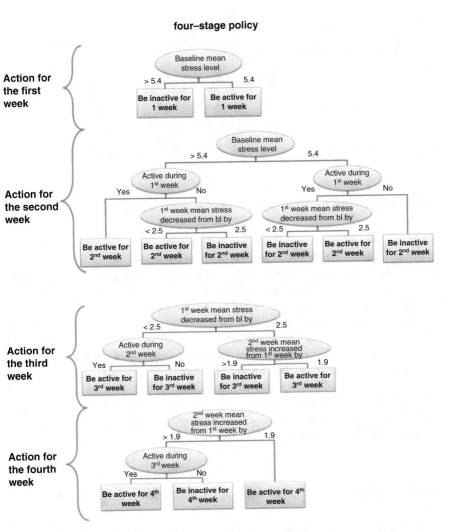

Fig. 13.2 From Fig. 13.3 in [69]. Estimated optimal policy for a four-stage study using the stress-activity data. Note: bl means baseline

data streams collected through mobile devices such as heart rate variability, sleep duration, and location for geofencing.

In summary, there exist various learning technologies that help us uncover the previously implicit cognitive map of bidirectional pathway between stress and exercise behavior for individual users. These insights serve as the foundation for a more progressive version of behavioral coaching and patient-centered care in the future.

Fig. 13.3 In [71] division of learning modules in the category daily care activities ($n = 243$)

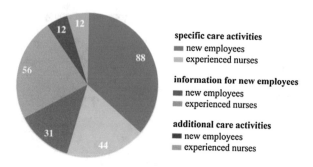

specific care activities
■ new employees
▬ experienced nurses

information for new employees
■ new employees
▬ experienced nurses

additional care activities
■ new employees
▬ experienced nurses

13.3 Knowledge Symbiosis Learning for Care Management

In the second theme of case studies, we introduce the potential of applying a smart learning framework that incorporates domain knowledge with data-driven learning into a fully automated patient education system. In particular, we focus on providing situational support to caregivers via mobile assistants that can learn from user feedback. We will discuss two major types of learning approaches and provide evidence on the superior performance of the hybrid approach. In this case study, we will describe a major German national cohort study, *Witra*-Care, and use it to illustrate how the Mobile Care Backup (MoCaB) system adopts a knowledge-based infrastructure to enable a mobile agent to act intelligently in response to a situation. As such, a variety of assessments would be deployed to monitor caregivers and their relatives. The knowledge-based infrastructure would then derive actionable insights from the assessments to check on patient compliance and provide the aforementioned situational support.

13.3.1 Application: AI in Intelligent Education for Healthcare

One of the central problems of the global healthcare system is the low health literacy at the population level. Take Germany for example. More than half of the German population have considerable difficulties in finding, understanding, assessing, or applying relevant information [72]. Based on the WHO estimates [73], an approximated burden of 15 billion euros per year is imposed on the German healthcare system. Similar trends are commonly observed across the globe. Increasing access to health-relevant knowledge is hence a highly important task for the society.

However, the solution is not straightforward in current system, wherein the strict scheduling of physicians only allows for limited time for patient education [74]. Over the years, with the increasing technical competence of e-health users, the barriers have decreased for the adoption of a technology-enabled solution for healthcare education. However, so far – to the authors' knowledge – there has not been a technology-enabled education system in healthcare yet. Meanwhile, the idea of building automated education systems already finds its application in many other fields, such as student education. In this case study, we will discuss the development of a hybrid smart learning framework that incorporates knowledge symbiosis learning in the implementation of similar applications.

13.3.2 Background in Intelligent Tutoring Systems

The intelligent tutoring system (ITS) is a typical example for educational systems. Here, teachers are supported by a computer system that can help structure a student's learning process. This sort of intelligent systems is expected to lend support to personalized tutoring and in turn reduce the teacher-student ratio. This would help grant easier access to quality education. ITS is different from computer-assisted instruction, where computers are utilized for administering the delivery of content with predetermined rules. Instead, ITS adapts the learning process individually for each student and therefore is referred to as intelligent. Typical applications of ITS include automated diagnosis of student weakness areas and automatic adjustment of education content according to the degree of difficulty.

The current development of ITS benefits from the foundation set by learning theories developed over the past few decades. For example, cognitivistic theories help inform ITS design in knowledge provisioning in relation to a learner's pre-existing knowledge [75, 76], whereas constructivistic theories help ITS to provide exercises in the learner's realistic settings and select learning tasks relevant to the learner's living experiences [77]. (For a review of the major learning theories, please refer to [78].)

A prerequisite of ITS is a lower barrier of access to technology. This was first reached in the 1980s by the introduction of personal computers leading to a boom of ITS. One example from that time is the LISPITS [79]. It was developed for teaching the programming language LISP. The system was able to detect errors and provide constructive feedback in real time [79, 80]. LISPITS has yielded positive evaluation, reducing the required time for learners to complete their assigned exercises while also improving on their scores [81].

The smart learning framework developed to support ITS consists of the following four models:

- Domain model
- Student model
- Tutoring model
- User interface model

First, the *domain model* or *expert model* represents the domain knowledge of an expert. More precisely as depicted in [82], it "contains the concepts, rules and problem-solving strategies of the domain to be learned. It can fulfill several roles: as a source of expert knowledge, a standard for evaluating the student's performance or for detecting errors, etc." To represent the domain models, both glass box model and the black-box model have been developed. In the glass box model, domain knowledge is represented by an expert system, which claims to be capable of representing the experts' problem-solving strategy and is accessible by the students. The black-box model, on the contrary, does not claim this [83].

Second, the *student model* represents a variety of properties of the students. The functionality of the ITS depends directly on the amount of information provided by the student model; therefore, the student model should be as comprehensive as possible. This includes all the factors that influence the student's learning behavior, e.g., the student's cognitive and affective state or learning progress. Based on the data collected, the student model infers new knowledge about the target student and to leverage such knowledge for selecting a pedagogical strategy. Sometimes it is called *diagnostic model* as it is used to diagnose the learning process. One common way to model the student's knowledge is to model it as a subset of the expert knowledge base or as the difference from the expert knowledge base.

Third, the *tutoring model* unites the knowledge from both the domain and the student model to adapt the tutoring strategies. It simulates the pedagogic behavior of a human teacher to select content and plan for interventions.

Finally, the interactions with the learner are realized by the *user interface model*. Because the presentation type often affects student's learning capabilities, ITS would need to account for different visualization strategies [82].

For the implementation of these models in the smart learning framework, various artificial intelligence approaches have been deployed. In particular, two main categories of methods emerge: symbolic AI and soft computing. The first category of symbolic AI methods is knowledge-driven and requires the explicit representation of knowledge bases, which means to explicitly express expert knowledge in formal forms using a representation language. The expatiating process incorporates significant cognitive work and therefore often results in high-quality knowledge bases.

Contrarily, the second category of soft computing methods is data driven and learns a desired behavior by investigating existent data. During the learning process, the parameters of a mathematical model are adapted according to the observation of learner performance. Generally, the learned internal representation is not directly accessible and not understandable by humans, as most soft computing models behave like black boxes.

Since both categories of method have their strengths and weaknesses, researchers have investigated into the possibility of combining methods from both categories into a knowledge symbiosis learning framework. (For comprehensive analysis, please refer to [84].)

For example, [85] developed an ITS that aims to leverage the everyday working knowledge of control center operators for incident analysis and diagnosis as well as for service restoration. The system consists of two parts, DiagTutor and CoopTutor. The first part is used for the training of fault diagnosis, and the second part is used for restoration training. During diagnosis, control center operators must analyze a combination of alarm messages and additional information, for example, temporal sequence of the alarms, in order to identify the root cause of the fault. Furthermore, the control center operators need to deal with incomplete and inconsistent information due to data loss or additional errors. The user is modeled by his conclusions about the faced problem, and as domain model an existing expert system is used. This expert system [86] is in use as a decision support system at a real power plant to identify operators' misconceptions and to assist in problem solving with hints by monitoring a student's learning process. The main components of this expert system were written in Prolog, a knowledge representation language that had a huge impact in AI.

The *tutoring module*, named as *curriculum planning module*, utilizes case-based reasoning to find a problem fitting the target student's needs. In order to increase the coverage of scenarios the ITS can handle, the knowledge base consists of both situations from a real control center and artificial cases. A rule-based system in the backend is implemented to adapt the level of difficulty of the content, taking the student's global knowledge level and the student's learning rate into account. The ITS also leverages an artificial neural network to calculate the distance for case-based reasoning. Given a student's knowledge level, it calculates the adequacy for every case in parallel. This setup is a good example of how the two different AI techniques can be integrated. The CoopTutor also adopts constrained-based modeling, a mathematical approach in which decisions have to be made in a rule-bound system, in order to accommodate the cooperation aspect in a multiagent system. All important roles of the control center are either simulated or taken by students. When educating a team of students, the interaction pattern is analyzed by the system to ensure everybody is participating in the training [85].

In terms of knowledge representation, typical techniques for *student modeling* include ontologies [87–89] and fuzzy logic [90, 91]. The same techniques are commonly used for the *domain model*. A few student models have used artificial neural networks [92]. ITS systems commonly apply the same technique for both model types [87–91]. This is due to the increased implementation effort when deploying multiple techniques in a single system.

An exception is provided in [93], where fuzzy logic is used for *domain modeling* and a combination of fuzzy logic and rules is used for *student modeling*. This approach exhibits no explicit tutor model. The tutor functionality in this ITS is implicitly implemented by the combination of the other two models. The *tutoring modeling* commonly adopts techniques such as fuzzy logic [87, 90, 91] and case-based reasoning [91, 94]. The extensive use of soft computing approaches is an indication of the elusiveness of this problem.

13.3.3 Challenges Facing the Development of ITS

The development of an intelligent tutoring system has confronted several challenges. First, the development process is usually a very long-term task. The development of the above-presented system by [85], for example, spanned longer than 10 years.

Such a long development process is cost-intensive and can hold a high risk of poor transferability in real world settings. As a result, its development often does not pass the prototype phase.

Second, the evaluation process of ITS is also challenging as ITS is often complex and composed of multiple system components. The combination of a variety of AI methods complicates the task as well. Although different evaluation techniques have been developed [95], there is still no one-fits-all approach; hence, no guarantee to the generalizability of the evaluation results. Generally, only the behavior of the complete system can be observed, and even if a fault is detected during evaluation, the localization of the fault inside the system can be difficult.

Third, the elusiveness of expert knowledge also poses challenges. When being inquired on the same question multiple times, the surveyed experts would often provide different answers to the same question. One approach to tackle this elusiveness is soft computing, in particular, fuzzy logic.

However, additional challenges arise when ITS is to be implemented in the healthcare domain to counter the health literacy issues of the population. For one, the target learners of ITS in this domain start with little domain knowledge. Therefore, a proper ITS must address a wide range of topics. But with the increasing range of topics, the domain and tutoring models must also be more comprehensive; yet, the features and information needed for doing so in the healthcare domain are highly heterogenous. A comprehensive assessment indicates a significant time commitment and burden for the patients, which results in the decrease of system usability.

Last but not least, healthcare experts often find it difficult to formalize the knowledge bases needed for building up ITS knowledge delivery strategies inside the domain and tutoring models. Furthermore, these are the experts who have very strict time constraints, and consequently the very short evaluation makes it difficult to properly evaluate the systems validity.

13.3.4 Case Study: Implicit Knowledge Learning for Nurses

Personal exchange has a strong influence on the learning process in nursing [96], but heavy workloads and high employee turnover show the necessity of on-the-job training. In this case study, we introduce the Witra Care (a German acronym for knowledge transfer) project, which aims to employ mobile technologies to support on-the-job training processes for nurses by recording, categorizing, and re-providing knowledge snippets in the form of micro-learning modules [97]. At Hannover Medical School (MHH), experienced nurses support new employees for several weeks by sharing their knowledge. In Witra Care, nine new employees and seven experienced nurses created micro-learning modules in a mixed form of text, images or recorded video and audio via a mobile application. As such, they collectively created 303 learning modules, which were structured around the following three categories: daily healthcare activities, health quality management, and general information for new employees. These categories are not disjunctive, and each has several subcategories. Most modules were allocated to the category of daily care activities. Among the learning modules, those regarding specific care activities were mostly created by new employees, whereas those regarding general information for new employees were mainly created by experienced nurses (see Fig. 13.1) [71].

Using these learning modules, the nurses' explicit and implicit knowledge could be visualized and utilized for instructing new employees to the ward. Furthermore, it is now possible to tell the differences between what new employees and experienced conversant nurses consider as important for acquisition. The findings could help improve the training process of new employees. A main mobile app-based interface was developed. Furthermore, speech-controlled smart glass-based interface was also developed as a hands-free device for tasks such as preparing for an operation. The nurses showed interest in the use of new technologies for supporting on-the-job training. However, there exist technical limitations, e.g., short battery lifetime, inherent to the wearable interface, which significantly limit the use of such wearable interfaces in real-life settings [97].

13.3.5 Case Study: Implicit Knowledge Learning for Caregivers

In Germany, there are about 2.9 million people in need of care, with 1.38 million being generally supplied by relatives only [98]. In this case study, *Mobile Care Backup (MoCaB)*, a support system for informal caregivers taking care of their relatives, is developed in the form of a mobile application. The situation of caring for a relative often comes spontaneous and surprising and so informal caregivers often complain about a lack of required knowledge. Therefore, the main component of the supportive system is a proactive knowledge provision. Besides this, the caregiving relative receives instructional exercises for self-relief and support with his or her organizational network. These knowledge-driven components account for the fact that the knowledge demand of a caregiver differs from professional nurses, and hence the knowledge provided must be comprehensive yet easy to interpret.

In this case study, experts in nursing science created 87 knowledge resources, each of which was presented as a textual dialog with a caregiver. The knowledge resources focus on fundamental topics and topics relevant to caregivers caring relatives suffering from stroke or dementia.

In addition to knowledge resources, the experts also formulated delivery strategies regarding the care recipient's disease, residential situation, and the duration of the care situation. These strategies are implemented as an ontology using the web ontology language (OWL). However, the experts were only able to formulate an explicit delivery strategy for a very small number of knowledge resources as this is not their main expertise. For the rest of knowledge resources, the experts were not able to formulate such delivery strategies, but still indicated that these were important topics to be covered.

This gap surfaces the need to handle both explicit and implicit knowledge. While the implicit knowledge is elusive and would often require the use of examples to make such knowledge interpretable, the example-oriented approach did not cover all the topics. Therefore, to characterize the rest of critical knowledge resources, the assessment forms from the *student model* were used.

In particular, the *student model* utilizes two comprehensive assessment instruments: (1) the caregiver burden inventory (CBI) [99] and (2) a German survey assessment (NBA) [100]. The first assessment instrument, as the name suggests, measures the burden of a caregiver. It includes 24 items, which are divided into five segments: time dependence, developmental burden, physical burden, social burden and emotional burden. The regular use case of the second instrument, the NBA, is the determination of the amount of financial relief due to persons' needs of assistance. It spans six different areas, including mobility, cognitive and communicative abilities, psychological problems, self-reliance and organizing everyday life. In the use case's student model, parts of it are used to infer from the patient's needs to the caregiver's issues. All items of the assessment tools used in the student model are represented as Boolean values.

To explicitly extract the implicit knowledge delivery strategies from the experts' minds, they were asked to weigh each assessment item with the score of 0–3 for every knowledge resource. Thereby, a zero weighting indicates that the item is not important for the provision of this knowledge resource and a weighting of three indicates that the item is highly important for the provisioning of the knowledge resource. To make a statement about the importance of a knowledge resource for a specific caregiver, the caregiver's profile and the experts' weighting must be matched. This is done by calculating a score based on formula (13.1), where n_i is the number of profile items marked with weight i and c_i is the number of profile items of weight i fulfilled by the caregiver:

$$s = c_1 * \frac{\frac{1}{6}}{n_1} + c_2 * \frac{\frac{1}{3}}{n_2} + c_3 * \frac{\frac{1}{2}}{n_3} \tag{13.1}$$

The weights assigned by the experts are dynamized by dividing a fixed portion by the number of items marked with this weight (n_i). The fixed portions follow the same distances as the weights assigned by the experts. The use of dynamical weights provides an edge, wherein the percentile of a weighting group i is independent from the number of items with this weight [101]. If the scores are calculated for all knowledge resources in a given profile, the knowledge resources then can be ordered by importance and delivered in descending order (from the most important to the least).

But even with the comprehensive assessment by CBI and NBA and formula (13.1), it is not possible to deliver all knowledge resources dynamically. A small number of knowledge resources about basic care knowledge are problematic. Since some of these knowledge resources include fundamental knowledge, they will be shown to all caregiving relatives before starting with the dynamical proactive delivery of knowledge. Thus, the delivery strategies of the *tutoring model* are

Fig. 13.4 Dialog-based user interface for knowledge delivery of the MoCaB mobile application translated from German

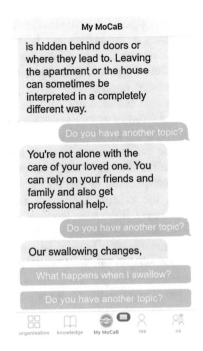

composed of two parts: the first part handles the rule-based delivery of basic care knowledge in the beginning, implemented as an OWL ontology, and the second part is based on a statistical scoring function. Thereby, the ontology excludes irrelevant knowledge resources and includes relevant knowledge resources, which are then rated by the scoring function.

The knowledge delivery user interface is based on a dialog with a hypothetical advisor, who guides the conversation. The interface is similar to messenger chat histories (see Fig. 13.4), commonly used by smartphone users. For system interactions, two buttons are provided. The first button above indicates the choice of getting more knowledge about a specific topic, and the second button below helps skip to the next topic. For this, the knowledge resources are all represented in an XML-based format. The first button plays the next system-side message with the next belonging user answer inside a knowledge resource, and the second button leads to the next dialog from the list ordered by the importance as determined by the *tutoring model*.

Furthermore, additional information about the relationship between the caregiver and the care recipient is used to personalize the dialogs. In order to emphasize on the personalization aspect of the knowledge delivery strategy, a direct addressing was used instead of the German polite form in the dialogs. This dialog-based presentation for content delivery received positive feedback in the usability evaluation study; the subjects were all able to operate the whole dialog without further intervention [102]. Besides the proactive delivery, all knowledge resources can be accessed independently from the user profile of the mobile application.

In another follow-up evaluation study, the subjects also provided positive feedback to the knowledge delivery strategies. In addition, another expert group was also consulted to assess the validity of the knowledge delivery strategies. In this follow-up evaluation study, the experts were asked to judge the orderings of knowledge resources for given profiles. The test scenarios used in this part of evaluation included not only the orders obtained by the *MoCaB* system but also artificial orders that were designed to be (partly) incorrect. This evaluation process was repeated a few weeks later with the same group of experts.

The evaluation results show that the experts were able to capture and correct inconsistencies in the implicit part of the *domain model*. But on the one hand, it is not possible to find all inconsistencies with this approach, and, on the other hand, the experts' judgments from the first and second repetition differed in some cases, illustrating the elusiveness of their implicit knowledge. Moreover, this evaluation only incorporates the opinions from a small number of experts, leaving its generalizability to be questionable.

However, in this case study, the focus is on educating informal caregivers by providing them personalized knowledge, and this group of users should be able to judge and provide feedback on whether the knowledge profile created for them fits their own expectation. By utilizing the additional information, the system can address the abovementioned problems by adding an algorithm-driven learning capacity. Since the internal structure of delivery strategy after feedback optimization is not known, the algorithm must have the capability to adapt. For example, an artificial neural network (ANN) is particularly suitable for its strength in capturing intricate relationship in high dimensional data.

Even though the knowledge delivery algorithms mentioned above seem to work quite well [101], it could be further improved by user feedback. To achieve this, both the *domain model* and the *tutoring model* should be integrated into the artificial neural network. There are multiple ways to implement this. The first option is to train the ANN for importance scoring. Because the goal of the overall task is to determine the ordering of available knowledge resources, the ANN could calculate the scores for all knowledge resources in parallel, which then must be ordered externally. Another option would be to train the ANN for both scoring and ordering. To enable the second option, a more complex network structure is needed, which would in turn result in a much longer training phase. The extra effort is non-trivial, compared to simply applying a common sorting algorithm.

The schematic representation of a working ANN structure is shown in Fig. 13.5. It first receives the input from the *student model* (via the CBI and NBA assessments) and returns a score for the importance of every knowledge resource, each represented by an output neuron. In this setup, the neural network must extract the implicit knowledge (e.g., experts' domain knowledge and the delivery strategies from the *tutoring model*) from observation data while training. The regression models themselves are each performed by a strain inside the network. The strains consist of three fully connected layers. Whereas one layer is for extracting the expert weightings while training, the other two layers are for regression itself. In the first training step, the artificially created profiles are used to teach the ANN

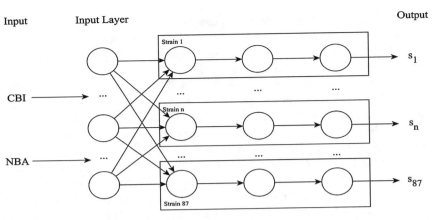

Fig. 13.5 Schematic representation of the artificial neural network used for user feedback utilization. The number of neurons inside the single strains are simplified and not representative

the expert's domain knowledge and existent delivery strategy, followed by a second training step utilizing the user feedback. In the second step, the user feedback will be consolidated to adjust the output of the ANN.

13.4 Continuous Learning for In-Field Decision-Making

Last but not least, we introduce the potential of learning from continuous monitoring over time, especially how to learn actionable insights from long-term monitoring not only to inform in-field operation decisions at the personal level but also to inform policy-making decisions at the population level. The case study of Blast Gauge, a small device currently being worn by US Army Soldiers in Afghanistan, would be given to illustrate how applying smart learning methods to learn from continuous monitoring data can help quickly assess the possible impact of explosive blasts, come to rapid treatment of the affected troops, and eventually lead to important longer-term policy decisions.

13.4.1 Application: Risk Inference for Traumatic Brain Injury

Traumatic brain injury (TBI) is a major cause of death and disability, contributing to more than 2.5 million ED visits, 282,000 hospitalizations, and 30% of injury deaths in United States alone [103]. In battlefields, TBI is the result from non-kinetic explosive effects on soldiers, yielding more than 200,000 injuries per year [104]. Previous studies such as DARPA's Preventing Violent Explosive Neurologic Trauma (PREVENT) program have shown that TBI caused by blast exposure has to be treated differently from typical overpressure injuries, hence needing a granular understanding of the relationship between the blast components and neurological injury. However, existing practice of medical personnel relies on visual signs and personal accounts of patients to alert them about the possibility of TBI.

In this case study, we introduce the DARPA Blast Gauge project, which aims to identify a quantitative means for measuring blast-related exposure and in turn provide a screening tool for medical personnel to better identify soldiers at risk for TBI. The Blast Gauge collects quantitative data along with the screening tool in order to uncover the mechanisms of TBI.

During the course of pilot testing over 2 years, the Blast Gauge has been used to check on soldiers who suffered shrapnel injuries (visible wounds) but did not report blast exposures to the medical personnel. By design, the Blast Gauge is a wristwatch-sized, self-contained system that continuously measures the amount of blast exposure a soldier has been exposed to. The color of the light (red, green, yellow) on the gauges display signals the level of exposure occurred during the engagement. Following the light indicator, different treatments would be prescribed according to the standard protocol. For example, if the light is yellow, which indicates moderate exposure, the medical personnel would download the data from the Gauge to evaluate the risk for TBI. If a mild TBI is detected, then follow-up treatment for TBI will begin immediately.

The pilot testing study consisted of two phases. The first phase of the pilot involved approximately 900 soldiers of an Army brigade in their active combat roles. The second phase of the pilot study deployed the Gauge to more than 10,000 soldiers in a variety of units across the military. The monitoring data collected by the Gauge not only provide accurate information collected by the Gauge are used for medical personnel to understand what their patients actually experienced during an exposure. This capability enables actionable insights being generated for in-field decision-making for medics and doctors and embedding such actionable insights directly in the care flow that provides immediate triaging to injured soldiers. For example, DARPA was using the data compiled from these devices to understand blast propagation, provide new insights into sources and causes of traumatic brain injury, and ultimately develop technologies that help minimize exposure and improve medical care.

The continuous measurements and visible signs for person-level in-field decision-making, but also provide a quantitative basis to enable policy-level decision-making and develop a long-term TBI diagnostic solution for the battlefield. This is especially important to the Army, in which soldiers are used to the mentality of not seeking medical aid for the "invisible" injuries, such as TBI. The Gauge not only stop at the diagnosis of the TBI risk, but also helps ensure the triage – even in the cases in which soldiers might downplay their own symptoms to avoid being taken out of fight. The positive results of the pilot study have led to the encouragement of the House Armed Services Committee in its 2013 National Defense Authorization Act Report [105].

13.5 Discussion

Generally, a smart learning framework consists of three major components – representation, evaluation, and optimization [37]. Using the above three themes of case studies, we provided a snapshot of three major applications using state-of-the-art smart learning methods in this framework on mobile and IOT e-health.

By incorporating N-of-1 analytics and actionable learning methods, the smart learning framework helps uncover the previously implicit behavioral pathways that depict the sequential bidirectional relationship between stress perception and exercise behavior. It also helps improve the experience of using the actionable and interpretable insights learned from mobile and IOT e-health data, leading to better integration of data science and science of care for patient-centered care.

By incorporating knowledge-enhanced learning methods, the framework can leverage the injected knowledge bases to provide a better learning performance with less bias and more interpretability. Table 13.1 lists a few examples of knowledge-enhanced learning methods in the framework.

First, for the representation component in the framework, knowledge injection can help refine input features in at least the following three ways: (1) adding knowledge-based features; (2) reconstructing features based on the hierarchical structure of knowledge bases; (3) discovering novel feature relationships from knowledge bases.

Second, for evaluation, knowledge injection can enhance our understanding of data-driven models or even help refine model parameters. Take clustering for example, treatment guidelines can help estimate how many sub-populations (clusters) could be identified by the algorithm. Take graph modeling as another example. Literature about disease stages can help determine the number of latent states. Finally, for optimization, we can perform model compression with knowledge distillation [56], filter knowledge-conflicting targets from recommendations directly, or perform knowledge-based regularization in learning networks [53]. In particular, logic rules have been used for knowledge-based post-regulation, e.g., in [53], to improve the deep neural networks such as convolutional neural networks (CNNs) and recurrent neural networks (RNNs). Similarly, incorporating logic rules into Latent Dirichlet Allocation has been introduced in [54, 55].

Table 13.1 The three components of learning frameworks and examples of knowledge injection-based enhancement

Representation	Evaluation	Optimization
Adding knowledge features;	Verify weighting of feature correlations;	Knowledge filtering;
Reconstructing features with knowledge-based hierarchy;	Determine the number of sub-populations to be clustered;	Post-regularization
Discovering novel feature relationships from knowledge	Determine the number of hidden states (e.g., disease stages)	

Besides, the framework could be developed with different smart learning principles, ranging from supervised learning to unsupervised learning and reinforcement learning. The data sources could be structured (e.g., electronic health records), unstructured (e.g., clinical notes), or semi-structured (e.g., medical images). The knowledge sources could be clinical guidelines (e.g., NICE guideline for type 2 diabetes management [106]), well-known risk models (e.g., Framingham Cardiovascular Disease 10-year risk model), and medical concept graphs (e.g., a knowledge graph of medical concepts for estimating feature correlations). The applications of such frameworks widely range from the determination of clinical trial eligibility [107] to risk assessment and treatment recommendations [56].

In retrospect of the *MoCaB* case study, quite a few lessons about the best practice and requirements for future implementation have been learned. For example, at the beginning of the study, the plan was to implement a fully rule-based system. However, because only a small portion of the experts' knowledge is explicitly expressible, this plan had to be adapted by finding an appropriate assessment to extract the implicit knowledge for the domain model. Yet this task has been proven to be time-consuming, since it involves an iterative process incorporating extensive cooperation between domain experts in healthcare and experts in computer science. This workload should be regarded when planning to develop such systems in the future.

Furthermore, the integration of the target users plays an important role for the development of an ITS system for care management. The choice of the system's target user group and recruiting of participants should account for not only their information and social need, but also the burden on their time commitment.

Our results indicate that AI-based systems can contribute to adaptive education of caregivers for care management. While this approach necessitates intensive efforts in terms of participatory system design and evaluation, its potential benefits may prove rewarding.

In this chapter, we have gathered requirements for evolving a smart learning framework to incorporate N-of-1 actionable insights and knowledge learning symbiosis in healthcare, aiming to generate real-world evidence in practice. In particular, this chapter focuses on promising applications and smart learning methods that help combine data with a priori knowledge for decision support. This chapter aims to share with leading practitioners and researchers in the field about the requirements for the next-generation, data-knowledge fusion systems, especially in areas where a large amount of evidence has been accumulated in knowledge bases.

We provided an overview for the principles and methodologies of the framework for better interpretability. To achieve this, there exist new challenges that would require breakthroughs in both continuous learning analytics and implementation science. While the former uncovers relationships between patient-specific characteristics, treatment trajectory and patient response to interventions, the latter provides guidance to embed actionable insights into care flows in practice for sense-making and decision support.

In the future, the smart learning framework for mobile/IOT e-health is expected to further help patients in chronic disease management scenarios, or monitor mental health conditions such as Parkinson's, Alzheimer's, Huntington's disease, post-traumatic stress disorder (PTSD), and behavioral conditions such as autism and ADHD. Further analysis of a patient's speech or written notes can identify additional indicators of meaning, syntax and intonation in language. Combining the various measurements from heterogeneous mobile and IOT e-health data with other clinical information sources such as magnetic resonance imagings (MRIs) and EEGs can paint a more complete picture of the individual-specific characteristics for health professionals to better identify, understand, and treat the underlying disease. The important part of mobile and IOT e-health is to make what were once invisible signs become clear signals of patients behavioral risk and how well they are responding to interventions, complementing regular clinical visits with EMA assessments in daily-life settings.

Acknowledgment Theme 2: The projects *Witra-Care* and *Mobile Care Backup (MoCaB)* are funded by German Federal Ministry of Education and Research (grant numbers *Witra-Care:* 16SV6380; *MoCaB*: 16SV7472).

References

1. E.K. Choe, N.B. Lee, B. Lee, W. Pratt, J.A. Kientz, in *Proceedings of the 32nd Annual ACM Conference on Human Factors in Computing Systems*. Understanding quantified-selfers' practices in collecting and exploring personal data. (Toronto, ON, Canada, Apr. 26–May 1). (ACM Press, New York, 2014), pp. 1143–1152

2. K. Shameer, M.A. Badgeley, R. Miotto, B.S. Glicksberg, J.W. Morgan, J.T. Dudley, Translational bioinformatics in the era of real-time biomedical, health care and wellness data streams. Brief. Bioinform. **18**(1), 105–124 (2017). https://doi.org/10.1093/bib/bbv118

3. B. Knowles, A. Smith-Renner, F. Poursabzi-Sangdeh, D. Lu, H. Alabi, Uncertainty in current and future health wearables. Commun. ACM **61**(12), 62–67 (2018). https://doi.org/10.1145/3199201

4. J.B. Bassingthwaighte, Strategies for the physiome project. Ann. Biomed. Eng. **28**(8), 1043–1058 (2000). https://doi.org/10.1114/1.1313771

5. X. Li, J. Dunn, D. Salins, G. Zhou, W. Zhou, S.M. Schüssler-Fiorenza Rose, et al., Digital health: Tracking physiomes and activity using wearable biosensors reveals useful health-related information. PLoS Biol. **15**(1), e2001402 (2017). https://doi.org/10.1371/journal.pbio.2001402

6. R. Voelker, Smart watch detects seizures. JAMA **319**(11), 1086 (2018). https://doi.org/10.1001/jama.2018.1809

7. G.H. Tison, J.M. Sanchez, B. Ballinger, A. Singh, J.E. Olgin, M.J. Pletcher, et al., Passive detection of atrial fibrillation using a commercially available smartwatch. JAMA Cardiol. (2018). https://doi.org/10.1001/jamacardio.2018.0136

8. J.E. Dimsdale, Psychological stress and cardiovascular disease. J. Am. Coll. Cardiol. **51**(13), 1237–1246 (2008)

9. D.M. Lloyd-Jones, Y. Hong, D. Labarthe, et al., Defining and setting national goals for cardiovascular health promotion and disease reduction: The American Heart Association's strategic impact goal through 2020 and beyond. Circulation **121**, 586–613 (2010)

10. B.H. Marcus, L.H. Forsyth, E.J. Stone, P.M. Dubbert, T.L. McKenzie, A.L. Dunn, S.N. Blair, Physical activity behavior change: Issues in adoption and maintenance. Health Psychol. **19**, 32–41 (2000)

11. P. Salmon, Effects of physical exercise on anxiety, depression, and sensitivity to stress: A unifying theory. Clin. Psychol. Rev. **21**(1), 33–61 (2001)

12. D. Scully, J. Kremer, M.M. Meade, R. Graham, K. Dudgeon, Physical exercise and psychological well-being: A critical review. Br. J. Sports Med. **32**(2), 111–120 (1998)
13. J.M. Jakicic, K.K. Davis, R.J. Rogers, W.C. King, M.D. Marcus, D. Helsel, et al., Effect of wearable technology combined with a lifestyle intervention on long-term weight loss. JAMA **316**(11), 1161 (2016). https://doi.org/10.1001/jama.2016.12858
14. S.S. Gollamudi, E.J. Topol, N.E. Wineinger, A framework for smartphone-enabled, patient-generated health data analysis. PeerJ **4**, e2284 (2016). https://doi.org/10.7717/peerj.2284
15. S.R. Steinhubl, J. Waalen, A.M. Edwards, L.M. Ariniello, R.R. Mehta, G.S. Ebner, E.J. Topol, Effect of a home-based wearable continuous ECG monitoring patch on detection of undiagnosed atrial fibrillation. JAMA **320**(2), 146 (2018). https://doi.org/10.1001/jama.2018.8102
16. A. Oguntimilehin, O.B. Abiola, O.A. Adeyemo, A clinical decision support system for managing stress. J. Emerg. Trends Comput. Inf. Sci **6**(8), 436–442 (2015)
17. P.M. Gollwitzer, P. Sheeran, Implementation intentions and goal achievement: A meta-analysis of effects and processes. Adv. Exp. Soc. Psychol. **38**, 69–119 (2006). https://doi.org/10.1016/S0065-2601(06)38002-1
18. A. Prestwich, M. Perugini, R. Hurling, Can implementation intentions and text messages promote brisk walking? A randomized trial. Health Psychol. **29**(1), 40–49 (2010a). https://doi.org/10.1037/a0016993
19. A. Prestwich, I. Kellar, How can the impact of implementation intentions as a behavior change intervention be improved? Eur. Rev. Appl. Psychol **64**(1), 35–41 (2014a). https://doi.org/10.1016/j.erap.2010.03.003
20. P. Pirolli, S. Mohan, A. Venkatakrishnan, L. Nelson, M. Silva, A. Springer, Implementation intention and reminder effects on behavior change in a mobile health system: A predictive cognitive model. J. Med. Internet Res. **19**(11), e397 (2017). https://doi.org/10.2196/jmir.8217
21. A. Prestwich, M. Perugini, R. Hurling, Can implementation intentions and text messages promote brisk walking? A randomized trial. Health Psychol. **29**(1), 40–49 (2010b Jan). https://doi.org/10.1037/a0016993
22. A. Prestwich, I. Kellar, How can the impact of implementation intentions as a behavior change intervention be improved? Eur. Rev. Appl. Psychol **64**(1), 35–41 (2014b Jan). https://doi.org/10.1016/j.erap.2010.03.003
23. D.C. Mohr, K. Cheung, S.M. Schueller, C.H. Brown, N. Duan, Continuous evaluation of evolving behavioral intervention technologies. Am. J. Prev. Med. **45**(4), 517–523 (2013)
24. C.M. Kennedy, J. Powell, T.H. Payne, J. Ainsworth, A. Boyd, I. Buchan, Active assistance technology for health-related behavior change: An interdisciplinary review. J. Med. Internet Res. **14**(3), 80 (2012)
25. C. Skinner, J. Finkelstein, in *Proceedings of the IASTED International Conference on Telehealth/Assistive Technologies*. Review of mobile phone use in preventive medicine and disease management. (ACTA Press, 2008), pp. 180–189
26. C.A. Depp, B. Mausbach, E. Granholm, V. Cardenas, D. Ben-Zeev, T.L. Patterson, D.V. Jeste, Mobile interventions for severe mental illness: Design and preliminary data from three approaches. J. Nerv. Ment. Dis. **198**(10), 715–721 (2010)
27. M. Lin, Z. Mahmooth, N. Dedhia, R. Frutchey, C.E. Mercado, D.H. Epstein, et al., Tailored, interactive text messages for enhancing weight loss among African American adults: The TRIMM randomized controlled trial. Am. J. Med. **128**(8), 896–904 (2015). https://doi.org/10.1016/j.amjmed.2015.03.013
28. L. Piwek, D.A. Ellis, S. Andrews, A. Joinson, The rise of consumer health wearables: Promises and barriers. PLoS Med **13**(2), e1001953 (2016)
29. P.J. Teixeira, E.V. Carraça, M.M. Marques, H. Rutter, J.-M. Oppert, I.D. Bourdeaudhuij, J. Lakerveld, J. Brug, Successful behavior change in obesity interventions in adults: a systematic review of self-regulation mediators. BMC Med **13**(1), 84 (2015)
30. B. Chen, S. Patel, L.D. Toffola, P. Bonato, *Proceedings of the 2nd Conference on Wireless Health*. Long-term monitoring of COPD using wearable sensors. (2011), p. 19
31. A. Lange, J.P. van de Ven, B. Schrieken, Interapy: Treatment of post-traumatic stress via the internet. Cogn. Behav. Ther **32**(3), 110–124 (2003)

32. A.C. King, E.B. Hekler, L.A. Grieco, S.J. Winter, J.L. Sheats, M.P. Buman, B. Banerjee, T.N. Robinson, J. Cirimele, Harnessing different motivational frames via mobile phones to promote daily physical activity and reduce sedentary behavior in aging adults. Plos ONE **8**(4), e62613 (2013)
33. S.J. Winter, J.L. Sheats, A.C. King, The use of behavior change techniques and theory in technologies for cardiovascular disease prevention and treatment in adults: A comprehensive review. Prog. Cardiovasc. Dis. **58**(6), 605–612 (2016)
34. D. Ben-Zeev, K.E. Davis, S. Kaiser, I. Krzsos, R.E. Drake, Mobile technologies among people with serious mental illness: Opportunities for future services. Adm. Policy Ment. Health Ment. Health Serv. Res. **40**(4), 340–343 (2013)
35. C.A. Christmann, A. Hoffmann, G. Bleser, Stress management apps with regard to emotion-focused coping and behavior change techniques: A content analysis. JMIR Mhealth Uhealth **5**(2), e22 (2017)
36. E.B. Litvin, A.M. Abrantes, R.A. Brown, Computer and mobile technology-based interventions for substance use disorders: An organizing framework. Addict. Behav. **38**(3), 1747–1756 (2013)
37. P. Domingos, A. Pedro, A few useful things to know about machine learning. Commun. ACM **55**(10), 78 (2012)
38. D. Castelvecchi, Can we open the black box of AI? Nature **538**(7623), 20–23 (2016)
39. S.A. Murphy, A generalization error for Q-learning. J. Mach. Learn. Res **6**(Jul), 1073–1097 (2005a)
40. Y.K. Cheung, B. Chakraborty, K.W. Davidson, Sequential multiple assignment randomized trial (SMART) with adaptive randomization for quality improvement in depression treatment program. Biometrics **71**(2), 450–459 (2015)
41. P.J. Schulte, A.A. Tsiatis, E.B. Laber, M. Davidian, Q-and A-learning methods for estimating optimal dynamic treatment regimes. Stat. Sci **29**(4), 640 (2014)
42. S.A. Murphy, An experimental design for the development of adaptive treatment strategies. Stat. Med. **24**(10), 1455–1481 (2005b)
43. L.I. Wagner, J. Duffecy, F. Penedo, D.C. Mohr, D. Cella, Coping strategies tailored to the management of fear of recurrence and adaptation for E-health delivery: The FoRtitude intervention. Cancer **123**(6), 906–910 (2017)
44. P. Klasnja, E.B. Hekler, S. Shiffman, A. Boruvka, D. Almirall, A. Tewari, S.A. Murphy, Microrandomized trials: An experimental design for developing just-in-time adaptive interventions. Health Psychol. **34**(Suppl), 1220–1228 (2015). https://doi.org/10.1037/hea0000305
45. R. Caruana, Y. Lou, J. Gehrke, P. Koch, M. Sturm, N. Elhadad, in *Proceedings of KDD*. Intelligible models for healthcare: Predicting pneumonia risk and hospital 30-day readmission. (2015)
46. Z.C. Lipton, in *Proceedings of ICML Workshop on Human Interpretability in Machine Learning (WHI 2016)*. The mythos of model interpretability. (2016)
47. P.S. Hsueh, S. Das, S. Dey, T. Wetter, in *Proceedings of MEDINFO*. Making sense of Patient Generated Health Data (PGHD) with better interpretability: The transition from more to better. (2017a)
48. X. Hu, P.-Y.S. Hsueh, C.-H. Chen, K.M. Diaz, Y.-K.K. Cheung, M. Qian, A first step towards behavioral coaching for managing stress: A case study on optimal policy estimation with multi-stage threshold Q-learning, in *AMIA ... Annual Symposium Proceedings. AMIA Symposium, 2017*, (2017), pp. 930–939. Retrieved from http://www.ncbi.nlm.nih.gov/pubmed/29854160
49. X. Hu, P.S. Hsueh, C. Chen, M. Qian, F.E. Parsons, I. Ensari, K. Daz, Y.K. Ceung, An interpretable health behavioral policy for mobile device users. IBM J. Res. Dev. **62**(1) (Jan 2018). https://doi.org/10.1147/JRD.2017.2769320
50. K.A. Bartholomew, The perspective of a practitioner, in *Knowledge Coupling*, (Springer, New York, 1991), pp. 235–277

51. P.S. Hsueh, H. Chang, S. Ramakrishnan, Next-generation wellness: A technology model for personalizing healthcare, in *Healthcare Information Management*, 4th edn., (Springer, Cham, 2016)

52. P.S. Hsueh, F. Martin-Sanchez, K. Kim, S. Peterson, S. Dey, B. Yang, Y-K. Cheung, T. Wetter (2017b), Secondary Use of Patient Generated Health Data (PGHD), IMIA Yearbook Review 2017

53. Z. Hu, X. Ma, Z. Liu, E. Hovy, E. Xing, Harnessing deep neural networks with logic rules, in *Proceedings of the 54th Annual Meeting of the Association for Computational Linguistics (Volume 1: Long Papers)*, (2016), pp. 2410–2420. https://doi.org/10.18653/v1/P16-1228

54. D. Andrzejewski, X. Zhu, M. Craven, B. Recht, in *IJCAI*. A framework for incorporating general domain knowledge into Latent Dirichlet Allocation using first-order logic. (2011a), pp. 1171–1177

55. D. Andrzejewski, X. Zhu, M. Craven, B. Recht, A framework for incorporating general domain knowledge into Latent Dirichlet Allocation using first-order logic. IJCAI **2011**, 1171–1177 (2011b)

56. J. Mei, H. Liu, X. Li, G. Xie, Y. Yu, in *MedInfo*. A decision fusion framework for treatment recommendation system. (2015), pp. 300–304

57. D.E. Warburton, C.W. Nicol, S.S. Bredin, Health benefits of physical activity: The evidence. Can. Med. Assoc. J. **174**(6), 801–809 (2006)

58. M.M. Burg, J.E. Schwartz, I.M. Kronish, K.M. Diaz, C. Alcantara, J. Duer-Hefele, K.W. Davidson, Does stress result in you exercising less? Or does exercising result in you being less stressed? Or is it both? Testing the bi-directional stress-exercise association at the group and person (N of 1) level. Ann. Behav. Med. **51**(6), 799–809 (2017)

59. W.T. Riley, D.E. Rivera, A.A. Atienza, W. Nilsen, S.M. Allison, R. Mermelstein, Health behavior models in the age of mobile interventions: Are our theories up to the task? Transl. Behav. Med. **1**(1), 53–71 (2011)

60. J.M. Smyth, S.A. Wonderlich, M.J. Sliwinski, R.D. Crosby, S.G. Engel, J.E. Mitchell, R.M. Calogero, Ecological momentary assessment of affect, stress, and binge-purge behaviors: Day of week and time of day effects in the natural environment. Int. J. Eat. Disord. **42**(5), 429–436 (2009)

61. S. Shiffman, A.A. Stone, M.R. Hufford, Ecological momentary assessment. Annu. Rev. Clin. Psychol. **4**, 1–32 (2008)

62. K.E. Heron, J.M. Smyth, Ecological momentary interventions: Incorporating mobile technology into psychosocial and health behavior treatments. Br. J. Health Psychol. **15**(1), 1–39 (2010)

63. Y.K. Cheung, P.-Y.S. Hsueh, M. Qian, S. Yoon, L. Meli, K.M. Diaz, et al., Are nomothetic or ideographic approaches superior in predicting daily exercise behaviors? Methods Inf. Med. **56**(06), 452–460 (2017)

64. R.S. Sutton, A.G. Barto, *Reinforcement Learning: An Introduction* (MIT press, Cambridge, 2018)

65. S.A. Murphy, M.J. van der Laan, J.M. Robins, et al., Marginal mean models for dynamic regimes. J. Am. Stat. Assoc. **96**(456), 1410–1423 (2001)

66. M. Zhang, D.E. Schaubel, Double-robust semiparametric estimator for differences in restricted mean lifetimes in observational studies. Biometrics **68**(4), 999–1009 (2012)

67. Y. Zhao, D. Zeng, A.J. Rush, M.R. Kosorok, Estimating individualized treatment rules using outcome weighted learning. J. Am. Stat. Assoc. **107**(499), 1106–1118 (2012)

68. X. Hu, P.-Y. Hsueh, C.-H. Chen, K.M. Diaz, F.E. Parsons, I. Ensari, et al., An interpretable health behavioral intervention policy for mobile device users. IBM J. Res. Dev. **62**(1), 4–1 (2018)

69. X. Hu, P.-Y.S. Hsueh, C.-H. Chen, K.M. Diaz, Y.-K.K. Cheung, M. Qian, in *AMIA Annual Symposium Proceedings*. A first step towards behavioral coaching for managing stress: A case study on optimal policy estimation with multi-stage threshold Q-learning. (American Medical Informatics Association, 2017), p. 930

70. Y. Cheung, P.-Y. Hsueh, I. Ensari, J. Willey, K. Diaz, Quantile coarsening analysis of high-volume wearable activity data in a longitudinal observational study. Sensors **18**(9), 3056 (2018)

71. M. Behrends, T. Kupka, R. Schmeer, I. Meyenburg-Altwarg, M. Marschollek, Knowledge transfer in health care through digitally collecting learning experiences – results of Witra care. Stud. Health Technol. Inform. **225**, 287–291 (2016)

72. D. Schaeffer, D. Vogt, E-M. Berens, K. Hurrelmann. Gesundheitskompetenz der Bevölkerung in Deutschland: Ergebnisbericht, Universität Bielefeld, Fakultät für Gesundheitswissenschaften (2016)

73. I. Kickbusch, *Health Literacy. The Solid Facts* (World Health Organization, Geneva, 2013)

74. G. Irving, A.L. Neves, H. Dambha-Miller, A. Oishi, H. Tagashira, A. Verho, J. Holden, International variations in primary care physician consultation time: A systematic review of 67 countries. BMJ Open **7**, e017902 (2017)

75. D.H. Schunk, *Learning Theories: An Educational Perspective* (Macmillan, New York, 1991)

76. T.M. Duffy, D. Jonassen, Constructivism: New implications for instructional technology? Educ. Technol. **31**(5), 3–12 (1991)

77. D.J. Cunningham, Assessing constructions and constructing assessments: A dialogue. Educ. Technol. **31**(5), 13–17 (1991)

78. P.A. Ertmer, T.J. Newby, Behaviorism, cognitivism, constructivism: Comparing critical features from an instructional design perspective. Perform. Improv. Q **26**, 43–71 (2013)

79. A.T. Corbett, J.R. Anderson, LISP intelligent tutoring system research in skill acquisition, in *Computer Assisted Instruction and Intelligent Tutoring Systems: Shared Goals and Complementary Approaches*, ed. by J. Larkin, R. Chabay, (Prentice-Hall Inc., Englewood Cliffs, New Jersey, 1992), pp. 73–110

80. J. Anderson, Skill acquisition and the LISP tutor. Cogn. Sci. **13**, 467–505 (1989)

81. J. Anderson, A.T. Corbett, K.R. Koedinger, R. Pelletier, Cognitive tutors: Lessons learned. J. Learn. Sci. **4**, 167–207 (1995)

82. R. Nkambou, J. Bourdeau, R. Mizoguchi (eds.), *Advances in Intelligent Tutoring Systems* (Springer, Berlin, 2010), p. 4

83. I. Goldstein, S. Papert, Artificial intelligence, language, and the study of knowledge. Cogn. Sci **1**, 84–123 (1977)

84. M. Minsky, Logical versus analogical or symbolic versus connectionist or neat versus scruffy. AI Mag. **12**, 34–51 (1991)

85. L. Faria, A. Silva, Z. Vale, A. Marques, Training control centers' operators in incident diagnosis and power restoration using intelligent tutoring systems. IEEE Trans. Learn. Technol **2**, 135–147 (2009)

86. Z.A. Vale, A. Machado, M. Fernanda Fernandes, C. Ramos, Sparse: An intelligent alarm processor and operator assistant. IEEE Expert **12**, 86–93 (1997)

87. G. Acampora, J.M. Cadenas, V. Loia, E.M. Ballester, A multi-agent memetic system for human-based knowledge selection. IEEE Trans. Syst. Man Cybern. A **41**, 946–960 (2011)

88. Z. Yu, Y. Nakamura, D. Zhang, S. Kajita, K. Mase, Content provisioning for ubiquitous learning. IEEE Pervasive Comput. **7**, 62–70 (2008)

89. F. Colace, M. de Santo, Ontology for E-learning: A Bayesian approach. IEEE Trans. Educ. **53**, 223–233 (2010)

90. P. Verma, S.K. Sood, S. Kalra, Student career path recommendation in engineering stream based on three-dimensional model. Comput. Appl. Eng. Educ. **25**, 578–593 (2017)

91. G. Tsaganou, M. Grigoriadou, T. Cavoura, D. Koutra, Evaluating an intelligent diagnosis system of historical text comprehension. Expert Syst. Appl. **25**, 493–502 (2003)

92. O. Taylan, B. Karagözoğlu, An adaptive neuro-fuzzy model for prediction of student's academic performance. Comput. Ind. Eng. **57**, 732–741 (2009)

93. K. Chrysafiadi, M. Virvou, Fuzzy logic for adaptive instruction in an E-learning environment for computer programming. IEEE Trans. Fuzzy Syst. **23**, 164–177 (2015)

94. M.L. Espinosa, N.M. Sánchez, Z.Z. García Valdivia, in *Proceedings of the 2007 Euro American conference on Telematics and information systems*, ed. by do R.P.C. Nascimento. Concept maps and case-based reasoning. (ACM, New York, NY, 2007), p. 1

95. A. Iqbal, R. Oppermann, A. Patel, A classification of evaluation methods for intelligent tutoring systems, in *Software-Ergonomie '99: Design von Informationswelten*, ed. by U. Arend, E. Eberleh, K. Pitschke, (Vieweg+ Teubner Verlag, Wiesbaden, 1999), pp. 169–181

96. A.-M. Kamin, *Beruflich Pflegende als Akteure in digital unterstützten Lernwelten* (Springer Fachmedien Wiesbaden, Wiesbaden, 2013)

97. M. Marschollek, C. Barthel, M. Behrends, R. Schmeer, I. Meyenburg-Altwarg, M. Becker, Smart glasses in nursing training - redundant gadget or precious tool? A pilot study. Stud. Health Technol. Inform. **225**, 377–381 (2016)

98. The Federal Statistical Office (2015), Pflegestatistik 2013 – Pflege im Rahmen der Pflegeversicherung, Wiesbaden. Online Available: https://www.destatis.de/DE/Publikationen/Thematisch/Gesundheit/Pflege/PflegeDeutschlandergebnisse5224001139004.pdf?__blob=publicationFile. Accessed on: Dez. 11 2018

99. M. Novak, C. Guest, Application of a multidimensional caregiver burden inventory. The Gerontologist **29**, 798–803 (1989)

100. Second Bill to Strengthen Long-Term Care (2015), https://www.bundesgesundheitsministerium.de/topics/long-term-care/second-bill-to-strengthen-long-term-care.html. Accessed 12 Dec 2018

101. D. Wolff, M. Behrends, M. Gerlach, T. Kupka, M. Marschollek, Personalized knowledge transfer for caregiving relatives. Stud. Health Technol. Inform. **247**, 780–784 (2018)

102. M. Rutz, M. Behrends, D. Wolff, T. Kupka, M-L. Dierks (2018) Hallo Du, ich bin Mo – Der Dialog als personalisierte Form der Wissensvermittlung in einem mobilen Assistenzsystem. In Zukunft der Pflege: Tagungsband der 1. Clusterkonferenz 2018, Boll S, Hein A, Heuten W & Wolf-Ostermann K., eds. ISBN 978-3-8142-2367-4

103. C.A. Taylor, J.M. Bell, M.J. Breiding, L. Xu, Traumatic brain injury–related emergency department visits, hospitalizations, and deaths — United States, 2007 and 2013. MMWR Surveill. Summ **66**(SS-9), 1–16 (2017). https://doi.org/10.15585/mmwr.ss6609a1

104. DVBIC. Defense and Veterans Brain Injury Center. DoD Worldwide Numbers for TBI (2017). https://dvbic.dcoe.mil/dod-worldwide-numbers-tbi

105. Full Text of H.R. 4310: National Defense Authorization Act for Fiscal Year 2013. GovTrack. Retrieved 13 Oct 2018

106. Behaviour change: individual approaches. Public health guideline (2014), https://www.nice.org.uk/guidance/ph49/resources/surveillance-report-2017-behaviour-change-individual-approches-2014-nice-guideline-ph49-4667934061/chapter/How-we-made-the-decision?tab

107. C. Weng, S.W. Tu, I. Sim, R. Richessond, Formal representation of eligibility criteria: A literature review. J. Biomed. Inform. **43**(3), 451–467 (2010)

Index

© Springer Nature Switzerland AG 2020
F. Firouzi et al. (eds.), *Intelligent Internet of Things*,
https://doi.org/10.1007/978-3-030-30367-9

Printed in the United States
by Baker & Taylor Publisher Services